THE QUANTUM IN CHEMISTRY:

An Experimentalist's View

THE QUANTUM IN CHEMISTRY:

An Experimentalist's View

Roger Grinter

School of Chemical Sciences and Pharmacy
University of East Anglia

John Wiley & Sons, Ltd

Other Wiley Editorial Offices

John Wiley & Sons Inc., 111 River Street, Hoboken, NJ 07030, USA

Jossey-Bass, 989 Market Street, San Francisco, CA 94103-1741, USA

Wiley-VCH Verlag GmbH, Boschstr. 12, D-69469 Weinheim, Germany

John Wiley & Sons Australia Ltd, 42 McDougall Street, Milton, Queensland 4064, Australia

John Wiley & Sons (Asia) Pte Ltd, 2 Clementi Loop #02-01, Jin Xing Distripark, Singapore 129809

John Wiley & Sons Canada Ltd, 22 Worcester Road, Etobicoke, Ontario, Canada M9W 1L1

Wiley also publishes its books in a variety of electronic formats. Some content that appears in print may not be available in electronic books.

Library of Congress Cataloging-in-Publication Data:

Grinter, Roger.
    The quantum in chemistry : an experimentalist's view / Roger Grinter.
        p. cm.
    Includes bibliographical references and index.
    ISBN-13: 978-0-470-01317-5 (cloth : acid-free paper)
    ISBN-13: 978-0-470-01318-2 (pbk. : acid-free paper)
    ISBN-10: 0-470-01317-6 (cloth : acid-free paper)
    ISBN-10: 0-470-01318-4 (pbk. : acid-free paper)
  1. Quantum chemistry.   2. Spectrum analysis.   3. Quantum measure theory.
I. Title.
    QD462. G75 2005
    541′.28 – dc22

                                                        2005009405

British Library Cataloguing in Publication Data

A catalogue record for this book is available from the British Library

ISBN-13 978-0-470-01317-5 (HB)   978-0-470-01318-2 (PB)
ISBN-10 0-470-01317-6 (HB)   0-470-01318-4 (PB)

Typeset in 10/12pt Times by Laserwords Private Limited, Chennai, India
Printed and bound in Great Britain by CPI Antony Rowe, Chippenham and Eastbourne
This book is printed on acid-free paper responsibly manufactured from sustainable forestry in which at least two trees are planted for each one used for paper production.

Contents

Preface... xiii

Chapter 1   The Role of Theory in the Physical Sciences 1

1.0 Introduction... 1
1.1 What is the role of theory in science?....................... 1
1.2 The gas laws of Boyle and Gay-Lussac....................... 3
1.3 An absolute zero of temperature.......................... 4
1.4 The gas equation of Van der Waals........................ 5
1.5 Physical laws....................................... 5
1.6 Laws, postulates, hypotheses, etc......................... 6
1.7 Theory at the end of the 19th century 7
1.8 Bibliography and further reading......................... 8

Chapter 2   From Classical to Quantum Mechanics................... 9

2.0 Introduction...................................... 9
2.1 The motion of the planets: Tycho Brahe and Kepler 10
2.2 Newton, Lagrange and Hamilton.......................... 11
2.3 The power of classical mechanics......................... 12
2.4 The failure of classical physics......................... 12
2.5 The black-body radiator and Planck's quantum hypothesis 13
    2.5.1 Planck's solution to the black-body radiation problem....... 15
    2.5.2 A qualitative interpretation of the form of the black-body
          emission curve in the light of Planck's hypothesis 16
    2.5.3 Quantisation in classical mechanics 18
2.6 The photoelectric effect 20
    2.6.1 Einstein's theory of the photoelectric effect confirmed
          experimentally.................................... 22
2.7 The emission spectra of atoms 23
    2.7.1 Bohr's theory of the structure of the hydrogen atom 24
    2.7.2 Comparison of Bohr's model with experiment 26
    2.7.3 Further development of Bohr's theory 26
2.8 de Broglie's proposal 26
2.9 The Schrödinger equation 28
    2.9.1 Eigenfunctions and eigenvalues 29

2.10 Bibliography and further reading.   30
      Problems for Chapter 2   34

Chapter 3 The Application of Quantum Mechanics.   37

3.0   Introduction.   38
3.1   Observables, operators, eigenfunctions and eigenvalues   38
3.2   The Schrödinger method.   39
3.3   An electron on a ring.   40
      3.3.1 The Hamiltonian operator for the electron on a ring   40
      3.3.2 Solution of the Schrödinger equation   41
      3.3.3 The acceptable eigenfunctions   41
3.4   Hückel's $(4N + 2)$ rule: aromaticity   44
3.5   Normalisation and orthogonality.   46
3.6   An electron in a linear box.   46
      3.6.1 The Hamiltonian operator for an electron in a linear box   46
      3.6.2 The acceptable eigenfunctions   47
      3.6.3 Boundary conditions   48
3.7   The linear and angular momenta of electrons confined within a
      one-dimensional box or on a ring   48
      3.7.1 The linear momentum of an electron in a box.   49
      3.7.2 The angular momentum of an electron on a ring   50
3.8   The eigenfunctions of different operators.   52
3.9   Eigenfunctions, eigenvalues and experimental measurements   53
3.10 More about measurement: the Heisenberg uncertainty principle   55
3.11 The commutation of operators   57
3.12 Combinations of eigenfunctions and the superposition of states   58
3.13 Operators and their formulation   59
      3.13.1 Position or co-ordinate, $\hat{x}$.   59
      3.13.2 Potential energy, $\hat{V}$ 59
      3.13.3 Linear momentum, $\hat{p}_x$ 60
      3.13.4 Kinetic energy, $\hat{W}$ 60
      3.13.5 Angular momentum, $\hat{L}$ 60
3.14 Summary.   60
3.15 Bibliography and further reading.   61
      Problems for Chapter 3   71

Chapter 4 Angular Momentum.   73

4.0   Introduction.   73
4.1   Angular momentum in classical mechanics   73
4.2   The conservation of angular momentum   75

4.3 Angular momentum as a vector quantity 75
4.4 Orbital angular momentum in quantum mechanics 76
    4.4.1 The vector model 77
4.5 Spin angular momentum 78
4.6 Total angular momentum 78
    4.6.1 The addition and conservation of angular momentum in
          quantum mechanics 79
    4.6.2 The laws of quantum-mechanical angular momentum 81
4.7 Angular momentum operators and eigenfunctions 82
    4.7.1 The raising and lowering, shift or ladder operators 82
4.8 Notation 83
4.9 Some examples 84
4.10 Bibliography and further reading 86
    Problems for Chapter 4 93

Chapter 5   The Structure and Spectroscopy of the Atom 95

5.0 Introduction 96
5.1 The eigenvalues of the hydrogen atom 96
5.2 The wave functions of the hydrogen atom 97
    5.2.1 The radial function, $R_{n,l}(r)$ 98
    5.2.2 The angular functions, $\Theta_{l,m}(\theta)$ and $\Phi_m(\phi)$ 99
5.3 Polar diagrams of the angular functions 100
    5.3.1 The s-functions 101
    5.3.2 The p-functions 101
    5.3.3 The d-functions 103
5.4 The complete orbital wave functions 104
5.5 Other one-electron atoms 104
5.6 Electron spin 105
5.7 Atoms and ions with more than one electron 105
    5.7.1 The self-consistent field 106
    5.7.2 Electron correlation 106
    5.7.3 The periodic table of the elements 107
5.8 The electronic states of the atom 107
    5.8.1 The five quantum numbers of a single electron 108
    5.8.2 Quantum numbers for the many-electron atom 108
    5.8.3 The assignment of term symbols 108
    5.8.4 Term energies and Hund's rules 110
5.9 Spin-orbit coupling 111
    5.9.1 Russell–Saunders or LS coupling 111
    5.9.2 jj coupling 112
    5.9.3 Intermediate coupling 113
    5.9.4 Inter-electronic spin-orbit coupling 115
5.10 Selection rules in atomic spectroscopy 115
    5.10.1 Angular momentum 115
    5.10.2 Parity 116

8.11 The Einstein coefficients   250
8.12 Bibliography and further reading   250
      Problems for Chapter 8   259

Chapter 9   Nuclear Magnetic Resonance Spectroscopy   261

9.0 Introduction   261
9.1 The magnetic properties of atomic nuclei   262
9.2 The frequency region of NMR spectroscopy   264
9.3 The NMR selection rule   264
9.4 The chemical shift   267
     9.4.1 The delta (δ) scale   268
     9.4.2 The shielding constant, σ (sigma)   270
9.5 Nuclear spin–spin coupling   270
9.6 The energy levels of a nuclear spin system   273
     9.6.1 First order spectra   274
     9.6.2 Second order spectra   275
9.7 The intensities of NMR spectral lines   276
9.8 Quantum mechanics and NMR spectroscopy   277
9.9 Bibliography and further reading   278
     Problems for Chapter 9   287

Chapter 10   Infrared Spectroscopy   289

10.0 Introduction   289
10.1 The origin of the infrared spectra of molecules   290
10.2 Simple harmonic motion   290
10.3 The quantum-mechanical harmonic oscillator   293
      10.3.1 Quantisation of the energy   293
      10.3.2 Zero-point energy   294
      10.3.3 Vibrational eigenfunctions   294
10.4 Rotation of a diatomic molecule   294
      10.4.1 Eigenfunctions of the rigid rotator   297
10.5 Selection rules for vibrational and rotational transitions   297
      10.5.1 A semi-classical view of the selection rules   301
      10.5.2 Infrared intensities   301
10.6 Real diatomic molecules   302
10.7 Polyatomic molecules   303
      10.7.1 Normal co-ordinates, normal vibrations, vibrational
              eigenfunctions and eigenvalues   303
      10.7.2 Vibrations of real polyatomic molecules   305
      10.7.3 Characteristic group frequencies   308
      10.7.4 Large molecules   308
10.8 Anharmonicity   309
      10.8.1 Fermi resonance   309
      10.8.2 Vibrational angular momentum and the Coriolis interaction   311
10.9 The *ab-initio* calculation of IR spectra   316

10.10 The special case of near infrared spectroscopy 317
10.11 Bibliography and further reading. 317
      Problems for Chapter 10. 324

Chapter 11 Electronic Spectroscopy 327

11.0 Introduction 327
11.1 Atomic and molecular orbitals 328
11.2 The spectra of covalent molecules. 329
    11.2.1 $\pi \rightarrow \pi^*$ transitions 329
    11.2.2 n $\rightarrow \pi^*$ transitions 330
    11.2.3 Transition-metal complexes 330
11.3 Charge transfer (CT) spectra. 330
11.4 Many-electron wave functions 332
11.5 The $1s^1 2s^1$ configuration of the helium atom; singlet and triplet states . . 333
    11.5.1 The energies of the 1s $\rightarrow$ 2s excited states of the helium atom . 335
    11.5.2 The one-electron energies; operator
             $-\frac{1}{2}\nabla_1{}^2 - \frac{1}{2}\nabla_2{}^2 - 2/r_1 - 2/r_2$ 335
    11.5.3 The two-electron, i.e. electron-repulsion, energy; operator $1/r_{12}$ 336
    11.5.4 The total energies of singlet and triplet state 338
    11.5.5 Electron repulsion in the triplet and singlet states of the excited
           helium atom: a diagrammatic illustration 339
    11.5.6 Summary of Section 11.5. 341
11.6 The π-electron spectrum of benzene 341
11.7 Selection rules 344
    11.7.1 Electron spin (multiplicity) and transition probability 344
    11.7.2 Spatial aspects of transition probability for an allowed electronic
           transition. 346
    11.7.3 The vibrational factor in the transition probability 347
11.8 Slater determinants (Appendix 6) 348
11.9 Bibliography and further reading. 348
      Problems for Chapter 11. 348

Chapter 12 Some Special Topics 351

12.0 Introduction 352
12.1 The Hückel molecular orbital (HMO) theory 352
    12.1.1 The basis of Hückel's approach 352
    12.1.2 The method. 353
    12.1.3 Hückel's assumptions 354
    12.1.4 Determination of HMO energies and AO coefficients 354
    12.1.5 Applications of HMO energies. 357
    12.1.6 Applications of HMO coefficients 361
    12.1.7 Some final comments on the Hückel theory 362
12.2 Magnetism in chemistry. 363
    12.2.1 Magnetic susceptibility: diamagnetism and paramagnetism 364

12.2.2 Magnetic susceptibility: ferromagnetism and antiferromagnetism   365
12.2.3 Magnetic fields and dipoles: some definitions   365
12.2.4 The magnetic effect of electronic orbital motion   366
12.2.5 The consequences of chemical bonding   368
12.2.6 The magnetic effect of electron spin   369
12.2.7 Magnetism in practice   370
12.2.8 Systems of interacting molecular magnets   374
12.2.9 A note of warning   376
12.2.10 An application   377
12.3 The band theory of solids   378
12.3.1 The tight binding approximation   378
12.3.2 The electron–gas (free-electron) approximation   381
12.3.3 Molecular and ionic solids   386
12.3.4 Applications   387
12.3.5 Metals, insulators and semiconductors   387
12.3.6 Optical properties of solids   390
12.3.7 Mechanical properties of solids   390
12.4 Bibliography and further reading   390
Problems for Chapter 12   397

Appendices . . . .

1 Fundamental Constants and Atomic Units   401
2 The Variation Method and the Secular Equations   403
3 Energies and Wave Functions by Matrix Diagonalisation   411
4 Perturbation Theory   417
5 The Spherical Harmonics and Hydrogen Atom Wave Functions   425
6 Slater Determinants   429
7 Spherical Polar Co-ordinates   431
8 Numbers: Real, Imaginary and Complex   433
9 Dipole and Transition Dipole Moments   435
10 Wave Functions for the ^3F States of d^2 using Shift Operators   439

Index   443

Preface

In his prologue to '*The Go-Between*' L.P. Hartley wrote: '*The past is a foreign country: they do things differently there.*' He might have been describing quantum mechanics, which goes some way to explaining why it is normally only the denizens of that strange land, the theoreticians, who dare to enter into print about it. So what is a self-confessed experimentalist doing there?

Of the many factors that have influenced the development of chemistry in the second half of the 20th century, none has been more important than the inexorable diffusion across the traditional subject boundaries of concepts, which have their origins in theoretical physics. Students of chemistry and research scientists alike, not only in chemistry, are under continual pressure to assimilate and apply these ideas; quantum mechanics forms a central part of many of them. But the learning is hard, and although there are many excellent books they tend to be written by authors who are by nature theoreticians and whose approach to the subject differs from that of the experimentalist in many ways. I know, because I have been in this position all my working life.

My original motivation, which has not changed with the years, was to interpret myself, in as exact a way as I was able, the experimental observations that I as a spectroscopist was making. In the course of my studies I spent a lot of time with most of the classic texts on the subject which are directed at chemists and some intended for physicists. Those in the English language that is. However, as excellent as they mostly are, two aspects of them often failed to satisfy me. Firstly, their authors are very good mathematicians and do not always make allowances for the lesser experience and ability in those realms of people like me. This also has the unfortunate side effect of encouraging students at all levels to believe that their difficulties are attributable to a lack of mathematical expertise when they are, in fact, more commonly due to a failure to grasp some essential quantum-mechanical concept. In my experience, once the concept has been grasped, the mathematical expression of it is much easier to understand. Secondly, expositions of quantum mechanics frequently lack examples of chemical applications. I am of the view that a real example, even when it is much simplified and gives a result which does not agree well with experiment, is always better than no example at all. The texts which do contain examples are usually those directed at a particular branch of experimental measurement, a particular branch of spectroscopy for example, and they, naturally, cannot concern themselves with a broader view of the quantum mechanics which they apply. In this book I have tried to bring the basic quantum-mechanical theory closer to real chemical examples and to make the inevitable mathematics involved subordinate to the understanding of the principles.

I owe a great deal to many friends with whom I have worked. My interest in the interpretation of physical–chemical measurements was first aroused by Stephen Mason and continued by Edgar Heilbronner. When I arrived at the University of East Anglia

(UEA) I had the great good fortune of entering into a period of collaborative research with Andrew Thomson who was as determined as I to learn some quantum mechanics. Together with some excellent post-doctoral fellows and graduate students, we spent many a lunch hour chewing literally on our packed lunches and figuratively on the theory of groups, ligand fields, angular momentum, irreducible tensors etc. This was for me a learning experience like no other and I am in debt to all who took part in the struggle; especially to Andrew.

Then, as this book began to grow, I was most grateful to Norman Sheppard when he expressed an interest in reading several of the early chapters and commenting on them with his usual, meticulous care. I have also benefited from discussions of particular problems with many other colleagues here at UEA. Positive and encouraging suggestions made by reviewers of sample chapters were also much appreciated and I have tried to take account of them.

In spite of the help I have received, it is too much to hope that no errors and/or misconceptions remain and I would like to expand a little on that subject. No error in a book from which a reader is trying to learn something can be trivial; even typographical errors can seriously mislead a student who is insecure in his/her knowledge of the subject. Readers will surely detect such errors and I would be most grateful if they would inform me of them. But misconceptions are an altogether more serious problem and in this connection I make a special appeal. In attempting to make quantum mechanics more approachable I have not hesitated to simplify and to draw on classical analogies wherever possible. In so doing I fear that I may have ventured out on a number of limbs too weak to support the weight which I have placed upon them. I hope that readers who detect this type of problem or who feel uneasy about statements which I have made will be good enough to point their concerns out to me.

A final word of thanks must go to four people. To my parents who, though they themselves had only elementary educations, did everything they could to further mine. And to my wife, Charlotte, and daughter, Rebecca, who have given me so much support and encouragement in my scientific endeavours. I trust that they will feel that this book was a worthwhile enterprise.

I also wish to thank Ivan Rodwell for provision of the three cartoons in this book.

Roger Grinter

http://www.grinter.org/quantum.html/

CHAPTER 1

The Role of Theory in the Physical Sciences

1.0 Introduction 1
1.1 What is the role of theory in science? 1
1.2 The gas laws of Boyle and Gay-Lussac 3
1.3 An absolute zero of temperature 4
1.4 The gas equation of Van der Waals 5
1.5 Physical laws 5
1.6 Laws, postulates, hypotheses, etc. 6
1.7 Theory at the end of the 19th century 7
1.8 Bibliography and further reading 8

1.0 INTRODUCTION

In this chapter we shall consider the role of theory in science and enquire what is meant when we speak of a 'law' in the physical sciences. We shall ask: what are these laws, how do they arise and what is their value or purpose within science? There are two reasons for approaching the subject of the quantum in chemistry in this way. Firstly, these are questions of interest in their own right to which, in my view, insufficient attention is paid in the teaching of science. This can result in a degree of confusion, especially in a subject as inherently complex as quantum theory, where newcomers to the subject are apt to think that if only they knew more mathematics they could derive results which, in fact, cannot be and never were derived. This leads to an undesirable focus of attention upon the mathematical rather than the conceptual aspects of the problem. Thus, the second reason for discussing scientific laws is an attempt to place the laws of quantum theory in a perspective in which their origin, value and meaning can be better appreciated by beginners in the field.

1.1 WHAT IS THE ROLE OF THEORY IN SCIENCE?

It is a matter of historical fact that man has been observing the natural world, and recording his observations, since ancient times. We may safely conclude that the verbal communication of observation is even older, as the times and routes of migrating animals, birds

The Quantum in Chemistry   R. Grinter
© 2005 John Wiley & Sons, Ltd

and fish are of crucial importance to people who live by hunting, while a knowledge of the seasons, of rainfall etc. is essential to food gatherers.[1] The importance of these and other natural phenomena made those whose knowledge of them was most extensive the leaders of their communities and exceptional status was accorded to those who had, or were believed to have, the ability to predict such events.

Historically, prediction has been attempted either through an appeal to the supernatural or by means of a reasoned extrapolation of facts already known. These two approaches to the same problem are not as different as they might appear. If an observed fact, the annual flooding of a river for example, is believed to be under the direct and immediate control of the gods, then it is quite rational to consult those gods about such events. Nor does the involvement of the supernatural necessarily conflict with the making and recording of observations. It has been suggested that a preoccupation with astrology may well account for the fact that the Mesopotamians of the first millennium BC excelled in astronomy.[2] Thus, the ancient fascination with the prediction of the future led not only to the use of rite and ritual but also to the recording and ordering of observations, a tendency which also received support from the widespread belief that there must be a system or order in the universe. The challenge of finding this system, and of demonstrating that one has found it by predicting the results of observations yet to be made, is the driving force of science and necessitates not only the collection of data but also the arrangement of that data within some conceptual framework that makes it easier to remember, understand and use.

We can distinguish two broad categories of such conceptual frameworks, or models. The first comprises models of an essentially descriptive nature in which the phenomenon in question is likened to objects of our everyday experience. An example of this type of model would be the description of the fundamental constituents of matter given by Lucretius, a Roman of the first century BC. In his view, all substances were composed of indestructible atoms and ... 'Things that seem to us to be hard and stiff must be composed of deeply indented and hooked atoms held firmly by their intertangling branches. ... Liquids, on the other hand, must owe their fluid consistency to component atoms that are hard and round, for poppy seed can be poured as easily as if it were water ....'[3] The molecular models used by modern chemists fall into this category; atoms are represented by coloured spheres and bonds by metal rods.

The second category of model is the mathematical model in which natural phenomena are represented by a set of symbols, the meanings of which have been defined, and which obey some particular rules of mathematical manipulation. A differential equation, for example, may be used to model a chemical reaction by providing a description of the rates of change with time of the concentrations of the reagents involved in the reaction. It is this kind of model to which we refer when we speak of the theoretical structure of the natural sciences. It is the type of model which is most useful to us when we are comparing numerical experimental data with theoretical predictions, i.e. in quantitative work.

A theoretical structure is essential to all the natural sciences. It provides the framework into which the pieces of the jigsaw of experimental data are fitted, thus revealing their inter-relationships and exposing gaps in our knowledge which need to be filled with the results of new experiments. In filling these gaps theory plays a leading role; it not only shows where new measurements are required but also tells the experimentalist what to expect when the experiments in question are performed. This is very important since, clearly, the apparatus must be designed so that it is capable of measuring the phenomena to be studied and the quantities used must be appropriate to the equipment.

For example, suppose we wish to determine the amount of silver in the waste solution from a photographic processing laboratory by precipitating insoluble silver chloride with hydrochloric acid and weighing it. Our theory is embodied in the chemical equation:

$$\text{AgX (in solution)} + \text{HCl} \longrightarrow \text{AgCl (precipitated)} + \text{HX}$$

The formula AgCl in this equation provides the information whereby the amount of silver in the precipitated silver chloride can be determined; provided, of course, that we know the relative atomic masses of silver and chlorine. The whole equation allows us to calculate how much silver chloride will be formed for a given amount of silver in the aqueous mixture. This information is required not only to determine the amount of silver, but also to plan the experiment in such a way that the precipitate to be weighed is of a mass appropriate for the chemical balance with which we propose to weigh it. To do this we may need to make a preliminary estimate of the amount of silver in the waste.

When, using a more elaborate example, we say that the length of the O–H bond in the water molecule is 95.7×10^{-12} m and the HOH bond angle $104.5°$, then these figures have been obtained using a theory, quantum mechanics in fact, which relates the measured absorption by water vapour of electromagnetic radiation in the microwave region (wavelengths of the order of $1-2$ cm) to the masses of the nuclei and the molecular geometry. Theory is not simply a substitute for experiment, it is a vital adjunct to it.

But theory is always a suggestion or *hypothesis*, the correctness, or otherwise, of which can only be tested against experimental fact. Therefore, theory must always be subordinate to experiment. If, after thorough checking for errors, the results of an experiment are found to differ from those predicted by theoretical calculations, then the theory must be amended, or perhaps even discarded. Thus, although the theoretical framework of science is an essential aspect which guides our progress towards a deeper understanding, we must always recognise that a current theory may one day prove to be inadequate and require replacement. These points can be illustrated by means of the gas laws.

1.2   THE GAS LAWS OF BOYLE AND GAY-LUSSAC

In 1662 Robert Boyle (1627–1691) published the results of a series of experiments on the compression of air in the closed, short arm of a J-shaped tube. Boyle observed that as the pressure, P, measured by the difference of height of the columns of mercury in the two arms of the tube, increased, the volume, V, of air in the closed end of the tube decreased. Further, he noted that *at constant temperature the volume is inversely proportional to the pressure*, a quantitative result which could be expressed in the simple equation:

$$P \propto 1/V \text{ or } PV = \text{constant} \tag{1.2.1}$$

The words in italics and Equation (1.2.1) are both expressions of what we now know as Boyle's law, and in answer to two of the questions posed above we may say that Boyle's law arose from of a series of experiments and that it expresses the results of those experiments in a convenient and precise mathematical form.

The dependence of the volume of a gas upon temperature, t, at constant pressure was studied by four French scientists, Guillaume Amontons (1663–1705), Jacques Charles

(1746–1823), Joseph Gay-Lussac (1778–1850) and Henri Regnault (1810–1878). The result of their labours is known as the law of Gay-Lussac, which was published in 1847 by Regnault. He refined the earlier experimental methods and expressed the law in the form:

$$V = V_0(1 + t/273) \qquad (1.2.2)$$

Where V_0 is the volume of the gas at $0\,^\circ$C. Here again we have a law which has been discovered by experimental measurements and, if we consider it carefully, we find that it says that, at constant pressure, the volume of a gas increases by 1/273 of its volume at $0\,^\circ$C for every degree rise in temperature. Clearly, Equations (1.2.1) and (1.2.2) are of great practical value. They can, for example, be used to calculate the volumes and pressures of gases at high temperatures. Such calculations are essential in the design of industrial plant for chemical processes, many of which take place at very high temperatures and pressures.

1.3 AN ABSOLUTE ZERO OF TEMPERATURE

But Equation (1.2.2) carries a far more fundamental message, as Amontons had realised. As we decrease the temperature of the gas below $0\,^\circ$C, i.e. when t becomes negative, the volume of the gas decreases. But this process must have a limit, since there is no such thing as a negative volume, and it is clear from Equation (1.2.2) that the limit of zero volume is reached at $t = -273\,^\circ$C. Therefore, $-273\,^\circ$C must be the lowest temperature which can be achieved. We are forced to a remarkable conclusion; although we can go up in temperature indefinitely, there is a clear lower limit. This surprising result has been substantiated experimentally and a more exact figure for the absolute zero of temperature is $-273.15\,^\circ$C, which is the origin of the scale of absolute temperature where temperature, T, is measured in degrees Kelvin or K:

$$T(\mathrm{K}) = t(^\circ\mathrm{C}) + 273.15 \qquad (1.3.1)$$

It is interesting to note here the importance of the accuracy of experimental data in formulating scientific laws. The mere observation that the volume of a fixed quantity of

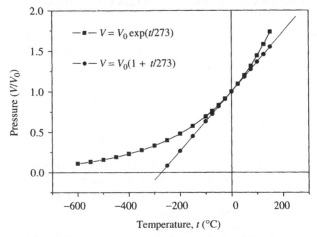

Figure 1.1 Exponential and linear temperature-volume relationships for a gas

gas decreased when the temperature was lowered might not have led to the concept of an absolute zero of temperature, much less to a value for it. Changes in physical quantities are frequently described by an exponential function and in Figure 1.1 it is shown that the equation:

$$V = V_0 \exp(t/273) \tag{1.3.2}$$

gives a temperature–volume relationship which is very similar to that given by Equation (1.2.2) in the region of temperature readily accessible to early researchers, i.e. -20 to $+100\,°C$. But according to Equation (1.3.2), the volume of the gas never reaches zero, no matter how low the temperature. So it was vitally important that the accuracy of Regnault's experimental measurements showed clearly that the relationship between the temperature and the volume of a gas was a linear rather than an exponential one.

1.4 THE GAS EQUATION OF VAN DER WAALS

Attempts to study Gay-Lussac's law down to very low temperatures failed because the gases all liquefied long before the absolute zero was reached. And, indeed, many gases were found to show marked deviations from the pressure–volume–temperature (PVT) behaviour described by Boyle and Gay-Lussac for what became known as ideal gases which obeyed the combined equation:

$$PV = nRT \tag{1.4.1}$$

where R is the ideal gas constant which has a value of $8.31\ \mathrm{J\,K^{-1}\,mol^{-1}}$ and n is the number of moles of gas in the sample (1 mole of gas contains 6.022×10^{23} molecules). These deviations of the PVT behaviour of real gases from the ideal could be represented by more complex gas laws such as that proposed by Johannes Diderik Van der Waals (1837–1923) in 1873:

$$(P + a/V^2)(V - b) = nRT \tag{1.4.2}$$

in which a and b respectively are parameters that allow for the attraction between the molecules and for the finite volumes of the molecules themselves. They have different values for different gases which are found by comparing the experimentally determined PVT behaviour of the gas with the Equation (1.4.2).

But to describe Van der Waals' equation in detail would deviate too far from the central theme of this book and we therefore turn to a summary of the answers to the questions about physical laws posed in Section 1.0.

1.5 PHYSICAL LAWS

A physical law expresses, either in words or in algebraic form, the result which is to be expected of a particular experiment. Thus, Equation (1.2.2) tells us what volume of gas we would have at any temperature, t, if its volume at $0\,°C$ is V_0. From Equation (1.2.1) we can calculate either pressure or volume, though we must first perform an experiment to determine the value of the constant for the particular gas. If it is an ideal gas then the constant is nRT, where n is the number of moles of gas in the sample (Equation (1.4.1)).

Laws arise, are discovered or deduced as a result of experimental observations. They are not derived by pure mathematical reasoning though they may be, and indeed very frequently are, expressed in an algebraic form. It follows from this that since laws are found as a result of experimentation they must also stand or fall by the results of any other experiments to which they relate. Thus the ideal gas law (Equation (1.4.1)) failed the test of application to all gases, especially in the region of low temperatures and/or high pressures. The law is therefore of limited applicability. But it is not useless and there are many circumstances in which it can provide valuable results.

A group of one or more laws normally underlies a theoretical model of some aspect of the real world as it is seen by science. Thus, the three laws of thermodynamics are the foundation stones, discovered by experiment, which allow us to describe quantitatively the inter-conversion of the various forms of energy; of heat into mechanical work for example. Together they form a mathematical model of all such processes. Newton's three laws perform the same function for mechanics, the study of motion; in the next chapter we shall examine them in more detail.

The value of laws to science and technology applies at various levels. At the most simple, but highly significant in applied science, there is the practical use of a law to predict a quantity which cannot be readily measured; the gas pressure in a novel chemical plant which is still at the planning stage, for example. A law also establishes the framework into which experimental results may be fitted as they become available. This not only provides the means of interpreting the results but also alerts the experimentalist when a new result cannot be reconciled with the current structure of the theory. This may mean that an experimental error has been made or that the law is flawed. At a deeper level, laws reveal new concepts which were not suspected at the time when the measurements which gave rise to the law were made. We may safely assume that the concept of an absolute zero of temperature was not the idea which stimulated the first experiments of Amontons; but it arose directly from them. Here lies the great importance of laws for the development of the natural sciences. As the full significance of the quantities related in a law is appreciated, scientists are led to a deeper understanding of their subject and to the formulation of new experiments to test the law at the more fundamental level. If the law fails, such a test it must be regarded as flawed and of limited use, but not necessarily totally useless. The important point is that experiment is the only test of the validity of a law. Laws summarise the results of a wealth of experimental data and present them in a condensed form suitable for application or for further study. But they are always subordinate to experiment.

1.6 LAWS, POSTULATES, HYPOTHESES, ETC.

The waters of the present discussion are sometimes muddied by the variety of terms used to describe the same thing. We speak, for example, of the gas *laws*, Planck's quantum *hypothesis*, the Pauli *principle* and the *postulates* of quantum mechanics. Each term highlighted in italics has essentially the same meaning; it is a statement, which we have here called a law, or a set of such statements, that summarises the results of experimental measurements. The different words express different aspects of the meaning of that statement. The terms *law* and *principle* emphasise the power and immutability; *postulate* and *hypothesis* the fact that this is a suggestion or a proposal which may later require modification. But in this book we shall regard all such expressions, and some others, as meaning laws.

1.7   THEORY AT THE END OF THE 19TH CENTURY

As the 19th century drew to a close the theoretical basis of the physical sciences appeared to be very mature and powerful. In particular, the three great structures of Maxwell's equations, which describe the behaviour of electromagnetic radiation, thermodynamics and mechanics, were remarkably successful in interpreting the experimental facts then known. Three examples show the range and power of these laws.

The laws of mechanics had been formulated by Isaac Newton (1643–1727) to model the motions of the planets and described these motions with remarkable accuracy. During the 1860s James Clerk Maxwell (1831–1879) and Ludwig Edward Boltzmann (1844–1906) used Newtonian mechanics to describe the motion of molecules in a gas, developing what we now call the kinetic theory of gases. The theory is in excellent agreement with the extensive experimental data encapsulated in the gas laws of Boyle and Gay-Lussac. The many deviations from the experimentally observed behaviour of real gases are due to the failure of other assumptions in the theory; that there is no attractive force between the molecules for example. Newtonian mechanics was thus shown to be applicable to bodies ranging in mass between 10^{-25} and 10^{+25} kg.

In the area of thermodynamics, the frequency of the chirping of the tree cricket, *Oecanthus*, has been found to depend upon the absolute temperature in strict conformity with the equation first put forward by Svante August Arrehnius (1859–1927).[4] The logarithm of the frequency of chirping is inversely proportional to the absolute temperature, showing that the tree cricket's chirping is quite involuntary and is controlled by its body chemistry which, in turn, is subject to the laws of thermodynamics. The same is true of the autonomous functions of the higher mammals; the human heart beat for example, though the temperature range available to the experimentalist is rather small in this case.

Our final example concerns electricity, magnetism and light. Michael Faraday (1791–1867) had shown that an electric current flowing in a coil produces a magnet and that when polarised light passes through a glass plate surrounded by the magnetic coil the plane of polarisation of the light is rotated. Thus, electricity, magnetism and light are related. An electric current can be measured by determining the magnetism it produces; the units of this measurement are called electromagnetic units, emu. An electric current can also be measured in terms of the flow of charge; the units of this type of measurement are electrostatic units, esu. In 1857, Gustav Kirchoff (1824–1887) showed experimentally that the ratio of the emu to the esu was equal to the velocity of light. These relationships between electricity, magnetism and light and between the emu and the esu were brilliantly and quantitatively interpreted by Maxwell with his mathematical model of electromagnetic radiation published in 1873. Maxwell's model, which is always referred to as *Maxwell's equations*, though they might equally well be called Maxwell's laws, showed that light and all wavelengths of electromagnetic radiation could be described in terms of a magnetic and an electric field that are orientated at right-angles to each other and oscillate with the frequency of the radiation. This led to the prediction that an oscillating electric spark would generate electromagnetic radiation, a prediction which was beautifully confirmed by Heinrich Hertz (1857–1894) in a series of experiments reported in 1886–1888. We shall explore Maxwell's description of electromagnetic radiation and the properties of polarised light further in Chapter 8.

With successes like these to its credit it seems scarcely surprising that some physicists apparently thought that there was little more to do in the field of theory other than to dot some i's and cross a few t's. Nevertheless, there were a small number of experiments the results of which defied interpretation in terms of the theories, i.e. laws, then available. It was the search for solutions to these problems which led to the revolutionary ideas of Max Planck (1858–1947) and Albert Einstein (1879–1955), and to quantum mechanics. In the next chapter we shall follow the history of mechanics to illustrate further the role which theory plays in science and to see how, early in the 20th century, some of the foundations of the structure of theoretical physics were found to be by no means as secure as they had once appeared.

1.8 BIBLIOGRAPHY AND FURTHER READING

1. C.S. Coon, *The Hunting Peoples*, Penguin Books, Harmondsworth, UK, 1976.
2. S.F. Mason, *A History of the Sciences*, Collier Books, New York, 1962.
3. Lucretius, *The Nature of the Universe*, Translation by R.E. Latham, Penguin Books, Harmondsworth, UK, 1951.
4. F.H. Johnson, H. Eyring and J.B. Stover, *The Theory of Rate Processes in Biology and Medicine*, Wiley, New York, 1974.

CHAPTER 2
From Classical to Quantum Mechanics

2.0   Introduction   9
2.1   The motion of the planets: Tycho Brahe and Kepler   10
2.2   Newton, Lagrange and Hamilton   11
2.3   The power of classical mechanics   12
2.4   The failure of classical physics   12
2.5   The black-body radiator and Planck's quantum hypothesis   13
      2.5.1   Planck's solution to the black-body radiation problem   15
      2.5.2   A qualitative interpretation of the form of the black-body emission
              curve in the light of Planck's hypothesis   16
      2.5.3   Quantisation in classical mechanics   18
2.6   The photoelectric effect   20
      2.6.1   Einstein's theory of the photoelectric effect confirmed
              experimentally   22
2.7   The emission spectra of atoms   23
      2.7.1   Bohr's theory of the structure of the hydrogen atom   24
      2.7.2   Comparison of Bohr's model with experiment   26
      2.7.3   Further development of Bohr's theory   26
2.8   de Broglie's proposal   26
2.9   The Schrödinger equation   28
      2.9.1   Eigenfunctions and eigenvalues   29
2.10 Bibliography and further reading   30
      Problems for Chapter 2   34

2.0   INTRODUCTION

In this chapter we shall first follow the historical course of the development of Newtonian or classical mechanics. We do this not only because classical mechanics is a limiting case of quantum mechanics, but also because the story of how classical mechanics developed is an excellent illustration of how a set of laws and the theoretical model which they sustain develops as a result of the interplay between experiment and theory. Having arrived with classical mechanics at the end of the 19th century, we shall investigate some particularly important experimental results which the laws of thermodynamics, mechanics and

The Quantum in Chemistry   R. Grinter
© 2005 John Wiley & Sons, Ltd

electromagnetic radiation that were available at that time were quite unable to interpret, despite their outstanding successes in many other applications. Finally, we shall see how Erwin Schrödinger and Werner Heisenberg cut the Gordian knot with the introduction of a new mechanics–quantum mechanics.

2.1 THE MOTION OF THE PLANETS: TYCHO BRAHE AND KEPLER

Before there can be any theory there must be experimental observations to be interpreted by the theory. (This proposition is not self-evident; the ancient Greeks developed theoretical concepts almost entirely without appeal to experimental evidence.) The collection of the experimental data which formed the basis for the first theory of mechanics may be traced back to the astronomical observations made, without a telescope, by the Danish nobleman Tycho Brahe (1546–1601), whose measurements were the most important and accurate of early modern times. Tycho, who is usually known by this latinised version of his Danish name, Tyge, was renowned for his fiery temperament and for most of his life he wore an artificial silver nose to replace the original appendage which had been sliced off in a duel at the age of 19. It is interesting to note that he thought it was not possible to make observations without the guidance of a theoretical system of the world, and he adopted a modified earth-centred system, now of course, known to be false. But the idea that measurements should be made in the light of a theory which, it is believed, can interpret them and fit them into the framework of our existing knowledge, is an important part of the modern scientific method; a point discussed in Chapter 1.

In 1599, Tycho Brahe moved from Copenhagen to Prague, where he was joined by German mathematical astronomer Johannes Kepler (1571–1630) in 1600. When he died in 1601, Tycho bequeathed his collection of data to Kepler who found, after many years of exacting work, that the observations of the planets could be interpreted in terms of the following three laws.

Kepler's Laws of Planetary Motion

1. The orbits of the planets are ellipses with the sun at one focus.

2. The line drawn from the sun to the planet sweeps out equal areas in equal times.

3. The square of the time required for a planet to complete its orbit is proportional to the cube of its mean distance from the sun, i.e. the length of the semi-major axis of the ellipse.

These laws are illustrated graphically in Figure 2.1.

In formulating his laws, Kepler brought the science of mechanics from the data-collection phase to the next important stage through which any developed scientific theory must pass. This is the stage at which the known experimental data are unified, in that it is shown that the data can be interpreted in terms of a small number of fundamental concepts: the laws. Clearly, if the laws are soundly based they have predictive power, i.e. the results of experiments or observations not yet made can be foretold. This is the only convincing test of the laws. A set of laws constitutes a theory, and when a theory fails systematically to predict correctly the result of experiments, then that theory must be amended or, in extreme cases, abandoned.

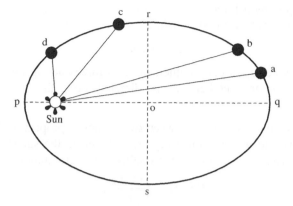

p–q is the major and r–s the minor axis of the ellipse which the path of the planet traces around the sun. The orbits of the planets of our solar system are much more like circles than that shown here which more closely resembles that of an asteroid, e.g. Icarus.

If the planet takes equal times to move from a to b and from c to d then the areas of the two sectors, a–sun–b and c–sun–d, of the ellipse are equal.

Figure 2.1   Kepler's laws of planetary motion

The predictive power of Kepler's laws is illustrated by his tables of astronomical data, published in 1627, which he calculated by means of his three laws and named the *Rudolphine Tables* in honour of his patron, the emperor Rudolph II. The planets Mercury and Venus, which have orbits of radius less than that of Earth, sometimes pass between the earth and the sun. At such times the planet in question, viewed from the earth, can be seen to move across the face of the sun. This is known as a transit of the planet. With the aid of his tables, Kepler correctly predicted the transits of Mercury (7th November, 1631) and of Venus (4th December, 1639 and 6th June, 1671) across the sun's disc. All these events actually took place as Kepler had said they would; but not before he died in 1630.

The next stage in this brief history of classical mechanics concerns a further phase through which scientific theories often pass; a phase in which a more fundamental, and hence more widely applicable, set of laws is sought. Kepler's laws described the motions of the planets – but why did they move in that way? An answer to that question might reveal that the motions of other bodies were governed by similar laws.

2.2   NEWTON, LAGRANGE AND HAMILTON

Isaac Newton (1643–1727) was not the only man of his time to grapple with this problem. His answer to it, together with accounts of many of his other scientific and mathematical achievements, appeared in his *Philosophiae Naturalis Principia Mathematica* in 1687. The four laws upon which he based his analysis were clearly of a much more fundamental, and therefore more general, nature than Kepler's.

Newton's Laws of Motion

1. A particle moves with a constant (perhaps zero) velocity (and therefore in a straight line) unless it is acted upon by a force.

2. A particle acted upon by a force (F) will move with an acceleration (a) proportional to that force, the constant of proportionality being the inverse of its mass (m).

$$a = F/m \qquad (2.2.1)$$

3. When two particles act upon one another, the two forces acting on the particles are equal and opposite to one another and are directed along the line joining the two particles.

4. The gravitational force of attraction between two bodies is proportional to the product of their masses (m_1 and m_2) and inversely proportional to the square of the distance (r) between them. The constant of proportionality is the gravitational constant (G).

$$F = G \times \left(\frac{m_1 m_2}{r^2} \right) \qquad (2.2.2)$$

The items or postulates 1 to 3 constitute the laws of what we now know as classical mechanics. During the following two centuries other formulations based on different postulates were developed, notably by Joseph-Louis Lagrange (1736–1813) and William Rowan Hamilton (1805–1865). However, these later formulations of classical mechanics were entirely consistent with Newton's and with each other; any one version could be derived mathematically from any other. The Lagrangian and Hamiltonian formulations found acceptance because they were more elegant in certain applications and the theory as a whole scored many truly remarkable successes.

2.3   THE POWER OF CLASSICAL MECHANICS

In Section 1.7 attention was drawn to the way in which Maxwell and Boltzmann developed the kinetic theory of gases by applying Newton's laws to the motion of the molecules of a gas, thereby obtaining a deeper understanding of the gas laws of Boyle, Charles and Gay-Lussac. The gas laws were to the study of gases what Kepler's laws were to mechanics; the first attempt to unify a mass of experimental data in a mathematical formalism. The work of Newton, Maxwell and Boltzmann brought these two strands of theoretical physics together and classical mechanics was thus shown to be largely applicable to bodies ranging in mass from 10^{-25} to 10^{+25} kg. Other examples of the ability of theory to interpret very diverse experimental observations have been cited in Chapter 1. Theoretical physicists at the end of the 19th century were fully justified in feeling that their science, which we may term classical theoretical physics, was capable of meeting any challenge the experimentalist might throw down. And yet experimental measurements already on record, or soon to be made, would be shown to be fundamentally at variance with the laws of classical physics and would, in due course, trigger a complete revolution in the subject. What were those measurements?

2.4   THE FAILURE OF CLASSICAL PHYSICS

It goes without saying that so cogent and successful a structure as classical physics did not fall easily or rapidly. If the laws are thought of as the belt of the structure then the blows that brought it down fell below that belt, striking at fundamental assumptions which were embodied in the laws but never explicitly stated. The fact that the structure needed to be replaced rather than simply repaired took many years to gain acceptance, and the replacement was scarcely viable until the late 1920s. The first 30 years of the 20th century was at one and the same time a period of turmoil and of intense excitement and creativity in the scientific world. It witnessed some of mankind's most outstanding intellectual achievements.

There was one major blow which we shall not discuss here. It fell in 1905 when Albert Einstein (1879–1955) showed that, because of the finite velocity of light, to say that two events take place simultaneously only has meaning if the events occur at the same place, and the concepts of space and time cannot therefore be separated as they are in classical physics. Einstein demonstrated that the velocity of light was a limiting velocity and that, although Newtonian mechanics was quite adequate for bodies moving with velocities only a fraction of that of light, it was not applicable to bodies moving at velocities approaching that limit. Nevertheless, we would not think of using relativistic mechanics to calculate the orbit of a communications satellite for example. Newtonian mechanics is much easier to apply and, for such a system, gives answers which are in practice quite indistinguishable from the relativistic results. This illustrates a very important point; although we know that classical mechanics is of limited applicability, it is still extremely useful and very widely used.

From the point of view of the development of quantum mechanics, a more important event had taken place four years before Einstein proposed his theory of relativity. In 1901 Max Karl Ernst Ludwig Planck (1858–1947) proposed a similarly radical solution to a problem that had been plaguing theoreticians throughout the 1890s; the energy spectrum of the black-body radiator.

2.5   THE BLACK-BODY RADIATOR AND PLANCK'S QUANTUM HYPOTHESIS

Any body at a temperature above the absolute zero emits electromagnetic radiation over a range of wavelengths. The energy required to produce this radiation, which is known as *thermal radiation*, comes from the thermal agitation of the particles of which the body is composed and, if the temperature of the body is above that of its surroundings, then, as the radiation is emitted, the body cools down. When the body reaches the temperature of its surroundings equilibrium is established and the body absorbs and emits radiation at exactly the same rates. Both the emission and the absorption processes are of interest to us. In 1879 Josef Stefan (1835–1893) discovered experimentally that the total energy per second (I_T) emitted from the surface of a body over all frequencies is proportional to the fourth power of its absolute temperature (T):

$$I_T \propto T^4$$

In 1884 Boltzmann deduced the constant of proportionality from thermodynamic principles and wrote the equation in the form:

$$I_T = \sigma e T^4 \tag{2.5.1}$$

where σ is the Stefan-Boltzmann constant ($\sigma = 5.67 \times 10^{-8}$ W m^{-2} K^{-4}) and e is a dimensionless constant called the *emissivity*, which lies between 0 and 1, depending upon the nature of the emitting surface.

The ability of a body to absorb electromagnetic radiation is defined by its *absorptivity*, *a*, which is the ratio of the total radiant energy falling upon its surface to the total energy absorbed by the surface. Gustav Kirchhoff showed by a powerful thermodynamic argument that *e* was equal to *a*, and this equality has also been substantiated by experiment. Thus, good absorbers of radiation are also good emitters, and vice versa. A body which is both a perfect emitter and a perfect absorber, i.e. one for which $e = a = 1.0$ is called a *black body* and is of particular interest. The best experimental realisation of a black body surface is a small hole in a cavity furnace with blackened inner walls. Radiation passing in through the hole has effectively no chance of being reflected back through it, so $a = 1.0$ and therefore $e = 1.0$ also.

The reason for the great interest in the properties of the perfectly emitting and absorbing black body can be seen by setting $e = 1.0$ in Equation (2.5.1). The total energy emitted is found to be dependent on the Stefan-Boltzmann constant and the fourth power of the absolute temperature but not upon any aspect of the body itself. Clearly, black-body radiation is a very fundamental phenomenon which demands an interpretation. Accurate experimental data were provided in 1899 by Otto Richard Lummer (1860–1925) and Ernst Pringsheim (1859–1917). They used a device not unlike a spectrometer to measure the way in which the intensity of the emitted radiation, $I(\lambda)$, depended upon the wavelength of that radiation, λ, and found curves such as those shown in Figure 2.2. The ordinate of Figure 2.2 is not $I(\lambda)$ but $\rho(\lambda)$, the energy density inside the cavity, which is equal to $4I(\lambda)/c$ where c is the velocity of light. More precisely, $\rho(\lambda)$ is the energy of the radiation in the wavelength interval λ to $\lambda + d\lambda$ in 1 m^3 of the cavity. Note how the peak of the curve moves to shorter wavelengths as the temperature is increased. This phenomenon is observed when the filament of an electric fire goes from dull red to almost white as the fire heats up. For a particular temperature, the total energy emitted, I_T, is proportional to the area under the curve and one can see how rapidly this rises with temperature, as is required by the factor T^4 in Equation (2.5.1).

Even before Lummer and Pringsheim's accurate data became available, the essential form of the relationship between energy and wavelength was known and had stimulated

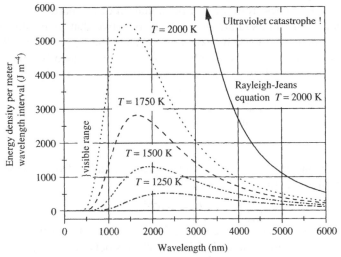

Figure 2.2   The emission spectra of a black-body radiator

many attempts to explain it. A particularly important theoretical study of black-body radiation was made around 1900 by John William Strutt, the third Lord Rayleigh (1842–1919) and James Hopwood Jeans (1877–1946). At the heart of this work lay the concept that the radiation was emitted (or absorbed) by many 'oscillators' which formed, or were contained within, the structure of the black body. According to electromagnetic theory, electric charges oscillating with frequency v emit radiation of the same frequency and from this the wavelength, λ, of the radiation can be readily calculated since:

$$v\lambda = c = \text{velocity of light} \tag{2.5.2}$$

However, although the equation obtained by Rayleigh and Jeans agreed with experiment at long wavelengths it was completely at variance with experiment at short wavelengths (Figure 2.2). It was especially disturbing that the equation predicted that the intensity of the emitted radiation rises to infinity as the wavelength approaches zero. This impossible result was known as the ultraviolet catastrophe.

2.5.1   Planck's solution to the black-body radiation problem

Max Planck was a man for whom honour and success in his professional life were accompanied by great personal tragedies; he outlived his first wife and all four of their children. He believed deeply that the only proper occupation for a theoretician was attempting to solve the most fundamental and challenging problems of the time. At the end of the 19th century there was no more fundamental a problem than that of black-body radiation and in 1901 Planck had already been working on it for about six years. Though he was by nature a very conservative man, he found himself forced to the conclusion that only an extremely radical change in the ideas which he and others had applied to the problem hitherto could lead to a relationship between $\rho(\lambda)$ and λ which would agree with experiment in the short-wavelength region. His revolutionary idea was that:

An oscillator with frequency v can possess only total energies, E_n, which satisfy the equation:

$$E_n = nhv \qquad (n = 0, 1, 2, 3, \ldots) \tag{2.5.3}$$

where h is a universal constant, now known as Planck's constant and n are the integers. Planck was proposing that energy came in 'pieces' of a certain size determined by the frequency (v). Just as the atomic theory requires that matter is not infinitely divisible, Planck's suggestion requires that the same is true of energy. When the possible energies of the oscillators were restricted in this way, Planck obtained a result in excellent agreement with the experimental data of Lummer and Pringsheim and a value for h of 6.55×10^{-34} J s. Planck's equation may be expressed in the form:

$$\rho(\lambda)\, d\lambda = \frac{8\pi hc}{\lambda^5} \cdot \frac{1}{(\exp\{hc/k\lambda T\} - 1)}\, d\lambda \tag{2.5.4a}$$

or on a frequency scale as:

$$\rho(v)\, dv = \frac{8\pi h v^3}{c^3} \cdot \frac{1}{(\exp\{hv/kT\} - 1)}\, dv \tag{2.5.4b}$$

where k is Boltzmann's constant and T the absolute temperature.

The boldness of Planck's hypothesis and the departure which it represented from the views accepted up to that time cannot be exaggerated. The assumption that energy was infinitely divisible into portions of any size underlay all formulations of classical mechanics and was so fundamental to them that it was never even stated. Now that assumption had been shown by Planck to be untenable. Planck coined the term *quanta* (singular *quantum*) for his little packets of energy ($h\nu$) and systems which can have only certain values of energy, in this or an analogous manner, are said to be *quantised*. The number n in Equation (2.5.3) which gives the number of quanta in the total energy (E_n) is called a *quantum number*.

It is interesting to note that the equation proposed by Planck for the possible total energies of an oscillator, Equation (2.5.3), had to be modified when the fully developed quantum theory of the harmonic oscillator became available about 25 years later. It was then shown that the correct formula for E_n is:

$$E_n = (n + \tfrac{1}{2})h\nu \qquad (n = 0, 1, 2, \ldots) \tag{2.5.5}$$

and each energy level is seen to be raised by $\tfrac{1}{2}h\nu$, the zero-point energy (see Chapter 10).

This uniform discrepancy in the energy levels caused no problem in Planck's work and in early applications of the developing 'old quantum theory' for which only energy differences were required. However, problems did arise when attempts were made to explain certain low-temperature phenomena, such as the specific heats of solids, and these difficulties were only overcome when the fully-fledged quantum theory revealed the presence of the zero-point energy.

2.5.2   A qualitative interpretation of the form of the black-body emission curve in the light of Planck's hypothesis

It is not immediately obvious that a change from a group of classical oscillators having all possible energies to a situation in which the energies of the oscillators are quantised can make such a dramatic difference to the predicted graph of $\rho(\lambda)$ versus λ. Since this is a question of the energy of the quantum and its dependence upon frequency, in Figure 2.3 the Planck function is plotted on a frequency rather than wavelength basis. Naturally, since ν is inversely proportional to λ, the plot looks very different in form though the essential features remain. To interpret this form we first note that in both the classical and quantum view the number of allowed frequencies in the frequency range between ν and $\nu + d\nu$, $N(\nu)\, d\nu$, is given by:[1]

$$N(\nu)\, d\nu = \frac{8\pi V \nu^2}{c^3}\, d\nu \tag{2.5.6}$$

where V is the volume of the cavity. The reason for the restriction on the number of allowed frequencies within the cavity is that each frequency is associated with a standing wave which must have nodes at the cavity walls. This requirement limits the possible wavelengths in just the same way as the possible wavelengths of a violin string are limited. Energy now enters our discussion since the value of $\rho(\nu)\, d\nu$ is obtained by multiplying $N(\nu)\, d\nu$ by, $\bar{\varepsilon}$, the mean energy of the oscillators. Thus there would appear to be a strong tendency for the graph of $\rho(\nu)$ versus ν to rise steeply with increasing ν. But there is another factor to be considered; the way in which the available energy is

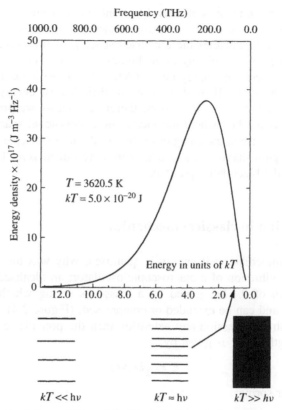

Figure 2.3   A qualitative interpretation of the black-body emission spectrum

distributed over the oscillators and, again, classical and quantum theories do not differ in that both use the Boltzmann distribution. The ratio of the numbers of oscillators (N_0 and N_1) in two energy levels (E_0 and E_1) is determined by the energy difference according to the equation:

$$\frac{N_1}{N_0} = \exp\left\{\frac{E_0 - E_1}{kT}\right\} \equiv \exp\left\{\frac{-\Delta E}{kT}\right\} \tag{2.5.7}$$

However, the crucial difference between the classical and quantum views is revealed when we determine the mean energy ($\bar{\varepsilon}$) of the oscillators. Classically, with a continuum of energy levels, we find:

$$\bar{\varepsilon} = kT \tag{2.5.8a}$$

whilst the quantised energy-levels give:

$$\bar{\varepsilon} = \frac{h\nu}{\exp(h\nu/kT) - 1} \tag{2.5.8b}$$

Noting that the exponential function may be expanded (see Appendix 8) as the series:

$$\exp(h\nu/kT) = 1 + h\nu/kT + (h\nu/kT)^2/2! + (h\nu/kT)^3/3! + \cdots$$

we see that the quantum value of $\bar{\varepsilon}$ tends to the classical value when $h\nu \ll kT$ and we take just the first two terms in the expansion. Thus, at the low-frequency end of the graph (Figure 2.3), where there is effectively a continuum of energy levels, the situation is almost classical and the classical and quantum theory plots are very similar. However, as we move to higher frequencies the quantised spacing of the energy levels begins to make itself felt. If we go immediately to the highest frequencies it is obvious that, at that limit, the spacing between the energy levels ($\Delta E = h\nu$) is so large that only the lowest energy level is occupied. If all oscillators are in their lowest energy state they cannot emit and $\rho(\nu)$ must fall to zero. It follows, therefore, that as we go from low to high frequencies a point is reached where the increasing energy-level spacing becomes more important than the increase in the energies of the oscillators themselves; the graph of $\rho(\nu)$ versus ν goes through a maximum and begins its long fall to zero. Eisberg[1] has a very good discussion of the black-body problem.

2.5.3 Quantisation in classical mechanics

A question arises immediately; if energy is quantised, why was this not noticed before 1901? Consider the vibration of a macroscopic oscillator; an idealised system consisting of two masses, each of 0.1 kg, joined by a mass-less spring which is 10 cm (0.1 m) long at equilibrium and can be extended or compressed; (Figure 2.4). If the spring obeys Hooke's law in both compression and extension then the potential energy, E, produced by a change of length of Δr is given by:

$$E = \tfrac{1}{2}k(\Delta r)^2$$

so for our particular spring:

$$E = \tfrac{1}{2}k(r - 0.10)^2 \ \text{(Joules)} \tag{2.5.9}$$

where k is the force constant measured in N m^{-1} and r is the length to which the spring has been extended or compressed. k represents the stiffness of the spring and if we pull the masses apart to extend the spring and then release them the model vibrates. In our idealised system there will be no losses of energy and the vibration will continue indefinitely. As the model vibrates energy is continually exchanged between kinetic and potential. At the maximum and minimum spring lengths all the energy is potential; at the point when the spring length is 10 cm all the energy is kinetic. The total energy, potential plus kinetic, is constant since energy must be conserved. This is a suitable model for an oscillating, homonuclear diatomic molecule, e.g. hydrogen (H_2) or oxygen (O_2).

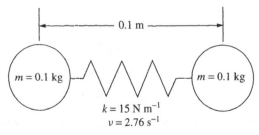

Figure 2.4 A macroscopic dumb-bell oscillator

The potential energy for a particular extension of the spring, and hence the total energy at any time, can be calculated from Equation (2.5.9). The results obtained using a force constant of 15 N m^{-1} are shown in Table 2.1. Since the energy is determined by $(r - 0.10)^2$, the figures apply to both extension and compression of the spring and they describe a parabola (Figure 2.5). Oscillations which arise as a result of a parabolic potential energy curve are simple harmonic oscillations and the frequency of oscillation (ν) can be calculated by means of the equation:

$$\nu = \frac{1}{2\pi} \cdot \sqrt{\frac{2k}{m}} \quad (\text{s}^{-1} \text{ or Hz}) \tag{2.5.10}$$

where m is the mass of each of the two equal masses which, in our case is 100 g = 0.1 kg. We find that the frequency is 2.757 Hz, i.e. the model performs 2.757 oscillations per second.

In this hypothetical model there is no limit on the potential energy which we can give the system; it simply depends upon the initial extension (or compression) which we choose to impose upon the spring. But note that although the energy may be different, the frequency of oscillation is always the same since it depends (Equation 2.5.10) only on m and k. But according to Planck the energy is quantised in units of $h\nu = 6.626 \times 10^{-34} \times 2.757$ J $= 1.827 \times 10^{-33}$ J. Comparing this result with the figures in Table 2.1, we see that the quantum of energy in this example is about 10^{28} times smaller than the energy of the oscillator. It is such a small quantity that there is no way in which we could measure it or detect its presence by observation of the vibrations of our model. As far as our model is concerned, energy is effectively continuous.

Table 2.1 The energy of the macroscopic model of a diatomic oscillator as a function of the distance between the masses

$(r - 0.1)$/m	0.000	0.005	0.010	0.015	0.020	0.025	0.030
E/J $\times 10^{-4}$	0.000	1.88	7.50	16.88	30.00	46.88	67.50

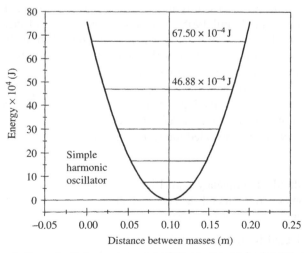

Figure 2.5 The parabolic potential energy curve of the macroscopic dumb-bell oscillator

But if we consider a molecule we shall find that two factors contribute to make quantisation a real observable phenomenon in a system of atomic dimensions. We take as our molecular model the hydrogen molecule (H_2). Experiment tells us that the mass of each atom in the molecule is 1.6738×10^{-27} kg and the frequency of oscillation (ν) is 1.3175×10^{14} Hz. Rearranging Equation (2.5.10) we have:

$$k = 2\pi^2 \nu^2 m \qquad (2.5.11)$$

which, with these data, gives a force constant of 573 N m^{-1}. And the quantum of energy for the vibrating hydrogen molecule is therefore $h\nu = 8.726 \times 10^{-20}$ J. This is much larger than that of the macroscopic model because of the great difference in the vibrational frequencies of the two systems.

The H–H bond length is 74.17 pm (1 pm $= 10^{-12}$ m) so, using Equation (2.5.9), a 10 % extension of the bond develops a potential energy of $(573/2) \times (7.4)^2 = 1.57 \times 10^{-20}$ J. This figure is approximately one fifth of the value of $h\nu$ for the hydrogen molecule calculated above and we must therefore expect that quantisation will play a much more important role in the vibrational behaviour of the hydrogen molecule than it does in the macroscopic model we discussed earlier. Thus, we anticipate that the effects of quantisation will be measurable in atoms and molecules, but not in macroscopic systems.

In summary, we may say that it is the magnitude of Planck's constant (h) which is the determining factor. Only when energies are so small or frequencies so high that $E \approx h\nu$ do the effects of quantisation become visible in our experimental measurements. Such conditions are found when we study the behaviour of atoms and molecules.

2.6 THE PHOTOELECTRIC EFFECT

Planck's solution of the problem of the energy distribution of black-body radiation was extremely difficult for scientists to accept; not least Planck himself who realised that, if it were correct, it would require the recasting of the whole of classical physics. The thought that energy, like matter, was not infinitely divisible struck at the very foundations of theoretical physics. What was desperately needed was some confirmation, some other experiment which could be successfully interpreted in terms of a quantisation of energy. The world of physics waited until 1905 for such evidence, which Einstein then provided with brilliant insight, though its true value was not recognised immediately. The experimental observation we shall discuss here is the *photoelectric effect*, but in his 1905 paper Einstein's discussion ranged much more widely.

In 1887 Hertz discovered that when a beam of ultraviolet light falls upon a metal surface particles capable of conducting electricity are emitted from the metal. In 1900 Philipp Eduard Anton Lenard (1862–1947) proved that the emitted particles were electrons and further experiments revealed the following salient facts about the photoelectric effect.

- There is a threshold of frequency (ν_0). Light having a frequency lower than ν_0 does not eject electrons from the metal surface, no matter what the intensity of the light. But when the frequency is increased above the threshold electrons appear immediately.

- The electrons ejected by radiation of frequency ν greater than ν_0 have kinetic energy and this kinetic energy increases as the frequency difference, $\nu - \nu_0$, increases.

- If the intensity of the light is increased at a constant frequency the kinetic energy of the ejected electrons remains the same, but they increase in number in direct proportion to the intensity.

These experimental observations are completely at variance with classical electromagnetic wave theory. If the energy of light is spread out in a wave a time delay would be expected, during which the electron accumulates energy from the wave, between the time at which the light source is switched on and the appearance of the first electrons. The delay was estimated theoretically to be about one minute whereas a delay of approximately 10^{-9} s was measured experimentally. Further, since in classical theory energy is accumulated from the light wave, any light should be capable of ejecting electrons; the process of accumulation would just take longer for less energetic light and there would be no other dependence upon v. Finally, for a given period of illumination, the kinetic energy of the ejected electrons is expected to increase with the *intensity* of the light, not with its *frequency*.

Einstein based his 1905 interpretation of the facts of the photoelectric effect on Planck's hypothesis, reasoning as follows. According to Planck, the oscillators in a light source can only have quantised energies given by nhv ($n = 0, 1, 2, 3, \ldots$). Therefore, as the oscillators change their energies from nhv to $(n-1)hv$ and emit radiation of frequency v and energy hv then, at the moment of emission, this radiation must emerge as a pulse within a short period. If this pulse of radiation then moves away from the source as a single entity rather than dispersing as a wave, light may be considered to be a stream of pulses of energy. (We now call these pulses of energy *photons*, though Einstein did not use that expression; see Chapter 8). If the stream of photons, each having energy hv, falls upon a metal surface one of two things may then occur. If the energy (hv) of the photon is less than the work function (W), which is the energy required to eject an electron from the metal surface, no photoelectron will be observed. If the photon has more energy than that required to eject the electron then the electron will be ejected and, because energy must be conserved, the excess of energy will be converted into the kinetic energy (KE) of the ejected electron. Einstein therefore proposed the following equations for the energetics of the photoelectric effect:

$$hv = W + KE \tag{2.6.1}$$

$$hv_0 = W \tag{2.6.2}$$

$$KE = hv - hv_0 = h(v - v_0) \tag{2.6.3}$$

These equations are in excellent agreement with the experimental facts as they are known today. But it is not widely appreciated that, at the time of Einstein's suggestion, no quantitative experimental data were available. Only the qualitative results outlined above were known to him and even these were not universally accepted. Thus, he was not able to calculate a value of Planck's constant with his new theory and his paper actually contains very few numerical results. It is concerned solely with describing a number of experimental observations and interpreting them in terms of a particle theory rather than a wave theory of light. The impact of this new hypothesis was not immediate, but when its significance was recognised it won Einstein the 1921 Nobel Prize for Physics. The lack of high-quality experimental data on which to test it was an important factor in the slow appreciation of Einstein's new concept. Eleven years were to pass before such data became available.

2.6.1 Einstein's theory of the photoelectric effect confirmed experimentally

The experimental work required to clinch Einstein's hypothesis, at least as far as the photoelectric effect is concerned, took many years to complete and turned out to be extremely difficult. Robert Andrews Millikan (1868–1953), who had earlier achieved fame for his determination of the charge on the electron, devoted about 11 years to this problem beginning in 1905. In 1916 he reported the definitive summary of his work on the photoelectric effect; his data for lithium and sodium are illustrated in Figure 2.6.

The horizontal axis of the figure gives the frequencies of the lines of the mercury discharge lamp used to excite the photoemission. The vertical axis records the minimum voltage required to prevent the photoelectrons leaving the metal. This is a measure of the kinetic energy of the photoelectrons. Negative values of the retarding voltage indicate that the electrons lacked sufficient energy to leave the metal surface and had to be drawn away from it. Each point is the result of a series of experiments in which the photocurrent for a range of more negative voltages is extrapolated back to zero current. The slope of the graph of retarding potential against frequency, when converted to appropriate units, gives the value of h/e and, using the value of e which he himself had determined, Millikan found values of $h = 6.584$ and 6.569×10^{-34} J s from his lithium and sodium data respectively. After a very careful consideration of errors Millikan concluded that $h = 6.57 \times 10^{-34}$ with an error of 0.5 %. When he recalculated h using Planck's original equation for black-body radiation with the most recent values of the other physical constants involved, Millikan found exactly the same value for h with an estimated error of 0.3 %. This result confirmed Einstein's ideas in the most convincing manner. It is not surprising that this brilliant and exacting experimental work was also cited by the Nobel Committee when they awarded the 1923 Nobel Prize for Physics to Millikan, largely, but not exclusively, for his determination of the charge on the electron.

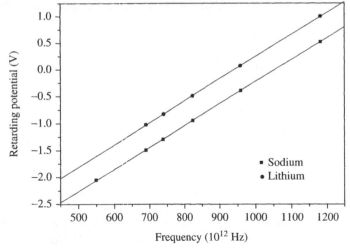

Reproduced with permission from R.A. Millikan, *Phys. Rev.*, **7**, 355 (1916).
© The American Physical Society.

Figure 2.6 Millikan's data for the photoelectric effect in lithium and sodium

We cannot leave this subject without emphasising again just how radical Einstein's photon hypothesis was and how small was the quantity of hard experimental evidence for it at the time. In the introduction to his paper, which did more than anything else to confirm it, Millikan himself described Einstein's proposition as a '... *bold, not to say reckless, hypothesis.*'; words which were quite justified in view of the dominance of the wave theory of light in 1905. But, in spite of its flimsy experimental base and the fact that quantitative experimental confirmation was such a long time coming, Einstein's photon hypothesis brought about a complete change in our view of light. Throughout the 18th and 19th centuries the wave theory of light had become increasingly dominant and had extended into the description of newly discovered forms of electromagnetic radiation such as ultraviolet, infrared and radio waves. Einstein's interpretation of the photoelectric effect demanded a complete reappraisal of the way in which light was regarded. He had extended Planck's revolutionary hypothesis while introducing an equally radical concept of his own.

2.7   THE EMISSION SPECTRA OF ATOMS

Of all the experimental data which classical theories appeared to find difficult to interpret, the greatest quantity of highly accurate data was contained in the photographs of atomic spectral emission. If a sample of like atoms is heated to a sufficiently high temperature, 4000 °C say, the atoms absorb energy from the furnace and re-emit it in the form of an atomic spectrum (Figure 2.7). The spectrum consists of a number, many hundreds in some cases, of very sharp lines of precise wavelength (frequency). The lines look like, and indeed are, a bar code for the atom and Robert Wilhelm Bunsen (1811–1899) and Gustav Kirchhoff (whose work on the black-body radiator has already been mentioned in Section 2.5) identified the previously unknown elements caesium (a sky-blue line found in 1860) and rubidium (a ruby-red line found in 1861) by measuring the spectra of minerals heated to incandescence by Bunsen's famous burner. Later, Kirchhoff showed that measurement of the absorption of light by atoms was similarly characteristic. Atomic emission and absorption spectra are among the most powerful qualitative and quantitative analytical tools currently available to us.

The lowest spectrum is that of copper and the one immediately above it is iron. The remainder are the spectra of ore samples. [The spectra are printed as 'positives' so that the emission lines appear white on a dark background.]

Figure 2.7   Some atomic emission spectra‡

---

‡ Christian, *Analytical Chemistry*, 5th edn, © Reprinted with permission of John Wiley & Sons, Inc.

But in the first decade of the 20th century the sharp lines of the atomic spectrum constituted a theoretical problem of great difficulty. Throughout this decade experimental evidence concerning the structure of the atom was accumulating and in 1911 Ernest Rutherford (1871–1937) proposed a model of the atom in which almost the whole mass of the atom was concentrated in a very small, positively charged nucleus at the centre while the negatively charged electrons occupied a much greater region of space surrounding the nucleus. Since atomic spectra were the result of emission of radiation by atoms, it should have been possible to draw conclusions about the structure of atoms from their detailed atomic spectra. But there was a problem. Since the electrons were negatively charged and the nucleus positively charged electrostatics predicted that the electrons would be drawn into the nucleus, destroying the atom immediately. A way in which such an event might be prevented would be if the electrons were to orbit the nucleus, like planets around a sun, with the force of the electrostatic attraction between electron and nucleus replacing the gravitational attraction between planet and sun. But even this was no solution. According to classical electromagnetic theory, a negative charge orbiting a positive charge constitutes an oscillating dipole which would radiate energy, just like Hertz's sparks (Section 1.7). Thus the atom should continuously lose energy until the electron fell into the nucleus and the radiation emitted would be of continuously varying wavelength (frequency), not the sharp lines actually observed.

2.7.1 Bohr's theory of the structure of the hydrogen atom

The above was the dilemma which faced Niels Henrik David Bohr (1885–1962) in 1913. He attacked it with that mixture of boldness and intuition which had characterised the work of Einstein and proposed four new laws.

Bohr's Laws of the Hydrogen Atom Structure

1. The electron in the hydrogen atom orbits the nucleus in a circular path, like a planet around the sun.

2. Of the infinite number of possible orbits, only those are allowed for which the orbital angular momentum of the electron is an integral multiple of Planck's constant divided by 2π.

3. Contrary to classical electromagnetic theory, electrons in these allowed orbits do not radiate energy.

4. When an electron changes its orbit a quantum of energy (photon) is emitted or absorbed in accordance with the equation $\Delta E = h\nu$, where ΔE is the difference in the energy of the two orbits.

In postulate 2 Bohr was following Planck's lead. Planck had proposed that the energies of the oscillators in a black body could have only certain quantised values; Bohr was suggesting a similar thing for the electron orbits of the hydrogen atom. But instead of energy he chose to quantise the angular momentum of the electron. It is not easy to say why Bohr chose to quantise angular momentum rather than energy, but it may be that it was because the units of Planck's constant (J s) are the same as those of angular

Table 2.2 The first eight Balmer lines ($2s \rightarrow np$; $n = 3, 4, \ldots, 10$) of hydrogen Balmer's formula: λ(calc.) $= 364.7n^2/(n^2 - 4)$

n	state energy[‡] ω/cm^{-1}	$\Delta\omega$/cm^{-1}	λ(obs.)/nm	λ(calc.)/nm
2	82 258.9			
3	97 492.3	15 233	656.5	656.5
4	102 823.8	20 565	486.3	486.3
5	105 291.5	23 033	434.2	434.2
6	106 632.2	24 373	410.3	410.3
7	107 440.4	25 182	397.1	397.1
8	107 965.0	25 706	389.0	389.0
9	108 324.7	26 066	383.6	383.6
10	108 582.0	26 323	379.9	379.9
∞	109 678.8	27 420	364.7	364.7

[‡]Charlotte E. Moore, *Atomic Energy Levels*, Vol I, US National Bureau of Standards, Circular 467, Washington DC, 1949.

momentum. However, following the discussion of Chapter 1, it should be clear that, as with the gas laws of Boyle and Gay-Lussac, there can be only one test of Bohr's laws and that is whether they reproduce the experimental findings.

Table 2.2 lists the wavelengths of a series of emission lines of the hydrogen atom which, in 1913, had been known for many years. They are called the Balmer lines because Johann Jakob Balmer (1825–1898), a Swiss school teacher, had shown in 1885 that the wavelengths of these lines (in nm $= 10^{-9}$ m) fitted the simple formula:

$$\lambda = \frac{364.7n^2}{(n^2 - 4)} \qquad (n = 3, 4, 5, \ldots) \qquad (2.7.1)$$

Balmer's equation, which predicts the wavelengths very accurately and was deduced from the experimental data alone, is illustrated with a graph in Figure 2.8. Thus, the test

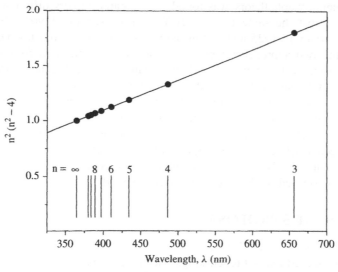

Figure 2.8 A graph of Balmer's equation for the hydrogen atom spectrum

of Bohr's laws is their ability to interpret Equation (2.7.1). The application of the Bohr model to Balmer's equation is described in detail in Box 2.1. In the next section the results derived there are used to test the Bohr theory against experiment.

2.7.2   Comparison of Bohr's model with experiment

As shown in Box 2.1, the Bohr model leads to Equation (B2.1.13) for the wavelengths of the atomic spectral lines of the hydrogen atom:

$$\lambda = \frac{32\varepsilon_0^2 ch^3}{e^4 m_e} \cdot \left(\frac{n^2}{n^2 - 4}\right) \qquad (B2.1.13)$$

On evaluating the factor $32\varepsilon_0^2 ch^3 / e^4 m_e$ using the values for the fundamental constants listed in Appendix 1, we obtain:

$$\frac{32\varepsilon_0^2 ch^3}{e^4 m_e} = \frac{32 \times (8.85419)^2 \times 2.99792 \times (6.62608)^3 \times 10^{118}}{(1.60218)^4 \times 9.10939 \times 10^{107}} = 364.507 \times 10^{-9} \text{ m}$$

This is in excellent agreement with Balmer, i.e. with experiment, and confirms the validity of Bohr's model for the hydrogen atom spectrum.

2.7.3   Further development of Bohr's theory

The excellent agreement of Bohr's theoretical result with the experimental data embodied in Balmer's equation was most gratifying and, at the time, it marked a very important step forward. But law 2, the arbitrary imposition of quantisation upon what was essentially a classical solution of the problem was to no one's liking, least of all to Bohr's. Furthermore, strenuous efforts by Bohr and others, notably Arnold Johannes Wilhelm Sommerfeld (1868–1951), to apply a similar, semi-classical approach to atoms with more than one electron failed to give results in agreement with experiment and it was clear that a fully acceptable theory of the structure of the atom was yet to be found. The search for a form of mechanics applicable to atoms and molecules was to last for a further 12 years. Amazingly, when a viable theory was found, two apparently very different theories were announced at almost the same time. Karl Werner Heisenberg (1901–1975) described his *matrix mechanics* in 1925 and Erwin Schrödinger (1887–1961) his *wave mechanics* in 1926. But the two approaches were soon shown to be different formulations of the same theory (compare the Newtonian, Lagrangian and Hamiltonian forms of classical mechanics), and we shall deal only with Schrödinger's version here.

But before we take our final step to quantum mechanics we must note a suggestion made in his PhD thesis by de Broglie which, though it was not substantiated by experiment until after the advent of quantum mechanics, showed very clearly that a profound difference was to be expected between the description of very small particles by classical and by quantum mechanics. It also had a great influence on Schrödinger who, in his famous paper of 1926, described de Broglie's thesis as 'inspired'.

2.8   de BROGLIE'S PROPOSAL

Louis Victor Pierre Raymond Prince de Broglie (1892–1987) descended from a noble family who had served many French kings. He studied history before taking up physics and in his doctoral dissertation (1924) he reasoned as follows:

According to Einstein's theory of relativity:

$$E = mc^2 \tag{2.8.1}$$

but Einstein, following Planck, also suggested that:

$$E = h\nu = hc/\lambda \tag{2.8.2}$$

therefore, if the momentum (p) of the photon as a particle is mc, then:

$$p = mc = E/c = h/\lambda \tag{2.8.3}$$

and if this result applies to other particles as well as to photons, then all moving particles should be associated with a wavelength (λ) given by the equation:

$$\lambda = h/p = h/mv \tag{2.8.4}$$

where m and v are the mass and velocity of the particle respectively.

This remarkable suggestion was confirmed for electrons by Clinton Joseph Davisson (1881–1958) and Lester Halbert Germer (1896–1971) who, in 1927, demonstrated that a beam of electrons directed at the surface of a nickel crystal were selectively diffracted at certain angles, just as X-rays are diffracted, by the regularly spaced layers of atoms in a crystal.

A similar experiment was performed independently by George Paget Thomson (1892–1975) in 1927. In Thomson's experiment, electrons were accelerated by potentials of between 10 and 60 kV as a result of which, if de Broglie is correct, their associated wavelengths would lie between 12 and 5 pm respectively. This wavelength range is some 20 to 50 times smaller than the spacing between atoms in a metal and Thomson fired a very narrow beam of his accelerated electrons through extremely thin (10^{-7} m) foils of gold and platinum. He placed a photographic plate on the far side of the foil and when he developed the plate found it to show the circular interference pattern well known from the corresponding experiments with light beams. Furthermore, measurement of the interference pattern and a knowledge of the velocity of the electrons and the spacing of the atoms in the metal foil provided a quantitative proof of de Broglie's equation. Many subsequent experiments with heavier particles have confirmed the early results.[‡]

The deep significance of this result is apparent when we consider a simplified version of Thomson's experiment (Figure 2.9) in which a beam of electrons is directed at a screen containing just two slits with a photographic plate to observe the results. From experiment we know that an interference pattern is observed on the plate only if there are two (or more) slits in the screen. With one slit no interference pattern is seen. Also, we can reduce the intensity of the electron beam to such a level that we may consider the final pattern on the plate to be the accumulation of the results of many experiments, each with an individual electron. But how can this be interpreted if the electron is a particle? A particle can only pass through one of the slits, but the experiment seems to imply that the presence of a further slit, through which the electron did not pass, has had an influence on the electron's trajectory!

We are on very treacherous ground here. Experience has shown that we cannot speak of the path of a particle in quantum mechanics in the same way as we do in classical mechanics. We can only speak of those things which we can observe. In this case we can

---

[‡] A nice anecdote is told about the Thomsons. The father (JJT) won the Nobel Prize for Physics in 1906 for showing that the electron was a particle. The son (GPT) won the same prize in 1937 for showing that the electron was a wave!

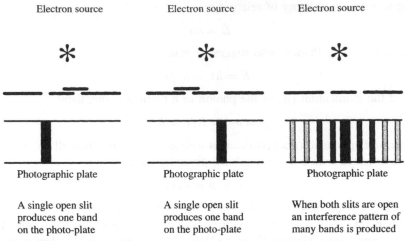

Figure 2.9   An idealised representation of G.P. Thomson's experiment

see the spot produced on the photographic plate by the electron, but we cannot observe the 'path' which it has taken between the electron source and the plate and we must not make inferences about what that path might, or might not, have been. A much more detailed discussion of this problem has been given by Feynman[2] and we shall return to the subject of what we can and cannot know in Chapter 3. At the moment we must continue on our difficult but well-defined path 'from classical to quantum mechanics', for we have almost reached our journey's end.

2.9 THE SCHRÖDINGER EQUATION

The title of the paper in which Schrödinger announced his new mechanics was '*Quantization as an Eigenvalue Problem*' and in the first paragraph he wrote:

'*In this communication I wish to show, first for the simplest case of the non-relativistic and unperturbed hydrogen atom, that the usual rules of quantisation can be replaced by another postulate, in which there occurs no mention of whole numbers. Instead, the introduction of integers arises in the same natural way as, for example, in a vibrating string for which the number of nodes is integral. The new conception can be generalised and I believe that it penetrates deeply into the true nature of the quantum rules.*[‡]

The title chosen for his paper by Schrödinger is very significant, and we shall return to it later. But let us first examine the opening paragraph quoted above. Schrödinger immediately makes clear that he is going to deal only with the simple case of the hydrogen atom. He is not going to include the effects of the theory of relativity in his discussion, nor will the atom be disturbed (perturbed) by any external forces. The second half of the first sentence shows that it is the arbitrary imposition of quantisation which is Schrödinger's

---

[‡] I have taken this quotation, *verbatim*, from Linus Pauling and E. Bright Wilson's '*Introduction to Quantum Mechanics*', published in 1935.[4] Like the book itself, the translation of the original German can scarcely be improved upon.

target and the second sentence reveals that he has found a theory in which quantisation arises naturally.

To illustrate the way in which integers can arise naturally in theoretical models of physical phenomena, Schrödinger refers to the vibration of a stretched string, which can only vibrate in such a way as to produce the fundamental frequency and the first, second, third, ... nth ... overtone, or harmonic, and for which the string has 0, 1, 2, 3, ... n ... nodes respectively between the fixed points at its ends. Here we must beware. Schrödinger's form of the new mechanics has become known as wave mechanics and he is here illustrating his new results with the well-known properties of waves on stretched strings. But this does not mean that his equation can be derived from such considerations. The Schrödinger equation is not derivable from classical mechanics, it is a new postulate as he himself makes clear in the quotation above. So, what is this new postulate? The question takes us back to consider the title of Schrödinger's paper.

2.9.1 Eigenfunctions and eigenvalues

German is a language with a remarkable ability for building compound words such as *Eigenfunktion* and *Eigenwert*.[3] It is easy to translate the second parts of the above German words '*Funktion*' = '*function*' and '*Wert*' = '*value*' (German nouns always begin with an upper-case letter), but it is very difficult to find a suitable translation for the qualifying word '*Eigen*'. The word 'special' has occasionally been used, but it does not imply an aspect of 'belonging to' which '*Eigen*' also has. Thus, an eigenvalue is not only a special value it also belongs to something, to its own special function and an operator in fact (see Box 2.2). The situation is best illustrated with an example of an eigenvalue-eigenfunction equation:

$$\frac{\partial^2}{\partial\vartheta^2}\,(\sin 3\vartheta) = \frac{\partial}{\partial\vartheta}\,(3\cos 3\vartheta) = -9\sin 3\vartheta$$

This equation is a rather unusual one. The mathematical operator, $\partial^2/\partial\vartheta^2$, operates on the function $\sin(3\vartheta)$ (we show it in two steps to make the process clear) to give exactly the same function back again, but multiplied by a number, -9 in this case.

There are rather few cases where an operator, e.g. $+$, $-$, $\times$, $\div$, integrate, differentiate, when applied to a function gives the function back again multiplied only by a number. When this unusual situation does occur the function is known as an eigenfunction of the operator and the number as the associated eigenvalue. It is this mathematical phenomenon which lies at the heart of Schrödinger's mechanics. He discovered a way of formulating an operator for the energy of the electron in the hydrogen atom and then determined the eigenfunctions and eigenvalues of that operator. The eigenvalues are then the possible energies of the electron in the hydrogen atom, as could be tested immediately by a comparison with the experimental spectral data. The eigenfunctions are functions which describe the possible ways in which the electron may be distributed around the nucleus of the hydrogen atom, each distribution being associated with a particular energy given by the associated eigenvalue. These hydrogenic eigenvalues and eigenfunctions will be discussed in detail in Chapter 5. At the moment we shall find it easier if we first consider the eigenvalues and eigenfunctions of some even simpler problems (Chapter 3). But simple though they are, we shall find that these problems have solutions with important chemical applications.

2.10   BIBLIOGRAPHY AND FURTHER READING

1. R.M. Eisberg, *Fundamentals of Modern Physics*, Wiley, New York, 1967.
2. R. Feynman, *Six Easy Pieces*, Penguin Books, London, 1998.
3. An example frequently quoted is: *Donaudampfschiffartsgesellschaftskapitänsfrau* = the wife (*frau*) of a captain (*kapitäns*) in the Danube (*Donau*) steam-ship (*dampfschiffarts*) company (*gesellschaft*).
4. L. Pauling and E.B. Wilson, *Introduction to Quantum Mechanics*, McGraw-Hill, New York, 1935.

BOX 2.1   Bohr's model of the hydrogen atom in detail

Motion in a circle

Bohr's model of the hydrogen atom pictures the electron orbiting the nucleus like a planet around the sun. The theory appeals to the results of the classical mechanics of motion in a circle so it will be useful if we derive those we require before we start.

Figure B2.1.1(a) shows a mass m moving with a constant speed (v) along a circular path of radius r centered at O. Though the speed is constant, the velocity, which is a vector quantity and therefore has direction as well as magnitude, is not. The magnitude of the velocity of the mass is its speed, which is constant, but its direction is continuously changing. The velocity vector (v) is always at right angles to the radial vector (r) and Figure B2.1.1(b) illustrates the situation at two closely spaced instants in time during which the mass has moved through a very small angle (dß). The distance moved along the arc, ds, is equal to $r \cdot$ **dß** if **dß** is measured in radians. Also, the speed along the arc is $v = ds/dt$. In Figure (c) the difference in the two velocity vectors, d$v = v_2 - v_1$, is found by means of vector subtraction. Since the magnitudes of v_2 and v_1 are equal to the constant v, for small angles where the arc of the circle and its chord are not appreciably different in length, d$v = v \cdot$ **dß**. Thus we have:

$$ds = r \cdot \textbf{dß} \quad \text{and} \quad dv = v \cdot \textbf{dß}$$

Eliminating **dß** gives: $dv/ds = v/r$.

The acceleration, a, of the mass can now be found because:

$$a = \frac{dv}{dt} = \frac{ds}{dt} \cdot \frac{dv}{ds} = v \cdot \frac{v}{r} = \frac{v^2}{r} \tag{B2.1.1}$$

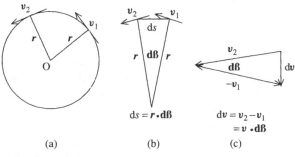

$$ds = r \cdot \textbf{dß} \qquad dv = v_2 - v_1$$
$$= v \cdot \textbf{dß}$$

        (a)   (b)   (c)

Figure B2.1.1   Motion in a circle

It is clear from figure (c) that for very small angles (**dß**), the change in velocity (dv) is at right angles to v and therefore directed along r towards O. Thus, if the change in velocity is directed towards O so also must be the acceleration.

The angular momentum (**l**) generated by the counter-clockwise motion illustrated in (a) is given by the vector product:

$$l = r \times p = r \times mv = rmv \qquad (B2.1.2)$$

The last equality follows because r and v are always at right angles to each other and the vector product of two vectors is the product of their magnitudes multiplied by the sine of the angle between them. The direction of the vector l is out of the paper towards the reader.

Angular momentum and energy

For the system described above all energy is kinetic energy which can be readily expressed in terms of the angular momentum:

$$E\text{(kinetic energy)} = \frac{mv^2}{2} = \frac{(mv)^2}{2m} = \frac{(l/r)^2}{2m} = \frac{l^2}{2mr^2} \qquad (B2.1.3)$$

The Bohr model

Consider an electron of mass m_e and charge $-e$ moving with velocity v in a circular orbit of radius r around a much heavier nucleus of charge $+e$ (Figure B2.1.2). The electron follows this path because of the coulombic force of attraction, $-e^2/4\pi\varepsilon_0 r^2$, pulling it towards the nucleus. ε_0 is the vacuum permittivity. The electron does not fall into the nucleus because of its velocity at right-angles to the line between the two particles, the radial vector. But, as shown above, the coulombic force does cause an acceleration of the electron towards the nucleus equal to $-v^2/r$, where the negative

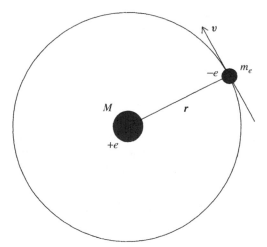

Figure B2.1.2   Bohr's model of the hydrogen atom

sign indicates that the acceleration, like the force which causes it, is in the direction of decreasing r. Now, according to Newton, force = mass × acceleration so:

$$\frac{e^2}{4\pi\varepsilon_0 r^2} = \frac{m_e v^2}{r} \quad \text{or} \quad r = \frac{e^2}{4\pi\varepsilon_0 m_e v^2} \tag{B2.1.4}$$

The discussion up to this point has been based entirely upon classical theory and therefore we find that no restriction is placed upon the radius of the electron's orbit (r). For any choice of r a velocity can be found to balance the above equation. Bohr's second law introduces quantisation with the postulate that only those orbits for which the orbital angular momentum is an integer multiple of $h/2\pi$ are allowed. The angular momentum is $m_e v r$ so Bohr's quantum condition is:

$$m_e v r = nh/2\pi \qquad n = 1, 2, 3, \ldots, \infty \tag{B2.1.5}$$

$n =$ zero is not permitted because it would imply either $v = 0$, giving rise to an infinite radius (Equation (B2.2.1)) or $r = 0$. If the allowed orbits satisfy Equation (B2.1.5) then the allowed values of the radius can be obtained by using Equation (B2.1.5) to eliminate v from Equation (B2.1.4) giving:

$$r = \frac{e^2}{4\pi\varepsilon_0 m_e} \times \frac{4\pi^2 m_e^2 r^2}{n^2 h^2} \quad \text{or} \quad m_e r = \frac{\varepsilon_0 n^2 h^2}{e^2 \pi} \tag{B2.1.6}$$

and using Equation (B2.1.5) again:

$$v = \frac{nh}{2\pi m_e r} = \frac{nh}{2\pi} \times \frac{e^2 \pi}{\varepsilon_0 n^2 h^2} = \frac{e^2}{2\varepsilon_0 nh} \tag{B2.1.7}$$

To obtain the total energy, E, of the atom we use:

$$E = \text{potential energy} + \text{kinetic energy}$$

so that:

$$\begin{aligned} E &= -\frac{e^2}{4\pi\varepsilon_0 r} + \frac{m_e v^2}{2} \\ &= -\frac{e^2}{4\pi\varepsilon_0}\left(\frac{e^2 \pi m_e}{\varepsilon_0 n^2 h^2}\right) + \frac{m_e}{2}\left(\frac{e^2}{2\varepsilon_0 nh}\right)^2 \\ &= \frac{e^4 m_e}{4\varepsilon_0^2 n^2 h^2}\left(-1 + \frac{1}{2}\right) = \frac{-e^4 m_e}{8\varepsilon_0^2 n^2 h^2} \end{aligned} \tag{B2.1.8}$$

The presence of the quantum number n on the right-hand side of Equation (B2.1.8) for E shows why n cannot be zero and that we have quantisation of the total energy of the hydrogen atom. We recognise this by writing the quantum number in brackets:

$$E(n) = \frac{-1}{n^2} \times \left(\frac{e^4 m_e}{8\varepsilon_0^2 h^2}\right) \tag{B2.1.9}$$

We now require to see if we can reproduce Balmer's Equation (Equation (2.7.1)). According to Bohr's fourth postulate, the energy of a spectral line is given by the difference between two of the above quantised energy values of the hydrogen atom, i.e. the energy calculated for two different quantum numbers, n_1 and n_2 say. Thus:

$$E(n_2) - E(n_1) = \Delta E = \frac{e^4 m_e}{8\varepsilon_0^2 n^2 h^2} \times \left(\frac{1}{n_1^2} - \frac{1}{n_2^2}\right) = \frac{e^4 m_e}{8\varepsilon_0^2 n^2 h^2} \times \frac{(n_2^2 - n_1^2)}{n_1^2 n_2^2}$$

(B2.1.10)

But the frequency of light (ν) and the wavelength (λ) are related by the velocity (c) in the equation $c = \nu\lambda$, so that:

$$\Delta E = h\nu = hc/\lambda$$

(B2.1.11)

Therefore, the wavelength of a line in the Balmer series is:

$$\lambda = \frac{8\varepsilon_0^2 c h^3}{e^4 m_e} \times \left(\frac{n_1^2 n_2^2}{n_2^2 - n_1^2}\right) = \frac{8 n_1^2 \varepsilon_0^2 c h^3}{e^4 m_e} \times \left(\frac{n_2^2}{n_2^2 - n_1^2}\right)$$

(B2.1.12)

A comparison of this equation with Equation 2.7.1 suggests that $n_1 = 2$ and $n_2 = n$, and if we make these substitutions our expression for λ can now be written:

$$\lambda = \frac{32\varepsilon_0^2 c h^3}{e^4 m_e} \times \left(\frac{n^2}{n^2 - 4}\right)$$

(B2.1.13)

The crucial test of Bohr's laws is now obtained by evaluating the factor $32\varepsilon_0^2 c h^3 / e^4 m_e$ in Equation (B2.2.13). This calculation is carried out in Section 2.7.2. where excellent agreement with Balmer, and therefore with experiment is found.

BOX 2.2   Operators

We can divide the symbols used in mathematics into two types, the *operators* and the *operands*. The operands are the numbers, or in the case of algebra, the letters which represent the numbers. The operators are the symbols which tell us what we have to do with the numbers. Thus, in the expression $2 + 5$, the operator, $+$, tells us that we are to add the number 5 to the number 2. Similarly, 2×5 indicates that 5 is to be multiplied by 2. The operator $\sqrt{}$ tells us to take the square root of the operand. As examples of more complicated operators we may take those for differentiation and integration which, as it happens, occur frequently in quantum mechanics. In a crude analogy we might say that the surgeon is the operator and the patient the operand.

Pairs of operators are linked by a very important property which is independent of the operand. This may be illustrated using the operators for multiplication by y, y×, multiplication by z, z×, and taking the square root, $\sqrt{}$. As the operand we shall use $4y^3$.

First we take the operators y× and z× and apply them to the operand in the order y× z×. We obtain:

$$y \times (z \times [4y^3]) = y \times (4zy^3) = 4zy^4$$

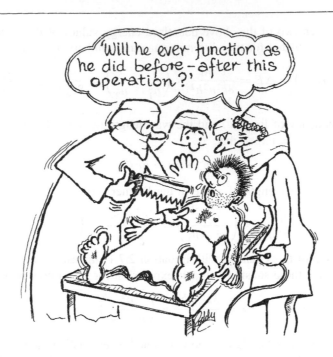

And if we apply the two operators in the reverse order we get the same result, i.e.:

$$z \times (y \times [4y^3]) = z \times (4y^4) = 4zy^4$$

But if we choose the operator pair $y\times$ and $\sqrt{\ }$, we find that the order in which we apply them is important because:

$$\sqrt{(z \times [4y^3])} = \sqrt{(4zy^3)} = 2z^{\frac{1}{2}}y^{\frac{3}{2}}$$

But:

$$z \times \sqrt{[4y^3]} = z \times (2y^{\frac{3}{2}}) = 2zy^{\frac{3}{2}}$$

We say that the operator pair $y\times$ and $z\times$ *commute*, but the pair $y\times$ and $\sqrt{\ }$ do not commute. The commutation or non-commutation of pairs of operators plays a very important role in quantum mechanics as we shall see in Chapter 3.

PROBLEMS FOR CHAPTER 2

1. For the special case of a planet with a circular orbit, deduce Kepler's third law from Newton's laws.

2. Calculate the radius of the circular orbit of a geostationary satellite, i.e. a satellite which remains in exactly the same position above the surface of the earth. Determine

its altitude and orbital velocity. Assume that the mass of the earth is 5.983×10^{24} kg, its radius is 6.378×10^6 m and the gravitational constant $G = 6.673 \times 10^{-11}$ m^3 kg^{-1} s^{-2}.

3. Estimate the work functions W(Na) and W(Li) from Figure 2.5.

4. The most powerful tennis players are reputed to serve the ball at a speed of about 200 km/hour. If the mass of the ball is 58 g calculate its wavelength.

5. Confirm the range of wavelengths quoted in Section 2.8 for Thomson's accelerated electrons.

6. By expanding the exponential in Equation 2.5.4b, show that $\rho(\nu) \propto \nu^2$ when $h\nu \ll kT$.

CHAPTER 3

The Application of Quantum Mechanics

3.0	Introduction	38
3.1	Observables, operators, eigenfunctions and eigenvalues	38
3.2	The Schrödinger method	39
3.3	An electron on a ring	40
	3.3.1 The Hamiltonian operator for the electron on a ring	40
	3.3.2 Solution of the Schrödinger equation	41
	3.3.3 The acceptable eigenfunctions	41
3.4	Hückel's $(4N + 2)$ rule: aromaticity	44
3.5	Normalisation and orthogonality (Box 3.2)	46
3.6	An electron in a linear box	46
	3.6.1 The Hamiltonian operator for an electron in a linear box	46
	3.6.2 The acceptable eigenfunctions	47
	3.6.3 Boundary conditions	48
3.7	The linear and angular momenta of electrons confined within a one-dimensional box or on a ring	48
	3.7.1 The linear momentum of an electron in a box	49
	3.7.2 The angular momentum of an electron on a ring	50
3.8	The eigenfunctions of different operators	52
3.9	Eigenfunctions, eigenvalues and experimental measurements	53
3.10	More about measurement: the Heisenberg uncertainty principle	55
3.11	The commutation of operators (Boxes 2.2 and 3.6)	57
3.12	Combinations of eigenfunctions and the superposition of states	58
3.13	Operators and their formulation	59
	3.13.1 Position or co-ordinate, $\hat{x}$	59
	3.13.2 Potential energy, $\hat{V}$	59
	3.13.3 Linear momentum, $\hat{p}_x$	60
	3.13.4 Kinetic energy, $\hat{W}$	60
	3.13.5 Angular momentum, $\hat{L}$	60
3.14	Summary	60
3.15	Bibliography and further reading	61
	Problems for Chapter 3	71

The Quantum in Chemistry   R. Grinter
© 2005 John Wiley & Sons, Ltd

3.0   INTRODUCTION

Consider the hydrogen molecule, H–H. The hydrogen atoms each have a mass of 1.674×10^{-27} kg and their nuclei are separated by a bond of length $\sim 74 \times 10^{-12}$ m. In actual fact, this is the equilibrium bond length of the molecule, for this minute entity is in constant vibration with a frequency of $\sim 132 \times 10^{12}$ oscillations per second. Just pause for a moment and try to picture the molecule. In the time it takes to blink, for example, it will have completed 132 million, million vibrations. Is it permissible to describe such a motion as a 'vibration' when we cannot really even imagine such frequencies? Masses as small as 10^{-27} kg and distances as short as 10^{-12} m are indescribably remote from the phenomena which we experience directly with our own senses. Quantum mechanics, which has now stood the test of more than 75 years of rigorous application, is the mathematical model which has been constructed to describe atomic and molecular systems and to interpret the results of the experiments which we perform upon them. But it should come as no surprise when the picture of this world of atoms and molecules, drawn by quantum-mechanical methods, sometimes seems very strange to us.

The early pioneers of the subject soon discovered that, in order to keep on course in interpreting of the results of quantum mechanics, we must be very careful to focus our attention upon the quantities which can be measured in a particular experimental system and refrain from speculation about those things which cannot. (Recall the interference of photons discussed in Section 2.8.) Thus, to return to the example of the hydrogen molecule, the equilibrium bond length of the molecule can be both measured and calculated to high degrees of accuracy. But there is no way in which the paths of the two electrons which form that bond can be followed or described. We can only give a figure for the probability of finding an electron (we cannot say which one!) at any particular point in space.

Thus, in this chapter we are concerned not only to illustrate the way in which quantum mechanics is applied, but also to make clear just what information can be obtained from quantum-mechanical calculations or by experiment. We shall find that, though the information which we can obtain is restricted when compared with that available in classical mechanics, this is not so surprising when we consider the fundamental limitations which must apply to measurements on atomic and molecular systems.

3.1   OBSERVABLES, OPERATORS, EIGENFUNCTIONS AND EIGENVALUES

Any phenomenon we can observe, and particularly one which we can measure and assign a numerical value to, e.g. energy or angular momentum, is known as an *observable*. It is the primary objective of quantum mechanics to provide a theory by means of which the values of such observables can be calculated. Just as classical mechanics provides an equation for determining the kinetic energy of a moving mass (k.e. $= \frac{1}{2}mv^2$), so quantum mechanics provides us with an equation by means of which the kinetic energy of a moving electron can be determined.

In Section 2.9 we saw that, when proposing his new wave mechanics, Erwin Schrödinger suggested that the mathematical concept of eigenfunctions and eigenvalues would be found to lie at the heart of any theory in which quantisation arose naturally, rather than being arbitrarily imposed. All subsequent work has demonstrated that Schrödinger's intuition was indeed correct and in the application of quantum mechanics today we formulate Schrödinger's method in terms of four entities: an *observable*, the particular quantum-mechanical *operator* associated with that observable and the *eigenfunctions* and *eigenvalues* of that operator. The operator, eigenfunction and eigenvalue are related in the equation:

$$\textit{Operator operating on Eigenfunction} = \textit{Eigenvalue} \times \textit{Eigenfunction} \qquad (3.1.1)$$

The usual mathematical formulation of the relationship is:

$$\hat{\mathcal{R}}\Psi_n = r_n\Psi_n \qquad (3.1.2)$$

and a specific example is: $\partial^2(\sin ax)/\partial x^2 = -a^2 \sin ax$. The operator, $\hat{\mathcal{R}}$ (designated as such with the caret mark or circumflex; $\partial^2/\partial x^2$ in the example), operates upon the eigenfunction, Ψ_n ($\sin ax$), which is a function of the co-ordinate system in which we place the atom or molecule we are studying, e.g. the Cartesian co-ordinates x, y and z. The result is the eigenfunction multiplied by the eigenvalue, r_n ($-a^2$). We shall see below that the fact that the eigenfunction is describing a real atomic or molecular system places important limitations upon the form which it may take. The observable appears in Equations (3.1.1) and (3.1.2) as the eigenvalue, r_n, which is simply a number that is the value of the observable in the units we have chosen for the calculation, e.g. an energy in J. The subscript, 'n', indicates that there are generally many eigenfunctions of $\hat{\mathcal{R}}$ and an equal number of corresponding eigenvalues, each characterised by a different value of n. We shall find that n is a quantum number, or more often a set of quantum numbers, e.g. those describing the electronic state of an atom.

3.2 THE SCHRÖDINGER METHOD

Schrödinger's method consists of three basic steps:

1. Determine the operator associated with the observable to be calculated.

2. Find the eigenvalues and eigenfunctions of the operator.

3. Ensure that the eigenfunctions are physically acceptable, in which case the corresponding eigenvalues will be the possible, quantised values of the observable.

It is the process of choosing the appropriate operator where a postulate is made which can only be tested against experiment. In formulating classical mechanics Newton made the three postulates described in Section 2.2. To interpret the motion of the planets he added a fourth postulate, gravitational attraction, and then demonstrated that these postulates were appropriate by showing that the data for the motion of the planets around the sun could be accurately reproduced by a mathematical model of planetary motion based upon them. In just the same way, Schrödinger postulated the form of the operator that would have as its eigenvalues the possible energies of the electron in a hydrogen atom and tested his postulate by calculating the hydrogen atom spectrum. At the same time he showed the way in which the operators required for all the other quantum-mechanical problems of interest in chemistry can be obtained (Section 3.13). In order to illustrate

Schrödinger's method we shall consider two problems which, though they are idealised and simple, also have interesting chemical applications. The problems concern the quantum mechanics of an electron confined to move within a one-dimensional box or around a ring.

3.3   AN ELECTRON ON A RING

Following Schrödinger's method, our first task is to determine the form of the energy operator. This operator is called the *Hamiltonian*, after William Rowan Hamilton (1805–1865) who developed a formulation of classical mechanics in which the relationship of the latter to quantum mechanics can be most clearly recognised. We shall write it as $\hat{\mathcal{H}}$.

3.3.1   The Hamiltonian operator for the electron on a ring

Consider an electron of mass m_e which is confined to move at velocity v_e in a circular path of radius r (Figure 3.1). There is no positive charge at the centre of the circle, as there is in the Bohr model of the atom, so the electron has only kinetic energy. If the electron obeyed the laws of classical mechanics its energy would be:

$$E = \tfrac{1}{2}m_e v_e^2 = (m_e v_e r)^2 / 2m_e r^2 \tag{3.3.1}$$

We write the expression for the classical kinetic energy in the above, more complicated, way because Schrödinger's recipe for finding a quantum-mechanical operator depends upon first writing down the classical mechanical expression for the observable in question, in this case the kinetic energy, and then transforming it to the required operator by means of a set of rules. In this case we note that the quantity $(m_e v_e r)^2$ is the square of the angular momentum of the electron (see Chapter 4) and Schrödinger's postulate for obtaining the operator corresponding to angular momentum squared, $\hat{l}^2$, is:

$$l^2 = (m_e v_e r)^2 \Rightarrow \hat{l}^2 = -(h/2\pi)^2 \partial^2/\partial\phi^2 \tag{3.3.2}$$

That is, we form the operator by replacing the classical expression for the square of the angular momentum by $-\hbar^2 \{\hbar \equiv h/2\pi\}$ times the second derivative of the wave-function with respect to ϕ, the angle defining the position of the electron as it travels around the ring. Rules for forming other operators will be given as the need arises. Thus, we require

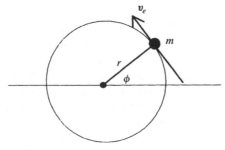

Figure 3.1   An electron on a ring

the eigenvalues (E_n) and eigenfunctions (Ψ_n) which solve the equation:

$$\hat{\mathcal{H}}\Psi_n \equiv -\frac{\hbar^2}{2m_e r^2} \cdot \frac{\partial^2 \Psi_n}{\partial \phi} = E_n \Psi_n \qquad (3.3.3)$$

As above, the subscript n will be used to characterise the eigenfunctions and the associated eigenvalues, of which there are usually a large number. It must be emphasised that the form of the operator is a postulate. Although Schrödinger's method makes use of a formula from classical mechanics as a stepping stone in obtaining the operator, the relationship between the classical expression and the quantum operator is *postulated*: the operator is not *derived* from classical mechanics. The proof of the validity of the postulate lies in the results which it gives and which we now determine.

3.3.2 Solution of the Schrödinger equation

In the second of Schrödinger's three steps, we now seek a function which, if it is to be an eigenfunction, must be such that when differentiated twice with respect to the variable ϕ it returns to its original form. Two functions with just this property spring readily to mind; $\sin(a\phi)$ and $\cos(a\phi)$, where a is a number (Box 3.1):

$$\partial^2(\sin a\phi)/\partial\phi^2 = a\partial(\cos a\phi)/\partial\phi = -a^2 \sin a\phi$$

$$\partial^2(\cos a\phi)/\partial\phi^2 = -a\partial(\sin a\phi)/\partial\phi = -a^2 \cos a\phi$$

Either of these functions is an eigenfunction of the operator with an eigenvalue of $a^2\hbar^2/2m_e r^2$ because:

$$-\frac{1}{2m_e r^2}\hbar^2 \cdot \frac{\partial^2 \sin(a\phi)}{\partial^2 \phi} = \frac{a^2\hbar^2}{2m_e r^2} \cdot \sin(a\phi) \qquad (3.3.4)$$

with a very similar result for the cosine function.

Therefore, the energy of our system is $a^2\hbar^2/2m_e r^2$. However, a moment's consideration shows that we are not yet home and dry. Though h and m_e are fixed, natural constants and the radius (r) would be fixed for any particular problem, no restriction has been placed upon the constant a. Consequently, any desired energy can be obtained with a suitable choice of the value of a; i.e. there is as yet no quantisation. We must now proceed to Schrödinger's third step and examine the eigenfunctions $\sin(a\phi)$ and $\cos(a\phi)$ to see if they are physically acceptable.

3.3.3 The acceptable eigenfunctions

The meaning of the eigenvalues of Schrödinger's equations was clear from the start and never appears to have presented the pioneers of quantum mechanics with any difficulties. The interpretation of the eigenfunctions, on the other hand, raised many philosophical problems about which argument and discussion still continue more than 75 years after Schrödinger's first publication of his new mechanics. However, we shall adopt the interpretation first suggested by Max Born (1882–1970) in 1926. This is the view taken in all chemical applications of quantum mechanics. Born proposed that the *square* of the eigenfunction, evaluated at any point in space, gives the *probability* of finding the particle described by that eigenfunction at that particular point. In interpreting the eigenfunction in this way Born took a very fundamental step. He abandoned the classical certainty of a precise position for the particle and introduced in its place the concept of the probability of finding a particle at any particular point in space. The significance of this departure from the certainty of classical mechanics cannot be exaggerated.

Thus, if a particular value of the angle ϕ is chosen and the eigenfunctions $\sin(a\phi)$ and $\cos(a\phi)$ of the particle on a ring at that point evaluated, then the square of the result gives the probability of finding the electron at that point in its endless journey around the ring. For a concrete example, if a value $a = 4.8$ is chosen, the functions $\sin(4.8\phi)$ and $[\sin(4.8\phi)]^2$ plotted against ϕ for values of ϕ from 0 to ca. 3π are shown in Figure 3.2. We note that the eigenfunction itself has regions of both positive and negative sign, i.e. it has the property of phase. But the probability function, the squared eigenfunction, is always positive as it must be if it represents probability. Using Figure 3.2 we can readily find the value of the eigenfunction for any value of ϕ and hence determine the probability that the electron can be found at that particular position around the ring. But the linear ϕ-axis of Figure 3.2 conceals, or rather does not make obvious, a fundamental deficiency in the eigenfunction $\sin(4.8\phi)$. Figure 3.3 in which the function is plotted on a dotted circle (positive values outside the circle, negative values inside) shows that when the angle ϕ has reached the value of 2π and the electron begins a new circuit of the ring another, different value of the eigenfunction is obtained for the same position on the ring. To be specific; $\sin(4.8\phi) = 0.0$ for $\phi = 0$ but for $\phi = 2\pi$, $\sin(4.8\phi) = -0.951$. On the Born model this would imply that the probability of finding the electron at the same place on the ring is proportional[‡] to two different numbers, 0.0 and $(-0.951)^2 = 0.904$. This

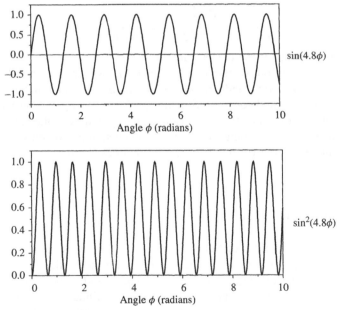

Figure 3.2　Linear plot of $\sin(4.8\phi)$ and $[\sin(4.8\phi)]^2$

[‡] We use the word 'proportional' because the wave function still has to be multiplied by a factor which we shall explain and determine in Section 3.5.

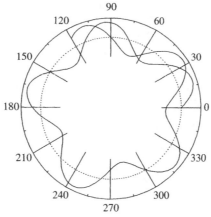

Figure 3.3 Circular plot of $\sin(4.8\phi)$, $\phi = 0 - \text{ca}.2.5\pi$

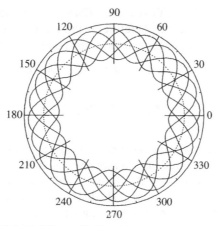

Figure 3.4 Circular plot of $\sin(4.8\phi)$, no limit on ϕ

cannot be so. Hence, though the eigenfunction is perfectly acceptable *mathematically*, it is *physically* quite unacceptable. To be physically acceptable, an eigenfunction must have only one value (*i.e. it must be single-valued*) at any point. Another way of looking at the result is shown in Figure 3.4. If we plot the eigenfunction against ϕ with no upper limit on ϕ, we discover that after five cycles around the ring the value of the eigenfunction begins to repeat itself and a rather pretty pattern results. A detailed examination reveals that at any value of ϕ there are five values of the eigenfunction which are such that their sum is equal to zero. Thus the phases of the eigenfunction cancel each other giving a total of zero and hence zero probability. In the language of waves, we would say that the function has been reduced to zero by destructive interference.

The only physically reasonable eigenfunctions are ones for which the value at $\phi = 0$ is equal to the value at $\phi = 2\pi, 4\pi, \ldots$ (Figure 3.5). What does this mean for our eigenfunctions and eigenvalues? How does this requirement restrict the value of a? We require that:

$$\sin(a\phi) = \sin(a[\phi + 2\pi]) = \sin(a\phi + 2a\pi) = \sin(a\phi) \cdot \cos(2a\pi) + \cos(a\phi) \cdot \sin(2a\pi)$$

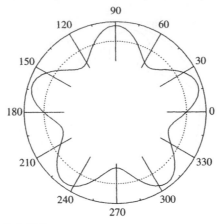

Figure 3.5 Circular plot of $\sin(5\phi)$

Clearly, the equality will be achieved if $\cos(2a\pi) = 1.0$ and $\sin(2a\pi) = 0.0$, which is only true when $a = 0, 1, 2, 3, \ldots$ Therefore, if the eigenfunctions are to be physically reasonable a must be zero or an integer and the corresponding eigenfunctions are given by:

$$\Psi_{\sin a} = \sin(a\phi) \quad \text{and} \quad \Psi_{\cos a} = \cos(a\phi) \tag{3.3.5}$$

and the energies by:

$$E_a = a^2\hbar^2/2m_e r^2 \tag{3.3.6}$$

for both eigenfunctions.

Thus, quantisation has arisen in the most natural way, as a result of the requirement that our eigenfunctions should be physically reasonable on Born's model. Figure 3.5 shows the function $\Psi = \sin(5\phi)$ in the circular form in which a whole number of cycles of the function fit exactly into the circular path. It is now natural to interpret our integer a as the number n, which we introduced in Equations (3.1.2) and (3.3.3) to allow for the fact that there may be many eigenfunctions and eigenvalues of the same operator. In this example the acceptability of the eigenfunction was determined by the requirement that it be single-valued. Since the probability must be smoothly continuous and finite, in order to be physically acceptable, the Born model also places the general requirements on eigenfunctions that they too must be continuous and finite everywhere.

It is interesting to note that the energy levels occur in pairs of equal energy (Figure 3.5), one level corresponding to the sine eigenfunction and the other to the cosine eigenfunction. In quantum-mechanical parlance, such levels are said to be *degenerate*. In Section 3.7.2 we shall discover the fundamental origin of this degeneracy. For $a = 0$ there is only one eigenfunction, the cosine function (since $\sin(0) = 0$), and therefore only one lowest energy level of zero energy.

3.4 HÜCKEL'S (4N + 2) RULE: AROMATICITY

It was stated above that, although the quantum-mechanical problems to be discussed in this chapter were simple ones, they nevertheless throw valuable light upon chemical problems. The quantum-mechanical description of the electron confined to move on a ring provides a basis for Hückel's $(4N + 2)$ rule for the aromatic hydrocarbons. Recall that the aromatic hydrocarbons, e.g. benzene, naphthalene, anthracene, phenanthrene etc., have $6, 10, 14, \ldots = (4N + 2)$ conjugated carbon atoms, where $N = 1, 2, 3, \ldots$ (Figure 3.7). It is generally accepted that the particular stability of these molecules depends upon the presence of a π-electron system, and that each carbon atom contributes one electron to this system, i.e. the number of π-electrons in an aromatic hydrocarbon is equal to the number of carbon atoms $(4N + 2)$. (The Hückel theory of π-electron systems is discussed in detail in Chapter 12.) In many of these molecules, especially the smaller ones, all the carbon atoms lie on the periphery of the carbon-atom skeleton and we can approximate the sequence of carbon atoms to a ring around which the delocalised π-electrons may circulate while the remainder of the bonding electrons are localised in the C–C and C–H bonds. In this model, the energy-level scheme for the π-electrons is that shown in Figure 3.6 and if we fill the levels with electrons, two per level (the Pauli principle),

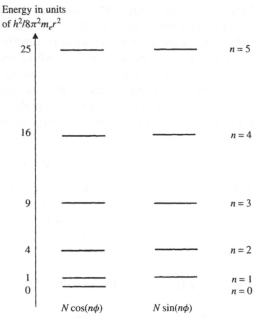

Figure 3.6   The energy levels for an electron on a ring

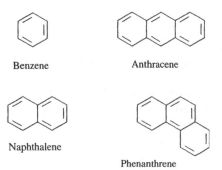

Figure 3.7   The structures of some of the smaller aromatic hydrocarbons

starting from the lowest (the aufbau principle), then it is easy to see that to produce a stable configuration in which all levels up to a particular energy are completely filled requires $(4N + 2)$ electrons.

This concept goes a long way towards explaining the particular stability of the aromatic hydrocarbons. In 1949 the idea was taken further by John R. Platt (1918–) who used the electron-on-a-ring model to account for and classify the electronic spectra of the aromatic hydrocarbons (see Problem 6). His classification is widely used by spectroscopists and photochemists.

3.5   NORMALISATION AND ORTHOGONALITY (BOX 3.2)

We have not yet completed the task of ensuring that our eigenfunctions are physically reasonable. If the square of an eigenfunction, evaluated for any value of ϕ, gives the probability of finding the electron at that position on the ring, then the sum of all such values around the ring must be 1.0 ($\equiv$ certainty) since, by definition, the electron is somewhere on the ring. Since the probability distribution is a continuous function of the co-ordinates, the sum becomes an integral over the complete range of all of them, in brief, *over all space*. The process of ensuring that an eigenfunction gives rise to a total probability of one is known as the *normalisation* of the eigenfunction. It is an additional requirement to ensure that the chosen mathematical functions are also acceptable physically. The final, normalised forms of our two types of eigenfunction with the quantum number now written as n are (see Box 3.2):

$$\Psi_{s,n} = (1/\sqrt{\pi}) \cdot \sin(n\phi) \quad \text{and} \quad \Psi_{c,n} = (1/\sqrt{\pi}) \cdot \cos(n\phi) \ (n = 1, 2, 3, \ldots) \quad (3.5.1)$$

When $n = 0$, $\cos 0 = 1$ and for this function the normalising constant is $1/\sqrt{(2\pi)}$.

The eigenfunctions of a quantum-mechanical operator have another interesting property, orthogonality, which is closely related to normalisation. The orthogonality of the above eigenfunctions is also demonstrated in Box 3.2.

3.6   AN ELECTRON IN A LINEAR BOX

In this section we examine the energy-levels of an electron confined within a one-dimensional box; an idealised situation which also casts significant light on chemical problems.

3.6.1   The Hamiltonian operator for an electron in a linear box

Suppose that we have an electron confined by infinite potential barriers within a one-dimensional linear box of length L. We shall again assume that there is no electrostatic potential present so that the only energy is kinetic energy. To construct the appropriate operator we follow Schrödinger's method and first write down the classical expression for the kinetic energy in terms of the momentum, p:

$$E = \tfrac{1}{2}m_e v^2 = (m_e v)^2/2m_e = p^2/2m_e$$

The postulate for the conversion of the classical expression for the square of the linear momentum to the appropriate operator is:

$$p_q^2 = (m_e v_q)^2 \Rightarrow \hat{p}_q^2 = -\hbar^2 \cdot \frac{\partial^2}{\partial q^2} \tag{3.6.1}$$

where $q = x$, y or z, the co-ordinate along the box. Thus, the expression for the Hamiltonian operator is:

$$\hat{\mathcal{H}} = -(\hbar^2/2m_e)\partial^2/\partial x^2 \tag{3.6.2}$$

Recalling the discussion of the problem of an electron on a ring, we can guess that the functions:

$$\Psi_{s,b} = \sin(bx) \quad \text{and} \quad \Psi_{c,b} = \cos(bx) \tag{3.6.3}$$

should be investigated since both are eigenfunctions of the operator $-(\hbar^2/2m_e)\partial^2/\partial x^2$ with eigenvalues of $b^2\hbar^2/2m_e$.

3.6.2 The acceptable eigenfunctions

But are these functions physically acceptable? We again consider the behaviour of the eigenfunctions and in particular their behaviour at $x = 0$ and $x = L$, the two ends of the box. Since the electron is, by definition, confined within the box, the probability of finding the electron must fall smoothly to zero at $x = 0$ and L. $\cos(bx) = 1.0$ at $x = 0$ and is therefore unacceptable on physical grounds. The sine function, however, goes smoothly to zero as x goes to 0, and the function must be such that this is also true at $x = L$. Since $\sin(n\pi) = 0$ when n is an integer we can satisfy the foregoing requirement if we choose b so that $bL = n\pi$ or $b = n\pi/L$. Thus, our only physically acceptable eigenfunctions of the energy operator are:

$$\Psi_{s,n} = N \sin(n\pi x/L) \quad (n = 1, 2, 3, \ldots) \tag{3.6.4}$$

where N is a normalising constant. The corresponding eigenvalues of the energy are:

$$E_n = \left(\frac{n^2\pi^2}{L^2}\right) \cdot \left(\frac{\hbar^2}{2m_e}\right) = \frac{n^2h^2}{8m_e L^2} \quad (n = 1, 2, 3, \ldots) \tag{3.6.5}$$

Note that n cannot be zero since $\sin(0) = 0$ and there is therefore no eigenfunction. Thus, the system *never has an energy of zero*, i.e. it always has some energy and the presence of this *zero-point energy* is an important point to which we shall return in Section 3.11.

In the eigenfunction, the letter b which previously had no particular significance has now been replaced by the quantum number n (multiplied by π/L). To find the normalising constant (N) we simply require that the probability of finding the electron at any point along the box, summed up, i.e. integrated, over the complete length of the box, is one, i.e. that:

$$\int_0^L \{N \sin(n\pi x/L)\}^2 \, \mathrm{d}x = 1.0$$

This leads to a value of N of $\sqrt{(2/L)}$ (Box 3.3), so our complete set of physically acceptable eigenfunctions is:

$$\Psi_{s,n} = \sqrt{(2/L)} \cdot \sin(n\pi x/L) \quad (n = 1, 2, 3, \ldots) \tag{3.6.6}$$

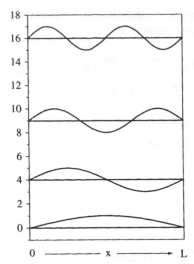

Figure 3.8 Energy levels and eigenfunctions for an electron in a linear box

These functions are illustrated for $n = 1$ to 4 in Figure 3.8. The π-electron spectra of linear conjugated polyenes can be interpreted rather well in the light of these results (Box 3.4).

3.6.3 Boundary conditions

In selecting a physically acceptable eigenfunction from the various possible mathematical solutions of the Schrödinger equation for an electron in a linear box, we made use of the fact that the eigenfunction must behave in an appropriate manner at the ends (boundaries) of the box. This led not only to the complete rejection of many eigenfunctions but also to the selection of particular members of the set of sine functions, i.e. those for which $b = n\pi/L$, where n is an integer. It was at this point that quantisation entered the problem. Quantisation is intimately associated with the behaviour of eigenfunctions at the boundaries of the system and we say that boundary conditions are imposed upon the eigenfunctions to determine those which are physically acceptable. In the case of the electron on a ring there was no obvious physical boundary; but the fact that the eigenfunctions were required to have the same value at $\phi = 0$ and 2π was the boundary condition. Boundary conditions of this type known as *periodic boundary conditions*.

3.7 THE LINEAR AND ANGULAR MOMENTA OF ELECTRONS CONFINED WITHIN A ONE-DIMENSIONAL BOX OR ON A RING

In determining the energy levels in the two idealised models described above, we have used the linear or angular momentum of the electron as an intermediate step in the construction the required energy operator, the Hamiltonian. In this section we return to consider the momenta themselves, and their corresponding eigenvalues and eigenfunctions. We take the linear system first.

3.7.1 The linear momentum of an electron in a box

In Section 3.6.1 we used the operator for the square of the linear momentum of the electron in a linear box, $-\hbar^2 \cdot \partial^2/\partial x^2$, in order to form the Hamiltonian operator so that the operator for linear momentum appears to be readily available as the square root of $-\hbar^2 \cdot \partial^2/\partial x^2$. But this presents two problems. Firstly, what is the square root of $\partial^2/\partial x^2$? This is not difficult; $\partial^2/\partial x^2$ means differentiate the function twice with respect to x, i.e. $(\partial/\partial x) \cdot (\partial/\partial x)$, so a single differentiation would represent the square root. The $-\hbar^2$ poses a greater problem in that we have to find the square root of a negative quantity! But we do not allow such problems to deter us; we simply follow the lead of the mathematicians who confronted this problem many years ago and solved it by inventing a new number, i, the square root of -1; i.e. $i^2 = -1$ (Appendix 8). Having done this we can express the square root of any negative number, e.g. $(\pm 2i)^2 = -4$, $(\pm 5.3i)^2 = -28.09$. Such numbers are called *imaginary* numbers and with their aid we can write the operator for linear momentum as:

$$\hat{p} = -i\hbar \cdot \frac{\partial}{\partial x} \qquad (3.7.1)$$

Of the two possible signs for the square root we choose the negative one for reasons which will become clear below.

If we apply this operator to our energy eigenfunctions $\Psi_{s,n}$ we see immediately that they are not eigenfunctions of the linear momentum operator because differentiation of sine gives cosine. What we need is a function which does not change when differentiated once. The exponential function, e^{nx} or $\exp(nx)$ is such a function (Box 3.1) and:

$$-i\hbar \cdot \frac{\partial}{\partial x}\{\exp(nx)\} = -i\hbar \cdot n\{\exp(nx)\} \qquad (3.7.2)$$

We now have a mathematical eigenfunction, but there is a serious problem here; the calculated value of our linear momentum, $-in\hbar$, contains a factor i and is therefore an imaginary quantity. This is quite unacceptable for a quantity which we can measure in the laboratory and which must therefore be real! But this can be overcome if we write $+in$ in place of n as the argument of the exponential function because then:

$$-i\hbar \cdot \frac{\partial}{\partial x}\{\exp(+inx)\} = -i\hbar \cdot in\{\exp(+inx)\} = n\hbar\{\exp(+inx)\} \qquad (3.7.3)$$

since $i^2 = -1$.

Thus, the function $\exp(+inx)$ is an eigenfunction of the linear momentum operator with an eigenvalue of $+n\hbar$, and by choosing the negative sign for the operator we have the result that the sign of the momentum is the same as that of the argument of the exponential function. We interpret this function as describing the motion of an electron in the positive x-direction with a positive value of p_x. Similarly, the function $\exp(-inx)$ is an eigenfunction of the linear momentum operator with an eigenvalue of $-n\hbar$ and describes an electron moving in the negative x-direction. Further, we should note that these two exponential functions are also eigenfunctions of the energy operator each having the same eigenvalue, i.e. $n^2\hbar^2/2m_e$.

But neither of the exponential functions is physically acceptable for a box of finite length L: both fail to satisfy the requirement that they must go smoothly to zero at $x = 0$ because $\exp(0) = 1.0$. Therefore, since the eigenfunctions of the linear momentum operator are not acceptable eigenfunctions for a box of finite length, no linear momentum can be measured for a particle confined within such a box. But if the box is infinitely long then there are, in effect, no boundaries and the two exponential functions are both physically acceptable. But then no restrictions are placed upon the mathematical eigenfunctions and there is therefore *no quantisation*. In such a case the eigenvalues of linear momentum corresponding to the two types of eigenfunction are:

$$\text{eigenfunction} \qquad \text{eigenvalue}$$
$$\Psi_{+b} = N \exp(+ibx) \quad p_{+b} = +b\hbar$$
$$\Psi_{-b} = N \exp(-ibx) \quad p_{-b} = -b\hbar$$

where b can have any value. The normalisation of the eigenfunction of an electron in an infinitely long box presents some difficulties and we shall not enter into that subject here (see Section 12.3 where periodic boundary conditions are used). The translational energy which results from this linear momentum is, of course, also unquantised. It is the form of these functions that is responsible for the name wave mechanics which is usually given to quantum mechanics in the Schrödinger form. The exponential functions describe electrons moving along x in a positive or negative direction and the energy calculated is the kinetic energy associated with that movement or momentum. The sine functions have no associated momentum since they are not eigenfunctions of the momentum operator. Inasmuch as they can be formed as the sum and difference of the two exponential functions (Appendix 8) they are standing waves which describe electrons moving in both directions along x with mutually cancelling momenta. But the energy is not zero because it is proportional to the square of the momentum.

3.7.2   The angular momentum of an electron on a ring

For an introduction to angular momentum see Chapter 4.

Since the electron is moving around the ring it has orbital angular momentum. To obtain the operator for angular momentum ($\hat{l}$) we follow the procedure of the last section and find:

$$\hat{l} = -i\hbar \cdot \partial/\partial\phi \qquad (3.7.4)$$

Where we again choose the negative square root for the same reason as in the linear case.

Here also we find that exponential functions are what we require and, in particular, the function $\exp(+in\phi)$ is an eigenfunction of the angular momentum operator with an eigenvalue of $+n\hbar$, while the function $\exp(-in\phi)$ is also an eigenfunction of $\hat{l}$ with an eigenvalue of $-n\hbar$. This looks encouraging, we expect to find equal and opposite values of the angular momentum since the electron can circle the ring in a clockwise (negative a.m.) or counter-clockwise (positive a.m.) direction.

But are the functions physically reasonable, are they single-valued? Is the value at $\phi = 0$, for example, equal to that at $\phi = 2\pi$? And are they normalised? To answer the first question we note the standard results (Appendix 8):

$$\exp(+in\phi) = \cos(n\phi) + i \sin(n\phi) \quad \text{and} \quad \exp(-in\phi) = \cos(n\phi) - i \sin(n\phi)$$

We know from our deliberations above that the sine and cosine functions are both single-valued if n is an integer or zero. Therefore, our exponential eigenfunctions of the angular momentum operator will also be single-valued if $n = 0, 1, 2, 3, \ldots$ When $n = 0$ the two eigenfunctions are one and the same and they have an eigenvalue of zero. For all other values of n the system has paired equal and opposite values of the angular momentum. Each value is an integral multiple of Planck's constant divided by 2π. Unlike the energy eigenvalues which are determined by n^2, the angular momentum eigenvalues are determined by n and are evenly spaced (Figure 3.9).

To normalise the functions we must make allowance for the fact that they contain both a real term, $\cos(n\phi)$ and an imaginary term $i \sin(n\phi)$. An expression which contains both real and imaginary parts is said to be *complex*. Born allowed for this possibility in defining the meaning of a quantum-mechanical eigenfunction and proposed that in the case of complex eigenfunctions the probability was given, not by the square of the function, but by the product of the function and its complex conjugate. The complex conjugate of any complex function is obtained simply by changing $+i$ to $-i$ and $-i$ to $+i$ everywhere (Appendix 8). Thus $\exp(+in\phi)$ is the complex conjugate of $\exp(-in\phi)$, and *vice versa*. Therefore, the probability, $P(\phi)$ of finding the electron at a particular angle, ϕ, around the ring is given by:

$$P(\phi) = \exp(+in\phi) \cdot \exp(-in\phi) = \exp(0) = 1.0$$

This result tells us two important facts. Firstly, the distribution of the electron does not depend upon the value of ϕ, it is the same at all points around the ring. But it also shows that the function requires normalisation (see Box 3.2) so that the sum of the probabilities all around the ring add up to 1.0. We find the normalising constant by integrating the expression above from 0 to 2π. We find:

$$\int_0^{2\pi} \exp(+in\phi) \cdot \exp(-in\phi)\, d\phi = \int_0^{2\pi} \exp(0)\, d\phi = \int_0^{2\pi} 1\, d\phi = [\phi]_0^{2\pi} = 2\pi$$

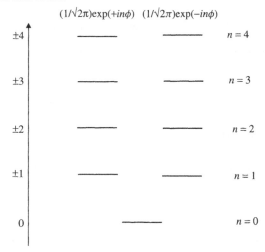

Figure 3.9 The angular momentum eigenvalues of an electron on a ring

which shows that each eigenfunction must be multiplied by $\sqrt{(1/2\pi)}$. Therefore, our final normalised wave functions are:

$$\Psi_{+n} = \sqrt{\frac{1}{2\pi}} \cdot \exp(+in\phi) \quad \text{and} \quad \Psi_{-n} = \sqrt{\frac{1}{2\pi}} \cdot \exp(-in\phi) \qquad (3.7.5)$$

3.8 THE EIGENFUNCTIONS OF DIFFERENT OPERATORS

We have determined the eigenvalues and eigenfunctions of the operators for the energy and angular momentum of an electron confined to move on a circle of radius r. In so doing we have noted that, although the functions $\Psi_{s,n} = \sqrt{(1/\pi)} \cdot \sin(n\phi)$ and $\Psi_{c,n} = \sqrt{(1/\pi)} \cdot \cos(n\phi)$, with n an integer, are eigenfunctions of the energy operator, or Hamiltonian, $\hat{\mathcal{H}}$:

$$\hat{\mathcal{H}} = \frac{-\hbar^2}{2m_e r^2} \cdot \partial^2/\partial^2\phi$$

they are not eigenfunctions of the angular momentum operator:

$$\hat{l} = -i\hbar \cdot \partial/\partial\phi$$

The reason for this is quite clear, the sine and cosine functions return to their original form only when differentiated twice with respect to the angle ϕ. Whereas an eigenfunction of $\hat{l}$ must be a function such as the exponential, the first derivative of which is the exponential function itself (Box 3.1). If we differentiate the exponential functions twice with respect to ϕ we find that they are also eigenfunctions of the energy operator with the same eigenvalues as the sine and cosine functions; e.g.:

$$-\left(\frac{1}{2m_e r^2}\right) \cdot \hbar^2 \cdot \sqrt{\frac{1}{2\pi}} \cdot \frac{\partial^2}{\partial \phi^2}\{\exp(in\phi)\} = \left(\frac{n^2\hbar^2}{2m_e r^2}\right) \cdot \sqrt{\frac{1}{2\pi}} \cdot \{\exp(in\phi)\}$$

The results are summarised in Table 3.1.

From these results we can draw the important conclusion that it is possible for a system obeying the laws of quantum mechanics to be in a state that is described by a function

Table 3.1 The energy and angular momentum of an electron on a ring

Observable	Energy	Angular momentum
Operator	$\hat{\mathcal{H}}$	$\hat{l}$
Function	Eigenvalue	Eigenvalue
$\frac{1}{\sqrt{\pi}} \cdot \sin(n\phi)$	$n^2\hbar^2/2m_e r^2$	not an eigenfunction
$\frac{1}{\sqrt{\pi}} \cdot \cos(n\phi)$	$n^2\hbar^2/2m_e r^2$	not an eigenfunction
$\frac{1}{\sqrt{2\pi}} \cdot \exp(+in\phi)$	$n^2\hbar^2/2m_e r^2$	$+n\hbar$
$\frac{1}{\sqrt{2\pi}} \cdot \exp(-in\phi)$	$n^2\hbar^2/2m_e r^2$	$-n\hbar$

which is an eigenfunction of one operator, but is not an eigenfunction of some other operator. The system may also be described by a function which is an eigenfunction of more than one operator. Since the eigenvalues of operators are the observables which we can measure experimentally, these results place important restrictions on what we can know about a system. This question will be taken up again in Section 3.10.

3.9 EIGENFUNCTIONS, EIGENVALUES AND EXPERIMENTAL MEASUREMENTS

We have seen that the eigenfunctions and eigenvalues which solve Schrödinger's equation for a particular system are, respectively, wave functions describing the distributions of the electron(s) involved in the problem and the values of the quantised observables of the electrons such as energy and momentum. There is a direct correspondence between the mathematical process of solving Schrödinger's equation to obtain the value of an observable by calculation and the measurement of that observable in an experiment. But we have also seen that it is possible for a quantum-mechanical system to be described by a function which is an eigenfunction of one operator but not of another. For example, the function $\Psi_{c,n} = (1/\sqrt{\pi}) \cdot \cos(n\phi)$ is an eigenfunction of the energy operator for an electron on a ring, but it is not an eigenfunction of the corresponding angular momentum operator. We might ask, therefore, what would be the result of attempting to measure the angular momentum of the electron on a ring when it is in the state described by the above cosine function? Quantum mechanics gives a clear answer to this question.

Suppose first that we measure an observable (r) in a system which is in a state described by an eigenfunction (Ψ_a) of the operator $(\hat{\mathcal{R}})$ that corresponds to this observable. We call such a state an eigenstate of $\hat{\mathcal{R}}$. Then the result of the (idealised and error-free) measurement will be the eigenvalue (r_a) corresponding to that eigenfunction and operator, i.e. the solution of the Schrödinger equation:

$$\hat{\mathcal{R}}\Psi_a = r_a \Psi_a \tag{3.9.1}$$

And, with the qualifications to be described below, we shall get the same answer, r_a, every time we repeat the measurement.

If we now attempt to measure the same observable in a system which is *not* in a state described by an eigenfunction of the operator $\hat{\mathcal{R}}$ corresponding to the observable r, then the result of the measurement will still be one of the possible eigenvalues of $\hat{\mathcal{R}}$, but we cannot say which. If we repeat the measurement we shall, in general, obtain a different result; another eigenvalue of $\hat{\mathcal{R}}$. If the measurement is repeated many times, the value measured each time will always be one of the eigenvalues of $\hat{\mathcal{R}}$, and there may be repetition of some values, but we cannot say what value any particular measurement will give. However, as the number of measurements increases, the average result of the measurements will tend to a value $(\bar{r})$ known as the mean value, which can be calculated using the formula:

$$\bar{r} = \int_{\text{all space}} \Psi_a^* \hat{\mathcal{R}} \Psi_a \, dv \bigg/ \int_{\text{all space}} \Psi_a^* \Psi_a \, dv \tag{3.9.2}$$

The division by $\int \Psi_a^* \Psi_a \, dv$ allows for the use of un-normalised wave functions and the * indicates that we have to take the complex conjugate of Ψ_a if it is a complex function (see Appendix 8). The integration over 'all space' means integration over the full range of co-ordinates available to the system, e.g. from $-\infty$ to $+\infty$ in x, y and z. Equation (3.9.2) is an important complement to Equation (3.9.1). It provides us with the means of calculating a value for an observable even when the Ψ_a's are not eigenfunctions of the operator $\hat{\mathcal{R}}$. $\bar{r}$ is then the average of a large number of measurements. It is the result which we would *expect* to get as the mean of a large number of measurements and is accordingly called the *expectation value* of r. Here probability again enters into the predictions of quantum mechanics; in such circumstances we cannot predict the result of an individual measurement; the best we can do is to give the value to which the mean of a large number of measurements will tend.

Before proceeding to an example, we note that in quantum mechanics the idea of repeating an experimental measurement is not a simple concept. It is one of the foundation stones of the philosophy of quantum mechanics that, in general, a system is affected by any measurement made upon it; its state, and the wave function which describes it, may not be the same after the measurement as they were before it. Therefore, after each measurement the system must be allowed to return to its original state before the next measurement can be made. This is a point which we shall discuss in detail in the next section. Since, in chemistry, a sample subjected to a measurement usually consists of a very large number ($\approx 10^{23}$) of identical molecules or atoms, we may consider in such cases that the result obtained is the mean value defined as above but that it is obtained as the result of the same measurement made simultaneously upon many identical systems.

By way of an example, in Box 3.5 we use Equation (3.9.2) to determine the expectation value for the angular momentum of an electron on a ring when it is in the cosine function state. The result of zero may at first appear unlikely, but it is quite easy to understand. According to Appendix 8:

$$\cos(n\phi) = \tfrac{1}{2} \cdot \{\exp(+in\phi) + \exp(-in\phi)\} \qquad (3.9.3)$$

Thus, the cosine function is composed of equal parts of the two exponential functions which are the eigenfunctions of the angular momentum operator. Since the two eigenfunctions have equal and opposite angular momenta (Table 3.1) the combination of the two gives an expectation value of zero. Thus, an experimental measurement of the angular momentum when the electron is in the cosine function state may give a value of $+n\hbar$ or of $-n\hbar$. But if many measurements are made the two values will occur with equal probability and the mean result will tend to the value of zero. Alternatively, as noted above, if the sample is a typical 'chemical' one and contains a very large number of atoms or molecules then the result of one measurement on the sample as a whole, i.e. simultaneously on many individual atoms (molecules), may be regarded as the average of many sequential measurements on an individual atom (molecule).

Equation (3.9.2) is of particular value in problems where we have difficulty in solving Schrödinger's equation and obtaining its eigenfunctions. Because of the complexity of many chemical systems, it is normally impossible to solve Equation (3.9.1) directly. Under such circumstances it is useful if we can postulate an approximate wave function for the system and use that to determine an approximation to the observable quantities by means of Equation (3.9.2). This type of calculation plays an essential role in the theory of the chemical bond as we shall see in Chapter 6.

3.10   MORE ABOUT MEASUREMENT: THE HEISENBERG UNCERTAINTY PRINCIPLE

In Section 3.9 we noted that it is a fundamental quantum-mechanical principle that no measurement can be made upon a system without affecting that system. This should not really surprise us. Suppose that we wish to measure the time at which a locomotive passes a particular point on a railway line, thereby fixing its position at that particular time. We might stretch a cotton thread attached to a timing mechanism across the track. There would be no concern that the collision of the locomotive with the thread would slow the train to any degree which could be measured. This being so, a second device of the same type placed a known distance down the track would enable us to fix the position of the locomotive for a second time and measure its average velocity between those two points on the line and, if we know its mass, its mean linear momentum.

"Professor Bees-knees, the eminent entomologist, timing gnats".

But what if the moving object to be timed was a fly? If the mechanism was sufficiently delicate we might be able to determine the time at which the fly struck the first timing device, but we certainly could not maintain that the fly's further progress would be unaffected by its contact with the thread. Thus, a measurement of its mean velocity or momentum would be invalidated since it would not be independent of the measurement itself. In the case of the fly, the problem could be overcome by using beams of light and photocells in place of the mechanical timing devices.

But what if the object to be studied was an electron? Electrons are not indifferent to photons; as we saw in Section 2.6, electrons can be ejected from metals by photons. Thus, we could not be sure if the impact of the 'measuring' photon had changed the position of the electron, its velocity, or both. There are no means by which events at the atomic and molecular level can be observed in such a way that those events are not influenced by the act of observation itself. Just as in sociology, where merely to ask a particular question has an influence upon the respondent's attitude to that subject, there is no uninfluential observation in quantum mechanics.

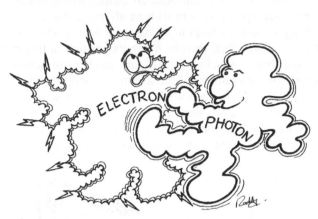

'A kick from a passing photon is not something which an electron can simply ignore'

To observe a system is to interact with it and change it.

In fact, the problem of the accuracy of measurement is even more subtle than the above discussion suggests. In 1927 Heisenberg showed that, in systems obeying the laws of quantum mechanics, there are pairs of related properties which cannot be measured exactly at the same time (Box 3.6). This situation is fundamental to quantum-mechanical systems and Heisenberg expressed the result in the form of his uncertainty principle, an example of which is:

$$\Delta p_x \cdot \Delta x \geq h/4\pi \qquad (3.10.1)$$

Δp_x is the uncertainty in the result of a measurement of the linear momentum of a particle along x and Δx is the uncertainty in the measurement of the x co-ordinate, and their product has to be greater than Planck's constant divided by 4π. Other pairs of observables linked by an uncertainty relationship are given in Box 3.6. It is important to note that this uncertainty of measurement has nothing to do with the quality and precision of the equipment available for the measurement, nor with the ability of the experimentalist. The limitation on the measurement is fundamental and Equation (3.10.1) assumes that perfect experimental devices are used by perfect experimentalists. Thus, to return to the problem of the fly, the problem of getting the fly to hit the timing device is quite irrelevant to the argument here.

The magnitude of Planck's constant and its presence in Equation (3.10.1) explains why the consequences of the uncertainty principle were never noticed in classical mechanics. Since the product of the two uncertainties is about one order of magnitude less than Planck's constant it is so small that the individual uncertainties in the values of position and momentum are much too small to be observed in an experiment with macroscopic bodies. If we divide both sides of Equation (3.10.1) by, m, the mass of the particle concerned, we obtain:

$$\Delta v_x \cdot \Delta x \geq h/4\pi m \qquad (3.10.2)$$

which shows clearly how the product of the two uncertainties becomes smaller as the mass of the particle increases, i.e. as we go from systems which require a quantum-mechanical description to those to which classical mechanics applies. Note that the quantity on the right-hand side of the inequality (3.10.1) is not always given as $h/4\pi$, it is quite common to find $h/2\pi$ for example. These differences are a consequence of the different ways in which the uncertainties Δp_x and Δx can be defined, but this in no way invalidates the fundamental uncertainty principle which Heisenberg first enunciated. The particular value of $h/4\pi$ applies to the case where the uncertainties are assumed to follow the Gaussian distribution law.

3.11 THE COMMUTATION OF OPERATORS (BOXES 2.2 AND 3.6)

In Section 3.10 we recognised that there are important limitations on what we can know about a system which obeys the laws of quantum mechanics. In particular we saw that the accuracy with which the two properties, momentum in the x direction and x co-ordinate, could be simultaneously measured, was restricted by the Heisenberg uncertainty principle. We discussed the matter from an essentially experimental point of view; now we look at the problem from the theoretical standpoint.

According to Section 3.9, every observable quantity has an associated operator by means of which either the exact value of the observable or a mean value can be calculated using Equations (3.9.1) or (3.9.2) respectively. How would the problem of *calculating* two properties be formulated? It seems natural to suggest that, in direct correspondence with experimental measurement, we first calculate the one and then the other. Thus, if we have a system in the state Ψ_a for which we wish to know the simultaneous values of two properties P and Q we could perform the calculation:

$$\hat{P}\hat{Q}\Psi_u = \hat{P}q_a\Psi_a = q_a\hat{P}\Psi_a = q_a p_a \Psi_a \quad \text{or} \quad \hat{Q}\hat{P}\Psi_a = \hat{Q}p_a\Psi_a = p_a\hat{Q}\Psi_a = p_a q_a \Psi_a$$

And since p_a and q_a are simply numbers we obtain the same result, regardless of the order of application of $\hat{P}$ and $\hat{Q}$. But these two calculations assume that Ψ_a is an eigenfunction of both $\hat{P}$ and $\hat{Q}$ which is by no means always the case. This can be illustrated using the momentum and position operators and the eigenfunction for a electron in an infinitely long box from Section 3.7.1: $\Psi_{+b} = N\exp(+ibx)$.

The operator for linear momentum along x is $-i\hbar\partial/\partial x$ and the operator for the x co-ordinate is x itself and we investigate the effect of applying these two operators consecutively to the above eigenfunction. We have either:

$$\hat{x}(-i\hbar)\frac{\partial}{\partial x}\Psi_{+b} = -i\hbar Nx\frac{\partial}{\partial x}\exp(+ibx) = -i\hbar Nxib\exp(+ibx) = \hbar bx\Psi_{+b}$$

or

$$-i\hbar\frac{\partial}{\partial x}\hat{x}\Psi_{+b} = -i\hbar N\frac{\partial}{\partial x}x\exp(+ibx) = -i\hbar N\{\exp(+ibx) + xib\exp(+ibx)\}$$

$$= -i\hbar\Psi_{+b} + \hbar bx\Psi_{+b}$$

The results are not the same, in fact:

$$(\hat{x}\,\hat{p}_x - \hat{p}_x\,\hat{x})\,\Psi_{+b} = i\hbar\Psi_{+b} \tag{3.11.1}$$

When the two results for the consecutive operation of two operators are not equal we say that the operators do not commute and the root of the problem lies in the fact that it is not possible to find a function which is simultaneously an eigenfunction of both operators. As a consequence, it is not possible to know, simultaneously, both the observables associated with the operators. Thus, Heisenberg's uncertainty principle is a manifestation of the fact that the operators corresponding to the two observables involved do not commute.

We can see some of the effects of Heisenberg's uncertainty principle in relation to position and momentum along the x co-ordinate using, once more, our pair of exponential wave functions for the electron in a linear box. The probability, $P(x')$, of finding an electron at a point having the co-ordinate x' is proportional to the product of the wave function and its complex conjugate evaluated at that point, i.e.:

$$P(x') = N^2 \exp(-ibx') \cdot \exp(+ibx') = N^2 \exp(0) = N^2$$

where N is the normalising constant. The important point here is that the probability is the same at every point along the box, since it does not depend upon x. Thus, for an infinitely long box the uncertainty in our knowledge of the position of the electron is infinite since it can be anywhere in the box. But it is only for a box of infinite length that the momentum has the exact values of $\pm bh/2\pi$. Thus, for the infinitely long box the linear momentum of the electron is known exactly but the uncertainty in its position is infinite. For any box of finite length the uncertainty in our knowledge of the position of the electron is finite, since then it must lie between 0 and L. Therefore, there must also be a finite uncertainty in our knowledge of the momentum of the electron. Further, since there is no potential energy term in the problem, the energy is all kinetic energy and therefore proportional to the square of the momentum so that the energy of the electron must also be uncertain. There could be no uncertainty if the energy was zero, so the fact that there is a zero-point energy (Section 3.6.2) is in accord with the Heisenberg uncertainty principle.

3.12   COMBINATIONS OF EIGENFUNCTIONS AND THE SUPERPOSITION OF STATES

In Section 3.9 we saw that a function which was not an eigenfunction of the angular momentum operator for the electron on a ring could be written as a sum of two eigenfunctions of that operator (Equation (3.9.3)). This is an important result which may be generalised further. Any function which obeys the boundary conditions (Section 3.6.3) of a problem can be expressed as a sum of the eigenfunctions of that problem. The value of this result lies in the way in which it allows us to address the following problem. We know that an atom (or molecule) may exist in an electronic eigenstate of the Hamiltonian operator. In this state it has a precisely measurable energy. We also know that if light of an appropriate frequency strikes the atom it may absorb a photon and make a transition to an eigenstate of higher energy and that this process takes a finite time. Estimates of the time required are discussed in Chapter 8, but we may say here that it is of the order of 10^{-18} s. But what is the appropriate description of the atom during the time in which it has left the initial eigenstate, ψ_i, but has not yet arrived in the final eigenstate, ψ_f? Since both ψ_i and ψ_f are eigenfunctions of the same Hamiltonian operator, any other state of the atom, which is not an eigenstate, may be described as a sum, suitably normalised, of these two states plus contributions from all the other eigenstates of the Hamiltonian operator for the atom. The technical term for such a description is a *superposition of states*. In this particular problem it is reasonable to assume that the contributions from the other eigenstates are vanishingly small so that at any moment during the transition the atom is in the state χ, where χ can be approximated as a superposition of just two states, i.e.:

$$\chi = c_i \psi_i + c_f \psi_f \qquad (3.12.1)$$

And, assuming that the functions ψ_i and ψ_f are orthogonal and normalised, χ will be normalised if:

$$c_i^2 + c_f^2 = 1.0 \qquad (3.12.2)$$

Before the photon strikes the atom, $c_i = 1.0$ and $c_f = 0.0$. When the process of absorbing the photon is complete, $c_i = 0.0$ and $c_f = 1.0$. At any point in time during the absorption process, c_i is decreasing and c_f increasing, subject always to the normalisation condition (Equation (3.12.2)). Thus the system undergoing a spectroscopic transition does not enter some form of limbo about which we know nothing; we have a description of the system throughout the process. This description of the absorption or emission of a photon is examined in more detail in Section 8.6.

3.13 OPERATORS AND THEIR FORMULATION

It has been emphasised above that the form of the operator for any particular observable is a postulate which has been tested against its success in calculations of measurable quantities using that operator. Since classical mechanics is a limiting case of quantum mechanics, we might expect that relationships between operators thus derived would reflect the relationships between the corresponding classical quantities. This is indeed the case and use can be made of that fact in the construction of operators. As far as the usual chemical applications of quantum mechanics are concerned, the operators for position (co-ordinate), potential energy, kinetic energy, linear momentum and angular momentum, play by far the most important roles. The relationships between these operators mirror the relationships of the corresponding classical quantities as the following will show.

3.13.1 Position or co-ordinate, $\hat{x}$

The operator for the position of a particle, x or r say, is simply $\hat{x}$ or $\hat{r}$. Where the caret mark ($\wedge$) is used to distinguish the operator from the observable. Therefore, the expectation value ($\bar{x}$) of the x co-ordinate of a particle described by the normalised wave function ψ is:

$$\bar{x} = \int\limits_{\text{all space}} \psi \hat{x} \psi \, dx \, dy \, dz \equiv \langle \psi | \hat{x} | \psi \rangle$$

using the notation introduced by Dirac.

3.13.2 Potential energy, $\hat{V}$

Since potential energy is invariably expressed in terms of the position of a particle in some field of force, e.g. the distance of the electron from the nucleus in the hydrogen atom or the angle between a nuclear magnet and an applied magnetic field, the essential operator in potential energy is the position operator. Therefore, using the result above, the classical expression for the potential energy (V) of two charges of $+e$ and $-e$ separated by a distance r is replaced by the operator $\hat{V}$ in the following way:

$$V = -e^2/r \Rightarrow \hat{V} = -e^2/\hat{r}$$

3.13.3   Linear momentum, $\hat{p}_x$

We have already seen that linear momentum in the x direction is replaced in the operator formalism by differentiation with respect to x multiplied by $-ih/2\pi$, i.e.:

$$p_x = mv_x \Rightarrow \hat{p}_x = -i\hbar \cdot \frac{\partial \psi}{\partial x}$$

3.13.4   Kinetic energy, $\hat{W}$

We deduce the form of the kinetic energy operator by expressing kinetic energy as the square of the momentum divided by mass and using the last result, i.e.:

$$W = \tfrac{1}{2}mv_x^2 = \frac{(mv_x)^2}{2m} \Rightarrow \hat{W} = -\frac{\hbar^2}{2m} \cdot \frac{\partial^2 \psi}{\partial x^2}$$

3.13.5   Angular momentum, $\hat{L}$

The angular momentum (see also Chapter 4) of a particle of mass m rotating about a centre is the vector product of the distance of the particle from the centre (r) and its linear momentum (p). If the rotation takes place in the xy-plane ($z = 0$, $p_z = 0$), then resolving the vectors r and p into their Cartesian components, we have our angular momentum operator as:

$$L = r \times p = xp_y - yp_x \Rightarrow \hat{L} = -i\hbar \left\{ \hat{x} \cdot \frac{\partial \psi}{\partial y} - \hat{y} \cdot \frac{\partial \psi}{\partial x} \right\}$$

3.14   SUMMARY

In this chapter we have studied the application of quantum mechanics to simple problems and have been particularly concerned to examine the exact details of the description of the world of atoms and molecules which the method provides. This differs from the description of the macroscopic world given by classical mechanics in a number of very important ways and, on the face of it, we appear to have less information available to us when we use quantum mechanics than we have when we use classical mechanics. With this in mind, it is essential that we focus our attention upon things which can be measured and do not attempt to grapple with problems for which quantum mechanics provides no answers. It is no more meaningful, for example, to attempt to calculate the *path* of an electron than it is to discuss its *colour*.

Thus, we can:

- Determine precise values of physical observables where the system concerned is in an eigenstate of the appropriate operator.

- Determine a mean value of an observable where the system is not in an eigenstate of the appropriate operator.

- Describe a system when it is not in an eigenstate in terms of appropriate superpositions of the set of eigenstates of the relevant operator.

With these theoretical methods, some of which are illustrated in a discussion of the properties of polarised light in Box 3.7, we can interpret almost all the experiments which are of interest to chemistry. Where such theoretical methods fail to give us a quantitatively accurate answer, as they frequently do, it is because of the difficulties which we have in solving the Schrödinger eigenvalue-eigenfunction Equation (3.9.1), or in evaluating the integral in Equation (3.9.2) in order to determine an expectation value, and not because of a failure of the basic theory. Fortunately, as we shall see in Chapters 6 and 7, approximate solutions are frequently quite sufficient for our purpose.

3.15 BIBLIOGRAPHY AND FURTHER READING

It is not easy to recommend books on quantum mechanics, there is a very wide choice and many of the classic texts are now very old, though their comprehensive treatment of the subject remains as relevant as it ever was.

There are two useful and comparatively recent introductory texts by Green:

1. N.J.B. Green, *Quantum Mechanics 1*, Oxford University Press, 1997.
2. N.J.B. Green, *Quantum Mechanics 2*, Oxford University Press, 1998.

And an additional volume in the same series.

3. P.A. Cox, *Introduction to Quantum Theory and Atomic Structure*, Oxford University Press, 1995.

Much more advanced but with many excellent, fully-worked applications is:

4. R.M. Golding, *Applied Wave Mechanics*, Van Nostrand, London, 1969.

One of the great classics, not for the beginner is:

5. P.A.M. Dirac, *Quantum Mechanics*, 4th edn, Oxford University Press, 1967.

Now also available from Dover Publications.

The following three also deserve the accolade 'classic'. They are all very comprehensive texts particularly aimed at chemists and they provide an invaluable source of reference about most of the quantum-mechanical problems and methods of interest to chemists.

6. L. Pauling and E.B. Wilson, *Introduction to Quantum Mechanics*, McGraw-Hill, New York, 1935.

7. H. Eyring, J. Walter and G.E. Kimball, *Quantum Chemistry*, Wiley, London, 1944.
8. W. Kauzmann, *Quantum Chemistry*, Academic Press, London, 1957.

More modern books with chemists in mind are:

9. P.W. Atkins and R.S. Friedman, *Molecular Quantum Mechanics*, Oxford University Press, 1997.
10. I.N. Levine, *Quantum Chemistry*, 5th edn, Prentice-Hall Inc., New Jersey, 2000.

The following are aimed more at physicists, but all present rather different approaches to the subject than the preceding volumes.

11. L.D. Landau and E.M. Lifshitz, *Quantum Mechanics*, English 2nd edn, Pergamon Press, Oxford, 1964.
12. J.J. Sakurai, *Modern Quantum Mechanics*, Benjamin/Cummings, Menlo Park, 1985.
13. A.S. Davydov, *Quantum Mechanics*, Addison-Wesley Publishing Company Inc., Reading, 1965.
14. N. Zettili, *Quantum Mechanics Concepts and Applications*, Wiley, Chichester, 2001.

BOX 3.1   The sine, cosine and exponential functions

We make frequent use of the sine, cosine and exponential functions, and their derivatives, in this chapter. The particular results which we require can be readily derived if we recall that each of these functions can be expressed as a power series:

$$\sin(ax) = ax - (ax)^3/3! + (ax)^5/5! - (ax)^7/7! + \cdots$$

$$\cos(ax) = 1 - (ax)^2/2! + (ax)^4/4! - (ax)^6/6! + \cdots$$

$$\exp(ax) \equiv e^{ax} = 1 + ax + (ax)^2/2! + (ax)^3/3! + (ax)^4/4! + \cdots$$

$$\exp(-ax) \equiv e^{-ax} = 1 - ax + (ax)^2/2! - (ax)^3/3! + (ax)^4/4! + \cdots$$

$n! = n(n-1)(n-2)(n-3)\cdots 1$ is called *factorial n*; e.g. $5! = 5 \times 4 \times 3 \times 2 \times 1 = 120$.

Differentiating $\sin(ax)$ term-by-term with respect to x we have:

$$d[\sin(ax)]/dx = a - 3a \cdot (ax)^2/3! + 5a \cdot (ax)^4/5! - 7a \cdot (ax)^6/7! + \cdots$$

$$= a[1 - (ax)^2/2! + (ax)^4/4! - (ax)^6/6! + \cdots] = a \cdot \cos(ax)$$

Similarly

$$d[\cos(ax)]/dx = -a \cdot \sin(ax)$$

$$d[\exp(ax)]/dx = a \cdot \exp(ax)$$

and

$$d[\exp(-ax)]/dx = -a \cdot \exp(-ax)$$

BOX 3.2   The Normalisation and orthogonality of eigenfunctions

Normalisation

The square of an eigenfunction (ψ) evaluated at any point in space gives the probability that the particle described by that eigenfunction will be found at that point. Since the probability of finding the particle somewhere within the whole of the region of space in which it is allowed to move must be one (certainty), it follows that the sum of the values of ψ^2 evaluated at every point in space must be one for a physically acceptable eigenfunction. This requirement is represented mathematically by an integration over all space:

$$\int_{\text{all space}} \psi^2 \, \mathrm{d}v = 1.0$$

We can illustrate the normalisation process using the eigenfunctions which we have found for the electron confined to a ring. For the cosine functions for example:

$$\psi_{c,a} = N \cdot \cos(a\phi)$$

Where $a = 0, 1, 2, \ldots$ and N is the normalising constant which we have to find. In this particular problem the 'whole of space', as far as the electron is concerned, is a complete circuit of the ring, i.e. all values of ϕ from 0 to 2π.

For the special case where $a = 0$, $\cos(a\phi) = 1.0$ and we have:

$$\int_0^{2\pi} \psi_{c,a}^2 \, \mathrm{d}\phi = N^2 \int_0^{2\pi} 1 \, \mathrm{d}\phi = N^2[\phi]_0^{2\pi} = 2\pi N^2$$

Therefore, for a normalised eigenfunction we must have $N = \pm 1/\sqrt{2\pi}$. We always take the positive sign, but the choice of sign can have no effect upon the calculation of any property which we can measure.

When a is not zero we have:

$$\int_0^{2\pi} \psi_{c,a}^2 \, \mathrm{d}\phi = N^2 \int_0^{2\pi} \{\cos(a\phi)\}^2 \, \mathrm{d}\phi = \frac{N^2}{2} \int_0^{2\pi} \{1 + \cos(2a\phi)\} \, \mathrm{d}\phi$$

$$= \frac{N^2}{2} \left[\phi + \frac{\sin(2a\phi)}{2a} \right]_0^{2\pi} = \frac{N^2}{2} \cdot 2\pi = N^2\pi$$

Thus, for these functions $N = \pm 1/\sqrt{\pi}$ and the correctly normalised eigenfunctions are written:

$$\psi_{c,a} = \left(\frac{1}{\sqrt{\pi}} \right) \cdot \cos(a\phi)$$

The sine functions can be normalised in the same way and the normalising constant is found to be $\sqrt{(1/\pi)}$ for these also.

Orthogonality

The eigenfunctions which solve Schrödinger's equation also have the property of orthogonality, i.e. when the product of two different eigenfunctions is integrated 'over all space' the result is zero. Thus, for two cosine eigenfunctions:

$$\frac{1}{\pi} \int_0^{2\pi} \cos(a\phi) \cdot \cos(b\phi) \, d\phi = 0 \quad \text{if } a \neq b$$

This is a very important property which expresses the fact that any one eigenfunction cannot be written in terms of the other eigenfunctions of the same Schrödinger operator. This is exactly the same as saying that a coordinate on the x-axis of a graph cannot be expressed in terms of a co-ordinate, or co-ordinates, on the y-axis of the same graph. This is because the two axes are orthogonal, which is another way of saying that they are at right angles to each other. We can demonstrate the orthogonality of two cosine functions simply by carrying out the required integration:

$$\frac{1}{\pi} \int_0^{2\pi} \cos(a\phi) \cdot \cos(b\phi) \, d\phi = \frac{1}{\pi} \int_0^{2\pi} \{\cos([a+b]\phi) + \cos([a-b]\phi)\} \, d\phi$$

$$= \frac{1}{\pi} \left[\frac{\sin\{(a+b)\phi\}}{(a+b)} + \frac{\sin\{(a-b)\phi\}}{(a-b)} \right]_0^{2\pi} = 0$$

Both terms are zero at $\phi = 0$ and 2π because a and b are integers.

BOX 3.3   Normalisation and orthogonality of the eigenfunctions for an electron in a linear box

Normalisation (See also Box 3.2)

The normalisation integral is:

$$\int_0^L \{N \sin(n\pi x/L)\}^2 \, dx = 1.0$$

but $\sin^2 \theta = \frac{1}{2}(1 - \cos 2\theta)$ and the integral can therefore be written as:

$$\frac{N^2}{2} \int_0^L \{1 - \cos(2n\pi x/L)\} \, dx = \frac{N^2}{2} \left[x - \frac{L}{2n\pi} \cdot \sin(2n\pi x/L) \right]_0^L$$

$$= \frac{N^2}{2}(L - 0 - 0 - 0)$$

since $\sin(2n\pi) = \sin(0) = 0$.

Therefore, $N^2 L/2 = 1$ or $N^2 = L/2$ and $N = \pm\sqrt{(2/L)}$.

We may take either sign for the square root since the phase of the wave function has no effect upon any quantity which we can measure.

Orthogonality (See also Box 3.2)

The wave functions corresponding to two different values of the quantum number, n, are orthogonal, i.e. the integral of their product over the length of the box is zero. We can prove this by integrating the product of two wave functions having quantum numbers n and m with normalising constants N and M respectively:

$$\frac{2\,NM}{L} \int_0^L \sin(n\pi x/L) \cdot \sin(m\pi x/L)\,dx$$

$$= \frac{NM}{L} \int_0^L \left\{ \cos\left([n-m]\frac{\pi x}{L}\right) - \cos\left([n+m]\frac{\pi x}{L}\right) \right\} dx$$

$$= \frac{NM}{L} \left[\frac{\sin([n-m]\pi x/L)}{[n-m]\pi/L} - \frac{\sin([n+m]\pi x/L)}{[n+m]\pi/L} \right]_0^L$$

$$= 0 \text{ because } n \text{ and } m \text{ are integers.}$$

BOX 3.4 Electronic spectra of conjugated polyenes

According to the theory of the chemical bonding in conjugated polyenes (Chapter 6), the π-electrons of these molecules occupy orbitals which are delocalised over the whole of the carbon-atom chain. These electrons should therefore behave rather like electrons in a linear box and, in particular, the electron-in-a-box model should provide approximate values for their energy levels. There is no doubt that the absorption band of longest wavelength, which moves into the visible spectral region as the length of the chain increases, is due to an electronic transition from the highest occupied to the lowest unoccupied π-electron energy level and the energy of this band has been correlated with the gap between these energy levels in many quantum-mechanical treatments of polyene spectra. (See, for example, S.F. Mason, *Quarterly Reviews of the Chemical Society*, **15**, 287, 1961.) Theory and experiment may be related in the following manner.

Consider a cyanine, a linear polyene, which has the general formula shown in Figure B3.4.1. There are $2k+2$ bonds in the conjugated system between the two nitrogen atoms, but it is well known that methyl groups extend the conjugated system. We therefore allow for this (hyperconjugative) effect by adding one bond for each of the two $H_3C-N-CH_3$ groups giving a total of $2k+4$ bonds in the conjugated system. The length, L, of the conjugated system can now be written as $(2k+4)l$, where l is the mean bond length. Each carbon atom contributes one π-electron and the two nitrogen atoms contribute three between them (1 from N+, 2 from N:), so that we have a total of $2k+4\pi$-electrons.

The energy levels of an electron in a box are given by the formula:

$$E_n = \frac{n^2 h^2}{8m_e L^2} \quad (n = 1, 2, 3, \ldots)$$

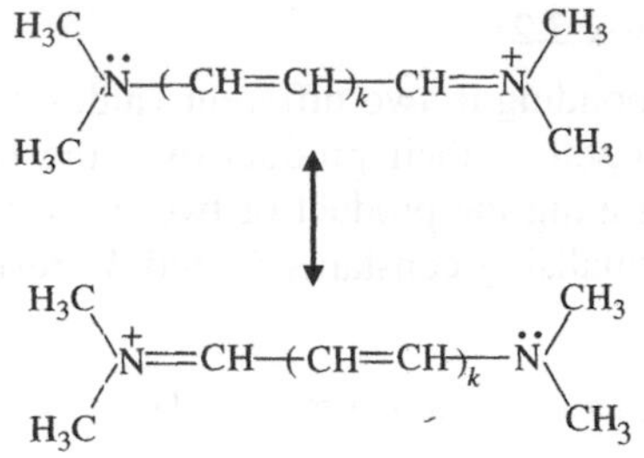

Figure B3.4.1   The resonance structures of cyanine dyes

Since the electrons occupy the energy levels in pairs, the quantum number (n_{hol}) of the highest occupied level is $(2k + 4)/2 = k + 2$ and the quantum number of the lowest unoccupied level, n_{lul}, is $k + 2 + 1 = k + 3$. Thus, using the above formula the energy difference (ΔE) between the lowest unoccupied and highest occupied levels is:

$$\Delta E = E_{lul} - E_{hol} = \frac{\{(k + 3)^2 - (k + 2)^2\}h^2}{8m_e(2k + 4)^2l^2} = \frac{(2k + 5)h^2}{8m_e(2k + 4)^2l^2}$$

and we therefore expect that a graph of ΔE against $(2k + 5)/(2k + 4)^2$ will be a straight line of gradient $h^2/8m_el^2$. To calculate the gradient we require an estimate of the mean bond length (l). Because of the resonance between the two cyanine tautomers, there is no bond alternation as in the polyenes and all the C–C bonds are equal in length. There are very few hydrocarbons for which this is the case, but benzene is one such which suggests that we set $l = 1.39 \times 10^{-10}$ m, so that the theoretical gradient is:

$$\frac{h^2}{8m_el^2} = \frac{(6.626 \times 10^{-34})^2}{8 \times 9.109 \times 10^{-31} \times (1.39 \times 10^{-10})^2} = 312 \times 10^{-20} \text{ J}$$

For comparison with experiment we use the data for the cyanines taken from Mason's review and given in Table B3.4.1. The data are plotted in Figure B3.4.2.

The experimental slope of 313×10^{-20} J is in remarkable agreement with the simple model of the electron in a box. But it must be noted that for other polyenes, where all C–C bonds are either formal double bonds, C=C, or formal single bonds, C–C, the resulting alternation of bond length and the consequent alternation of the electric field of the σ-electrons, in which the π-electrons move, makes the simple theory less applicable. These additional effects are discussed by Mason. Nevertheless, this example shows that even the most simple quantum-mechanical model is capable

Table B3.4.1   The electronic spectra of the cyanines

k	$(2k + 5)/(2k + 4)^2$	λ_{max}/nm	$\Delta E/\text{J} \cdot 10^{-20}$
1	0.1944	312.5	63.55
2	0.1406	416.0	47.74
3	0.1100	519.0	38.26
4	0.0903	625.0	31.78
5	0.0765	734.5	27.04
6	0.0664	848.0	23.42

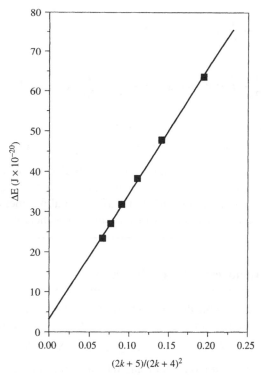

Figure B3.4.2 ΔE *versus* chain-length parameter for cyanine dyes

of giving a quantitative interpretation of real chemical data and thereby enhancing our understanding of the fundamental properties of conjugated molecules.

BOX 3.5 Calculation of the expectation value of the angular momentum of an electron on a ring

We consider the electron in the state having the wave function:

$$\psi_a = \frac{1}{\sqrt{\pi}} \cos(a\phi)$$

which is *not* an eigenfunction of the angular momentum operator, $-i\hbar\partial/\partial\phi$. The expectation value of the angular momentum, $\bar{l}$, for this eigenfunction is calculated as follows:

$$\bar{l} = \int_0^{2\pi} \frac{1}{\sqrt{\pi}} \cos(a\phi) \left(-\frac{i\hbar\partial}{\partial\phi} \right) \frac{1}{\sqrt{\pi}} \cos(a\phi) \, d\phi$$

$$= -\frac{i\hbar}{\pi} \int_0^{2\pi} \cos(a\phi) \frac{\partial}{\partial\phi} \cos(a\phi) \, d\phi$$

$$= \frac{ia\hbar}{\pi} \int_0^{2\pi} \cos(a\phi)\sin(a\phi)\,\mathrm{d}\phi$$

$$= \frac{ia\hbar}{2\pi} \int_0^{2\pi} \sin(2a\phi)\,\mathrm{d}\phi = \frac{-ia\hbar}{2\pi} \cdot \frac{1}{2a}[\cos(2a\phi)]_0^{2\pi} = 0.$$

Bearing in mind (Appendix 8) the fact that:

$$\cos(ax) = \tfrac{1}{2}\{\exp(+iax) + \exp(-iax)\}$$

we see that the cosine function is a sum of the two angular momentum eigenfunctions, $\exp(+iax)$ and $\exp(-iax)$. Thus, if we measured the angular momentum of the system we would expect to find values of $+a\hbar$ and $-a\hbar$ in equal numbers which would lead to a mean value of zero as the number of measurements increased.

BOX 3.6   Heisenberg uncertainty relationships

The best known of these, and the most important as far as chemistry is concerned, are those involving momentum and position along the same coordinate:

$$\Delta p_x \cdot \Delta x \geq h/4\pi \quad \Delta p_y \cdot \Delta y \geq h/4\pi \quad \Delta p_z \cdot \Delta z \geq h/4\pi$$

and those involving two components of angular momentum, i.e.:

$$\Delta l_x \cdot \Delta l_y \geq h/4\pi \quad \Delta l_y \cdot \Delta l_z \geq h/4\pi \quad \Delta l_z \cdot \Delta l_x \geq h/4\pi$$

In these relationships it is assumed that there is a Gaussian distribution of the uncertainties, i.e. Δp is the root-mean-square of the δp of the individual measurements:

$$\Delta p = \sqrt{\sum_i (\delta p_i)^2}$$

As we have seen in this chapter, the uncertainty relationships are intimately related to the non-commutation of the operators involved. In the case of the angular momentum operators, this is expressed in the following useful equalities derived in Box 4.1:

$$[\hat{l}_x, \hat{l}_y] = i\hbar\hat{l}_z; \quad [\hat{l}_y, \hat{l}_z] = i\hbar\hat{l}_x; \quad [\hat{l}_z, \hat{l}_x] = i\hbar\hat{l}_y$$

where $[a, b] \equiv a \cdot b - b \cdot a$.

There is also an uncertainty relationship which connects energy and time:

$$\Delta E \cdot \Delta t \geq h/4\pi$$

It is very valuable and we use it in Chapter 8. However, it must be clearly understood that, since time is not an observable, the relationship has an origin and significance quite different from the others quoted here. The distinction is discussed in more advanced texts, e.g. J.J. Sakurai, *Modern Quantum Mechanics*, Benjamin/Cummings, Menlo Park, 1985.

There is a very good discussion of the Heisenberg uncertainty principle in M. Born, *Atomic Physics*, Blackie & Son, 3rd Edn, London 1944, Appendix XXII.

BOX 3.7   Polarised light and quantum mechanics

We can use well-known results from classical optics to illustrate several of those aspects of quantum mechanics which appear puzzling when we first encounter them. The experiments and their interpretation have been discussed by Dirac in '*The Principles of Quantum Mechanics*'. The first edition of this famous book appeared in 1930 and was one of the most important contributions to the mathematical and philosophical foundations of quantum mechanics. The experiments in questions are the following. (See Chapter 8 for further information about polarised light.)

A beam of light propagating in the z direction and polarised in the xz plane is directed onto a polariser. The latter is an optical device which transmits incident light if the light is polarised along a particular direction which we here call $t - t'$, but which absorbs all the incident light if it is polarised along $a - a'$, the direction at right angles to $t - t'$ (Figure B3.7.1). We assume the polariser to be perfect.

Both $a - a'$ and $t - t'$ are perpendicular to the direction of propagation of the light. When the incident light is polarised at some intermediate angle, β, to $t - t'$, experiment shows that if I_0 is the intensity of the incident polarised light and I_t the intensity of the transmitted light then:

$$I_t = I_0 \cos^2 \beta \quad \text{and} \quad I_a = I_0 \sin^2 \beta$$

where $I_a = I_0 - I_t$, the intensity of the light lost by absorption. The transmitted light is also polarised but along $t - t'$ and not along its original direction of polarisation at an angle of β to $t - t'$.

Using a technique known as photon counting, the experiments described above can be carried out with a beam of light of such a low intensity that the photons passing through the polariser can be individually recorded. For light polarised along $t - t'$ every photon is transmitted while for light polarised along $a - a'$ every photon is absorbed and this presents us with little conceptual difficulty. But what of the experiments with light polarised at an angle of β to $t - t'$? If we now consider the fate of each individual photon we are unable to say definitely whether it will be transmitted or absorbed by the polariser. We can only say that there is a probability of $\cos^2 \beta$ that it will be transmitted and a probability of $\sin^2 \beta$ that it will be absorbed

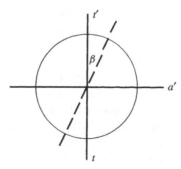

Figure B3.7.1   Plane of polarisation of incident light with respect to directions of transmission, $t - t'$, and absorption, $a - a'$

since this will give us the known experimental result if a large number of photons are subjected to the polarisation test. We find that we can only make statements about probabilities in respect of the results of a measurement on a single particle which obeys the laws of quantum mechanics. Thus, the first aspect of quantum mechanics illustrated by the above example is the statistical nature of our knowledge concerning the behaviour of individual particles and the results of individual measurements.

The second important point which emerges above is that the making of the measurement has a profound influence upon the particle subjected to that measurement. The experimental result is that all the photons which pass through the polariser are polarised along $t - t'$ whereas those which are absorbed must be polarised along $a - a'$. But all the incident photons were polarised at an angle of β to $t - t'$ so that the polarisation of every photon subjected to measurement has been changed. Therefore, if we make a further measurement on the photons which pass through the polariser the objects of the second measurement are not exactly the same as the objects of the first measurement, their observable property of polarisation has been changed. The photons have been 'forced' into eigenstates of the operator by the measurement process. In general, there is no such thing as an experimental measurement which has no effect upon the system measured, though there would be no apparent effect if each incident photon was polarised along $t - t'$.

The concept of polarisation as an observable property leads us to ask what the corresponding operator, eigenvalues and eigenfunctions are. There are two eigenstates of polarisation in the above experiment. If a photon is in eigenstate ψ_t then it is transmitted by the polariser, if it is in the eigenstate ψ_a then it is absorbed. If we call the 'polarisation analysing' operator $\hat{A}$ then:

$$\hat{A} \, \psi_t = 1 \times \psi_t \quad \text{and} \quad \hat{A} \, \psi_a = 0 \times \psi_a$$

and for a beam of light polarised along $t - t'$ every measurement gives the same result: photon transmitted. For light polarised along $a - a'$ we always get the result: photon absorbed and annihilated.

If we now consider the case of photons polarised at an angle of β to $t - t'$ then, according to quantum mechanical principles, this state of polarisation may be described as a linear combination of the possible eigenstates of polarisation, i.e.:

$$\psi_\beta = C_t \psi_t + C_a \psi_a$$

where the coefficients C_t and C_a give the contributions of each of the two eigenstates to the state ψ_β. The coefficients must be normalised so that:

$$C_t^2 + C_a^2 = 1.0 \tag{B3.7.1}$$

because the wave function ψ_β describes a single photon and the total probability of it being either absorbed or transmitted must be 1.0. Furthermore, and again according to quantum-mechanical principles, the outcome of the determination of the polarisation of a photon in the state ψ_β must be such that the probability that the photon is polarised along $t - t'$ and is transmitted is C_t^2 and the probability that the photon is polarised along $a - a'$ and is absorbed is C_a^2. But we know these probabilities from experiment to be $\cos^2 \beta$ and $\sin^2 \beta$ respectively, so that:

$$C_t = \pm \cos \beta \quad \text{and} \quad C_a = \pm \sin \beta$$

and the fact that $\sin^2 \beta + \cos^2 \beta = 1.0$ is exactly what is required to satisfy the normalisation condition Equation (B3.7.1). This last point illustrates how the concepts of the superposition of states and expectation value allow us to extract useful information from measurements on systems which are not, at the time immediately before the measurement, in eigenstates of the operator corresponding to the observable property which we are measuring. It also shows that eigenfunctions (eigenstates) and eigenvalues are fundamental but not immutable things; they are determined by the measurements which we choose to make. If we rotated the polariser by Θ^o then a new pair of eigenstates would be appropriate for a discussion of the experiment above.

PROBLEMS FOR CHAPTER 3

1. Confirm the entries in columns three and four of the following table.

Operator, $\hat{P}$	Function, Ψ	Is Ψ an eigenfunction of $\hat{P}$?	If 'yes', what is the eigenvalue?
$\dfrac{\partial^2}{\partial\theta^2}$	$\sin\theta$	yes	-1
$\dfrac{\partial}{\partial\theta}$	$\sin^2\theta$	no	$-$
$\left(\dfrac{\partial}{\partial x} + \alpha\right)$	$\exp(-ax)$	yes	$-a + \alpha$
$\left(\dfrac{x\partial}{\partial x} + \alpha\right)$	x^2	yes	$2 + \alpha$
$\left(\dfrac{x\partial}{\partial x} + \dfrac{y\partial}{\partial y}\right)$	xy	yes	2
$xy\left(\dfrac{\partial}{\partial x} + \dfrac{\partial}{\partial y}\right)$	xy	no	$-$

2. Show that the functions $\exp(+inx)$ and $\exp(-inx)$ are eigenfunctions of the energy operator for an electron in a linear box, each having the same eigenvalue of $n^2\hbar^2/2m_e$.

3. A deeper understanding of the following problem should be possible after Chapters 4 and 5 have been studied. But no knowledge of the material covered there is required to complete the present exercise.

   The following, normalised functions are functions of the polar coordinate, θ, only, apart from ψ_s which is simply a constant:

$$\psi_s = \sqrt{(1/2)} \qquad \psi_d = \sqrt{(5/8)}(3\cos^2\theta - 1)$$
$$\psi_p = \sqrt{(3/2)}\cos\theta \qquad \psi_f = \sqrt{(7/8)}(5\cos^3\theta - 3\cos\theta)$$

Show that these functions are eigenfunctions of the angular momentum operator $\hat{L}_\theta^2$:

$$\hat{L}_\theta^2 = \cot\theta \cdot \partial/\partial\theta + \partial^2/\partial\theta^2$$

The eigenvalues of $\hat{L}_\theta^2$ give the square of the orbital angular momentum in units of $-(h/2\pi)^2$. Show that the eigenvalues are $-l(l+1)$, where $l = 0, 1, 2$ and 3 for s, p, d and f respectively.

4. The functions $\Psi_a = N_a\sqrt{x(L-x)}$ and $\Psi_b = N_b x(L-x)$, where N_a and N_b are normalising constants, are not eigenfunctions for the electron in a one-dimensional box of length L, but they do satisfy the boundary conditions of that system.
   a) Show that $N_a = 2\sqrt{3}/L^2$, $N_b = \sqrt{(30/L^5)}$ and that the functions are not orthogonal.

   b) Using the position operator (Section 3.13.1) and the expression for an expectation value (Equation (3.9.2)) calculate the expectation value of x, $\bar{x}$, for both wave functions, Ψ_a and Ψ_b. Sketch graphs of the two functions [set $L = 1.0$] to explain the difference in the two values of $\bar{x}$.

   c) Show that $(\langle\Psi_b|x|\Psi_b\rangle)^2 \neq \langle\Psi_b|x^2|\Psi_b\rangle$. This is equivalent to saying that the square of the mean of a set of numbers is not equal to the mean of the squares of those numbers.

5. Calculate the expectation value for the energy, E_b, using the function Ψ_b from question 4, the kinetic energy operator (Equation (3.6.2)) and the expression for an expectation value (Equation (3.9.2)). Write down the lowest exact energy eigenvalue, E_1, for the electron-in-a-box problem (Equation (3.6.5)) and show that $E_b = 1.0132E_1$. The fact that the expectation energy calculated with the approximate wave function is greater than the exact lowest energy is an example of the variation theorem (Appendix 2).

Note: Be careful to distinguish between h and $\hbar = h/2\pi$; Equation (3.6.5) uses the former and Equation (3.6.2) the latter.

6. Use Equation (3.3.6) to draw quantitative energy-level schemes for the π-electrons of benzene (C_6H_6), naphthalene ($C_{10}H_8$) and anthracene($C_{14}H_{10}$). Estimate the radius of the ring by assuming that its circumference is the number of peripheral C–C bonds × 139 pm. Place the available electrons in the lower energy levels of your energy-level schemes and calculate the energy required for a transition from the highest occupied to the lowest unoccupied level for each molecule. Compare your results with the experimental data for the *para*-band which is found in the electronic spectra of these molecules at 9.55, 6.87 and 5.24 × 10^{-19} J for benzene, naphthalene and anthracene respectively.

   The agreement of theory and experiment is good, considering the simplicity of the model. Among the most important aspects of the problem which we have neglected are the effects of the bonds across the ring in naphthalene and anthracene and the strong configuration interaction (see Chapter 11) between the four possible excited states of the same energy.

CHAPTER 4
Angular Momentum

4.0   Introduction   73
4.1   Angular momentum in classical mechanics   73
4.2   The conservation of angular momentum   75
4.3   Angular momentum as a vector quantity   75
4.4   Orbital angular momentum in quantum mechanics   76
      4.4.1   The vector model   77
4.5   Spin angular momentum   78
4.6   Total angular momentum   78
      4.6.1   The addition and conservation of angular momentum in quantum
              mechanics   79
      4.6.2   The laws of quantum-mechanical angular momentum   81
4.7   Angular momentum operators and eigenfunctions   82
      4.7.1   The raising and lowering, shift or ladder operators   82
4.8   Notation   83
4.9   Some examples   84
4.10 Bibliography and further reading   86
      Problems for Chapter 4   93

4.0   INTRODUCTION

In the application of the quantum theory to chemistry, angular momentum is arguably as important as energy. It plays a vital part in spectroscopy and in the theory of atomic structure. It has even been suggested that angular momentum plays a more important role than energy in determining the outcome of intermolecular collisions.[1] Before considering the quantum theory of angular momentum we shall describe the phenomenon in classical terms.

4.1   ANGULAR MOMENTUM IN CLASSICAL MECHANICS

Consider a mass (m) moving with velocity (v) in a circular path (radius r) (Figure 4.1). At any moment in time the velocity of the body is directed at right angles to a line joining the body to the centre of the circle. By definition, the linear momentum (p) of the body

The Quantum in Chemistry   R. Grinter
© 2005 John Wiley & Sons, Ltd

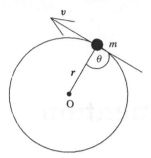

Figure 4.1 Angular momentum generated by an orbiting mass, *m*

is the product of the mass and the velocity:

$$\boldsymbol{p} = \boldsymbol{m}\boldsymbol{v} \tag{4.1.1}$$

and the angular momentum (*a*) is the vector product of *p* and *r*,:

$$\boldsymbol{a} = \boldsymbol{r} \times \boldsymbol{p} = \boldsymbol{r}\boldsymbol{p}\sin\theta = \boldsymbol{r}\boldsymbol{m}\boldsymbol{v} \tag{4.1.2}$$

where θ is the angle between *r* and *p*. The quantities in bold type, *r*, *v*, *p* and *a* are all vector quantities, i.e. they have both magnitude and direction. The vector *a* is located at the centre of the circle about which the mass *m* rotates and, because it is the vector product of *r* and *p*, its direction is at right angles to the plane containing *r* and *p*. In order to define a positive and negative direction for *a* we first define a right-handed co-ordinate system (Figure 4.2), which is one where a rotation from $+x$ to $+y$ advances a right-handed screw along the positive direction of the *z*-axis. Then, if *v* is directed from $+x$ towards $+y$ the resulting angular momentum is positive, whereas if the direction of motion is from $+y$ to $+x$ *a* is negative. Thus, by rotating a wheel on an axle we can generate either positive or negative angular momentum, depending upon the direction of rotation. The angular momentum so generated is represented by a vector which lies along the axle. If the rotating wheel is viewed from $+z$, clockwise (Figure 4.3(a)) and

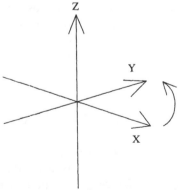

In a right-handed co-ordinate system, a rotation
from +X to +Y advances a right-handed screw
in the positive Z-direction

Figure 4.2 Definition of a right-handed co-ordinate system

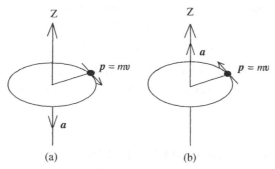

Figure 4.3 Angular momenta produced by clockwise and counter-clockwise rotations

counter-clockwise (Figure 4.3(b)) rotations produce negative (*a* directed away from the observer) and positive (*a* directed towards the observer) angular momentum respectively.

4.2 THE CONSERVATION OF ANGULAR MOMENTUM

Like energy, angular momentum is a conserved quantity and, being a vector quantity, both the magnitude and direction of *a* are maintained if they are not influenced by external factors. The constancy of direction is exploited in the gyroscope which, prior to satellite technology, was widely used in navigational equipment. It is the property which keeps a spinning top upright and is surprisingly strong. Take a cycle wheel, hold the spindle, one end in each hand, and start the wheel rotating gently. Now move the wheel vertically up and down or horizontally from side to side. You should feel no resistance to these movements since you are not attempting to change the direction of the angular momentum vector. Now try to tip the wheel by raising one hand and lowering the other; the resistance to this movement arises because the direction of the angular momentum is conserved. If you do this while seated on a swivelling office chair you will find that the angular momentum is conserved by a rotation of the chair.

The conservation of angular momentum makes the spinning bullets from a gun with a rifled barrel much more stable in flight than those from a smooth-bored weapon, greatly increasing the range of accuracy and hence lethality. The introduction, by the Union, of the Springfield rifled-bore musket into the American Civil War (1861–1865) caused the Confederacy dreadful losses before they adopted new tactics and obtained similar weapons themselves. This effect on projectiles was known to the ancient Greeks who, around 500 BC, used wooden javelins with a length of string wound around the centre in athletic competitions. When the javelin was thrown the athlete would hold on to the string, causing the javelin to rotate thereby increasing the length of his throw.

4.3 ANGULAR MOMENTUM AS A VECTOR QUANTITY

Since it is a vector quantity, angular momentum has a component in any direction in space which we choose to specify. It is useful to be able to specify the components along the x, y and z axes of a Cartesian co-ordinate system and this can be easily done. If we know the angles α, β and γ between the angular momentum vector (a) and the x, y, and z axes respectively, then the required equations are:

$$a_x = a \cos\alpha, \quad a_y = a \cos\beta, \quad a_z = a \cos\gamma \qquad (4.3.1)$$

The classical angular momentum which has been described above is also called orbital angular momentum since it arises as a result of a mass executing an orbit around a fixed point.

4.4   ORBITAL ANGULAR MOMENTUM IN QUANTUM MECHANICS

Orbital angular momentum is also found in quantum mechanics (Box 4.1) but the quantum and classical orbital angular momenta differ in two aspects related to quantisation. Firstly, in quantum mechanics the total angular momentum is quantised with possible magnitudes given by the equation:

$$a = \sqrt{d(d+1)} \cdot \frac{h}{2\pi} \equiv \sqrt{d(d+1)} \cdot \hbar \qquad (4.4.1)$$

where h is Planck's constant and d is a quantum number which can take all positive integer values, including zero, i.e. $d = 0, 1, 2, 3, \ldots$ A molecule rotating freely in space has angular momentum of this type. Note that the quantity $h/2\pi$ occurs so frequently and is so important that it has its own symbol ($\hbar$) which is spoken 'h-cross'.

Secondly, though the x, y and z components of a classical angular momentum vector can always be specified as in Equation (4.3.1), quantum mechanics places very important restrictions on the components of angular momentum which can be specified. The first of these restrictions is that the component in only one direction in space may be specified. Thus, if the z component is given, no x or y components may be specified. Furthermore, the components in the direction which can be specified, and there is no restriction in our choice of this one direction, are quantised. Whichever the direction chosen, z in Equation (4.4.2), the allowed quantised components are given by the equation:

$$a_z = m_d h/2\pi \equiv m_d \hbar \qquad (4.4.2)$$

where m_d is a quantum number which can take all integer values from $-d$ to $+d$:

$$m_d = -d, -(d-1), -(d-2), \ldots -1, 0, +1, \ldots +(d-1), +d \qquad (4.4.3)$$

giving $2d + 1$ values for m_d. The results expressed in Equations (4.4.1), (4.4.2) and (4.4.3) are deduced from the angular momentum commutation relationships in Box 4.1.

Thus, for $d = 0,$ $m_d = 0$

$\qquad\qquad d = 1,$ $m_d = -1, 0, +1$

$\qquad\qquad d = 2,$ $m_d = -2, -1, 0, +1, +2$ etc.

In Equation (4.4.2), the z axis has been selected as the one for which the components of angular momentum are specified. This is purely a matter of choice but it is a convention which has been universally adopted. The symbol, m_d, for the quantum number above was chosen because of the close relationship between this quantum number and the magnetic properties of atoms. The origins of this relationship will be described in Chapter 12.

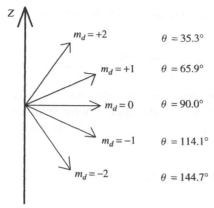

Figure 4.4 The five z-components of an angular momentum characterised by an angular momentum quantum number of 2

Thus, in quantum mechanics, orbital angular momentum is described by two quantum numbers, d which gives the value of the quantised total angular momentum through Equation (4.4.1) and m_d which gives the allowed components of the angular momentum in the z direction through Equation (4.4.2) and is restricted in its values by Equation (4.4.3). This form of quantisation is known as space quantisation because the spatial orientation of the angular momentum with respect to the z-axis is restricted by the quantisation of its z component. Note that the maximum component, a_z is $d\hbar$ and is therefore always smaller than the total angular momentum, $\sqrt{[d(d+1)]}\hbar$. This means that the quantised component (a_z) never lies exactly along the total angular momentum vector (a) (Figure 4.4). This is a further consequence of the uncertainty principle. If a_z were aligned along a we would know not only the values of a and a_z, but also those of a_x and a_y, because they would both be zero. Thus, we would know everything that there is to know about the angular momentum of the system and in quantum mechanics that is not permitted, as we have seen in Section 3.10. Figure 4.4 illustrates this point.

4.4.1   The vector model

We have a situation in which the length of a vector and its projection on the z axis are defined, but nothing more about its orientation can be said. Since there is no reason to treat the x and y axes differently, we must assume that the projection of the vector in the xy-plane points in all possible directions with equal probability. This concept forms the basis of the *vector model* of angular momentum in which the vector d is envisaged as precessing about the z axis so that it describes a cone about that axis in which the half-angle at the apex of the cone (θ) is given by Equation (4.4.4):

$$\cos\theta = \frac{m_d}{\sqrt{d(d+1)}} \qquad (4.4.4)$$

We should not leave this aspect of angular momentum without taking advantage of the opportunity it provides of illustrating the way in which classical mechanics may be regarded as a limiting case of quantum mechanics. The fact that there is always a finite angle (θ) between the total quantum-mechanical angular momentum vector and its maximum z component is a non-classical result. Since the maximum value of m_d is d, $\cos(\theta[\min])$ is given by:

$$\cos(\theta[\min]) = \frac{m_d[\max]}{\sqrt{d(d+1)}} = \frac{d}{\sqrt{d(d+1)}} = \sqrt{\frac{d}{d+1}} \qquad (4.4.5)$$

Thus, as d becomes very large $\cos(\theta[\min])$ tends to 1, $\theta[\min]$ tends to zero and the maximum possible z component of the angular momentum approaches the value of the total angular momentum, which is what it would be in classical mechanics.

4.5   SPIN ANGULAR MOMENTUM

As we have seen in Section 2.7, in his model of the hydrogen atom Bohr quantised electronic orbital angular momentum thirteen years before the birth of quantum mechanics. In 1925, further studies of the spectra of atoms and the effect of magnetic fields upon them led Samuel Abraham Goudsmit (1902–1978) and George Eugene Uhlenbeck (1900–1988) to postulate that, in addition to its orbital angular momentum, an electron also possesses another type of angular momentum which they called spin angular momentum. Like the orbital angular momentum, the spin angular momentum also plays a role in the magnetic properties of atoms, though the relationship between the angular momentum and the magnetism is not exactly the same in both cases (see Section 12.2).

Wolfgang Pauli (1900–1958), who had used the concept of electron spin to formulate his exclusion principle in 1925, introduced spin into the new quantum mechanics two years later; but in a phenomenological manner. In 1928 Dirac placed the concept on a much firmer footing when he found that the electron-spin quantum number arose naturally when the quantum mechanics of the hydrogen atom was made compatible with the theory of relativity. Detailed study then revealed that the spin angular momentum obeyed Equations (4.4.1), (4.4.2) and (4.4.3) exactly as orbital angular momentum did, but with one important difference. For spin the quantum number d could take positive half-integer as well as integer values. Thus, in addition to the d and m_d values given above, for spin we can also have:

$$d = \tfrac{1}{2}; \qquad m_d = -\tfrac{1}{2}, +\tfrac{1}{2}$$
$$d = \tfrac{3}{2}; \qquad m_d = -\tfrac{3}{2}, -\tfrac{1}{2}, +\tfrac{1}{2}, +\tfrac{3}{2}$$
$$d = \tfrac{5}{2}; \qquad m_d = -\tfrac{5}{2}, -\tfrac{3}{2}, -\tfrac{1}{2}, +\tfrac{1}{2}, +\tfrac{3}{2}, +\tfrac{5}{2} \quad \text{etc.}$$

4.6   TOTAL ANGULAR MOMENTUM

The total angular momentum of an atom or molecule is usually composed of both spin and orbital parts and both of these contributions may arise from several sources. For example, the spin angular momentum may be electron spin, but the spins of the nuclei are also important in many branches of atomic and molecular spectroscopy. Orbital angular momentum may be due to orbital motion of electrons. It may also arise as a result of the rotation in space of the molecule as a whole. These various angular momenta are not independent; they interact with each other by mechanisms of which we shall give some examples in the chapters on atomic structure and spectroscopy. It is important therefore that we have a way of adding angular momenta together.

4.6.1 The addition and conservation of angular momentum in quantum mechanics

In classical mechanics the addition of two angular momenta is quite straightforward. The tails of the two vectors are brought to the same point and the addition parallelogram constructed to give their sum. In Figure 4.5a, $a + b = c$. Thus, for any pair of vectors a and b, each of fixed magnitude (length) and direction (orientation), there can be one, and only one, resultant (c). Angular momentum is conserved and c must be the vector sum of a and b. Figure 4.5b shows an alternative form of the addition diagram.

The addition of angular momentum in quantum mechanics is not as simple because all contributions to the total angular momentum are themselves quantised and the addition process, also known as *coupling*, must ensure that the sum itself and also its components are quantised. It can be shown (Box 4.2)[2] that if two angular momenta characterised by the angular momentum quantum numbers d_1 and d_2 are added to form a total angular momentum characterised by the quantum number d, then d can take all the values given by the Equation (4.6.1); also known as the Clebsch–Gordan series:

$$d = d_1 + d_2, \; d_1 + d_2 - 1, \; d_1 + d_2 - 2, \ldots |d_1 - d_2| \tag{4.6.1}$$

e.g. $d_1 = 3, d_2 = 1 : d = 4, 3, 2.$ $d_1 = 2, d_2 = \frac{3}{2} : d = \frac{7}{2}, \frac{5}{2}, \frac{3}{2}, \frac{1}{2}.$

(Note how the series terminates at the modulus of the difference between d_1 and d_2.)

Although d_1 and d_2 are defined as completely as is possible in quantum mechanics, according to Equation (4.6.1) $2d_< + 1$, where $d_<$ is the smaller of d_1 and d_2, resultants of the addition of d_1 and d_2 are permissible, rather than the single resultant which we would have in classical mechanics, (Figure 4.5). It is tempting to attempt to draw two vectors having lengths characterised by their values of d_1 and d_2 and orientations with respect to the z axis determined by their values of m_{d1} and m_{d2} and to add these in the classical manner, comparable to Figure 4.5, to obtain a vector characterised by d and m_d given by the rules for the addition of quantum-mechanical angular momentum quoted above. This is not possible because the plane containing the z axis and d_1 cannot be assumed to be coincident with that containing the z axis and d_2. The rules apply to the quantum numbers and the angular momentum vectors constructed using them cannot be added as in classical mechanics. But the z components of d_1 and d_2 both lie along the z axis and can be directly added and each component of the combined vector, $|d, m_d\rangle$, is a sum of the components of the combining vectors, $|d_1, m_{d1}\rangle$ and $|d_2, m_{d2}\rangle$, which includes all those which satisfy the requirement that $m_d = m_{d1} + m_{d2}$. For example, one can show using the raising and lowering operators (Section 4.7.1 and Box 4.1) that in the

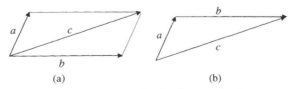

(a)   (b)

Figure 4.5 The addition of angular momenta in classical mechanics

case of the second of the above examples the resultant vector $|\frac{5}{2}, +\frac{1}{2}\rangle$ with $d = \frac{5}{2}$ and $m_d = +\frac{1}{2}$, is:

$$|\tfrac{5}{2}, +\tfrac{1}{2}\rangle = 3\sqrt{(3/70)}|2, -1\rangle|\tfrac{3}{2}, +\tfrac{3}{2}\rangle + \sqrt{(6/70)}|2, 0\rangle|\tfrac{3}{2}, +\tfrac{1}{2}\rangle$$
$$- 5/\sqrt{(70)}|2, +1\rangle|\tfrac{3}{2}, -\tfrac{1}{2}\rangle - 2\sqrt{(3/70)}|2, +2\rangle|\tfrac{3}{2}, -\tfrac{3}{2}\rangle$$

The numerical coefficients in the above sum are known as Clebsch–Gordan coefficients and formulae for calculating them from d, m_d, d_1, m_{d1}, d_2 and m_{d2} can be found in many sources.[2-5] They are also extensively tabulated.[6]

In spite of the obvious difficulty of illustrating the quantum-mechanical addition of vectors, one frequently encounters diagrams of that type and they are very helpful in understanding complex coupling processes such as are found in molecular electronic spectroscopy. The basic situation is illustrated in Figure 4.6. In Figure 4.6(a) the two vectors, d_1 and d_2, are not coupled. Each precesses about the z axis, independently with constant z components (m_1 and m_2) but not in phase. d_1, d_2, m_1 and m_2 are *good quantum numbers* and the corresponding angular momenta and z components are separately observable. In Figure 4.6(b) d_1 and d_2, are coupled to form d, about which they precess together and in phase. They no longer have constant z components; d_1, d_2, m_1 and m_2 are not good quantum numbers and the corresponding angular momenta and components are not observable. d precesses about z with a constant z component, m, the quantum numbers d and m are good quantum numbers and the corresponding angular momentum and its z component are experimental observables. This point is illustrated with an example in Section 4.6.2.

One other consequence of the fact that, in quantum mechanics, there can be more than one result for the sum of two vector quantities should be mentioned here. We shall investigate it further in Chapter 8 when we consider the question of the conservation of angular momentum in connection with spectroscopic selection rules. At this point we need only say that if, before a particular event, a system consists of two angular momenta characterised by the angular momentum quantum numbers d_1 and d_2 and if, after the event, the system has an angular momentum characterised by d then if d satisfies *any* of the $2d_< + 1$ Equations (4.6.1), angular momentum has been conserved.

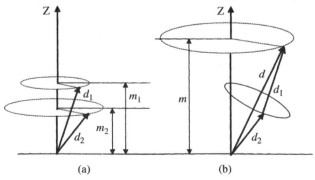

Figure 4.6 The addition (coupling) of two angular momenta

4.6.2 The laws of quantum-mechanical angular momentum

Once we have a formula for the addition of two angular momenta we can add any number of contributions by repeated applications of the formula. (But note how rapidly the number of possible values for the total angular momentum rises!)

And although there may be many contributions to the total angular momentum of a molecule, we can make three general statements about angular momentum which each individual contribution to the total and the total itself must obey. They are:

1. The square of the angular momentum, a^2, is given in terms of the quantum number d in the form of Equation (4.4.1):

$$a^2 = [d(d+1)]\hbar^2$$

   where d can take any positive integer or half-integer value, including zero.

2. The components of the angular momentum, $\boldsymbol{a}_z$, in any specified direction (conventionally the z-axis) are given by Equation (4.4.2):

$$\boldsymbol{a}_z = m_d\hbar$$

   where the quantum number m_d can take all $2d + 1$ values from $-d$ to $+d$ in unit steps, Equation (4.4.3).

3. If the angular momentum is purely *orbital* angular momentum the quantum number d, and hence m_d, can take only integer values. When spin is involved half-integer values are also possible.

These three statements might be termed the laws of quantum-mechanical angular momentum and two important points about them should be made. Firstly, it is rare to observe the angular momentum directly, usually it is the *transitions* between states of angular momentum which give rise to observable experimental data. Secondly, in cases where two or more angular momenta are combined, or *coupled* (Section 4.6.1), the observable experimental data are related to changes in the *total* angular momentum and must be interpreted in terms of the corresponding *total* quantum numbers. But the individual quantum numbers for the contributing angular momenta play an important role in describing the system theoretically and it is in that sense that we can say that the individual contributions to the total angular momentum obey the above laws, even when they are not experimentally observable.

It might be helpful to make this point clear with an example. In a heavy atom, the orbital angular momentum of an electron (quantum number l) may couple with the spin angular momentum of the same electron (quantum number s) to give a total angular momentum characterised by the quantum number j. (See Section 5.8 for more details on this phenomenon which is called *spin-orbit coupling*.) Experimental observations are interpretable in terms of j which is said to be a good quantum number. The orbital and spin angular momenta corresponding to the quantum numbers l and s only feature in the theoretical description of the atom, they are not observed directly in the experimental measurements and, consequently, l and s are not good quantum numbers.

4.7 ANGULAR MOMENTUM OPERATORS AND EIGENFUNCTIONS

In Chapter 3 we discussed the basic application of Schrödinger's theory of observables, operators, eigenfunctions and eigenvalues using explicit mathematical functions for the operators for energy and angular momentum in our examples. We consider these ideas again here because, for angular momentum, the concept takes an extremely simple form in which, for many applications, we do not need to give full mathematical expressions for either the operators or the eigenfunctions. This is very fortunate, not only on account of the resulting simplicity but also because the operators for spin angular momentum cannot be obtained from the corresponding classical expressions because there is no classical analogue of quantum-mechanical spin.

In Box 4.1 eigenfunction-eigenvalue equations for the first two laws of Section 4.6 (Equations (B4.1.22a) and (B4.1.22b)) are derived. We repeat those equations here with $\hat{d}$ as the operator for total angular momentum and $\hat{d}_z$ that for its z component:

$$\hat{d}|d, m_d\rangle = \sqrt{d(d+1)} \cdot \hbar |d, m_d\rangle \tag{4.7.1}$$

and

$$\hat{d}_z|d, m_d\rangle = m_d \cdot \hbar |d, m_d\rangle \tag{4.7.2}$$

Here we are adopting a notation, first suggested by Dirac, in which we specify the eigenfunction $|d, m_d\rangle$, solely by giving the quantum numbers which define it and determine its eigenvalues. Angular momentum eigenfunctions are characterised by two quantum numbers. The first (d) gives the total angular momentum, in units of $h/2\pi$, according to Equation (4.7.1). The second (m_d) gives the z component of that angular momentum, also in units of $h/2\pi$, according to Equation (4.7.2). The corresponding eigenfunction is exactly identified, as far as its angular momentum properties are concerned, if the values of these two quantum numbers are given, and that is frequently sufficient for our purposes. The symbol $|d, m_d\rangle$ is therefore all that we need to write Equations (4.7.1) and (4.7.2), which satisfy the requirement that the operator operating on the eigenfunction is equal to the eigenfunction multiplied by a number, the eigenvalue. The eigenvalues of the angular momentum operators are the observable, quantised values of the total angular momentum and its z component given in units of $h/2\pi$ by the quantum numbers.

4.7.1 The raising and lowering, shift or ladder operators

Since we cannot know the values of the x and y components of the angular momentum if we already know the z component, we cannot write eigenfunction-eigenvalue equations like Equation (4.7.2) for $\hat{d}_x$ and $\hat{d}_y$. However, there are two operators derived from $\hat{d}_x$ and $\hat{d}_y$ which we use in Box 4.1. Although they are not eigenfunction-eigenvalue equations, they are of great value to us. The operators are the raising and lowering operators, $\hat{d}_+$ and $\hat{d}_-$ defined by the equations:

$$\hat{d}_+ = \hat{d}_x + i\hat{d}_y \tag{4.7.3}$$

and

$$\hat{d}_- = \hat{d}_x - i\hat{d}_y \tag{4.7.4}$$

In these two equations, $i = \sqrt{(-1)}$; for more information about i see Appendix 8. The two operators have very important properties which are deduced in Box 4.1 (Equations (4.1.31a) and (4.1.31b)) and repeated here as Equations (4.7.5) and (4.7.6):

$$\hat{d}_+|d, m_d\rangle = \sqrt{[d(d+1) - m_d(m_d + 1)]} \cdot \hbar|d, m_d + 1\rangle \qquad (4.7.5)$$

$$\hat{d}_-|d, m_d\rangle = \sqrt{[d(d+1) - m_d(m_d - 1)]} \cdot \hbar|d, m_d - 1\rangle \qquad (4.7.6)$$

The application of either operator does not change the value of the total angular momentum (the quantum number d remains the same) but each changes the value of the z component of the angular momentum by one unit of $h/2\pi$. $\hat{d}_+$ raises the value of d_z, i.e. $(m_d \rightarrow m_d + 1)$, and $\hat{d}_-$ lowers it $(m_d \rightarrow m_d - 1)$. For example:

$$\hat{d}_+|2, 1\rangle = \sqrt{[(2 \times 3) - (1 \times 2)]} \cdot \hbar|2, 2\rangle = 2\hbar|2, 2\rangle$$

$$\hat{d}_-|2, 1\rangle = \sqrt{[(2 \times 3) - (1 \times 0)]} \cdot \hbar|2, 0\rangle = \sqrt{6}\hbar|2, 0\rangle$$

These are not eigenfunction-eigenvalue equations because the functions on the left- and right-hand sides of Equations (4.7.5) and (4.7.6) are not the same. Rather, these two operators take us up ($\hat{d}_+$) and down ($\hat{d}_-$) a set of m_d values, one step at a time. They are therefore known as the *raising and lowering, ladder* or *shift* operators. Note that if we try to raise an m_d value which is already at its maximum, i.e. $m_d = d$, we get a result of zero. The same is true if we try to lower a value of m_d which is already at its minimum value, i.e. $m_d = -d$.

4.8 NOTATION

The quantum numbers d and m_d in the above laws represent any form of angular momentum. It has become customary to denote the various types of angular momentum with a particular choice of symbols for d as the quantum number and this practice is so widespread and consistent that it should be stated here.

The orbital angular momentum of an electron in an atom:	l and m_l.
The spin angular momentum of an electron:	s and m_s.
The combined spin and orbital angular momentum of an electron in an atom:	j and m_j.
Where the angular momentum results from more than one electron this is indicated by using the upper case letters:	L, M_L, S, M_S, J, M_J.
The spin angular momentum of an atomic nucleus:	I and M_I.
The orbital angular momentum of a rotating diatomic molecule:	J and M_J.
The orbital angular momentum of a molecule rotating in space:	R and M_R.

(In the last three cases the upper case letter is always used, even for one nucleus or one molecule.)

4.9   SOME EXAMPLES

A single electron has $s = \frac{1}{2}$ and therefore $m_s = -\frac{1}{2}$ and $+\frac{1}{2}$. Thus, a single electron can be found in two different spin states. When two electrons occupy the same orbital (spatial region) in an atom or molecule one has $m_s = -\frac{1}{2}$ and the other $m_s = +\frac{1}{2}$. They are said to be spin-paired or, simply, paired.

The s, p, d, ... atomic orbitals of atoms (see Chapter 5) have orbital angular momentum quantum numbers, $l = 0, 1, 2, ...$ respectively. Associated with these there are m_l values of 0 for s, $-1, 0, +1$ for p and $-2, -1, 0, +1, +2$ for d, etc. Thus, there is just one s orbital, but there are three p orbitals, five d orbitals, seven f orbitals, ... The element boron consists of two isotopes, 19.6 % of ^{10}B and 80.4 % of ^{11}B. The nuclear spin quantum numbers of the two isotopes are 3 and $\frac{3}{2}$ respectively and we therefore have:

$$^{10}BI = 3 \qquad M_I = -3, -2, -1, 0, +1, +2, +3$$

$$^{11}BI = \tfrac{3}{2} \qquad M_I = -\tfrac{3}{2}, -\tfrac{1}{2}, +\tfrac{1}{2}, +\tfrac{3}{2}$$

The rigidity and simplicity of the rules of quantum-mechanical angular momentum make more complex systems quite easy to handle, in principle, though frequently tedious in practice. Suppose that we have one electron in a p orbital. Then, $s = \frac{1}{2}$ and the possible m_s values are $-\frac{1}{2}$ and $+\frac{1}{2}$, while $l = 1$ and the possible m_l values are -1, 0 and $+1$. Recall that the m_s and m_l values give the components of the angular momenta along the z axis. Therefore, adding the z components of the orbital and spin vectors we have six possibilities for the z component of the combined momenta, m_j:

m_l	m_s	m_j	m	m_s	m_j	m_l	m_s	m_j
$-1 + (-\frac{1}{2}) = -\frac{3}{2}$			$0 + (-\frac{1}{2}) = -\frac{1}{2}$			$+1 + (-\frac{1}{2}) = +\frac{1}{2}$		
$-1 + (+\frac{1}{2}) = -\frac{1}{2}$			$0 + (+\frac{1}{2}) = +\frac{1}{2}$			$+1 + (+\frac{1}{2}) = +\frac{3}{2}$		

The result of adding the z components of orbital and spin angular momenta are six m_j values which represent the z components of the combined angular momenta. And we know from the above laws, which angular momentum *always* obeys, that the values of these components must go from $-m_j$ to $+m_j$ in unit steps. Therefore, if there is an m_j value of $-\frac{3}{2}$ it must be associated with further m_j values of $-\frac{1}{2}, +\frac{1}{2}$ and $+\frac{3}{2}$. Furthermore, we also know that the maximum m_j value will be the same as the j value for the total angular momentum, so we have discovered that our single p electron can have a state of angular momentum characterised by the quantum numbers:

$$j = \tfrac{3}{2} \quad \text{and} \quad m_j = -\tfrac{3}{2}, -\tfrac{1}{2}, +\tfrac{1}{2}, +\tfrac{3}{2}$$

We remove these four values from our list above (it does not matter which of the two $+\frac{1}{2}$ and $-\frac{1}{2}$ values we choose; this is merely a book-keeping exercise) and there remain just two m_j values, $-\frac{1}{2}$ and $+\frac{1}{2}$. Clearly, there is a second angular momentum state for our p electron which is characterised by the quantum numbers:

$$j = \tfrac{1}{2} \quad \text{and} \quad m_j = -\tfrac{1}{2}, +\tfrac{1}{2}$$

These results for the possible values of j should be compared with those obtained by adding the spin and orbital angular momenta according to the rule given in Section 4.6.1.

We can apply a similar method to determine the spin states of a two-electron system. We first tabulate all the possible combinations of m_{s1} and m_{s2} and their total M:

m_{s1}	m_{s2}	M	m_{s1}	m_{s2}	M	m_{s1}	m_{s2}	M
$+\frac{1}{2}$	$+\frac{1}{2}$	$+1$	$+\frac{1}{2}$	$-\frac{1}{2}$	0	$-\frac{1}{2}$	$-\frac{1}{2}$	-1
			$-\frac{1}{2}$	$\frac{1}{2}$	0			

Clearly, the value $M = +1$ must be associated with two other components, $M = 0$ and $M = -1$. Thus, there is a state with a total spin of $S = 1$ which, because it has three components $|1, +1\rangle, |1, 0\rangle$ and $|1, -1\rangle$, is called a *triplet*. A *singlet*, $|0, 0\rangle$, with $S = 0$ and $M = 0$ remains.

To find the eigenfunctions we first use the lowering operator $(\hat{S}_-)$ on the combined spin function $|1, +1\rangle$:

$$\hat{S}_-|1, +1\rangle = \sqrt{1(1+1) - 1(1-1)}|1, +1\rangle = \sqrt{2}|1, 0\rangle$$

We now apply the lowering operators $(\hat{s}_-(1)$ and $\hat{s}_-(2))$ to the spin functions of the individual electrons:

$$\{\hat{s}_-(1) + \hat{s}_-(2)\}| + \tfrac{1}{2}(1), +\tfrac{1}{2}(2)\rangle = \hat{s}_-(1)| + \tfrac{1}{2}(1), +\tfrac{1}{2}(2)\rangle + \hat{s}_-(2)| + \tfrac{1}{2}(1), +\tfrac{1}{2}(2)\rangle$$

$$= | - \tfrac{1}{2}(1), +\tfrac{1}{2}(2)\rangle + | + \tfrac{1}{2}(1), -\tfrac{1}{2}(2)\rangle$$

Equating the two results:

$$|1, 0\rangle = \sqrt{\tfrac{1}{2}}\{| - \tfrac{1}{2}(1), +\tfrac{1}{2}(2)\rangle + | + \tfrac{1}{2}(1), -\tfrac{1}{2}(2)\rangle\}$$

Further lowering operations on both sides of the above equation give:

$$|1, -1\rangle = | - \tfrac{1}{2}(1), -\tfrac{1}{2}(2)\rangle$$

The singlet state, $|0, 0\rangle$, must have $M = 0$ and be orthogonal to the state $|1, 0\rangle$. These two conditions determine its eigenfunction to be:

$$|0, 0\rangle = \sqrt{\tfrac{1}{2}}\{| - \tfrac{1}{2}(1), +\tfrac{1}{2}(2)\rangle - | + \tfrac{1}{2}(1), -\tfrac{1}{2}(2)\rangle\}$$

As a final example we consider the spin properties of the hydrogen atom which have some interesting applications. The nucleus of the hydrogen atom, the proton, has a spin of $I = \frac{1}{2}$ and $M_I = -\frac{1}{2}$ and $+\frac{1}{2}$. The single electron has $s = \frac{1}{2}$ and $m_s = -\frac{1}{2}$ and $+\frac{1}{2}$. Therefore, there are four possible spin states of the hydrogen atom:

M_I		m_s		M	M_I		m_s		M
$-\frac{1}{2}$	$+$	$(-\frac{1}{2})$	$=$	-1	$-\frac{1}{2}$	$+$	$\frac{1}{2}$	$=$	0
$+\frac{1}{2}$	$+$	$(-\frac{1}{2})$	$=$	0	$+\frac{1}{2}$	$+$	$\frac{1}{2}$	$=$	$+1$

As always, we note that we have values of $M_I + m_s$ which run from -1 to $+1$ in unit steps and we pick out these as the three components of a state characterised by a total angular momentum quantum number of 1. A spin state of zero total angular momentum remains. Now, because the spins of the nucleus and the electron are each associated with a magnetic moment (see Chapter 5), the two particles behave as if they were minute bar magnets and the energy of the atom depends upon the relative orientation of the two magnets. The energy of the state in which the two spins are aligned parallel, e.g. $M_I = +\frac{1}{2}, m_s = +\frac{1}{2}$ is greater than that in which the two spins are aligned anti-parallel, e.g. $M_I = -\frac{1}{2}, m_s = +\frac{1}{2}$, and the energy difference is 0.9412×10^{-30} J. Some further discussion of the nature of this interaction can be found in Section 9.6. Using the Bohr-Einstein equation, $E = h\nu$, (Section 2.7.1) this energy difference corresponds to electromagnetic radiation of frequency 1420.4 MHz or wavelength 0.2111 m. Thus, if the higher energy state should change into the lower energy state, i.e. if the relative orientations of the two spinning particles were to change from parallel to anti-parallel, then electromagnetic radiation of the above frequency/wavelength would be emitted.

In 1942, Hendrik Van de Hulst (1918–) suggested that, although this event would occur only very infrequently (approximately once every 11 million years in any one hydrogen atom), since there are vast amounts of hydrogen distributed as atoms throughout space, one might expect to detect this radiation if a sufficiently sensitive receiver was available. Following World War II, the technological advances achieved in pursuit of military objectives made the rapid development of radio astronomy possible and the 'song of hydrogen' was detected and used to map the galaxy.

Later, in the 1960s, when it was proposed that we might listen for radio transmissions from intelligent beings in outer space the question naturally arose as to what frequency of signal the listening device should be tuned to receive. It was persuasively argued that, since hydrogen is the most abundant element in the universe, the emission/absorption of hydrogen atoms would be known to all intelligent life and that this was therefore the frequency which should be used. The search has now occupied more than 10 000 hours of radio-telescope time with the star systems *Epsilon Eridani* and *Tau Ceti* (11 and 12 light years distant respectively) as the main targets. All results to date have apparently been negative.

In 1994 the 0.2111 m emission of hydrogen atoms was used to discover a new galaxy, Dwingeloo 1, which is about 10^7 light years from earth and lies in the plane of our own galaxy, the Milky Way. Star systems which lie within the Milky Way are very difficult for us to see because we have to look through the plane of the galaxy and the scattering of light by dust particles within the plane obscures the image. But the scattering of light is inversely proportional to the fourth power of the wavelength (Rayleigh's Law) so that the scattering of the long-wavelength hydrogen emission is scarcely affected by the dust. By searching for the 0.2111 m waves from within the galactic plane, astronomers in the Netherlands were able to detect the emission from Dwingeloo 1 and pin-point its position. When this had been done it was possible to confirm the discovery by means of a rather poor image obtained with an optical telescope. The search for 0.2111 m emission from other galaxies within the Milky Way continues.

4.10   BIBLIOGRAPHY AND FURTHER READING

1. A.J. McCaffery, Z.T. Alwahabi, M.A. Osborne and C.J. Williams, *J. Chem. Physics*, **98**, 4586, 1993.
    The comparatively recent book by Zare[2] covers a very large area of angular momentum theory and highlights many applications in physics and chemistry. It can be highly recommended to those who wish to study angular momentum theory and to apply it to the interpretation of physical measurements. Earlier, classic texts on the subject are those by Rose,[3] Edmonds[4] and by Brink and Satchler.[5] Bouten[7] has pointed out that there are, unfortunately, some inconsistencies in the treatment of rotations by both Rose and Edmonds, but one has to be an advanced user of the theory to appreciate or be affected by this problem.

2. R.N. Zare, *Angular Momentum*, Wiley, New York, 1988.
3. M.E. Rose, *Elementary Theory of Angular Momentum*, Wiley, New York, 1957.
4. A.R. Edmonds, *Angular Momentum in Quantum Mechanics*, Princeton University Press, 1957.
5. D.M. Brink and G.R. Satchler, *Angular Momentum*, 2nd Edn, Oxford University Press, 1962.
6. R.D. Cowan, *The Theory of Atomic Structure and Spectra*, California University Press, Berkeley, 1981.
7. M. Bouten, *Physica*, **42**, 572, 1969.

BOX 4.1 Angular momentum operators and commutation rules

The orbital angular momentum (***L***) of a single particle about an origin is the vector product of the distance from the origin (***r***) and the linear momentum (***p***):

$$L = r \times p \qquad (B4.1.1)$$

In Cartesian co-ordinates the components of ***L*** are:

$$L_x = yp_z - zp_y \qquad (B4.1.2a)$$

$$L_y = zp_x - xp_z \qquad (B4.1.2b)$$

$$L_z = xp_y - yp_x \qquad (B4.1.2c)$$

Note how the successive components can be obtained by a cyclic permutation of $x \to y \to z \to x \ldots$

If we replace the p's by the corresponding quantum-mechanical operators we obtain the operators for angular momentum in units of $h/2\pi \equiv \hbar$:

$$\hat{L}_x = -i\hbar \{ y\partial/\partial z - z\partial/\partial y \} \qquad (B4.1.3a)$$

$$\hat{L}_y = -i\hbar \{ z\partial/\partial x - x\partial/\partial z \} \qquad (B4.1.3b)$$

$$\hat{L}_z = -i\hbar \{ x\partial/\partial y - y\partial/\partial x \} \qquad (B4.1.3c)$$

These three operators do not commute with each other, for example:

$$\hat{L}_x\hat{L}_y\Psi = -\hbar^2 \{ y\partial/\partial z - z\partial/\partial y \}\{ z\partial/\partial x - x\partial/\partial z \}\Psi$$

$$= -\hbar^2 \{ y\partial/\partial z[z\partial/\partial x] - y\partial/\partial z[x\partial/\partial z] - z\partial/\partial y[z\partial/\partial x] + z\partial/\partial y[x\partial/\partial z] \}\Psi$$

The terms in square brackets must be differentiated as products so that we have:

$$= -\hbar^2 \{ y\partial/\partial x + yz\partial^2/\partial z\partial x - yx\partial^2/\partial z^2 - z^2\partial^2/\partial y\partial x + zx\partial^2/\partial y\partial z \}\Psi$$

Similarly:

$$\hat{L}_y\hat{L}_x\Psi = -\hbar^2 \{ z\partial/\partial x - x\partial/\partial z \}\{ y\partial/\partial z - z\partial/\partial y \}\Psi$$

$$= -\hbar^2 \{ z\partial/\partial x[y\partial/\partial z] - z\partial/\partial x[z\partial/\partial y] - x\partial/\partial z[y\partial/\partial z] + x\partial/\partial z[z\partial/\partial y] \}\Psi$$

$$= -\hbar^2 \{ zy\partial^2/\partial x\partial z - z^2\partial^2/\partial x\partial y - xy\partial^2/\partial z^2 + x\partial/\partial y - xz\partial^2/\partial z\partial y \}\Psi$$

The co-ordinates x, y and z commute with each other and so do the operators for partial differentiation with respect to those co-ordinates. Therefore, $yz\partial^2/\partial z\partial x = zy\partial^2/\partial x\partial z$ etc. and:

$$\{ \hat{L}_x\hat{L}_y - \hat{L}_y\hat{L}_x \}\Psi = -\hbar^2 \{ y\partial/\partial x - x\partial/\partial y \}\Psi = -i\hbar \cdot i\hbar \{ x\partial/\partial y - y\partial/\partial x \}\Psi$$

$$= i\hbar \hat{L}_z\Psi$$

Thus: $\{\hat{L}_x\hat{L}_y - \hat{L}_y\hat{L}_x\}\Psi \equiv [\hat{L}_x, \hat{L}_y]\Psi = i\hbar\,\hat{L}_z\Psi$ (B4.1.4a)

and similarly: $\{\hat{L}_y\hat{L}_z - \hat{L}_z\hat{L}_y\}\Psi \equiv [\hat{L}_y, \hat{L}_z]\Psi = i\hbar\,\hat{L}_x\Psi$ (B4.1.4b)

and: $\{\hat{L}_z\hat{L}_x - \hat{L}_x\hat{L}_z\}\Psi \equiv [\hat{L}_z, \hat{L}_x]\Psi = i\hbar\,\hat{L}_y\Psi$ (B4.1.4c)

These are the commutation relationships between the operators for the Cartesian components of angular momentum. The symbol $[a, b] = ab - ba$, is called the commutator of a and b.

But $\hat{L}^2 = \hat{L}_x^2 + \hat{L}_y^2 + \hat{L}_z^2$ commutes with each of its components, e.g. [dropping the explicit inclusion of the operand, Ψ, to simplify the notation]:

$$\hat{L}^2\hat{L}_z - \hat{L}_z\hat{L}^2 = \hat{L}_x^2\hat{L}_z + \hat{L}_y^2\hat{L}_z + \hat{L}_z^3 - \hat{L}_z\hat{L}_x^2 - \hat{L}_z\hat{L}_y^2 - \hat{L}_z^3$$

$$= \hat{L}_x\hat{L}_x\hat{L}_z - \hat{L}_x\hat{L}_z\hat{L}_x + \hat{L}_x\hat{L}_z\hat{L}_x - \hat{L}_z\hat{L}_x\hat{L}_x + \hat{L}_y\hat{L}_y\hat{L}_z$$

$$- \hat{L}_y\hat{L}_z\hat{L}_y + \hat{L}_y\hat{L}_z\hat{L}_y - \hat{L}_z\hat{L}_y\hat{L}_y$$

$$= \hat{L}_x(-i\hbar\hat{L}_y) + (-i\hbar\hat{L}_y)\hat{L}_x + \hat{L}_y(i\hbar\hat{L}_x) + (+i\hbar\hat{L}_x)\hat{L}_y = 0$$

In the second line above we have subtracted and then added $\hat{L}_x\hat{L}_z\hat{L}_x$ and $\hat{L}_y\hat{L}_z\hat{L}_y$. In summary:

$$[\hat{L}^2, \hat{L}_x] = [\hat{L}^2, \hat{L}_y] = [\hat{L}^2, \hat{L}_z] = 0$$ (B4.1.5)

[$\hat{L}$ also commutes with each of its components but it is simpler to work with $\hat{L}^2$].

From Equations (B4.1.5) we know (Section 3.11) that there is a set of functions which are simultaneously eigenfunctions of both $\hat{L}_z$ and $\hat{L}^2$. If we use the corresponding quantum numbers, M and L, to characterise these functions writing them in the form $|L, M\rangle$, we have the following two eigenvalue-eigenfunction equations, where K_L and K_M are the eigenvalues for the square of the total angular momentum, in units of $\hbar^2$, and its z component, in units of $\hbar$, respectively:

$$\hat{L}^2|L, M\rangle = K_L|L, M\rangle$$ (B4.1.6)

and $\hat{L}_z|L, M\rangle = K_M|L, M\rangle$ (B4.1.7)

Starting from these equations we can now deduce the most important properties of the angular momentum eigenvalues:

$$\hat{L}^2|L, M\rangle = \{\hat{L}_x^2 + \hat{L}_y^2 + \hat{L}_z^2\}|L, M\rangle = K_L|L, M\rangle$$ (B4.1.8)

and $\hat{L}_z\hat{L}_z|L, M\rangle = \hat{L}_z K_M|L, M\rangle = (K_M)^2|L, M\rangle$ (B4.1.9)

Subtracting Equation (B4.1.9) from Equation (B4.1.8) we obtain:

$$\{\hat{L}_x^2 + \hat{L}_y^2\}|L, M\rangle = \{\hat{L}^2 - \hat{L}_z^2\}|L, M\rangle = \{K_L - (K_M)^2\}|L, M\rangle$$ (B4.1.10)

Thus, the functions $|L, M\rangle$ are also eigenfunctions of $\hat{L}_x^2 + \hat{L}_y^2$ and the eigenvalues of the sum of the squares of two angular momentum components must be positive so that:

$$K_L \geq (K_M)^2$$ (B4.1.11)

We now define two new operators, the raising and lowering, ladder or step operators:

$$\hat{L}_+ = \hat{L}_x + i\hat{L}_y \tag{B4.1.12a}$$

and

$$\hat{L}_- = \hat{L}_x - i\hat{L}_y \tag{B4.1.12b}$$

Using the commutation rules B4.1.4 one can readily show that:

$$\hat{L}_z(\hat{L}_x + i\hat{L}_y) = (\hat{L}_x + i\hat{L}_y)(\hat{L}_z + \hbar) \tag{B4.1.13a}$$

and

$$\hat{L}_z(\hat{L}_x - i\hat{L}_y) = (\hat{L}_x - i\hat{L}_y)(\hat{L}_z - \hbar) \tag{B4.1.13b}$$

If the first of these two sequences of operations is applied to the functions $|L, M\rangle$ we have:

$$\hat{L}_z(\hat{L}_x + i\hat{L}_y)|L, M\rangle = (\hat{L}_x + i\hat{L}_y)(\hat{L}_z + \hbar)|L, M\rangle = (\hat{L}_x + i\hat{L}_y)(K_M + \hbar)|L, M\rangle$$

$$= (K_M + \hbar)(\hat{L}_x + i\hat{L}_y)|L, M\rangle = (K_M + \hbar)\hat{L}_+|L, M\rangle \tag{B4.1.14a}$$

Thus, $(\hat{L}_x + i\hat{L}_y)|L, M\rangle$ is found to be an eigenfunction of $\hat{L}_z$ with an eigenvalue of $(K_M + \hbar)$. Since $\hat{L}^2$ commutes with $\hat{L}_x$ and $\hat{L}_y$ it also commutes with $\hat{L}_+$ and the eigenfunction $(\hat{L}_x + i\hat{L}_y)|L, M\rangle$ remains an eigenfunction of $\hat{L}^2$ with eigenvalue K_L. Similarly, we find that $(\hat{L}_x - i\hat{L}_y)|L, M\rangle$ is an eigenfunction of $\hat{L}_z$ with an eigenvalue of $(K_M - \hbar)$ while remaining an eigenfunction of $\hat{L}^2$ with eigenvalue K_L:

$$\hat{L}_z(\hat{L}_x - i\hat{L}_y)|L, M\rangle = (K_M + \hbar)(\hat{L}_x - i\hat{L}_y)|L, M\rangle = (K_M - \hbar)\hat{L}_-|L, M\rangle \tag{B4.1.14b}$$

Therefore, we have a series of eigenfunctions of $\hat{L}^2$ and $\hat{L}_z$, all with the eigenvalue of K_L for $\hat{L}^2$ but with the following sequence of eigenvalues of $\hat{L}_z$:

$$\ldots (K_M - 3\hbar), (K_M - 2\hbar), (K_M - \hbar), K_M, (K_M + \hbar), (K_M + 2\hbar), (K_M + 3\hbar) \ldots$$

The series must terminate at both ends because $K_L \geq (K_M)^2$. Thus, if we denote the highest value by K_M'' and the lowest by K_M' the full sequence is:

$$K_M' \ldots (K_M - 2\hbar), (K_M - \hbar), K_M, (K_M + \hbar), (K_M + 2\hbar) \ldots K_M'' \tag{B4.1.15}$$

Therefore, if the corresponding eigenfunctions are $|L, M''\rangle$ and $|L, M'\rangle$ we must have:

$$\hat{L}_+|L, M''\rangle = 0 \tag{B4.1.16a}$$

and

$$\hat{L}_-|L, M'\rangle = 0 \tag{B4.1.16b}$$

Applying $\hat{L}_-$ to Equation (B4.1.16a) we have:

$$\hat{L}_-\hat{L}_+|L, M''\rangle = (\hat{L}_x - i\hat{L}_y)(\hat{L}_x + i\hat{L}_y)|L, M''\rangle$$

$$= \{\hat{L}_x^2 + \hat{L}_y^2 + i(\hat{L}_x\hat{L}_y - \hat{L}_y\hat{L}_x)\}|L, M''\rangle = \{\hat{L}_x^2 + \hat{L}_y^2 - \hbar\hat{L}_z\}|L, M''\rangle$$

$$= \{\hat{L}^2 - \hat{L}_z^2 - \hbar\hat{L}_z\}|L, M''\rangle = \{K_L - (K_M'')^2 - \hbar K_M''\}|L, M''\rangle = 0$$

Thus,

$$K_L = (K_M'')^2 + \hbar K_M'' \tag{B4.1.17a}$$

By operating with $\hat{L}_+$ on Equation (B4.1.16b) we can also show that:

$$K_L = (K_M')^2 - \hbar K_M' \qquad \text{(B4.1.17b)}$$

In order to satisfy Equations (B4.1.17a) and (B4.1.17b) and to be consistent with the assumption that $K_M' < K_M''$ we must have $K_M'' = -K_M'$. Furthermore, the sequence (Equation (B4.1.15)) requires that K_M'' be greater than K_M' by an integral number of units of $\hbar$. Thus, K_M'' must be of the form $n\hbar$ where n is an integer, including 0, or a half-integer. Thus far we have attached no particular significance to L. Therefore, since K_M'' depends only on L we may set $n = L$ and we have:

$$K_M'' = L\hbar \qquad \text{(B4.1.18)}$$

Using Equation (B4.1.17a) $\qquad K_L = (L\hbar)^2 + \hbar L\hbar$

or $\qquad\qquad\qquad\qquad K_L = L(L+1)\hbar^2 \qquad \text{(B4.1.19)}$

Now, the possible values of K_M are:

$$-L\hbar, -(L-1)\hbar \dots (L-1)\hbar, L\hbar \qquad \text{(B4.1.20)}$$

And M, to which we have attached no particular significance yet, can be used to characterise this sequence of K_M values:

$$K_M = M\hbar, \quad L \geq M \geq -L \qquad \text{(B4.1.21)}$$

If L is an integer M is an integer, if L is half-integral so too is M.
Equations (B4.1.6) and (B4.1.7) may now be written:

$$\hat{L}^2|L, M\rangle = L(L+1)\hbar^2|L, M\rangle \qquad \text{(B4.1.22a)}$$

and $\qquad\qquad \hat{L}_z|L, M\rangle = M\hbar|L, M\rangle \qquad \text{(B4.1.22b)}$

These results have been derived solely from the commutation relationships and they therefore apply to any properties for which the corresponding operators have the same commutation rules. It is fortunate that they also apply to the property which we call 'spin'. Since spin is a non-classical quantity, there was no reason to assume that the rules deduced here, based on the classical mechanics of orbital angular momentum, would apply. However, it can be shown using a method which does not depend on classical mechanics that the commutation rules deduced above also apply to spin. Treatments of this type are based on the theory of infinitesimal rotations and can be found in any of the references 2–5 cited in the bibliography of Chapter 4. The mathematics used is not as daunting as the words 'infinitesimal rotations' may make it appear.

The primary task of this Box, the deduction of the commutation rules for orbital angular momentum and from them Equations (B4.1.22a) and (B4.1.22b), has now been accomplished. But in the process we have also 'discovered' the raising and lowering operators, $\hat{L}_+$ and $\hat{L}_-$, which we find extremely useful throughout this book. We are now in a position to examine their properties a little more closely. We found that, in units of $\hbar$, when $\hat{L}_+$ is applied to $|L, M\rangle$ we obtain an eigenfunction of $\hat{L}_z$ with an eigenvalue of $(M + 1)$ which is still an eigenfunction of $\hat{L}^2$ with the same value of the quantum number L. Similarly, we found that $\hat{L}_-|L, M\rangle$ is an eigenfunction of

$\hat{L}^2$, with the same value of L, and of $\hat{L}_z$ with an eigenvalue of $(M - 1)$. Equations (B4.1.23a) and (B4.1.23b) summarise these statements:

$$\hat{L}_+|L, M\rangle = C_+|L, M + 1\rangle \tag{B4.1.23a}$$

$$\hat{L}_-|L, M\rangle = C_-|L, M - 1\rangle \tag{B4.1.23b}$$

C_+ and C_- are numbers which may result from the operation by $\hat{L}_+$ and $\hat{L}_-$ on $|L, M\rangle$. The statements immediately above would not be invalidated if the resulting eigenfunctions were multiplied by such constants and we need to determine their values. The complex conjugate (Appendix 8) of Equation (B4.1.23a) is:

$$\{\hat{L}_+|L, M\rangle\}^* = \{C_+|L, M + 1\rangle\}^* \tag{B4.1.24}$$

But $\{\hat{L}_+\}^* = \hat{L}_-$, $|L, M\rangle^* = \langle L, M|$ and $|L, M + 1\rangle^* = \langle L, M + 1|$, and Equation (B4.1.24) can be written:

$$\langle L, M|\hat{L}_- = C_+^*\langle L, M + 1| \tag{B4.1.25}$$

If we now multiply the right- and left-hand sides of Equation (B4.1.23a), on the left, by the same sides of Equation (B4.1.25) we obtain Equation (B4.1.26):

$$\langle L, M|\hat{L}_-\hat{L}_+|L, M\rangle = C_+C_+^*\langle L, M + 1|L, M + 1\rangle = C_+C_+^* \tag{B4.1.26}$$

The right-hand side reduces to $C_+C_+^*$ because $\langle\alpha|\beta\rangle$ means integrate the product $\alpha\beta$ over all space and therefore $\langle L, M + 1|L, M + 1\rangle = 1$ if the eigenfunctions are normalised, which we certainly require them to be. The operator product $\hat{L}_-\hat{L}_+$ can be evaluated using Equations (B4.1.4) and (B4.1.12) and we find, in units of $\hbar$, that:

$$\hat{L}_-\hat{L}_+ = \hat{L}^2 - \hat{L}_z^2 - \hat{L}_z \tag{B4.1.27}$$

Therefore:

$$\langle L, M|\hat{L}_-\hat{L}_+|L, M\rangle = \langle L, M|\hat{L}^2 - \hat{L}_z^2 - \hat{L}_z|L, M\rangle$$

$$= \langle L, M|L(L + 1) - M^2 - M|L, M\rangle = L(L + 1) - M(M + 1) \tag{B4.1.28}$$

since $\langle L, M|L, M\rangle = 1$ for normalised eigenfunctions. Finally, comparing Equations (B4.1.26) and (B4.1.28) we have:

$$C_+C_+^* = L(L + 1) - M(M + 1)$$

Therefore, since there is no reason to assume that C_+ is anything other than real:

$$C_+^* = C_+ = \{L(L + 1) - M(M + 1)\}^{\frac{1}{2}} \tag{B4.1.29}$$

An analogous development starting from Equation (B4.1.23b) gives:

$$C_-^* = C_- = \{L(L + 1) - M(M - 1)\}^{\frac{1}{2}} \tag{B4.1.30}$$

and the complete forms of Equations (B4.1.23a) and (B4.1.23b) are:

$$\hat{L}_+|L, M\rangle = \{L(L + 1) - M(M + 1)\}^{\frac{1}{2}}|L, M + 1\rangle \tag{B4.1.31a}$$

$$\hat{L}_-|L, M\rangle = \{L(L + 1) - M(M - 1)\}^{\frac{1}{2}}|L, M - 1\rangle \tag{B4.1.31b}$$

Readers may have wondered why the positive sign was chosen for the square root in Equation (B4.1.29). There is definitely an element of choice here and, since the raising and lowering operators link all the $2L + 1$ eigenfunctions of the manifold of M values having the same value of L, it is important. This choice of phase does not affect the calculation of any measurable physical property, but it does affect intermediate algebraic results which can be very important when these results have been tabulated for general use, as many have been. The positive sign was first chosen by Condon and Shortley[1] and is now effectively universal.

In the light of the above a reader might well ask, 'If we can find out so much about angular momentum simply by studying the commutation properties of the associated operators can we not approach other quantum-mechanical problems in the same way?' The short answer to that question is 'yes', the procedure is sometimes called *second quantisation*. The best-known example of it is the deduction of the energy levels of the harmonic oscillator.[2] In that problem the raising operator is known as the *creation* operator because each application generates an additional vibrational quantum. The lowering operator, each application of which removes a vibrational quantum, is called the *annihilation* operator. These concepts have very wide applicability and form the basis of the description of photons in the theory of quantum electrodynamics; see Section 8.9 and reference 2. Some less well-known applications have been described by Newmarch and Golding.[3]

1. E.U. Condon and G.H. Shortley, *The Theory of Atomic Spectra*, Cambridge University Press, 1935.
2. N. Zettili, *Quantum Mechanics Concepts and Applications*, Wiley, Chichester, 2001.
3. J.D. Newmarch and R.M. Golding, *American J. Phys.*, **46**, 658–660 1978.

BOX 4.2   The Clebsch–Gordan series

The mathematical proof of the Clebsch–Gordan formula for the addition of angular momenta in quantum mechanics is really only a rigorous argument along the following, more descriptive, lines.

Suppose that the two angular momenta to be added are characterised by total angular momentum quantum numbers d_1 and d_2. We wish to show that the quantum numbers of the resulting vectors, d, are given by the Clebsch–Gordan series:

$$d = d_1 + d_2, d_1 + d_2 - 1, d_1 + d_2 - 2, \ldots |d_1 - d_2|$$

The z-components of the resulting vectors, characterised by their m_d values, must simply be sums of the z-components of d_1 and d_2 and there are $(2d_1 + 1) \times (2d_2 + 1)$ such sums. We first arrange the possible sums in a rectangular array.

	d_1	$d_1 - 1$	$d_1 - 2$	$d_1 - 3$	$\cdots$	$-d_1 + 1$	$-d_1$
d_2	$d_1 + d_2$	$d_1 + d_2 - 1$	$d_1 + d_2 - 2$	$d_1 + d_2 - 3$	$\cdots$	$d_2 - d_1 + 1$	$d_2 - d_1$
$d_2 - 1$	$d_1 + d_2 - 1$	$d_1 + d_2 - 2$	$d_1 + d_2 - 3$	$d_1 + d_2 - 4$	$\cdots$	$d_2 - d_1$	$d_2 - d_1 - 1$
$d_2 - 2$	$d_1 + d_2 - 2$	$d_1 + d_2 - 3$	$d_1 + d_2 - 4$	$d_1 + d_2 - 5$	$\cdots$	$d_2 - d_1 - 1$	$d_2 - d_1 - 2$
$\vdots$	$\vdots$	$\vdots$	$\vdots$	$\vdots$		$\vdots$	$\vdots$
$-d_2 + 1$	$d_1 - d_2 + 1$	$d_1 - d_2$	$d_1 - d_2 - 1$	$d_1 - d_2 - 2$	$\cdots$	$-d_2 - d_1 + 2$	$-d_2 - d_1 + 1$
$-d_2$	$d_1 - d_2$	$d_1 - d_2 - 1$	$d_1 - d_2 - 2$	$d_1 - d_2 - 3$	$\cdots$	$-d_2 - d_1 + 1$	$-d_2 - d_1$

We now go through a procedure which is repeated several times in the examples in Section 4.9. At the top left corner of the array we find the maximum possible m_d value of $d_1 + d_2$ which means that there must be a value of d equal to $d_1 + d_2$ with $2(d_1 + d_2) + 1$ z-components running in unit steps from $d_1 + d_2$ to $-(d_1 + d_2)$. We can find this sequence of m_d values along the top row and down the far right column of our array and we tick them off. The maximum m_d value remaining in the array is now $d_1 + d_2 - 1$, again in the top left corner, and we can tick off the other $2(d_1 + d_2 - 1)m_d$ values which we find in the second row of the array and the column second from the right.

This process can be repeated until we have one of two possible scenarios. If $d_1 = d_2$ the last element remaining in the array is the bottom left corner, $d_1 - d_2$, which clearly corresponds to a vector with $d = 0$ and $m_d = 0$. If $d_1 > d_2$ and $d_1 - d_2 = n$ the number of colums exceeds the number of rows by $2d_1 + 1 - (2d_2 + 1) = 2(d_1 - d_2) = 2n$. Then the final situation is that $2n + 1$ elements remain at the left-hand end of the last row of the array, $d_1 - d_2$ being the first of these and $d_1 - d_2 - 2n = -(d_1 - d_2)$ the last. These are the m_d values of a vector with $d = d_1 - d_2$.

PROBLEMS FOR CHAPTER 4

1. Read Box 4.2 and, following the general description of the addition of two angular momenta, work through the specific example $d_1 = 2$ and $d_2 = \frac{3}{2}$.

2. Use Equation (4.1.2) to derive an equation for the angular momentum of a rotating diatomic ^{16}O molecule in terms of the mass of the ^{16}O atom (26.56019×10^{-27} kg), the bond length (1.20752×10^{-10} m) and the frequency, v, of the rotation. The angular momentum (L) must be quantised according to the equation $L = (h/2\pi)(J[J + 1])^{\frac{1}{2}}$. Find the rotational frequency for $J = 1$.

3. Use Equations (4.7.5) and (4.7.6) to determine the effects of the raising and lowering operators on the seven f-orbital functions, $|3, m_d\rangle$, $m_d = -3$ to $+3$.

4. The three equivalent protons of a methyl group couple to give eight ($2 \times 2 \times 2$) nuclear spin functions. Use an extension of the method described in Section 4.9 to show that the coupled spins form a state having $S = \frac{3}{2}$ and two states having $S = \frac{1}{2}$. Why do you think that these states are sometimes described as 'a quartet and two doublets'? Repeat the exercise using successive applications of Equation (4.6.1).

5. In problem 4 the spin states, $|S, M_S\rangle$, of three equivalent coupled protons have been determined. A more detailed description of the system requires the exact combinations of the eight nuclear spin functions which form a particular spin state. The raising and lowering operators may also be used to solve this problem.
   a) Starting with the eigenfunction $|\frac{3}{2}, +\frac{3}{2}\rangle$, apply the lowering operator, $\hat{S}_-$, successively to obtain the other three eigenfunctions with $S = \frac{3}{2}$, e.g.:

$$\hat{S}_-|\tfrac{3}{2}, +\tfrac{3}{2}\rangle = \sqrt{3}\,|\tfrac{3}{2}, +\tfrac{1}{2}\rangle \tag{1}$$

The multiplying factors which arise when each eigenfunction is generated from the one above are important.

The only way in which the eigenfunction $|\frac{3}{2}, +\frac{3}{2}\rangle$ can be generated from three nuclear spins with $s = \frac{1}{2}$ is for all three z-components to be parallel, i.e.:

$$|\tfrac{3}{2}, +\tfrac{3}{2}\rangle = |+\tfrac{1}{2}(1), +\tfrac{1}{2}(2), +\tfrac{1}{2}(3)\rangle$$

where only the m_s values of the individual protons have been indicated. To operate on the eigenfunction formulated as it is on the right of the above equation we write $\hat{S}_-$ in terms of the individual spin operators, $\hat{s}_-(a)$, $a = 1, 2, 3$, as:

$$\hat{S}_- = \hat{s}_-(1) + \hat{s}_-(2) + \hat{s}_-(3)$$

Each individual spin operator operates only on a particular spin so that we have:

$$\{\hat{s}_-(1) + \hat{s}_-(2) + \hat{s}_-(3)\}|+\tfrac{1}{2}(1), +\tfrac{1}{2}(2), +\tfrac{1}{2}(3)\rangle = \hat{s}_-(1)|+\tfrac{1}{2}(1), +\tfrac{1}{2}(2), +\tfrac{1}{2}(3)\rangle$$

$$+ \hat{s}_-(2)|+\tfrac{1}{2}(1), +\tfrac{1}{2}(2), +\tfrac{1}{2}(3)\rangle + \hat{s}_-(3)|+\tfrac{1}{2}(1), +\tfrac{1}{2}(2), +\tfrac{1}{2}(3)\rangle$$

$$= |-\tfrac{1}{2}(1), +\tfrac{1}{2}(2), +\tfrac{1}{2}(3)\rangle + |+\tfrac{1}{2}(1), -\tfrac{1}{2}(2), +\tfrac{1}{2}(3)\rangle$$

$$+ |+\tfrac{1}{2}(1), +\tfrac{1}{2}(2), -\tfrac{1}{2}(3)\rangle \tag{2}$$

Comparing the right-hand sides of equations 1 and 2 we see that:

$$|\tfrac{3}{2}, +\tfrac{1}{2}\rangle = \{1/\sqrt{3}\}\{|-\tfrac{1}{2}(1), +\tfrac{1}{2}(2), +\tfrac{1}{2}(3)\rangle$$

$$+ |+\tfrac{1}{2}(1), -\tfrac{1}{2}(2), +\tfrac{1}{2}(3)\rangle + |+\tfrac{1}{2}(1), +\tfrac{1}{2}(2), -\tfrac{1}{2}(3)\rangle\}$$

b) Continue the process to determine the eigenfunctions $|\frac{3}{2}, -\frac{1}{2}\rangle$ and $|\frac{3}{2}, -\frac{3}{2}\rangle$ in terms of the individual spin functions. Alternatively, start with $|\frac{3}{2}, -\frac{3}{2}\rangle$ and use $\hat{S}_+$.

c) To find the two doublets we require two expressions for $|\frac{1}{2}, +\frac{1}{2}\rangle$ which are normalised and orthogonal to each other and to $|\frac{3}{2}, +\frac{1}{2}\rangle$. Try:

$$\{1/\sqrt{6}\}\{2|-\tfrac{1}{2}(1), +\tfrac{1}{2}(2), +\tfrac{1}{2}(3)\rangle - |+\tfrac{1}{2}(1), -\tfrac{1}{2}(2), +\tfrac{1}{2}(3)\rangle$$

$$- |+\tfrac{1}{2}(1), +\tfrac{1}{2}(2), -\tfrac{1}{2}(3)\rangle\}$$

and

$$\{1/\sqrt{2}\}\{|+\tfrac{1}{2}(1), -\tfrac{1}{2}(2), +\tfrac{1}{2}(3)\rangle - |+\tfrac{1}{2}(1), +\tfrac{1}{2}(2), -\tfrac{1}{2}(3)\rangle\}$$

Show that these two functions have the required properties of normalisation and orthogonality and find the corresponding two functions $|\frac{1}{2}, -\frac{1}{2}\rangle$ by applying the lowering operators to the two $|\frac{1}{2}, +\frac{1}{2}\rangle$ functions.

CHAPTER 5

The Structure and Spectroscopy of the Atom

5.0	Introduction	96
5.1	The eigenvalues of the hydrogen atom	96
5.2	The wave functions of the hydrogen atom	97
	5.2.1 The radial function, $R_{n,l}(r)$	98
	5.2.2 The angular functions, $\Theta_{l,m}(\theta)$ and $\Phi_m(\phi)$	99
5.3	Polar diagrams of the angular functions	100
	5.3.1 The s-functions	101
	5.3.2 The p-functions	101
	5.3.3 The d-functions	103
5.4	The complete orbital wave functions	104
5.5	Other one-electron atoms	104
5.6	Electron spin	105
5.7	Atoms and ions with more than one electron	105
	5.7.1 The self-consistent field	106
	5.7.2 Electron correlation	106
	5.7.3 The periodic table of the elements	107
5.8	The electronic states of the atom	107
	5.8.1 The five quantum numbers of a single electron	108
	5.8.2 Quantum numbers for the many-electron atom	108
	5.8.3 The assignment of term symbols	108
	5.8.4 Term energies and Hund's rules	110
5.9	Spin-orbit coupling	111
	5.9.1 Russell–Saunders or LS coupling	111
	5.9.2 jj coupling	112
	5.9.3 Intermediate coupling	113
	5.9.4 Inter-electronic spin-orbit coupling	115
5.10	Selection rules in atomic spectroscopy	115
	5.10.1 Angular momentum	115
	5.10.1.1 Electric quadrupole transitions	116
	5.10.1.2 Magnetic dipole transitions	116
	5.10.2 Parity	116

The Quantum in Chemistry   R. Grinter
© 2005 John Wiley & Sons, Ltd

5.11   The Zeeman effect 117

       5.11.1   The normal Zeeman effect 118

       5.11.2   The anomalous Zeeman effect 120

5.12   Bibliography and further reading 121

       Problems for Chapter 5 129

5.0   INTRODUCTION

In 1926, the structure of the hydrogen atom, and an interpretation of the wavelengths of the visible spectral lines of the atom, was probably the most immediate problem facing the new quantum mechanics. In his paper of 1926, in which he inaugurated that formulation of quantum mechanics which we now call wave mechanics, Schrödinger proposed the correct form of the Hamiltonian operator for the problem, derived the required eigenvalues and eigenfunctions and showed that the former corresponded to the energy levels of the Bohr model of the hydrogen atom and were therefore in agreement with the experimental data. These results play a central role in our understanding of the structure of all atoms. The solution of the Schrödinger equation for the hydrogen atom is a straightforward but time-consuming task and we shall not enter into it here. Full details can be found in many texts.[1,2] We shall be much more concerned, in this chapter and in others, with the following aspects of the problem and their implications for our understanding of atomic structure, atomic spectroscopy, the periodic table and the nature of the chemical bond.

- What are the eigenvalues and how are they determined by the quantum numbers?

- What are the forms of the corresponding eigenfunctions and what do they tell us about the distribution of the electron in space around the nucleus of the atom?

- How are the eigenvalues and eigenfunctions of the hydrogen atom, a one-electron atom, modified by inter-electronic repulsion in atoms which have more than one electron?

- What are the magnetic properties of electrons and how do they manifest themselves in the structures and spectra of atoms?

5.1   THE EIGENVALUES OF THE HYDROGEN ATOM

If we assume that the nucleus has an infinite mass (see Section 5.5 for further details), then the eigenvalues of the hydrogen atom are given by a simple formula which depends upon just one quantum number (n) which is known as the *principal* quantum number. It can take any positive integer value from 1 to ∞:

$$E_n = -m_e e^4/8h^2\varepsilon_0^2 n^2 \text{ in J} \quad \text{or} \quad -m_e e^4/8ch^3\varepsilon_0^2 n^2 \text{ in cm}^{-1} \quad (5.1.1)$$

m_e is the mass of the electron, e its charge, c the velocity of light (in cm s^{-1}), h Planck's constant and ε_0 the permittivity of a vacuum. Note that the energy is a negative quantity. This derives from the fact that our energy zero is defined to be the state in which the proton and electron are infinitely far apart and motionless so that they have neither potential nor kinetic energy. As the two particles approach each other, the energy of the forming atom decreases because of the negative coulombic energy of the opposite charges; so much is quite obvious. What is not so obvious is the fact that the two particles can only be kept apart if they are in motion and this motion gives rise to a kinetic energy which must be positive and increases as the average distance between the electron and the nucleus decreases. However, it is a fundamental consequence of the laws of classical and quantum mechanics that the kinetic (T) and potential (V) energies are always in exactly the same ratio which, for a system where the potential energy results from electrostatic attraction, is $V = -2 \times T$. Therefore, for any eigenvalue, n, of the hydrogen atom the total energy may be expressed as:

$$E_n = T_n + V_n = T_n - 2T_n = -T_n = V_n/2$$

This relationship between T and V is known as the *virial theorem*. Because of it, we can be sure that the negative potential energy of the forming hydrogen atom will always exceed the positive kinetic energy.

The expression for the energy, Equation (5.1.1), was first deduced by Schrödinger and it provided an exact, quantitative interpretation of the Balmer formula. Naturally, all the other lines in the hydrogen atom spectrum can also be fitted to this formula (Box 5.1). As n increases n^2 increases very rapidly so that values of E_n become closer and closer together as they converge upon the value of zero. We see the experimental evidence for this in the positions of the lines in the spectra of the hydrogen atom, Figure B5.1.1 and Figure 2.8. Consider the Lyman series (Box 5.1). The energy required to raise the electron from the energy level, $n = 1$, to successively higher levels, $n = 2, 3, 4, \ldots$ increases, but by a smaller amount, for each increase in the higher quantum number. Finally, when $n = \infty$ the electron leaves the atom which is then said to have been *ionised*. For the Lyman series this ionisation limit is found at $109\,677.6$ cm^{-1} which corresponds to a wavelength of 91.2 nm. Since the Balmer series starts from the level $n = 2$ its ionisation limit is correspondingly smaller at $27\,419.4$ cm^{-1} or 364.7 nm.

5.2 THE WAVE FUNCTIONS OF THE HYDROGEN ATOM

The wavefunctions of the hydrogen atom are not easy to describe in a Cartesian co-ordinate system of three, mutually perpendicular axes. The polar co-ordinate system with the nucleus at the origin is much more suitable. The polar system is described and the two systems compared in Appendix 7.

Though at first sight the Schrödinger equation for the hydrogen atom looks more formidable in polar than in Cartesian co-ordinates, the apparent complexity conceals the fact that a most important simplification is possible. The Schrödinger equation in the three variables (r, θ and ϕ) can be *separated* into three equations in each of which only one of the variables occurs. This makes the solution of the equation and, more importantly from our point of view, the interpretation of the results much simpler. Because the equation can be separated, the total wavefunction can be written as a product of three functions, each containing only one co-ordinate variable, i.e.:

$$\Psi_{n,l,m}(r, \theta, \phi) = R_{n,l}(r) \cdot \Theta_{l,m}(\theta) \cdot \Phi_m(\phi) \tag{5.2.1}$$

Where the subscripts, n, l and m, indicate the quantum numbers upon which that particular part of the wave function is found to depend. Each of the functions $R(r)$, $\Theta(\theta)$

and $\Phi(\phi)$ can be written in a very general form and these expressions, especially those for $R(r)$ and $\Theta(\theta)$, are rather complicated. However, since we are interested only in a comparatively small number of them, we shall write those out explicitly (Appendix 5) and refer readers who require the general formulae to more advanced texts.[1,2]

A complete wave function may be thought of as an exact description of a region mapped out in space which can be occupied by one electron, or by two electrons with opposite spins. The probability of finding an electron at any point in that space is found by evaluating the wavefunction at that point and multiplying the result by its complex conjugate, i.e. $\Psi^*\Psi$. If the product $\Psi^*\Psi$ is integrated [summed up] over all the space then the result is one because the wavefunctions are normalised (Section 3.5). Since we now know that we cannot think of the motion of an electron in just the same way as we think of the motion of a planet around the sun, the expression 'orbit' is not appropriate. To make the difference clear, while at the same time reflecting the historical origins of the concept, the term *orbital* has been coined and is now universally used to describe the function Ψ.

5.2.1   The radial function, $R_{n,l}(r)$

The function $R_{n,l}(r)$ is known as the radial function because it contains only r, the radial co-ordinate of the polar co-ordinate system. Equation (5.2.1) shows that it is the only part of the wave function in which the principal quantum number, n, occurs. When we study the radial functions (Appendix 5) we see that, apart from a normalising factor, they consist of an exponential function, $\exp(-Zr/na_0) \equiv \exp(-\rho/2)$, and a polynomial in r. Because its argument is negative, the exponential factor ensures that the value of the wave function always goes to zero at large distances from the nucleus. This means that the square of the wave function, and therefore the probability of finding the electron, falls to zero as we move further from the centre of the atom. For the hydrogen atom $Z = 1$, but if the value of Z is larger, as it is for example in the helium (He^+) ion, then the exponential part of the radial function falls off much faster than when $Z = 1$ reflecting the way in which a larger nuclear charge holds the electron closer to the nucleus. The polynomial factor causes the radial function to change sign as the value of r changes as shown in Figure 5.1. Those values of r for which any particular radial function is zero are known as the *radial nodes* and we must note that such a node has the form of a sphere, enveloping the whole atom and with the nucleus at its centre. Not all radial functions have nodes. Of those listed in Appendix 5, $R_{2,0}(r)$ has one node at $\rho = 2.0$ ($r = 2.0a_0$; a_0 = Bohr radius), $R_{3,1}(r)$ has one node at $\rho = 4.0$ ($r = 6.0a_0$) and $R_{3,0}(r)$ has two nodes at $\rho = 1.268$ ($r = 1.90a_0$) and 4.732 ($r = 7.10a_0$). The number of nodes in an eigenfunction is intimately connected with the corresponding eigenvalue and we shall consider the subject of nodes again when we have described the *angular* functions $\Theta_{l,m}(\theta)$ and $\Phi_m(\phi)$.

Apart from plots of $R(r)$ itself, two other representations of the radial function are frequently used to illustrate the way in which the probability of finding an electron, i.e. the electron density, behaves as the distance from the nucleus increases. Since this probability is proportional to the square of the wave function, graphs of $R^2(r)$ against r are frequently used for this purpose. Another form of diagram is obtained in the following manner. Since the surface area of a sphere of radius r is $4\pi r^2$, the volume element

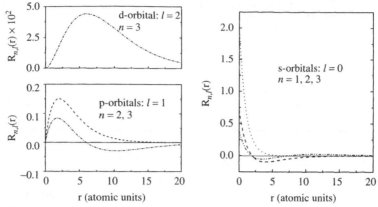

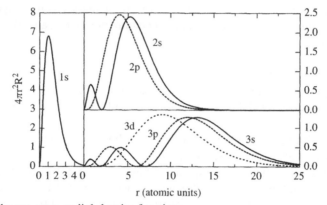

Figure 5.1 Hydrogen-atom radial functions

Figure 5.2 Hydrogen-atom radial density functions

enclosed between two spheres of radii r and $r + \delta r$ is $4\pi r^2 \delta r$. Therefore, the proportion of the hydrogen-atom electron which lies between two such spheres, i.e. at distance r from the nucleus, is $4\pi r^2 \cdot R^2(r)\delta r$. So a graph of $4\pi r^2 \cdot R^2(r)$ against r is a measure of the probability of finding an electron at a distance r from the nucleus (Figure 5.2) but it can easily be misleading. Because the function r^2 increases so rapidly with increasing r, the function $4\pi r^2 \cdot R^2(r)$ has one or more maxima which must be interpreted with care. For example, if we were able to stand at the hydrogen nucleus and experience the electron density of the 1s orbital there in the way in which we experience a fog we would find that the fog was very dense at the nucleus. As we walked away from the nucleus the thickness of the fog would decrease exponentially and we would not see any increase in the region of $r = a_0$ where the graph of $4\pi r^2 \cdot R^2(r)$ against r shows a maximum. The maximum is a consequence of the factor r^2. Of course, where there are radial nodes the electron density must fall to zero at these values of r and increase again at greater r.

5.2.2 The angular functions, $\Theta_{l,m}(\theta)$ and $\Phi_m(\phi)$

We might well have expected that an electron circulating around a nucleus would have some orbital angular momentum associated with it, and this is indeed the case. Further, we

Table 5.1   The possible quantum number combinations for the lower values of n

n	l	m	n	l	m
1	0	0	2	0	0
				1	-1 0 $+1$
3	0	0	4	0	0
	1	-1 0 $+1$		1	-1 0 $+1$
	2	-2 -1 0 $+1$ $+2$		2	-2 -1 0 $+1$ $+2$
				3	-3 -2 -1 0 $+1$ $+2$ $+3$

know from Chapter 4 that we can always characterise two aspects of angular momentum, the square of its total value and its z-component. The quantum numbers l and m do just that and therefore the quantum number for the square of total angular momentum, l, and the quantum number for its z-component, m, are related by the expression; $m = 0, \pm 1, \pm 2, \ldots, \pm l$.[‡] But the detailed solution of Schrödinger's equation for the hydrogen atom reveals that the value of l is itself restricted to integer values lying between 0 and $(n-1)$. In summary, the three spatial quantum numbers can take the following, interrelated values:

n, the principal quantum number $= 1, 2, 3, \ldots, \infty$
l, the orbital angular momentum quantum number $= 0, 1, 2, \ldots, (n-1)$
m, the quantum number for the z-component of the orbital a. m. $= 0, \pm 1, \pm 2, \ldots, \pm l$

The possible quantum number combinations for the lower values of n, which are the only ones of interest to us, are as shown in Table 5.1.

We find it easier to remember the quantum numbers and their significance if we use a letter instead of a number for l, and for reasons which stretch back into the history of atomic spectroscopy we use the letters s (sharp), p (principal), d (diffuse) and f (fundamental) for $l = 0, 1, 2$ and 3 respectively. Therefore, there are one s-orbital, three p-orbitals, five d-orbitals and seven f-orbitals.

The functions $\Theta_{l,m}(\theta) \cdot \Phi_m(\phi)$ occur in many problems which, like the atom, have spherical symmetry. They are known as the *spherical harmonics* (Appendix 5) and they are frequently written as a single function, $Y_{l,m}(\theta, \phi)$. The forms of these angular functions are very important, especially so since the distribution of electrons in space around an atomic nucleus is a major factor which affects the strength of chemical bonds and the geometry of molecules. Illustrations of the angular functions are therefore very important in the study of chemical bonding.

5.3   POLAR DIAGRAMS OF THE ANGULAR FUNCTIONS

The most useful illustrations of the angular parts of the wavefunctions of the hydrogen atom are obtained by plotting them in polar co-ordinates as described in detail below.

[‡] Our notation here is not quite consistent. If we were to follow the convention on angular momentum quantum numbers established in Chapter 4, then if l is the quantum number characterising total angular momentum, m_l would be the form of the quantum number for the z-component of that angular momentum. The subscript is omitted to conform with the usual notation for the atomic quantum numbers since no confusion can result from this and complexity in the notation is also reduced.

Since the quantum numbers m and l are related by $-l \leq m \leq +l$, we must be careful to combine only those functions $\Theta_{l,m}(\theta)$ and $\Phi_m(\phi)$ for which this condition is satisfied.

5.3.1 The s-functions

We start with the s-functions of which there is only one; $\Theta_{0,0}(\theta) \cdot \Phi_0(\phi) = \frac{1}{2}\sqrt{(1/\pi)}$. Clearly, this function has no dependence upon either of the two angular co-ordinates. It is a function which is exactly the same in all directions around the atom, i.e. it is spherically symmetrical. We can represent it by drawing a spherical boundary surface centred at the nucleus and of a radius such that a given proportion of the electron is enclosed within it, 95 % say. All s-orbitals are represented by a sphere, but the radius will depend upon the radial function with which the s-type angular function is combined in any particular case. At the risk of stating the obvious, it is emphasised that, after the last radial node, the electron density falls off exponentially with r so the boundary surface is chosen arbitrarily and has no physical reality.

5.3.2 The p-functions

The p-functions, $\Theta_{1,m}(\theta) \cdot \Phi_m(\phi) \equiv Y_{1,m}(\theta, \phi)$, are of three forms depending upon the value of m, i.e. $m = -1$, 0 or $+1$. For purposes of illustration it is best to avoid using the complex functions which contain $\exp(\pm mi\phi)$ and to use the alternative functions of $\cos(m\phi)$ and $\sin(m\phi)$. The difference between the two sets of functions is that the complex functions are eigenfunctions of $\hat{l}_z$, the operator for the z-component of the orbital angular momentum, whereas the sine and cosine functions are not (see Chapters 3 and 4). But, if we are not specifically interested in the orbital angular momentum of the electron we loose nothing by this. The simplest p-function is:

$$Y_{1,0} = \Theta_{1,0}(\theta) \cdot \Phi_0(\phi) = \sqrt{(3/4\pi)} \cdot \cos(\theta)$$

and to plot it we proceed as follows (Figure 5.3).

We draw the z-axis with the nucleus, N, at $z = 0$ and a radial co-ordinate from the nucleus in a direction which makes an angle θ with that axis. Recalling that θ is the angle between the z-axis and the radial co-ordinate (r), we mark off a length Np equal to $\cos(\theta)$ on the radial co-ordinate. We repeat this process for a number of values of θ and we note that, because of the way θ is defined, all angles in which p has a positive z co-ordinate are in the range $0 \leq \theta \leq \pi/2$ and $\cos(\theta)$ is therefore positive. All values of θ for which p has a negative z co-ordinate (e.g. p' in Figure 5.3) lie in the range $\pi/2 \leq \theta \leq \pi$ and

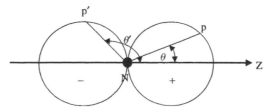

Figure 5.3 Plotting the angular function for a p_z-orbital

$\cos(\theta)$ is therefore negative. When we have marked out sufficient points, on both sides of the z-axis, we find that p and p' describe two perfect circles which touch at N. This is a *polar* graph of $\cos(\theta)$. Since the function $Y_{1,0}$ has no dependence upon the angle ϕ, an identical result would be obtained for any plane which contains the z-axis. Therefore, in three dimensions this representation of $Y_{1,0}$ consists of two spheres in contact and could be obtained by rotating the two circles of Figure 5.3 about the z-axis, i.e. through all possible values of ϕ.

The angular wavefunction has a value of zero at $\theta = 90°$, i.e. in the xy-plane, where it changes phase. This is an *angular* node in the wave function and all p-functions have one such planar angular node as we shall see when we consider the other two functions:

$$Y_{1,c} = \frac{\sqrt{3}}{2} \cdot \sin(\theta) \cdot \frac{1}{\sqrt{\pi}} \cdot \cos(\phi) \quad \text{and} \quad Y_{1,s} = \frac{\sqrt{3}}{2} \cdot \sin(\theta) \cdot \frac{1}{\sqrt{\pi}} \cdot \sin(\phi)$$

To plot these functions is slightly more complicated. First we fix a value for θ and a convenient choice is $90°$ since $\sin(90°) = 1$. This choice of θ means that we are varying ϕ in the xy-plane and we plot it in exactly the same way as we plotted θ above. It is clear that for $Y_{1,c}$ we must obtain the two circles in contact since we are again plotting a cosine function. That the result for the sine function should also be the same, but rotated through $90°$, may not be immediately obvious, but it can be readily demonstrated by a few minutes of plotting.

In Figure 5.4, the point p_x traces out $\cos(\phi)$ and the function $Y_{1,c}$ and p_y traces out $\sin(\phi)$ and the function $Y_{1,s}$. As with the function $Y_{1,0}$, the signs of the two lobes of the function are determined by the sign of $\sin(\phi)$ or $\cos(\phi)$ in the appropriate quadrant. Remember that ϕ is always the angle between the x-axis and the radius vector, $N - p(p')$. To obtain the complete angular functions we must now consider the result of varying θ over its range of 0 to π. The function in question is $\sin(\theta)$ which for any constant ϕ gives two circles in contact in the plane defined by the z-axis and the radius vector $N - p(p')$. The diameter of each circle is $N - p(p')$. Thus, if we fix a value of ϕ at say, zero, we obtain the polar plot of $Y_{1,c}$ as two circles in contact in the xz-plane. The two functions

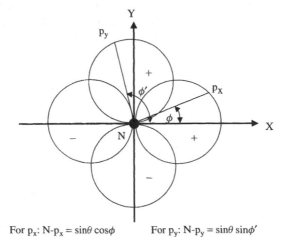

For p_x: N-$p_x = \sin\theta \, \cos\phi$　　　For p_y: N-$p_y = \sin\theta \, \sin\phi'$

Figure 5.4　Plotting the angular functions for p_x- and p_y-orbitals

$Y_{1,c}$ and $Y_{1,s}$ turn out to be exactly like $Y_{1,0}$ consisting of two spheres in contact but aligned along the x- and y-axes respectively.

Accordingly, these three angular functions are known as the p_x, p_y and p_z functions. It is important to note that a sum of the squares of the x-, y- and z-functions has no dependence upon θ or ϕ and is therefore spherically symmetrical. Consequently, if each p-function is occupied by one or by two electrons the total electron distribution is spherically symmetrical. This is clearly the case for the single s-function and it is also true of the five d-functions, the seven f-functions and so on.

A further point should be made with regard to the forms of the squared functions. If we make a polar plot of $Y^2_{1,0}$, $Y^2_{1,c}$ or $Y^2_{1,s}$ we obtain results similar to those obtained above but with an important difference; the two spheres become two pear-shaped lobes. But the directional properties of the functions remain quite unchanged and either form of plotting is suitable for a discussion of the spatial characteristics of p-functions.

5.3.3 The d-functions

Polar diagrams of the five d-functions can be obtained in the manner described above. They are shown in Figure 5.5. Again, for purposes of illustration the real sine and cosine functions have been chosen. Reflecting their orientation with respect to the Cartesian axes, the d-functions are named, d_{xy}, d_{xz} and d_{yz} for the three functions which are directed between the axes indicated, and $d_{x^2-y^2}$ for the function with its lobes lying along the x and y axes. The function $d_{z^2} \equiv Y_{2,0}$, is clearly different and the reason for this can be a source of difficulty. In actual fact, it is not as different from its companions as it appears since it can be written as a combination of two functions, $d_{z^2-x^2}$ and $d_{z^2-y^2}$, which have exactly the same form as the other four d-orbitals, i.e. four lobes with their axes at right angles and lying in a plane. We have to write the d_{z^2} function in the form given because

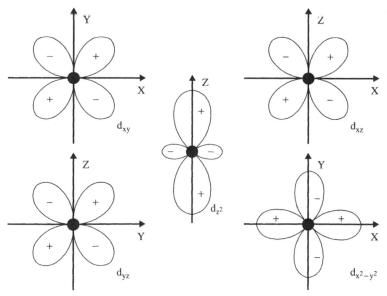

Figure 5.5 d-orbital angular functions

if we used the functions $d_{z^2-x^2}$ and $d_{z^2-y^2}$ we would have six d-functions which were not mutually orthogonal and therefore not independent of each other. To use such a set of functions would be like trying to plot a graph on a normal piece of graph paper on which three, rather than two, axes had been drawn.

The phases of the d-functions show that they each have two angular nodes which, for all except the d_{z^2}, consist of two mutually perpendicular planes. The angular nodes of the d_{z^2} function take the form of two cones with their apexes at the nucleus.

The seven f-functions are important in the chemistry of the lanthanides and actinides, but it would take us too far from the main theme of this chapter to discuss them here. Details may be found in reference[3].

5.4   THE COMPLETE ORBITAL WAVE FUNCTIONS

A complete wave function, $\Psi_{n,l,m}(r, \theta, \phi)$, which forms a solution of Schrödinger's equation for the hydrogen atom, is therefore a product of the three functions $R_{n,l}(r) \cdot \Theta_{l,m}(\theta) \cdot \Phi_m(\phi)$ in which we must recognise that the three quantum numbers, n, l and m are not entirely independent of each other.

The number of nodes in the complete wave function is the sum of the radial and angular nodes and rises with increasing orbital energy. The total number of nodes is $n - 1$ (Table 5.2). But the actual form of the angular functions does not change as the quantum number n increases, they simply reach out further from the nucleus, because of the increasing extension of the radial function, and are intersected by an increasing number of radial nodes. The occupation of the energy levels and the form of the associated wave functions provides the key to understanding the chemical bond and molecular geometry. These subjects are discussed in much more detail in Chapter 6.

5.5   OTHER ONE-ELECTRON ATOMS

Though hydrogen and its isotopes, ^{2}H and ^{3}H, are the only elements which have one electron in the neutral state, many positive ions which have just one electron are known and their spectra are frequently observed in high-energy discharges where electrons are

Table 5.2   The nodes of the hydrogen atom wave functions

Orbital $n\ l$	Number of nodes			Orbital $n\ l$	Number of nodes		
	radial	angular	total		radial	angular	total
1s	0	0	0	4s	3	0	3
				4p	2	1	3
2s	1	0	1	4d	1	2	3
2p	0	1	1	4f	0	3	3
3s	2	0	2	5s	4	0	4
3p	1	1	2	5p	3	1	4
3d	0	2	2	5d	2	2	4
				5f	1	3	4
				5g	0	4	4

Table 5.3 Values of the Rydberg constant (R) for hydrogen and some one-electron ions

Atom/ion	Z	R(cm^{-1})	Atom/ion	Z	R(cm^{-1})
H	1	109 677.59	^{2}H	1	109 707.42
^{3}H	1	109 717.35	He$^+$	2	109 722.26
Li^{++}	3	109 728.72	Be^{+++}	4	109 730.62

stripped away from atoms to form multiply charged ions. The energy levels and wave functions of these one-electron ions are given by the formulae for the hydrogen atom with the appropriate modification of the value of the nuclear charge, Z. Table 5.3 gives the values of the Rydberg constant, R (see Box 5.1), determined from the spectra of H, He$^+$, Li^{++} and Be^{+++}. The small change in the value of R as we go from H to Be^{+++} is due to the change in the mass of the nucleus which enters the formula for the energy levels when it is written in the strictly correct form:

$$E_n = -Z^2\mu e^4/8h^2\varepsilon_0^2 n^2 \text{ in J } \quad \text{or} \quad -Z^2\mu e^4/8ch^3\varepsilon_0^2 n^2 \text{ in cm}^{-1}$$

where $\mu = m_e \cdot m_N/(m_e + m_N)$ is the reduced mass of the electron, m_e, and the nucleus, m_N. For further discussion of the reduced mass see Chapter 10 and Box 10.1.

This variation of R with nuclear mass is not simply an arcane theoretical point. In 1932, the observation of a very weak line on the higher-energy side of each of the Balmer lines was important proof of the existence of a hydrogen isotope with an atomic mass of 2.

5.6 ELECTRON SPIN

Thus far, this account of the structure of the atom has been focused on the energy of the electron and its distribution in the space surrounding the atomic nucleus. The angular momenta involved are orbital angular momenta. But immediately prior to Schrödinger's publication of the new wave mechanics, Goudsmit and Uhlenbeck had proposed that the nuances of atomic spectral data, especially the Zeeman effect (see Section 5.11), could best be interpreted on the assumption that the electron had an intrinsic 'spin' angular momentum with $s = \frac{1}{2}$ and $m_s = \pm\frac{1}{2}$. At almost exactly the same time, Pauli proposed his exclusion rule (Section 5.7.2) according to which electrons with opposed m_s values were paired in the spatial orbitals described in the earlier part of this chapter. Thus, although he did not examine the electron spin functions, Schrödinger was undoubtedly well aware that spin was a property which would have to form a part of any comprehensive theory of atomic structure. We also know that, in 1928, the whole concept of electron spin was placed on a much firmer footing by Dirac's relativistic treatment of the hydrogen atom. But we need only note that the property of spin adds two more quantum numbers, s and m_s, to those required to fully specify an electron in an atom. The further consequences of electron spin will be taken up in Section 5.8 onwards.

5.7 ATOMS AND IONS WITH MORE THAN ONE ELECTRON

The Hamiltonian operator for a many-electron atom contains many terms. Each kinetic-energy term and the electron-nucleus attraction term involves only one electron. These one-electron terms present little difficulty to the would-be calculator of atomic structure. But the terms describing the mutual repulsion of the electrons each involve two electrons and though they are simple to write down they are very difficult to deal with computationally. The problem is this. In the helium atom for example, how can we know the energy of electron 1 when we do not know how electron 2 is distributed around the atom? We need to know the distribution of electron 2 in order to calculate the energy of the 1–2 inter-electronic repulsion. There is no closed algebraic solution of this *three-body problem* (2 electrons + 1 nucleus), neither in classical nor in quantum mechanics. However, there is a way forward if we are prepared to carry out a long cyclic calculation. The procedure is as follows, where we take the lithium (Li) atom as an example.

5.7.1   The self-consistent field

We have to determine a wave function for each of the three lithium electrons. Because every atom is spherically symmetrical, we know that the angular functions will be the same as those of the hydrogen atom orbitals and that we shall therefore have wavefunctions which we may describe as 1s and 2s, the first containing two electrons and the second one. The problem is to determine the appropriate radial functions which are strongly influenced by the presence of the inter-electronic repulsion. We start by guessing wave functions for electrons 2 and 3 basing our guess upon whatever information we have from experiment or theory. Having done this we solve the Schrödinger equation for electron 1 moving in the averaged electrostatic field of the nucleus and electrons 2 and 3. Next we solve the Schrödinger equation for electron 2 using the guessed wave function for electron 3 and the calculated wave function for electron 1. We now solve the equation for electron 3 using the wave functions calculated for electrons 1 and 2. The resulting three wavefunctions and total electronic energy are not the final answer since they are based upon the original guesses. But they do provide better approximations to the true answers and if we repeat the cycle of calculations we can make a further improvement in our wave functions and energy. If the cyclic calculation is continued the changes in the wave functions and energy become smaller with every cycle until, within the limits of accuracy required for that particular calculation, there is no further change in them. When this point is reached the electronic wave functions are said to be consistent with the electrostatic field from which they were calculated and which the electrons themselves produce. The process is known as a *self-consistent field* (SCF) calculation. Nowadays, computers make this task rather easy, but the first calculations of the type were done the hard way by the Hartrees, father and son, in the 1930s.[‡] Notable contributions to the subject were also made by Vladimir A. Fock (1898–1974) and wave functions calculated by this method are often called Hartree–Fock functions.

5.7.2   Electron correlation

However, even an SCF calculation does not provide an exact radial wavefunction. Because the electrons repel each other their motions are *correlated*, i.e. they keep out of each other's

---

[‡] The father (WH) was a retired school teacher with time to spare. His son, the physicist and mathematician Douglas Rayner Hartree (1897–1958), set him to work on the long, demanding SCF calculations with which the family name is now synonymous.

way, and this effect is not fully allowed for in the Hartree-Fock method which assumes that the electrons produce a *static*, spherical field and fails to allow for the *dynamic* effects of the inter-electronic repulsion which is known as *electron correlation*. This subject will be discussed further in Section 6.8.

5.7.3 The periodic table of the elements

Today, SCF wave functions are readily available. The angular functions are exactly the same as those of the hydrogen atom, but the radial functions are not. Nor can they be expressed as algebraic functions of r but are given as tables of values. Apart from the increase of energy which the inter-electronic repulsion engenders, it also has the consequence that the electrons can no longer be considered as individual, independent entities since their motions depend upon those of their fellows, and vice versa. Therefore, the wave function and energy of any one electron must depend upon the distributions and hence the wave functions of all the other electrons in the atom, and one cannot truly speak of the wave function of a single electron; the wave function for an atom is a function of the co-ordinates of *all* the electrons.

Nevertheless, chemists have found it possible, and extremely useful, to think in terms of individual electrons occupying individual orbitals in many-electron atoms and in terms of building up the atom (the *aufbau principle*) by filling the orbitals in order of increasing energy with two spin-paired electrons (the *Pauli principle* or *exclusion rule*). Thus we ascribe many of the particular properties of the transition metals and their ions to electrons occupying 3d (Ti-Zn), 4d (Zr-Cd) and 5d (Hf-Hg) orbitals and the electron which is so readily lost by the alkali metal atoms is a 2s (lithium), 3s (sodium), 4s (potassium) ... electron. The empty or partially filled orbitals play a similarly useful role. We envisage that the 3s electron lost by sodium enters the singly occupied 3p orbital of chlorine when the sodium and chlorine atoms combine to form sodium chloride.

In every case these are electrons or orbitals in the outer reaches of the particular atom, i.e. they are *valence* orbitals or *valence* electrons and have the highest, or nearly the highest, energy and value of n. Thus, a quantum-mechanical view of the periodic table is an arrangement of the elements in which those having the same electronic structure in their outermost, normally incomplete, shell are placed in columns. The fact that the structure of the incomplete outer shell of an element controls its reactivity, the number of chemical bonds which it forms (valency) and the geometry of the molecules of which it forms a part (see Chapter 6) explains the very similar chemical properties which were the first stimulus to the construction of the table.

5.8 THE ELECTRONIC STATES OF THE ATOM

Though individual orbitals are very useful for a qualitative or semi-quantitative description of the way in which a many-electron atom forms chemical bonds to other atoms, a more exact description is required if we wish to account quantitatively for the energetics of bond formation or atomic spectroscopy. For these purposes we require a description in which all the electrons in the atom are included; i.e. the electronic state of a many-electron atom must be characterised by means of quantum numbers that describe the total angular momentum of the atom. The exact form in which these quantum numbers are given depends upon the atom in question for reasons which we shall discover below. We start out by describing a hydrogen-like or one-electron atom.

5.8.1   The five quantum numbers of a single electron

These are the:

principal quantum number	n
orbital angular momentum quantum number	l
quantum number for the z-component of orbital angular momentum	m_l
spin angular momentum quantum number	s
quantum number for the z-component of the spin angular momentum	m_s

It is clear that, with the exception of the first, all these quantum numbers characterise an aspect of the angular momentum of the electron. (The subscript l has been re-introduced in m_l in order to distinguish it from m_s).

5.8.2   Quantum numbers for the many-electron atom

As noted above, the proper description of a many-electron atom must be a description which includes all the electrons. Since the spherical symmetry of the atom is maintained as more electrons are added, the angular momentum quantum numbers remain valid and are central to this description. In analogy with those which refer to individual electrons, we introduce four new quantum numbers, L, M_L, S and M_S, which describe the orbital and spin angular momenta of all the electrons, collectively. Extending the notation, lower case letters for one electron, upper case letters for two or more electrons, we say that atoms with values of $L = 0, 1, 2, 3, \ldots$ are in $S, P, D, F, \ldots$ states.

The value of S (total spin quantum number) is important because it characterises the multiplicity of the spin state, i.e. the number of M_S values, which is $2S + 1$ (Section 4.4), and the multiplicity is normally added to the symbol for the state as a left superscript, e.g. for $S = \frac{1}{2}$ and $L = 2$ we have a ^{2}D state (spoken doublet D), for $S = 2$ and $L = 1$ we have a ^{5}P state (quintet P). The values of M_S and M_L are not usually specified. Symbols such as ^{2}D and ^{5}P used to describe the states of atoms are known as *term symbols* and in the next section a method of finding the possible L and S values for a particular arrangement of electrons in orbitals, i.e. a *configuration*, is described.[‡]

5.8.3   The assignment of term symbols

Given a particular atomic configuration, i.e. an assignment of electrons to hydrogen-like atomic orbitals, the determination of the possible states or terms of the atom is quite

---

[‡] It is most unfortunate that in this notation, which is almost 100 years old and therefore quite impossible to change, the symbol S stands for two completely different angular momentum properties. Fortunately, the possibilities of error due to a confusion between the two quantities represented by S are very few.

straightforward. Firstly we recognise that if every one of the n^2 orbitals corresponding to a particular principal quantum number, n, is doubly occupied, all components of spin and orbital angular momenta cancel to give a total angular momentum of zero. Therefore, all such *closed shells* can be ignored in the determination of term symbols. The treatment of the partially filled shells is best carried out by constructing a table as illustrated in the following example in which the configuration p^2 is analysed. We concentrate our attention upon the possible z-components of orbital and spin angular momenta, since these can be simply added because they are all aligned along z, and for the p^2 problem the maximum $M_L = m_{l1} + m_{l2}$ value is $1 + 1 = 2$. Similarly, the maximum value of M_S is $\frac{1}{2} + \frac{1}{2} = 1$. Since we know (Section 4.4) that the z-components, M_X, of angular momentum, X, always take all values between $+X$ and $-X$ in unit steps, our M_L values must run from $+2$ to -2. Similarly, our M_S values must be $+1$, 0 and -1, and we arrange these data as shown in Table 5.4. The symbol in brackets gives the m_l and m_s of electron 1 and electron 2, i.e. $(m_{l1}m_{s1}; m_{l2}m_{s2})$, which add to give the M_L and M_S values at the side and head of the table. There is no entry in the position $M_L = +2$ and $M_S = +1$ because this could only be achieved if both electrons occupied the same p orbital ($m_l = +1$) and had the same spin ($m_s = +\frac{1}{2}$) which is forbidden by the Pauli principle. The bracketed symbols are the possible *microstates* of the system and we can readily check that we have found the correct number of them: 15. A p electron can have six combinations of the quantum numbers m_l and m_s. Therefore, we have a choice of six possibilities for our first electron. Since no two electrons can have the same combination of the five quantum numbers, we have only five possibilities for the second electron making $6 \times 5 = 30$ possibilities in total. But electrons are indistinguishable so that, for example, $(+1 + \frac{1}{2}; 0 - \frac{1}{2}) \equiv (0 - \frac{1}{2}; +1 + \frac{1}{2})$ and there are therefore $6 \times \frac{5}{2} = 15$ possible microstates of the system.

To find the terms of p^2 we first focus our attention upon the entry in the table which has the highest M_L value and, if there are more than one of these, upon that which has both the highest M_L and the highest M_S value. In Table 5.4 the entry with the highest M_L value is $(+1 + \frac{1}{2}; +1 - \frac{1}{2})$. The fact that there is a microstate with an M_L value of $+2$ tells us immediately that it must be accompanied four other microstates with M_L values of $+1$, 0, -1 and -2 and that the five together form a state with $L = 2$, i.e. a D state. The same reasoning tells us that since $M_S = 0$ there are no other spin components of this state which has $S = 0$ and $2S + 1 = 1$, i.e. we have found a ^1D term. We therefore tick off one micro state in each box having $M_S = 0$ and $M_L = +2 \ldots -2$. We do not need

Table 5.4 The microstates of the electron configuration p^2

		M_S		
		$+1$	0	-1
	$+2$		$(+1 + \frac{1}{2}; +1 - \frac{1}{2})$	
	$+1$	$(+1 + \frac{1}{2}; 0 + \frac{1}{2})$	$(+1 + \frac{1}{2}; 0 - \frac{1}{2})\ (+1 - \frac{1}{2}; 0 + \frac{1}{2})$	$(+1 - \frac{1}{2}; 0 - \frac{1}{2})$
M_L	0	$(+1 + \frac{1}{2}; -1 + \frac{1}{2})$	$(+1 + \frac{1}{2}; -1 - \frac{1}{2})\ (+1 - \frac{1}{2}; -1 + \frac{1}{2})$ $(0 + \frac{1}{2}; 0 - \frac{1}{2})$	$(+1 - \frac{1}{2}; -1 - \frac{1}{2})$
	-1	$(-1 + \frac{1}{2}; 0 + \frac{1}{2})$	$(-1 + \frac{1}{2}; 0 - \frac{1}{2})\ (-1 - \frac{1}{2}; 0 + \frac{1}{2})$	$(-1 - \frac{1}{2}; 0 - \frac{1}{2})$
	-2		$(-1 + \frac{1}{2}; -1 - \frac{1}{2})$	

to identify which of the microstates, in a group with more than one, belongs to ^1D, the ticking process is merely one of 'keeping account'.

We now identify the microstate among those remaining which has the highest M_L and M_S values; it is $(+1 + \frac{1}{2}; 0 + \frac{1}{2})$. Since it has $M_L = +1$ and $M_S = +1$ this microstate must be one of the nine which form a ^3P term and we tick off microstates in Table 5.4 accordingly. A single microstate with $M_S = 0$ and $M_L = 0$ remains and this must be the single component of a ^1S term. So our result is that the configuration p^2 gives rise to three terms; ^1D (5 components), ^3P (9 components) and ^1S (1 component) = 15 components in total = number of microstates in Table 5.4.

It should be admitted here that the above method of determining the terms of a particular configuration is more valuable as an illustration of the meaning of a term symbol and of the subtle way in which the spin and orbital angular momenta of two electrons combine than it is as a practical research tool. A configuration with several f electrons, such as we find in the rare earths, is effectively impossible to analyse with this method (e.g. f^4 has 1001 microstates!) and some extremely sophisticated mathematical techniques have been developed to deal with this type of problem. Each term in a configuration has a different energy because of the different contributions from inter-electronic repulsion in each one. But the $(2S + 1)(2L + 1)$ components within each term have almost exactly the same energy. In fact they would be degenerate were it not for an effect known as *spin-orbit coupling* which splits each term into a number of levels of different energy, as shown in Section 5.9.

5.8.4 Term energies and Hund's rules

As we have seen, a single electron configuration can give rise to several terms and, in general, all of these have different energies because of the different contributions from the inter-electronic repulsion in each one. Table 5.5 gives the energies of some examples of the configuration np^2. (The energy of the $^3P_0$ state has been arbitrarily set to zero.) The energies of the ^3P states are not all equal because of spin-orbit coupling (see Section 5.9), so a weighted mean energy is given in the fourth column of the table.

The electron spin plays an important role in determining the inter-electronic repulsion and it does so in a rather subtle way, the details of which depend upon the connection between the electron spin and spatial wave functions and the Pauli principle. This is explored in detail in Chapter 11. Here it will be sufficient to recognise that, in order to conform with the Pauli principle, different electron spin functions must be combined with different electron spatial functions. Thus, since the inter-electronic repulsion is directly determined by the spatial distribution of the electrons, the energies of the various terms are different. This fact was noted in 1927 by Friederich Hund (1896–), largely as a result

Table 5.5 The energies (in eV) of the terms arising from the configuration np^2

	$^3P_0$	$^3P_1$	$^3P_2$	^3P(mean)	$^1D_2$	$^1S_0$
Ge $4p^2$	0.0	0.0691	0.1748	0.1201	0.8834	2.0293
Sn $5p^2$	0.0	0.2098	0.4250	0.3060	1.0679	2.1279
Pb $6p^2$	0.0	0.9695	1.3205	1.0568	2.6605	3.0955

of experimental studies of atomic spectra. On the basis of his observations Hund proposed two rules for equivalent electrons:

1. The terms of highest multiplicity lie lowest in energy.

2. If two terms have the same multiplicity the lowest is that with the greatest value of L.

The data in Table 5.5 illustrate Hund's rules. The ^3P states are the lowest and one of them, $^3P_0$, is the ground state. Also, ^1D is lower than ^1S.

5.9 SPIN-ORBIT COUPLING

The origin of spin-orbit coupling is easy to understand. To an electron orbiting an atomic nucleus the nucleus appears to be an orbiting positive charge. This orbiting charge generates an electric current which, in turn, produces a magnetic field at the electron. This magnetic field interacts with the magnetic field due to the spin of the electron and the energy of interaction is known as the spin-orbit coupling energy. Theory shows that the effect is proportional to Z^4, where Z is the nuclear charge, and it therefore increases very steeply with increase in atomic number. The coupling of the spin and orbital angular momenta of an electron, which results from this interaction, mixes those quantities together so that, especially for heavy atoms, it is possible to speak of L and S only as approximate quantum numbers; the only exact angular momentum quantum number being J, which is a combination of the two. There are two ways of determining J, one of which is appropriate for light atoms and one for heavy.

5.9.1 Russell–Saunders or *LS* coupling

For light atoms the values of J can be obtained by adding together the values of L and S for the total orbital and spin angular momenta of the electrons determined as in Section 5.8.3. We add the angular momenta using the formula (Section 4.6):

$$J = L + S, L + S - 1, L + S - 2, \ldots, |L - S|$$

The values of J are written as a right subscript on the term symbol, e.g. for p^2:

$$^1D \ (S = 0, L = 2) \implies {}^1D_2$$
$$^3P \ (S = 1, L = 1) \implies {}^3P_2, {}^3P_1 \text{ and } {}^3P_0$$
$$^1S \ (S = 0, L = 0) \implies {}^1S_0$$

Each state having a particular value of J has $2J + 1$ components with M_J values running from $+J$ to $-J$ and in the absence of a magnetic field, *vide infra*, all are degenerate. Thus, as a result of spin-orbit coupling, a ^3P term which was nine-fold degenerate, is split into a five-fold degenerate $^3P_2$, a three-fold degenerate $^3P_1$ and a singly degenerate $^3P_0$ term. The five-fold degeneracy of the ^1D term is not removed and the $^1S_0$ term has only one component.

When we formulate spin-orbit coupling in this way we call it *LS* coupling because we combine values of L and S to give J. It is also known as Russell-Saunders coupling after two astronomers who first used it to interpret the atomic spectra of stars in 1925, before the advent of quantum mechanics as we know it.

Thus, in order to obtain a model of the electronic structure of a light, many-electron atom, we build up our description in three stages.

1. We first think of a many-electron atom as being composed of a number of electrons each moving in an orbital characterised by quantum numbers n, l, m_l, s and m_s. We recognise that the radial parts of these orbitals will depend upon the charge on the nucleus and that the electrons with low values of n will therefore be very close to the nucleus while electrons in orbitals with higher values of n will reach out much further into the space surrounding the atom. Those with the highest value of n are the valence electrons which are the most important in the formation of chemical bonds and in atomic spectroscopy.

2. Next we take account of the fact that, although the individual n values are unchanged, the detailed forms of the radial functions will be changed by the inter-electronic repulsion. Furthermore, because the electronic motions are correlated and because we cannot distinguish between individual electrons in a many-electron system, the individual values of l and s are no longer meaningful and must be replaced by L and S which describe the orbital and spin angular momenta of the electrons as a whole.

3. Finally, we recognise that the effect of spin-orbit coupling requires that we combine L and S to give a total angular momentum quantum number J.

It should be noted that the three steps above are associated with three physical effects of diminishing energy. In step 1 we have the kinetic and potential energies of the electrons moving in the field of the charged nucleus. In the second step, we add the smaller energy due to the inter-electronic repulsion. In step 3 we add the spin-orbit coupling energy which, for atoms with atomic numbers less than 20, is even smaller. When carried out with proper mathematical precision, this process of examining the largest contribution to the energy first and then adding in the effects of further, smaller energy terms in order of decreasing energy is called *perturbation theory* (Appendix 4). It has been known since the time of Lord Rayleigh (1842–1919) and has extensive uses in both classical and quantum mechanics. It is important that the energy contributions are taken in order of decreasing magnitude and in the case of very heavy atoms the spin orbit coupling, which goes as Z^4, exceeds the inter-electronic repulsion and must therefore be considered first.

5.9.2   *jj* coupling

Since we require to consider spin-orbit coupling before inter-electronic repulsion for atoms in the later rows of the periodic table, in step 2 we first combine the values of l and s, adding the angular momentum quantum numbers as described in Section 4.6, to make a j value for each individual electron, i.e. in the case of p^2:

$$\text{for each p electron } l = 1 \text{ and } s = \tfrac{1}{2} \Longrightarrow j_p = \tfrac{3}{2} \text{ and } \tfrac{1}{2}$$

Now, in step 3, we must add the j values together to give J for the whole atom. Where we have two inequivalent electrons, e.g. $p^1 d^1$, then we simply combine all the possible m_j values for the d-electron with all those for the p-electron, in pairs. But in the

Table 5.6 The M_J values of the microstates of np^2 in jj coupling

		$p_{\frac{3}{2}}$				$p_{\frac{1}{2}}$	
		$\frac{3}{2}$	$\frac{1}{2}$	$-\frac{1}{2}$	$-\frac{3}{2}$	$\frac{1}{2}$	$-\frac{1}{2}$
$p'_{\frac{3}{2}}$	$\frac{3}{2}$	—	2	1	0	2	1
	$\frac{1}{2}$	—	—	0	−1	1	0
	$-\frac{1}{2}$	—	—	—	−2	0	−1
	$-\frac{3}{2}$	—	—	—	—	−1	−2
$p'_{\frac{1}{2}}$	$\frac{1}{2}$	—	—	—	—	—	0
	$-\frac{1}{2}$	—	—	—	—	—	—

case of two equivalent electrons, e.g. p^2 for which we determined the term symbols in Section 5.8.3, we experience again the problem that we had in constructing Table 5.4, i.e. we must be aware of the Pauli principle. For the configuration np^2 we have possible j values of $1 + \frac{1}{2} = \frac{3}{2}$ and $1 - \frac{1}{2} = \frac{1}{2}$ and $4 + 2 = 6$ possible m_j values. A square, 6×6 array of all possible combinations of two m_j values (Table 5.6), would give 36 microstates. But there can only be $6 \times \frac{5}{2} = 15$ states for two p-electrons and the following states must be rejected: all six states on the diagonal of the array, because if two electrons in the same shell have the same j value and the same m_j value then they are identical and must be occupying the same atomic wave function with the same spin. This is contrary to Pauli's principle. Further, since electrons are indistinguishable, an off-diagonal array element of the form $[j\,m_j(1);\ j'\,m'_j(2)]$ cannot be distinguished from its 'mirror-image' across the diagonal $[j'\,m'_j(1);\ j\,m_j(2)]$. These two microstates *together* $\{[j\,m_j(1);\ j'\,m'_j(2)] - [j'\,m'_j(1);\ j\,m_j(2)]\}/\sqrt{2}$ describe a state of the system so that 15 of the 30 off-diagonal microstates must be neglected in the state-counting process. We arbitrarily choose the 15 microstates below the diagonal leaving $36 - 6 - 15 = 15$ microstates (Table 5.6). Following our usual method, we now start with the largest M_J value, remove the remaining $2J$ M_J values associated with it and repeat the process until all M_J values have been accounted for. The $p'_{\frac{3}{2}}p_{\frac{3}{2}}$ triangle gives $J = 2$ and 0, the $p'_{\frac{3}{2}}p_{\frac{1}{2}}$ rectangle gives $J = 2$ and 1 and the $p'_{\frac{1}{2}}p_{\frac{1}{2}}$ triangle gives $J = 0$. The total number of states calculated from these J values is $5 + 1 + 5 + 3 + 1 = 15$ confirming the correctness of our working.

5.9.3 Intermediate coupling

We have noted above that LS or Russell-Saunders coupling applies very well to light atoms and jj coupling to heavy atoms. But what about atoms of intermediate mass, or Z value; there is no sudden change from LS to jj coupling? This is a problem typical of the physical sciences and especially so of quantum mechanics. We have solutions for the extremes of our problem but not for the intermediate region for which, in fact, no simple solution is possible. The root of our difficulty lies in the application of the steps of the perturbation theory outlined in Section 5.9.1. It is important to take the larger terms before the smaller ones and for light atoms this means taking inter-electronic repulsion before spin-orbit coupling. For heavy atoms the order must be reversed. For the intermediate case the only correct approach is to take inter-electronic repulsion and spin-orbit coupling together at the same time, since they are of comparable magnitudes. This cannot be done in as simple a manner as in the examples described above, even for two electrons. A more complicated calculation is required (see Box 5.2). But with the help of the digital computer the calculation of the terms or electronic energy states of atoms presents little difficulty today.

For systems with a small number of electrons such as np^2 a correlation diagram showing the transition from LS, through the intermediate region, to jj coupling is useful in the assignment of atomic spectra. Figure 5.6, calculated with data from Condon and Shortley,[4] shows the np^2 correlation diagram. χ is proportional to the ratio of the spin-orbit coupling to inter-electronic repulsion so that the zero on the left-hand end of the abscissa represents pure LS coupling while the zero on the right represents pure jj. The assignment of the np^2 electronic states of germanium ($n = 4$), tin ($n = 5$) and lead ($n = 6$) is also included and the move from LS to jj coupling with increase of nuclear charge is very clear. The energy scales at the two ends of the diagram are not equal, they have been modified so that the total spread of the states is the same; 15 energy units. The details of this modification[4] need not concern us here. However, we should not leave this subject without remarking upon the value of the correlation diagram in chemistry. Because, as chemists, we are often interested in groups of similar species, e.g. the aromatic hydrocarbons or the atoms in a particular row or column of the periodic table, we wish to understand how the properties of a particular group of atoms or molecules vary as their fundamental physical properties vary across the group. This information can sometimes be displayed in diagrammatic form, the most well-known example of this being the periodic table itself. But there are many other important examples in physics and chemistry.

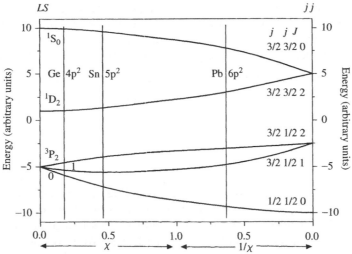

Reprinted with permission, E.V.Condon and G.H. Shortley, "*The Theory of Atomic Spectra*", 1935, Cambridge University Press.

Figure 5.6 Spin-orbit coupling in np^2

5.9.4   Inter-electronic spin-orbit coupling

The orbital magnetic moment of an electron can also interact with the spin magnetic moment of another electron, and vice versa. However, the energies concerned are very small compared with those due to the intra-electronic effects which we have described above and they are usually neglected.

5.10   SELECTION RULES IN ATOMIC SPECTROSCOPY[4,5,6]

In many places in this chapter reference has been made to the information which atomic spectroscopy provides about atomic structure. But we must not think that every energy state of the atom can be reached from every other state by the emission or absorption of radiation. The process is subject to restrictions, known as selection rules, which result from the requirements that energy and angular momentum must be conserved. It is precisely because these restrictions exist that so much detailed information has been obtained. The conservation of energy is embodied in the Bohr-Einstein selection rule:

$$\Delta E = h\nu$$

The change in energy of the atom must be exactly equal to the energy of the photon absorbed or emitted.

5.10.1   Angular momentum

The photon of electric dipole radiation has an angular momentum characterised by the angular momentum quantum number 1, but with only two z-components, $+1$ and -1, parallel and antiparallel to the direction of propagation. (This subject is discussed in more detail in Section 8.4 and the treatment here will therefore be brief.) Thus, since angular momentum is conserved, in one-electron atoms atomic transitions between atomic orbitals with angular momentum quantum numbers l and l' must be such that $l' = l + 1, l$ or $l - 1$, and the z-component of the angular momentum must also be conserved. The electron spin does not interact with the oscillating *electric* field of the radiation so it is not changed. (Note however, that the spin does interact with the radiation's *magnetic* field and this interaction is very important in magnetic resonance spectroscopy and in magnetic dipole transitions, Section 5.10.1.2.)

In summary, the selection rules for dipolar radiation in one-electron atoms are:

1. $\Delta s = 0$
2. $\Delta l = 0, \pm 1$, but $\Delta l = 0$ is forbidden by parity (see Section 5.10.2)
3. $\Delta m_l = \pm 1$

There is no restriction on the change in the value of the principal quantum number, Δn, which accounts for series of lines such as the Balmer series.

When we consider many-electron atoms the position becomes more complicated and, in general, the interaction of spin and orbital angular momenta must be considered. In the limit of pure LS coupling we have for electric dipole radiation:

1. $\Delta S = 0$
2. $\Delta L = 0, \pm 1$, but $L = L' = 0$ is forbidden

3. $\Delta M_L = \pm 1$
4. $\Delta J = 0, \pm 1$, but $J = J' = 0$ is forbidden

As we move away from LS coupling, into the intermediate region and towards jj coupling the quantum numbers S and L become increasingly poorly defined and the selection rules progressively less definitive. Selection rules expressed in terms of changes of state quantum numbers are the results of application of angular-momentum theory to the coupling between atom and radiation.[4,5] They apply only to idealised situations, e.g. *pure LS* coupling. If exact information on a particular transition is required the detailed wave functions for the states in question must be determined and the transition moment integral evaluated. This subject is discussed in much more detail in Box 5.2. Sections 8.7 and 11.7 and Appendix 9 also contain relevant information.

5.10.1.1 Electric quadrupole transitions

In some atoms and ions, e.g. O^{2+}, very weak transitions which are forbidden by the selection rules described above, but not induced by spin-orbit coupling, have been observed. These transitions satisfy selection rules in which the photon involved appears to carry two units of angular and they are ascribed to quadrupolar photons.[4,5] However, photons in a beam of light have not, to this author's knowledge, been shown to carry two quanta of angular momentum; in contrast to the case of a beam of circularly polarised light, the photons of which have been shown by Beth[7] to carry $h/2\pi$ units of angular momentum each. It would appear that a definitive analysis of the conservation of angular momentum in quadrupolar transitions is not yet available.

5.10.1.2 Magnetic dipole transitions

The magnetic fields arising from the orbital and spin angular momentum of the electron can also interact with the magnetic field of an impinging photon. The transitions which result are known as magnetic dipole transitions. They are of the order of 10^{-7} times weaker than electric dipole transitions.

5.10.2   Parity

In order to understand the concept of parity it is best if we first consider a wavefunction of an atom, $\Psi(x, y, z)$, and perform upon it the operation of multiplying the three Cartesian co-ordinates by -1. We call this operation the parity operation and it can be represented by the equation:

$$\hat{\Pi}\Psi(x, y, z) \Longrightarrow \Psi(-x, -y, -z),$$

where $\hat{\Pi}$ is the parity operator.

If we perform this operation on the hydrogen 1s orbital (Appendix 5) then:

$$\hat{\Pi}\Psi_{1s} = \hat{\Pi}\frac{1}{\sqrt{\pi}}\left(\frac{Z}{a_0}\right)^{\frac{3}{2}} \cdot \exp(-\rho/2) = \frac{1}{\sqrt{\pi}}\left(\frac{Z}{a_0}\right)^{\frac{3}{2}} \cdot \exp(-\rho/2),$$

and the function is unchanged because ρ is proportional to the positive square root of $(x^2 + y^2 + z^2)$. Consequently, Ψ_{1s} does not change sign when the signs of the Cartesian co-ordinates are changed and is therefore said to be *even*. All s orbitals are even. But if we apply the parity operator to the $2p_z$ wave function then:

$$\hat{\Pi}\Psi_{2pz} = \hat{\Pi}\frac{1}{2\sqrt{2\pi}}\left(\frac{Z}{a_0}\right)^{\frac{5}{2}} \cdot z\exp(-\rho/2) = \frac{-1}{2\sqrt{2\pi}}\left(\frac{Z}{a_0}\right)^{\frac{5}{2}} \cdot z\exp(-\rho/2),$$

and we find that the function changes its sign (because $\hat{\Pi}z = -z$). It is *odd* as are all p functions. But note that the wave function is still an eigenfunction of the Hamiltonian operator with the same energy eigenvalue as before.

All atomic wave functions can be classified into two groups. Those which are unchanged under the parity operation, the s, d, g, ... functions, are said to be even and they are sometimes denoted by g (from the German 'gerade' = even). Functions which change sign under the parity operation, the p, f, h, ... functions, are said to be odd and denoted by u (from the German 'ungerade'). The rules for combining parities are just the same as those for multiplying the numbers $+1$ and -1. Thus, if two functions of the same parity, even or odd, are combined the result is a function of even parity, whereas if two functions of different parities are combined the resulting function is of odd parity.

Electric quadrupole and magnetic dipole photons, Sections 5.10.1.1 and 5.10.1.2, are of even parity. But the great majority of experimental observations in atomic spectroscopy involve dipolar photons which are of odd (u) parity. This means that the initial and final states of a transition must be of opposite parity so that, (parity of initial state) $\times$ u = (parity of final state), and parity is conserved. Alternatively we may argue (Appendix 9) that, since the dipolar photon is of u symmetry, the product of the parities of the initial and final states must also be u so that the transition dipole moment integral (Appendix 9) as a whole is g. By both arguments the parity selection rule is found to be; u $\rightarrow$ g and g $\rightarrow$ u transitions allowed, u $\rightarrow$ u and g $\rightarrow$ g forbidden. This is known as the Laporte rule.

5.11 THE ZEEMAN EFFECT

Apart from interacting with each other, the magnetic fields due to the spin and orbital motion of an electron can also interact with an externally applied magnetic field. The effect of the interaction is to lift degeneracy and split spectral lines, as was first observed by Pieter Zeeman (1865–1943) in 1896. The quantum numbers which determine the angular momenta of the electron, l, m_l, s and m_s determine the associated magnetic moments according to the following relationships. If we choose, as we invariably do, the z-axis as that along which the angular momentum is quantised, the orbital motion of an electron gives rise to a magnetic moment in the z-direction. The relationship between the quantum number for the z-component of the orbital angular momentum (m_l) and the magnetic moment (μ_z) is derived from the classical equations for the magnetic field due to a circulating electron of charge $-e$ and mass m_e in Section 12.2.4; it is:

$$\mu_z^l = \frac{-e\hbar}{2m_e}m_l \equiv -\mu_B m_l \quad \text{where} \quad \mu_B = \frac{-e\hbar}{2m_e}$$

μ_B is the Bohr magneton and the negative sign shows that the magnetic moment is antiparallel to the orbital angular momentum.

Such a simple analysis is not possible for the magnetic moment due to the spin angular momentum. But using the relativistic Dirac equation it is found that:

$$\mu_z^S = -g_e\mu_B m_S = -2.0023\mu_B m_S \approx -2\,\mu_B m_S$$

g_e is known as the electron g-factor and the approximation, $g_e = 2$, is quite adequate for all but the most exacting experiments or calculations.

If we now impose an external magnetic field **(B)** upon the electron there will be an energy change (ΔE) given (Box 12.2) by:

$$\Delta E = -(\mu_z^l + \mu_z^S) \cdot B = (m_l + 2m_S)\mu_B B_z \equiv g\mu_B B_z \qquad (5.11.1)$$

where B_z is the component of the field in the z-direction. In this expression a useful experimental parameter, g, has been introduced. The g-value deduced from the experimentally observed splitting of spectral lines provides an immediate number for comparison with theoretical estimates of $m_l + 2m_s$. Where the true atomic electronic states are complex mixtures of the simple microstates as, due to effects like spin-orbit coupling they normally are, the value of g will not be the simple integer or half-integer which Equation (5.11.1) might lead us to expect. Since $\mu_B = 9.273 \times 10^{-24}$ J T^{-1} the energy change, or Zeeman splitting, is very small. Converting to wave numbers, we find that $\mu_B = 0.467$ cm^{-1} T^{-1} so that a field of the order of two Tesla is required to produce a splitting of 1 cm^{-1}. This explains why Faraday searched, without success, for the effect of a magnetic field upon spectral lines some 30 years before Zeeman found it.

5.11.1 The normal Zeeman effect

In 1896 when Zeeman first observed his eponymous effect there was no concept of electron spin. But orbital angular momentum and the associated magnetic effects were readily understandable with classical electrodynamics and early studies of the Zeeman effect therefore gave a mixed picture. Some results were easy to interpret and these cases became known as examples of the *normal* Zeeman effect. The remaining results were only understood when electron spin and its magnetic effects were postulated by Goudsmit and Uhlenbeck in 1925; prior to that they were said to be examples of the *anomalous* Zeeman effect. There is also an important relationship between the Zeeman effect and the polarisation of light which is best explained diagrammatically (Figure 5.7). The properties of polarised light are discussed in detail in Section 8.2 and that discussion will not be pre-empted here. In discussing Figure 5.7 we shall simply make a number of bald statements which will be elucidated in Chapter 8.

A $^1S_0$ to $^1P_1$ transition in the absence of a magnetic field is illustrated in Figure 5.7(a). An example would be the Xe6s^2 to Xe6s^{1}6p^1 transition of barium which is found at $\lambda = 553.70$ nm and is responsible for the bright green colour of the flame test for barium. The spectrum appears as a single line since, although there are three upper states $(M_J = +1, 0, -1)$, they are all of equal energy. In Figure 5.7(b) our atom sample has been placed at the centre of a solenoid so that it is subject to a magnetic field (B), which is parallel to the light beam. The magnetic field lifts the degeneracy of the $^1P_1$ state as required by Equation (5.11.1) so that it is now possible, in principle, to observe three transitions from the ground to the excited state. Furthermore, the upper state energy levels have angular momentum characterised by the quantum numbers $J = 1$ and $M_J = 0, \pm1$.

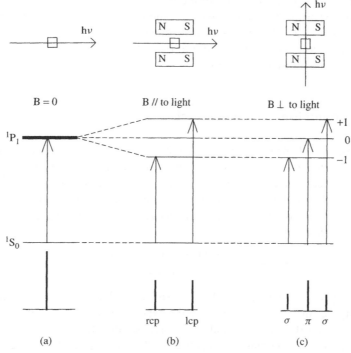

Figure 5.7 The normal Zeeman effect in the $^1S_0$ to $^1P_1$ transition of barium

Therefore, in order to conserve angular momentum, the photon must provide the angular momentum required for transitions from ground state to the M_J levels of the excited state, since the lower state has no angular momentum. Circularly polarised light has exactly the properties we require; an angular momentum of 1 and for right circularly polarised (rcp) photons a z-component of -1 and for lcp photons $+1$. Thus, if we use circularly polarised light in the Zeeman experiment we shall be able to see the transitions to the ± 1 states but the transition to $M_J = 0$ will not be observed since the z-component of the angular momentum cannot be conserved.

But if we bore a hole in our solenoid and illuminate the atom sample with a light beam perpendicular to the direction of the magnetic field (Figure 5.7(c)), a new polarisation possibility emerges. The oscillating electric and magnetic fields, which are the fundamental carriers of the polarisation property, can only oscillate in the plane perpendicular to the direction of propagation of the light. With the experimental arrangement of Figure 5.7(c) the light can be polarised in the direction of the field and if we use plane polarised light transitions to all three excited sates can be seen. The transition to $M_J = 0$ is energised by light polarised parallel to B and is denoted by π (Greek p for parallel). The transitions to $M_J = \pm 1$ are energised by light polarised perpendicular to B and are denoted by σ (Greek s for German senkrecht = perpendicular). The total probability of the transitions in each of the three figures is the same. If we set it to four arbitrary units in Figure 5.7(a), then we have two for each of the transitions in Figure 5.7(b) and one for each σ and 2 for π in Figure 5.7(c). Very important practical use of the Zeeman effect is made in quantitative analytical atomic spectroscopy where the ability to move an absorption line by applying a magnetic field enables the background of the line to be measured and subtracted. Finally, we note that the experimentally observed separation between the outer lines gives a value of g of 1.02 which is 2% larger than the theoretical value of 1.0. This probably arises because of mixing, brought about by spin-orbit coupling, with states having $S \neq 0$, which brings us to the anomalous Zeeman effect.

5.11.2   The anomalous Zeeman effect

When Goudsmit and Uhlenbeck proposed that electrons had an intrinsic spin angular momentum characterised by the angular momentum quantum number of $\frac{1}{2}$ but with a g-value of 2.0, twice as large as the g-value for orbital angular momentum, the puzzling splitting patterns observed in many spectra were rapidly elucidated and a formula expressing the g-value in terms of the quantum numbers L, S and J was derived by Alfred Landé (1888–1975), see Section 12.2:

$$g = 1 + \frac{J(J+1) + S(S+1) - L(L+1)}{2J(J+1)} \tag{5.11.2}$$

This formula gives the g-values of 0.500, 1.167 and 1.333 for $^3D_1$, $^3D_2$ and $^3D_3$ respectively. These figures may be compared with the experimental data for corresponding states of calcium which are 0.501, 1.162 and 1.329. The sodium D-lines provide a good illustration of the anomalous Zeeman effect (Figure 5.8). The values of $M_J \times g$ in Figure 5.8 are simply the product of the M_J value of the state and the its g-value as determined by Equation (5.11.2) and they are directly related to the spacing of the spectral lines; e.g. for $^2P_{\frac{3}{2}}$ $S = \frac{1}{2}$ $L = 1$, $J = \frac{3}{2}$ giving $g = \frac{4}{3}$ whence $M_J \times g = \pm 2$ and $\pm \frac{2}{3}$.

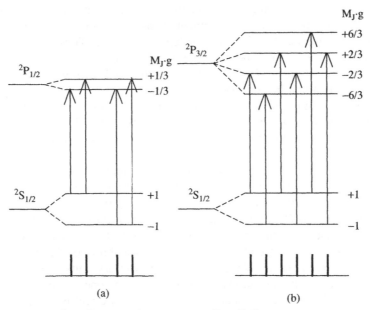

Figure 5.8   The anomalous Zeeman effect on the sodium D lines

5.12 BIBLIOGRAPHY AND FURTHER READING

1. L. Pauling and E.B. Wilson, *Introduction to Quantum Mechanics*, McGraw-Hill, New York, 1935, and Dover.
2. H. Eyring, J. Walter and G.E. Kimball, *Quantum Chemistry*, Wiley, New York, 1944.
3. Shapes and polar plots of f-orbitals. H.G. Friedman, Jr., G.R. Choppin and D.G. Feuerbacher, *J. Chem. Education*, **41**, 354–358, 1964. C. Becker, *J. Chem. Education*, **41**, 358–360, 1964.
4. E.U. Condon and G.H. Shortley, *The Theory of Atomic Spectra*, Cambridge University Press, 1935.
5. R.D. Cowan, *The Theory of Atomic Structure and Spectra*, California University Press, Berkeley, 1981.
6. J.C. Slater, *Quantum Theory of Atomic Structure*, Vols. I and II, McGraw-Hill, New York, 1960.
7. R.A. Beth, *Phys. Rev.*, **48**, 471 (1935); **50**, 115 (1936).

BOX 5.1 The spectral series of the hydrogen atom

In 1890 Johannes Robert Rydberg (1854–1919) wrote the Balmer formula for the lines in the emission spectrum of the hydrogen atom in the form which is most common today:

$$\tilde{\nu} = R \left\{ \frac{1}{r^2} - \frac{1}{s^2} \right\}$$

Here, $\tilde{\nu}$ is the value in wave numbers (cm^{-1}) of the spectral line, R the Rydberg constant ($109\,677.578\ cm^{-1}$) and r and s the principal quantum numbers of the lower and upper states connected by the transition respectively. Using this formula the five best-known series of hydrogen lines may be characterised as in the following table.

Series	r	s	first line		last line	
			$\tilde{\nu}$ (cm^{-1})	λ (nm)	$\tilde{\nu}$ (cm^{-1})	λ (nm)
Lyman (1906)	1	2, 3, …∞	82 258.2	121.6	109 677.6	91.2
Balmer (1885)	2	3, 4, …∞	15 232.9	656.5	27 419.4	364.7
Paschen (1908)	3	4, 5, …∞	5 331.6	1 875.6	12 186.4	820.1
Brackett (1922)	4	5, 6, …∞	2 467.8	4 052.2	6 854.9	1 458.8
Pfund (1924)	5	6, 7, …∞	1 340.5	7 459.9	4 387.1	2 279.4

The last line in each series corresponds to a transition in which the upper state has the principal quantum number of infinity in each case (Figure B5.1.1). In this state the electron has the highest energy it can have and still be bound to the nucleus. At higher energies the electron leaves the atom, i.e. *ionisation* takes place. Therefore, all series have a common upper level for their transition of highest energy; the ionisation limit. But the energy required to reach it diminishes since each series starts from a higher level. Schrödinger's result for the eigenvalues of the hydrogen atom (Equation (5.1.1)) shows that, in cm^{-1}:

$$R = \frac{me^4}{8ch^3\varepsilon_0^2}$$

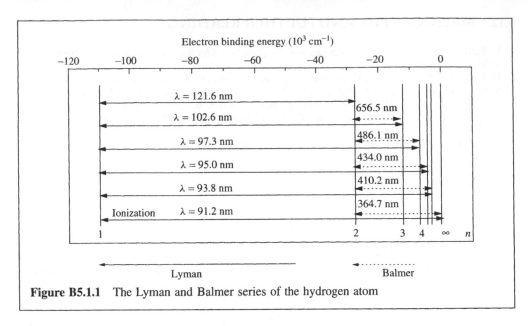

Figure B5.1.1   The Lyman and Balmer series of the hydrogen atom

BOX 5.2   Inter-electronic repulsion, spin-orbit coupling (SOC) and their effect upon transition probability and selection rules

The purpose of this Box is to provide a fairly detailed example of the way in which the relative magnitudes of inter-electronic repulsion and SOC affects not only state energies, as illustrated in Figure 5.6, but also wavefunctions and hence transition probabilities. This will also offer an opportunity to illustrate how transition probabilities are calculated from atomic wavefunctions. The example will be based on the configuration p^2 and a transition to that configuration from $s^1 p^1$. We begin with a discussion of p^2.

The energies of the p^2 configuration

We have found the terms of p^2 in the LS and jj limits in Section 5.9, but the methods used there do not reveal the combinations of microstates, i.e. the wavefunctions, which are associated with each term. These can be obtained for the limiting cases by use of the raising and lowering operators in the way illustrated for d^2 in Appendix 10, but for the intermediate case which concerns us here we require to diagonalise (Appendix 3) simultaneously the matrices of the operators for SOC, $\zeta \hat{l} \cdot \hat{s}$, and inter-electronic repulsion, e^2/r_{12}.

The basis states or microstates may be written in the form $(m_{l1} m_{s1}, m_{l2} m_{s2})$ with the m_s values represented by a superscript plus or minus sign, e.g. $(+1^+, 0^-)$, it being always understood that the microstate is actually a Slater determinant (Appendix 6), i.e.:

$$(+1^+, 0^-) = (1/\sqrt{2})\{|+1^+(1)0^-(2)\rangle - |0^-(1) + 1^+(2)\rangle\}$$

[If the two electrons are not both p-electrons it would usually be necessary to include the quantum number l in the notation.] For p^2 there are 15 microstates and in the matrices which follow we shall consider 10 of them, those having M_J values of $+2$, $+1$ and 0. The matrix elements for $M_J = -2$ and -1 are, of course, very similar to those for $M_J = +2$ and $+1$. We shall not go into the details of the calculation of the electron-repulsion matrix elements. Full treatments of this subject can be found in numerous sources.[1-3] The required integrals can be expressed as a sum of products of angular and radial parts the first of which can be obtained in closed form and tabulated for any combination of spherical harmonics (Appendix 5), i.e. for all atoms. The radial integration, on the other hand, depends upon the radial functions which differ from atom to atom and they must be determined individually, either theoretically or by comparison with experiment, for each atom. However, the results can be expressed in terms of a small number of integrals, F^0 and F^2 in the present case, where:

$$F^k(n^a l^a, n^b l^b) = e^2 \int_0^\infty \int_0^\infty (r_<^k/r_>^{k+1}) R_1^2(n^a l^a) R_2^2(n^b l^b) r_1^2 r_2^2 \, dr_1 \, dr_2$$

$R(nl)$ is the radial part of an atomic wavefunction function, $r_<$ is the smaller and $r_>$ the larger of r_1 and r_2, and the electron repulsion is obtained as a sum of two terms, $A^0 F^0 + A^2 F^2$, where A^0 and A^2 are the results of the angular integrations.

The inter-electronic repulsion matrix for the microstates of p^2 having $M_J = +2$ and $+1$ is:

	$(+1^+, +1^-)$	$(+1^+, 0^+)$	$(+1^-, 0^+)$	$(+1^+, 0^-)$	$(+1^+, -1^+)$
$(+1^+, +1^-)$	$F^0 + F^2/25$	0	0	0	0
$(+1^+, 0^+)$	0	$F^0 - F^2/5$	0	0	0
$(+1^-, 0^+)$	0	0	$F^0 - 2F^2/25$	$-3F^2/25$	0
$(+1^+, 0^-)$	0	0	$-3F^2/25$	$F^0 - 2F^2/25$	0
$(+1^+, -1^+)$	0	0	0	0	$F^0 - F^2/5$

The eigenvalues of this part of the matrix are $F^0 - F^2/5$ (3×) and $F^0 + F^2/25$ (2×).

The inter-electronic repulsion matrix for the microstates of p^2 having $M_J = 0$ is:

	$(+1^+, -1^-)$	$(+1^-, 0^-)$	$(+1^-, -1^+)$	$(0^+, 0^-)$	$(-1^+, 0^+)$
$(+1^+, -1^-)$	$F^0 + F^2/25$	0	$-6F^2/25$	$+3F^2/25$	0
$(+1^-, 0^-)$	0	$F^0 - F^2/5$	0	0	0
$(+1^-, -1^+)$	$-6F^2/25$	0	$F^0 + F^2/25$	$-3F^2/25$	0
$(0^+, 0^-)$	$+3F^2/25$	0	$-3F^2/25$	$F^0 + 4F^2/25$	0
$(-1^+, 0^+)$	0	0	0	0	$F^0 - F^2/5$

The eigenvalues of this part of the matrix are $F^0 - F^2/5$ (3×), $F^0 + F^2/25$ (1×) and $F^0 + 2F^2/5$ (1×).

The remaining five states of the electron-repulsion matrix give eigenvalues of $F^0 + F^2/25$ (2×) and $F^0 - F^2/5$ (3×) so that in total we have: $F^0 - F^2/5$ (9×), $F^0 + F^2/25$ (5×) and $F^0 + 2F^2/5$ (1×). Since F^0 appears on each diagonal element and nowhere else it simply sets the absolute energy scale and can usually be set to zero. Thus, these are the states ^3P, ^1D and ^1S which, with $F^0 = 0$ and $F^2 = 25$ arbitrary units, correspond in energy and degeneracy to the extreme left of Figure 5.6. The corresponding wavefunctions can be obtained as combinations of microstates by diagonalising the blocked out matrices. The energy matrix for p^2 in the absence of SOC is seen to be a function only of F^2 which can be treated as a parameter and its value obtained by comparison of calculated and experimental energy levels.

The matrix elements of the SOC are more readily obtained than those of electron repulsion. We express the operator in the form of a sum over (in this case two) electrons:

$$\zeta \sum_i \hat{l}_i \cdot \hat{s}_i = \zeta \sum_i \left\{ \frac{1}{2}(\hat{l}_{i+}\hat{s}_{i-} + \hat{l}_{i-}\hat{s}_{i+}) + \hat{l}_{iz}\hat{s}_{iz} \right\}$$

It is useful to draw up the following table:

	$|{+}1^+\rangle$	$|0^+\rangle$	$|{-}1^+\rangle$	$|{+}1^-\rangle$	$|0^-\rangle$	$|{-}1^-\rangle$
$\hat{l}_+\hat{s}_-$	0	$\sqrt{2}|{+}1^-\rangle$	$\sqrt{2}|0^+\rangle$	0	0	0
$\hat{l}_-\hat{s}_+$	0	0	0	$\sqrt{2}|0^+\rangle$	$\sqrt{2}|{-}1^+\rangle$	0
$\hat{l}_z\hat{s}_z$	$\frac{1}{2}|{+}1^+\rangle$	0	$-\frac{1}{2}|{-}1^+\rangle$	$-\frac{1}{2}|{+}1^-\rangle$	0	$\frac{1}{2}|{-}1^-\rangle$
$\hat{l}\cdot\hat{s}$	$\frac{1}{2}|{+}1^+\rangle$	$\sqrt{\frac{1}{2}}|{+}1^-\rangle$	$\sqrt{\frac{1}{2}}|0^-\rangle - \frac{1}{2}|{-}1^+\rangle$	$\sqrt{\frac{1}{2}}|0^+\rangle - \frac{1}{2}|{+}1^-\rangle$	$\sqrt{\frac{1}{2}}|{-}1^+\rangle$	$\frac{1}{2}|{-}1^-\rangle$

From which, for example:

$$\zeta \sum_i \hat{l}_i \cdot \hat{s}_i |{+}1^+0^+\rangle = \zeta \frac{1}{2}|{+}1^+0^+\rangle + \zeta\sqrt{\frac{1}{2}}|{+}1^+ + 1^-\rangle$$

giving rise to an on-diagonal matrix element of $\frac{1}{2}\zeta$ and an off-diagonal element of $\sqrt{\frac{1}{2}}\zeta$ with $\langle{+}1^+ + 1^-|$. The SOC matrix for the microstates of p^2 having $M_J = +2$ and $+1$ is:

	$({+}1^+, {+}1^-)$	$({+}1^+, 0^+)$	$({+}1^-, 0^+)$	$({+}1^+, 0^-)$	$({+}1^+, {-}1^+)$
$({+}1^+, {+}1^-)$	0	$\zeta/\sqrt{2}$	0	0	0
$({+}1^+, 0^+)$	$\zeta/\sqrt{2}$	$\zeta/2$	0	0	0
$({+}1^-, 0^+)$	0	0	$-\zeta/2$	0	0
$({+}1^+, 0^-)$	0	0	0	$\zeta/2$	$\zeta/\sqrt{2}$
$({+}1^+, {-}1^+)$	0	0	0	$\zeta/\sqrt{2}$	0

The eigenvalues of this part of the matrix are $+\zeta$ (2×) and $-\frac{1}{2}\zeta$ (3×).

The SOC matrix for the microstates of p^2 having $M_J = 0$ is:

	$(+1^+, -1^-)$	$(+1^-, 0^-)$	$(+1^-, -1^+)$	$(0^+, 0^-)$	$(-1^+, 0^+)$
$(+1^+, -1^-)$	ζ	0	0	0	0
$(+1^-, 0^-)$	0	$-\zeta/2$	$\zeta/\sqrt{2}$	$\zeta/\sqrt{2}$	0
$(+1^-, -1^+)$	0	$\zeta/\sqrt{2}$	$-\zeta$	0	$-\zeta/\sqrt{2}$
$(0^+, 0^-)$	0	$\zeta/\sqrt{2}$	0	0	$-\zeta/\sqrt{2}$
$(-1^+, 0^+)$	0	0	$-\zeta/\sqrt{2}$	$-\zeta/\sqrt{2}$	$-\zeta/2$

The eigenvalues of this part of the matrix are $+\zeta$ ($2\times$), $-\frac{1}{2}\zeta$ ($2\times$) and -2ζ ($1\times$).

When the results for $M_J = -1$ and -2 are included the 15 eigenvalues of the spin-orbit matrix are found to be -2ζ ($1\times$), $-\frac{1}{2}\zeta$ ($8\times$) and $+\zeta$ ($6\times$) which corresponds to the extreme right of Figure 5.6 with $\zeta = 5$ arbitrary units.

Inspection of the matrices of $\zeta \hat{l} \cdot \hat{s}$ and e^2/r_{12} shows that they are blocked out, though not in exactly the same way for both operators. The inter-electronic repulsion connects, i.e. has an off-diagonal matrix element between, microstates which have the same value of M_L and the same value of M_S whereas the SOC links microstates with the same value of $M_J (= M_L + M_S)$. However, not all pairs of states which satisfy these requirements are connected by an operator.

In cases of intermediate coupling, where the SOC and inter-electronic repulsion are of comparable magnitudes, we must sum the two matrices and diagonalise the result.

A similar analysis of the energetics of the s^1p^1 configuration might be made, but this is not necessary for our present purposes and it suffices to say that the following terms are possible; ^1P and ^3P in the absence of SOC and $^1P_1$, $^3P_2$, $^3P_1$ and $^3P_0$ when it is present.

Atomic transition probabilities

First, some preliminary observations. Apart from the Bohr-Einstein condition, the selection rules of atomic spectroscopy are couched in terms of the quantum numbers which describe particular aspects of the angular momentum of the atom. We shall assume that the Bohr-Einstein condition is satisfied and concentrate on the conditions on the quantum numbers. Very general derivations of the selection rules which rely on advanced angular momentum theory can be found in the specialist texts on atomic spectroscopy.[1-3] Here we shall confine our attention to specific examples which illustrate the general results.

The intensity of an electronic transition from an initial orbital, ϕ_i, to a final orbital, ϕ_f, in a one-electron atom is proportional to the square modulus of the transition moment integral (Appendix 9 and Section 8.7), i.e. $I \propto |\langle \phi_f | e\mathbf{r} | \phi_i \rangle|^2$ where e is the electronic charge and $\mathbf{r}$ the vector distance between the electron and the nucleus. For a many-electron atom $I \propto |\langle \psi_f | \Sigma_k e\mathbf{r}_k | \psi_i \rangle|^2$ where the sum over k runs over all the electrons and ψ_f and ψ_i represent electronic states. The transition moment operator is a *one-electron* operator which means that for a non-zero TM integral the states represented by ψ_i and ψ_f can only differ by one orbital because, for example, of the n terms in the following sum:

$$\langle\psi_f|\Sigma_k er_k|\psi_i\rangle = \langle\phi_a(1)\phi_b(2)\phi_c(3)\ldots\phi_n(n)|\Sigma_k er_k|\phi_a(1)\phi_b(2)\phi_x(3)\ldots\phi_n(n)\rangle$$

only:

$$\langle\phi_c(3)|e\mathbf{r}_3|\phi_x(3)\rangle \cdot \langle\phi_a(1)|\phi_a(2)\rangle \cdot \langle\phi_b(2)|\phi_b(2)\rangle\ldots\langle\phi_n(n)|\phi_n(n)\rangle$$

may have a value, all the others being zero because of the orthogonality of the atomic orbitals ϕ_c and ϕ_x.

Two approaches are open to us when we seek to evaluate a TM integral such as $\langle\phi_c(3)|e\mathbf{r}_3|\phi_x(3)\rangle$, one emphasises angular momentum and the other geometrical factors. Atomic orbitals can be formulated in a manner such that they are eigenfunctions of the angular momentum operators $\hat{l}$ and $\hat{l}_z$. In order to exploit this property we write the operator $e\mathbf{r}$ in a form appropriate for left- or right-circularly polarised light (Section 8.6), using the common notation $\mathbf{m}_-$ and $\mathbf{m}_+$ respectively:

$$\text{lcp} \sim \mathbf{m}_- = (e/\sqrt{2})(x - iy) = (e/\sqrt{2})(r\sin\theta\cos\phi - ir\sin\theta\sin\phi)$$

$$= (er\sin\theta/\sqrt{2})\exp(-i\phi)$$

$$\text{rcp} \sim \mathbf{m}_+ = (e/\sqrt{2})(x + iy) = (e/\sqrt{2})(r\sin\theta\cos\phi + ir\sin\theta\sin\phi)$$

$$= (er\sin\theta/\sqrt{2})\exp(+i\phi)$$

Since the radial part of an atomic orbital is different for every orbital and every atom, only angular integration is appropriate at this point. Radial integration can only be carried out when the radial functions of the orbitals concerned are specified. We find that the only non-zero matrix elements between s and p orbitals in this formulation are:

$$\langle p_-|\mathbf{m}_-|s\rangle = er/\sqrt{3} \quad \text{and} \quad \langle p_+|\mathbf{m}_+|s\rangle = er/\sqrt{3}$$

This form of TM integral is useful when we wish to interpret spectra measured in magnetic fields, as in the Zeeman effect and magnetic circular dichroism, where the levels are split according to their m_l values (Section 5.11 and Figures 5.7 and 5.8).

If we express the p-orbitals as p_x, p_y and p_z and write $e\mathbf{r}$ as:

$$\mathbf{m}_x = er\sin\theta\cos\phi, \quad \mathbf{m}_y = er\sin\theta\sin\phi \quad \text{and} \quad \mathbf{m}_z = er\cos\theta,$$

we find that the only non-zero TM integrals are:

$$\langle p_x|\mathbf{m}_x|s\rangle = \langle p_y|\mathbf{m}_y|s\rangle = \langle p_z|\mathbf{m}_z|s\rangle = er/\sqrt{3}$$

This formulation is appropriate for spectra in which the atom is in a non-spherical environment, e.g. when subjected to an electric field.

We can now return to the question of transitions between states of the configurations s^1p^1 and p^2. Since the microstates $(+1^+, +1^-)$ and $(+1^+, 0^+)$ appear at the top and to the left of Table 5.4, they must form components of the states ^1D and ^3P of p^2 respectively. Also, the matrix of inter-electronic repulsion shows that in the absence of SOC there is no mixing between these states, nor do they interact with any others.

Thus when $\zeta = 0$, the following wave functions are true descriptions of components of the ^{1}D and ^{3}P states of p^2:

$$^1D(p^2) = (1/\sqrt{2})\{|+1^+(1) + 1^-(2)\rangle - |+1^-(1) + 1^+(2)\rangle\}$$

$$^3P(p^2) = (1/\sqrt{2})\{|+1^+(1)0^+(2)\rangle - |0^+(1) + 1^+(2)\rangle\}$$

Note that the above wave functions are obtained by the expansion of the Slater determinants (Appendix 6) represented by the microstates $(+1^+, +1^-)$ and $(+1^+, 0^+)$. Similar considerations show that for the configuration s^1p^1 we can write:

$$^1P(s^1p^1) = (1/2)\{|s^+(1) + 1^-(2)\rangle - |+1^-(1)s^+(2)\rangle$$

$$- |s^-(1) + 1^+(2)\rangle + |+1^+(1)s^-(2)\rangle\}$$

$$^3P(s^1p^1) = (1/2)\{|s^+(1) + 1^-(2)\rangle - |+1^-(1)s^+(2)\rangle$$

$$+ |s^-(1) + 1^+(2)\rangle - |+1^+(1)s^-(2)\rangle\}$$

We now calculate some transition moment integrals between the above states, using the notation $\mathbf{M} = \mathbf{m}_1 + \mathbf{m}_2 = e\mathbf{r}_1 + e\mathbf{r}_2$.

$$\langle ^1D(p^2)|\mathbf{M}|^1P(s^1p^1)\rangle = (\tfrac{1}{2}\sqrt{2})\{\langle\langle+1^+(1) + 1^-(2)| - \langle+1^-(1) + 1^+(2)||\mathbf{M}|$$

$$|s^+(1) + 1^-(2)\rangle - |+1^-(1)s^+(2)\rangle - |s^-(1) + 1^+(2)\rangle + |+1^+(1)s^-(2)\rangle\}\}$$

$$= (\tfrac{1}{2}\sqrt{2})\{\langle+1^+(1) + 1^-(2)|\mathbf{M}|s^+(1) + 1^-(2)\rangle$$

$$+ \langle+1^+(1) + 1^-(2)|\mathbf{M}|+1^+(1)s^-(2)\rangle + \langle+1^-(1) + 1^+(2)|\mathbf{M}| + 1^-(1)s^+(2)\rangle$$

$$+ \langle+1^-(1) + 1^+(2)|\mathbf{M}|s^-(1) + 1^+(2)\rangle\}$$

The above reduction from eight integrals to four is possible because those terms in which the m_s value of the electron is different on either side of the operator, e.g. $\langle+1^+(1) + 1^-(2)|\mathbf{M}|+1^-(1)s^+(2)\rangle$, are zero because of the integration over the spin co-ordinates. As an example of a non-zero contribution to the transition moment we can evaluate:

$$\langle+1^+(1) + 1^-(2)|\mathbf{M}|s^+(1)+1^-(2)\rangle = \langle+1^+(1) + 1^-(2)|\mathbf{m}_1 + \mathbf{m}_2|s^+(1) + 1^-(2)\rangle$$

$$= \langle+1^+(1) + 1^-(2)|\mathbf{m}_1|s^+(1) + 1^-(2)\rangle + \langle+1^+(1) + 1^-(2)|\mathbf{m}_2|s^+(1) + 1^-(2)\rangle$$

$$= \langle+1^+(1)|\mathbf{m}_1|s^+(1)\rangle\langle+1^-(2)|+1^-(2)\rangle + \langle+1^+(1)|s^+(1)\rangle\langle+1^-(2)|\mathbf{m}_2|+1^-(2)\rangle$$

$$= \langle+1^+(1)|\mathbf{m}_1|s^+(1)\rangle$$

because of the orthonormality of the atomic orbital wave functions. All four integrals give the same contribution so that $\langle ^1D(p^2)|\mathbf{M}|^1P(s^1p^1)\rangle = \sqrt{2}\langle+1^-|\mathbf{m}|s^-\rangle$. Note that this is a transition in which $\Delta S = 0$, $\Delta L = 0$ and $\Delta M_L = +1$; although we have a two-electron atom we find that the selection rules for one-electron atoms apply because electron repulsion does not mix states with different S, L or M_L values.

[The reader may wish to confirm, simply by changing a few signs in the above equations, that:

$$\langle ^1D(p^2)|\mathbf{M}|^3P(s^1p^1)\rangle = 0, \text{ as expected since } \Delta S = 1]$$

But, as the above SOC matrix shows, SOC mixes states which have different S, L or M_L values. We can illustrate the effect of this by considering the two states in the top left-hand corner of the electron-repulsion and SOC matrices which now interact with each other but with no other states. Adding the two matrices we have the matrix on the left:

	$(+1^+, +1^-)$	$(+1^+, 0^+)$		$(+1^+, +1^-)$	$(+1^+, 0^+)$
$(+1^+, +1^-)$	$F^0 + F^2/25$	$\zeta/\sqrt{2}$	$\Longrightarrow$	0	$\zeta/\sqrt{2}$
$(+1^+, 0^+)$	$\zeta/\sqrt{2}$	$F^0 - F^2/5 + \zeta/2$		$\zeta/\sqrt{2}$	$\zeta/2 - 4F^2/5$

Since only the difference between the two diagonal matrix elements is important here, we may subtract $F^0 + F^2/25$ from each diagonal element and obtain the right-hand matrix above.

If we now suppose that $\zeta/2 = 4F^2/5$, which corresponds closely to the position of lead in Figure 5.6, the diagonal elements are equal and there is an equal mixing of $(+1^+, +1^-)$ and $(+1^+, 0^+)$. We can now no longer speak of pure $^1D(p^2)$ and $^3P(p^2)$ states; the eigenstates of the atom are equal mixtures of the two. Therefore, transitions from $^1P(s^1p^1)$ to the states first described as $^1D(p^2)$ or $^3P(p^2)$ will be equally allowed because of their component of $^1D(p^2)$. We see that the apparent failure of the selection rules is better described as a failure in the correct designation of the states involved. The selection rules on one-electron matrix elements are always rigorously obeyed, in the absence of perturbations, and the problem lies with our knowledge of the exact wavefunctions of the states involved in the transition. It should also be pointed out that the large SOC will also cause the $^1P(s^1p^1)$ state to mix with the $^3P(s^1p^1)$ which may further complicate matters. Of course, it might be argued that in this case of a high degree of mixing the situation would be clear; but if we move across Figure 5.6 to the position of tin, for example, we find that $F^2 = 2.847$ eV, $\zeta = 0.260$ eV and $\chi = 5\zeta/F^2 = 0.46$. The energy matrix is then:

	$(+1^+, +1^-)$	$(+1^+, 0^+)$
$(+1^+, +1^-)$	0	0.260
$(+1^+, 0^+)$	0.260	-2.148

for which the mixed wave functions $^1D'(p^2)$ or $^3P'(p^2)$ are:

$$^1D'(p^2) = 0.993\,^1D(p^2) + 0.119\,^3P(p^2)$$

and

$$^3P'(p^2) = 0.119\,^1D(p^2) - 0.993\,^3P(p^2)$$

For the tin atom, therefore, it seems quite reasonable to describe the states as $^1D(p^2)$ and $^3P(p^2)$, though we have to recognise that the selection rules will not be strictly obeyed because of the mixing induced by SOC. For germanium, for which $F^2 = 3.152$ eV, $\zeta = 0.109$ eV and $\chi = 0.17$ the mixing is negligible and the limiting LS selection rules should hold.

It is worth noting that, although it is the SOC which causes the mixing of states of different multiplicity, the degree of mixing depends upon the energy-separation between the mixing states and therefore upon the inter-electronic repulsion. In germanium, though the mixing of states is very small, the SOC makes a measurable impact on the state energies through the diagonal matrix element with $(+1^+, 0^+)$. Where only diagonal elements of SOC are significant in the comparison of theory and experiment we speak of *first-order* SOC.

REFERENCES

1. E.U. Condon and G.H. Shortley, *The Theory of Atomic Spectra*, Cambridge University Press, 1935.
2. J.C. Slater, *Quantum Theory of Atomic Structure*, Vols. I and II, McGraw-Hill, New York, 1960.
3. R.D. Cowan, *The Theory of Atomic Structure and Spectra*, California University Press, Berkeley, 1981.

PROBLEMS FOR CHAPTER 5

1. Predict the frequencies and wavelengths of a further series in the electronic spectrum of the hydrogen atom.

2. Calculate the wavelength of the first 'Balmer' line in the spectra of the deuterium atom and the Be^{3+} cation.

3. Show that the number of states, N_{states}, in a configuration l^p, where l is the orbital angular momentum quantum number and p is the number of electrons, is:

$$N_{states} = 2l!/\{p! \times (2l - p)!\}$$

Hence, confirm the fact that there are 1001 states of $f^4 \cdot [n! = n(n-1)(n-2)\ldots 1]$

4. Draw up a table analogous to Table 5.4 for the configuration sd and determine the states of the system.

5. Draw up a table analogous to Table 5.6 for the configuration sd with jj coupling and determine the states of the system.

6. The normalised radial wave functions for the hydrogen atom are given in Appendix 5. Show, using the functions for $n = 3$, that although they are different all those having the same value of n give the same result for the integral:

$$-\int_0^\infty R_{3l}(r)\frac{Ze^2}{4\pi\varepsilon_0 r}R_{3l}(r)r^2\,dr = -\frac{Z^2e^2n^2}{324\pi\varepsilon_0 a_0}.$$

This result tells us that the potential energy of the electron is the same in all three types of d orbital, despite the differences in their radial functions. Because of the virial theorem, this means that the kinetic and total energies will also be the same. Determine the kinetic and total energy of a hydrogen 3d electron and compare your answer with Equation (5.1.1).

7. Use the information on s ↔ p transition probabilities in Box 5.2 to confirm the relative intensities of the spectral lines of barium quoted in Section 5.11.1 and illustrated in Figure 5.7.

CHAPTER 6
The Covalent Chemical Bond

6.0   Introduction   132
6.1   The binding energy of the hydrogen molecule   133
6.2   The Hamiltonian operator for the hydrogen molecule   134
6.3   The Born–Oppenheimer approximation   136
6.4   Heitler and London: The valence bond (VB) model   137
6.5   Hund and Mulliken: the molecular orbital (MO) model   139
6.6   Improving the wave functions   140
    6.6.1   The value of Z   140
    6.6.2   Polarisation   140
6.7   Unification: Ionic structures and configuration interaction   141
6.8   Electron correlation   143
6.9   Bonding and antibonding MOs   145
6.10 Why is there no He–He Bond?   146
6.11 Atomic orbital overlap   146
    6.11.1 σ (sigma) overlap   147
    6.11.2 π (pi) overlap   147
    6.11.3 δ (delta) overlap   148
    6.11.4   Non-bonding overlap   148
6.12 The homonuclear diatomic molecules from lithium to fluorine   149
6.13 Heteronuclear diatomic molecules   151
6.14 Charge distribution   153
6.15 Hybridisation and resonance   153
    6.15.1   Hybridisation: Pauling 1931   153
    6.15.2   Hybridisation and the valence bond theory   156
    6.15.3   Hybridisation of carbon AOs   156
    6.15.4   The choice of hybrid orbitals   161
    6.15.5   The properties of hybrid-orbital bonds   162
6.16 Resonance and the valence bond theory   163
6.17 Molecular geometry   163
    6.17.1   The valence-shell electron-pair repulsion (VSEPR) model   164
    6.17.2   The VSEPR model and multiple bonds   165
6.18 Computational developments   167
6.19 Bibliography and further reading   168
    Problems for Chapter 6   176

The Quantum in Chemistry   R. Grinter
© 2005 John Wiley & Sons, Ltd

6.0 INTRODUCTION

There can be no doubt that the most important contribution of quantum mechanics to the science of chemistry is the way in which it has enabled us to understand the nature of the chemical bond, especially the covalent bond. The long and complex story of the theory of valency and the chemical bond before the advent of quantum mechanics has been well told by Palmer[1] and by Russell[2] and will not be repeated here. We set the objectives of this chapter by drawing from the first few pages of a book by one of the pioneers in the application of quantum mechanics to chemistry, Charles Alfred Coulson (1910–1974). In his best-selling undergraduate text, 'Valence',[3] Coulson discussed the major experimental phenomena which any valid theory of chemical bonding must explain. He raised, in particular, the following questions:

- Why do molecules form at all? Why is it that two hydrogen atoms form the very stable diatomic molecule H_2 while two helium atoms do not form He_2?

- Why is there saturation of valency, i.e. if H_2 is stable, why not H_3 or H_4 and so on? Similarly, why do we have CH_4 but not CH_5 or CH_6?

- What are the reasons for the shapes of molecules? Why, for example, is the carbon dioxide molecule (CO_2) linear but sulfur dioxide (SO_2) bent, with an O–S–O bond angle of $119°$?

Further, the theory must link the interpretations of the above experimental facts together, viewing them as aspects of a comprehensive theory of chemical structure as a whole which (and here I add to Coulson's requirements) should embrace not only covalent compounds but also ionic solids and metals.

In 1919, Max Born (1882–1970) and Fritz Haber (1868–1934) independently published the energetic cycle, now known as the Born–Haber cycle, by means of which the energy of an ionic lattice can be expressed in terms of the properties of the elements forming the lattice and the coulombic forces between the ions involved. With that work the essential features of the ionic bond were quantitatively understood, though it must be said that the excellent agreement of theory and experiment is a little fortuitous due to the selfcompensating nature of the ionic model, i.e. the mutual cancellation of errors.[4]

But the energetic basis of the covalent bond remained a complete mystery and its explanation was one of the first objectives of the new quantum mechanics. It is amazing to note that, without the aid of a computer or even an electronic calculator, Heitler and London produced a calculation of the binding energy of the hydrogen molecule in 1927, just one year after the publication of Schrödinger's wave equation! Though the calculation was not quantitatively very accurate, ways in which it might be improved were obvious and it was clear that quantum mechanics was on the right track as far as chemical bonding was concerned. By 1933, a calculation of the binding energy of H_2 correct to better than 0.6 % was reported; a formidable achievement when we consider the complexity of the problem, which will soon become apparent. This work is described now, giving a roughly chronological account of the significant steps in the application of quantum mechanics to the chemical bond in the hydrogen molecule; this exemplifies almost all the essential aspects of the theory. What it lacks, the interpretation of molecular geometry, i.e. bond angles, is described in Section 6.17.

6.1 THE BINDING ENERGY OF THE HYDROGEN MOLECULE

Before embarking upon a discussion of the calculation of the bond energy of H–H, we should be clear just what is to be calculated and how the result is to be compared with experiment. We first recall (Section 5.1) that the zero of energy is, by definition, that state in which all particles, two electrons and two protons in this case, are infinitely separated and at rest so that they have neither kinetic nor potential energy. If we now allow two hydrogen atoms to form at rest, the energy of each atom lies 13.605 eV below the zero of energy and we say that the electronic energy of the hydrogen atom is −13.605 eV. The energy of two H atoms is $2 \times -13.605 = -27.210$ eV (-43.596×10^{-19} J) (Figure 6.1; note the two breaks in the scale).

If the two hydrogen atoms approach sufficiently close to each other to form a bond then there is a further decrease in energy and there will be a particular value of the H–H internuclear distance for which the energy is a minimum, i.e. is the most negative. This distance is known as the equilibrium internuclear distance, R_e, and is found experimentally to be 74.1 pm. The corresponding experimental energy is 4.747 eV (7.6056×10^{-19} J) below the energy of the two, non-interacting hydrogen atoms. This is the energy which we shall seek to calculate by determining the difference between the electronic energy of two independent hydrogen atoms and that of the hydrogen (H–H) molecule with a bond length of R_e. The symbol for this energy difference is D_e. Since we know the electronic energy of the hydrogen atom very accurately, from both experiment and theory, the problem is simply that of calculating the electronic energy of the molecule.

We emphasise the electronic energy because the molecule also has other forms of energy that are not of interest to us in the present context. It may, for example, be moving

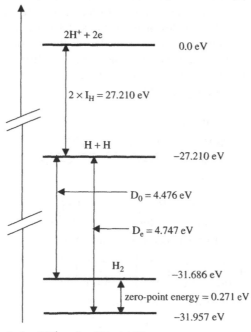

Figure 6.1 Energy levels for $2H^+ + 2e$, 2H and H_2

through space and rotating giving rise to kinetic energy contributions to its energy which we do not require in our current calculation. The molecule will also be vibrating and we cannot ignore this *zero-point vibrational energy* (see Chapter 10 for more details) since it impinges upon the comparison of theory with experiment. Experimentally, the energy required to dissociate the hydrogen molecule into two hydrogen atoms is found to be 4.476 eV. This is the energy D_0; the energy required to dissociate a real hydrogen molecule which has zero-point energy, as opposed to the theoretical value of D_e which is the dissociation energy of an hydrogen molecule which has no zero-point energy (Figure 6.1). Thus, the energy which we seek to calculate (D_e) is given by the equation:

$$D_e = D_0 + \text{zero-point energy} \qquad (6.1.1)$$

The zero-point energy of H–H is known from experiment to be 0.271 eV so that, experimentally:

$$D_e = 4.476 + 0.271 = 4.747 \text{ eV} \qquad (6.1.2)$$

In the following we shall compare theoretical and experimental values of D_e.

6.2 THE HAMILTONIAN OPERATOR FOR THE HYDROGEN MOLECULE

Following Schrödinger's method as in earlier chapters, our first step must be to write down an expression for the energy of the hydrogen molecule in classical terms. Having done that we then apply the rules for converting the classical expressions to the appropriate operators. A hydrogen molecule in which the nuclei have been labelled A and B and the electrons 1 and 2 is represented in Figure 6.2. The fact that we have so labelled the electrons and nuclei should not be taken to mean that we can actually distinguish between two electrons or two nuclei. Nothing could be further from the truth; it is fundamental to quantum mechanics that all particles of the same species are totally indistinguishable. The labels are used only to ensure that we consider all the possible terms in the Hamiltonian; the attraction between every nucleus and every electron, for example. With the aid of Figure 6.2 we can identify the following energy terms, their classical-mechanical expressions and the corresponding operators:

1. The kinetic energy of the nuclei:

$$\frac{(MV_A)^2}{2M} + \frac{(MV_B)^2}{2M} \Rightarrow \frac{-h^2(\nabla_A^2 + \nabla_B^2)}{8\pi^2 M} \qquad (6.2.1)$$

2. The kinetic energy of the electrons:

$$\frac{(mv_1)^2}{2m} + \frac{(mv_2)^2}{2m} \Rightarrow \frac{-h^2(\nabla_1^2 + \nabla_2^2)}{8\pi^2 m} \qquad (6.2.2)$$

3. The potential energy due to nucleus-electron attraction:

$$\frac{-e^2}{4\pi\varepsilon_0}\left[\frac{1}{r_{A1}} + \frac{1}{r_{B1}} + \frac{1}{r_{A2}} + \frac{1}{r_{B2}}\right] \Rightarrow \frac{-e^2}{4\pi\varepsilon_0}\left[\frac{1}{r_{A1}} + \frac{1}{r_{B1}} + \frac{1}{r_{A2}} + \frac{1}{r_{B2}}\right] \qquad (6.2.3)$$

4. The potential energy due to internuclear repulsion:

$$\frac{e^2}{4\pi\varepsilon_0} \cdot \frac{1}{R_{AB}} \Rightarrow \frac{e^2}{4\pi\varepsilon_0} \cdot \frac{1}{R_{AB}} \qquad (6.2.4)$$

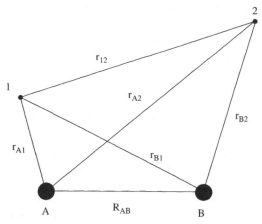

Figure 6.2   The Hamiltonian for the hydrogen molecule

5. The potential energy due to interelectronic repulsion:

$$\frac{e^2}{4\pi\varepsilon_0} \cdot \frac{1}{r_{12}} \Rightarrow \frac{e^2}{4\pi\varepsilon_0} \cdot \frac{1}{r_{12}} \tag{6.2.5}$$

In these expressions e is the *magnitude* of the charge on the protons and the electrons, m and M are the masses of the electron and proton respectively and ε_0 is the vacuum permittivity. The symbol ∇^2, which is spoken 'del squared', is defined by:

$$\nabla^2 \equiv \partial^2/\partial x^2 + \partial^2/\partial y^2 + \partial^2/\partial z^2 \tag{6.2.6}$$

Our total Hamiltonian operator is therefore the sum of ten terms and we immediately recognise that it is unlikely that a solution of Schrödinger's eigenvalue-eigenfunction equation can be found, even for this, the most simple of molecules. Therefore, if the theory is to be applicable, some way, or ways, of simplifying the problem must be found. Such methods have been found and the story of the theory of the chemical bond since 1926 is one of finding approximate methods that enable us to draw out the essence of the quantum-mechanical description, even when we are very far from finding an exact solution of Schrödinger's equation. The fact we are unable to solve Schrödinger's equation 'head-on' so to speak is by no means a disaster as two eminent theoretical physicists, Eugen P. Wigner and Frederick Seitz, recognised in 1955. Their opening lines in a review[5] of the theory of the cohesion (bonding) in metals read as follows:

'If one had a great calculating machine, one might apply it to the problem of solving the Schrödinger equation for each metal and obtain thereby the interesting physical quantities, such as the cohesive energy, the lattice constant and similar parameters. It is not clear, however, that a great deal would be gained by this. Presumably the results would agree with the experimentally determined quantities and nothing vastly new would be learned from the calculation. It would be preferable instead to have a vivid picture of the behaviour of the wave functions, a simple description of the essence of the factors which determine cohesion and an understanding of the origins of variation in properties from metal to metal.'

If we replace the words metal, cohesive energy and lattice constant by molecule, binding energy and bond length respectively, then this statement applies equally well to the quantum-mechanical study of molecules. We now have 'great calculating machines' which give us much assistance in the application of quantum mechanics, but we are still unable to solve Schrödinger's equation in most cases of chemical interest. Wigner and Seitz's criteria remain as valid today as they were in 1955, and we shall bear them very much in mind in this chapter. And so to our first simplifying approximation.

6.3   THE BORN–OPPENHEIMER APPROXIMATION

Max Born and J. Robert Oppenheimer (1904–1967) were among the first to consider the problem of applying Schrödinger's new equation to molecules. They recognised that the mass of the proton is 1836 times that of the electron and that the nuclei therefore move very much more slowly than the electrons. Because of this it should not be necessary to treat both types of particle in exactly the same manner as the terms in our Hamiltonian operator above require (compare Equation (6.2.1) with Equation (6.2.2) and Equation (6.2.4) with Equation (6.2.5)). In 1927 they showed with a rigorous mathematical analysis that only a very small error is introduced into the calculation of the electronic energy for any particular internuclear distance (bond length) if the nuclei are held stationary at the bond length in question. This removes the two nuclear kinetic energy terms (Equation (6.2.1)) from the Hamiltonian and the internuclear repulsion becomes a simple constant which can be added in at the end. The calculation can then be repeated for a number of bond lengths and a graph of bond length against energy plotted in order to find the minimum in the curve and the associated equilibrium bond length (R_e).

The type of graph we expect to obtain by this process is shown in Figure 6.3. On the right of the figure, where R is large, the curve of energy against R is horizontal. This is the region of two non-bonded hydrogen atoms where a change of the internuclear distance

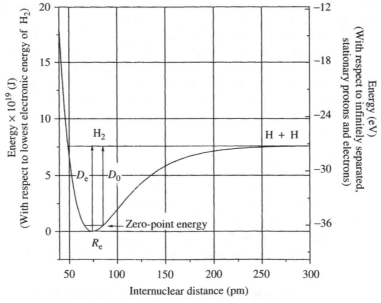

Figure 6.3   The hydrogen molecule: energy *versus* internuclear distance

makes no difference to the electronic energy, which is simply that of the two hydrogen atoms. The energy at this point is -27.210 eV, as explained in Section 6.1.

If a stable molecule is to be formed, there must be a decrease in energy as we move towards smaller values of R and this process will continue until the mutual repulsion of the nuclei and of the electrons begins to make itself felt and the energy begins to rise again; normally quite steeply. The energy difference between the minimum at R_e and the horizontal curve at large R is D_e and the whole curve can be constructed by calculating the electronic energy of two hydrogen atoms at a series of values of R. The fact that we are able to proceed in this way, i.e. the fact that the Born–Oppenheimer approximation is such an excellent one, has a significance which extends far beyond our present task and touches every aspect of chemistry. One of the most significant advances in chemistry in the 20th century was the way in which we now think of molecules as three-dimensional spatial entities. The concept of molecular shape and structure is now fundamental to our understanding of the way in which molecules interact and react chemically, and these properties are also recognised to be vital to advances in subjects such as materials science and molecular biology. But if the Born–Oppenheimer approximation could not be made and the nuclei and electrons in a molecule had to be treated even-handedly, then the motions of the two sorts of particles would be inextricably entwined. And there would then be no way in which we could draw the complex structural formulae upon which we so much rely and which are nothing more than representations of the equilibrium positions of the nuclei of the molecule. If the Born–Oppenheimer approximation was not possible, then there would also be no such thing as structural chemistry as we know it!

Vitally important though it is, the Born–Oppenheimer approximation still leaves us with a lot of work to do. The Hamiltonian still contains seven terms:

$$\hat{\mathcal{H}} = \frac{-h^2(\nabla_1^2 + \nabla_2^2)}{8\pi^2 m} - \frac{e^2}{4\pi\varepsilon_0}\left[\frac{1}{r_{A1}} + \frac{1}{r_{B1}} + \frac{1}{r_{A2}} + \frac{1}{r_{B2}} - \frac{1}{r_{12}} - \frac{1}{R_{AB}}\right] \quad (6.3.1)$$

and a direct solution of the Schrödinger equation is still impossible.

At this juncture we recall the discussion, in Section 4.9, of the problem of calculating an observable quantity for a system which is not in an eigenstate of the appropriate operator. According to Equation (3.9.2), a mean or expectation value of the energy which we require can be found for a normalised trial wave function (Ψ) using the equation:

$$\overline{E} = \iint \Psi^* \hat{\mathcal{H}} \Psi \, dv_1 \, dv_2 \quad (6.3.2)$$

Note that, since we now have two electrons to consider, we have to integrate over the three spatial coordinates of both; a total of six integrations. This is not a trivial calculation! But, although we have the Hamiltonian operator, we have no wave function. How can we proceed?

6.4 HEITLER AND LONDON: THE VALENCE BOND (VB) MODEL

In the first successful calculation of the energy of a covalent bond, Walter Heitler (1904–1981) and Fritz London (1900–1954) proposed that an approximate wave function could be obtained from our knowledge of the wave functions of the hydrogen atom in the following manner. We first suppose that our two hydrogen atoms are far apart with electron 1 in ϕ_A, the normalised 1s atomic orbital around nucleus A, and electron 2 in ϕ_B, the 1s orbital around nucleus B. The wave function, ψ, for this system of two separated hydrogen atoms will be:

$$\psi = \phi_A(1) \cdot \phi_B(2) \tag{6.4.1}$$

Note that the two 1s functions are combined as a product because each is a probability function and the probability of two events taking place is the *product* of the probabilities of each of the individual events taking place.

Now imagine that the two atoms move close enough together to form a bond. Then we might assume, to a first approximation, that the above wave function remains valid. This is what Heitler and London suggested; but with one important addition. When the two atoms form a bond the electrons occupy a common region of space and can no longer be thought of as confined to one particular nucleus. Furthermore, it is fundamental to the quantum theory that electrons are indistinguishable. Thus, the alternative wave function:

$$\psi = \phi_B(1) \cdot \phi_A(2) \tag{6.4.2}$$

is equally suitable, and for a correct description of the molecule both are required:

$$\Psi_{vb} = N\{\phi_A(1) \cdot \phi_B(2) + \phi_B(1) \cdot \phi_A(2)\} \tag{6.4.3}$$

We should also note that when the two hydrogen atoms are brought together so that the electrons occupy a common region of space, then the Pauli principle requires that the spins of the electrons must be paired and the above wave function represents the interaction of two hydrogen atoms with opposite electron spins. The inter-relationship between the spin and the spatial distribution of electrons raises many extremely important questions but, fortunately, we need only the spatial parts of the wave functions for our present

Table 6.1 A summary of some of the calculations on the hydrogen molecule

Type of wave function	$D_e{}^a$		$R_e{}^b$
	/eV	/J $\cdot 10^{-19}$	/pm
Simple valence bond (Z = 1)	3.14	5.03	86.9
Simple molecular orbital (Z = 1)	2.68	4.29	85.0
VB with Z = 1.166	3.78	6.06	74.3
MO with Z = 1.197	3.49	5.59	73.2
VB + polarisation (Z = 1.17, $\gamma = 0.123$)	4.02	6.44	74.0
χ (covalent) + $\lambda\chi$ (ionic) ($\lambda \approx 0.25$)	4.05	6.49	74.9
Best calculation without explicit inclusion of electron correlation	4.27	6.84	74.0^c
James and Coolidge; 13-term function	4.72	7.56	74.0
Kolos and Roothaan; 50-term function	4.7467	7.6047	74.1^d
Experimental values	4.7468 ±0.0007	7.6048 ±0.001	74.1

^{a}Dissociation energy per molecule
^{b}Equilibrium bond length
^{c}Calculated for this value of R only.
^{d}In 1968, Kolos and Wolniewicz reported a calculation using a 100-term function.
Generally, but not always because of the heavy calculations involved, each successive, improved calculation incorporates the improvements made in earlier work. An extensive review of the early calculations on the hydrogen molecule may be found in the paper by A.D. McLean, A. Weiss and M. Yoshimine, *Rev. Mod. Phys.*, **32**, 211 (1960).

purposes so, for the moment, we postpone the additional complication of spin. One aspect of the space-spin problem is explored in Box 6.4 and the subject is discussed more thoroughly in Section 11.5. The factor N has been introduced to normalise the wave function Equation (6.4.3) (Box 6.1). Inserting the wave function Ψ_{vb} into Equation (6.3.2) enabled Heitler and London to calculate the expectation value of the energy, $\overline{E}$, and by varying R, to find an energy minimum and a value for the equilibrium bond length of the hydrogen molecule. The calculated energy of 3.14 eV and bond length of 86.9 pm do not compare very well with the experimental values of 4.74 eV and 74.1 pm (Table 6.1). However, the result represented an enormous advance on the previous position and, moreover, it was immediately clear how the wave function could be refined, thereby improving the result. But before we consider the possible improvements we should introduce another form of wave function, a concept which was suggested by Friederich Hund (1896–) and Robert Sanderson Mulliken (1896–1986) among others.

6.5   HUND AND MULLIKEN: THE MOLECULAR ORBITAL (MO) MODEL

Hund and Mulliken approached the problem of finding a suitable wave function for H–H in a rather different way. The solutions of Schrödinger's equation for the hydrogen atom, the atomic orbitals, may be regarded as regions of space which can be occupied by one or two, but not more, electrons. Hund and Mulliken suggested that one might envisage similar regions of space surrounding not one but two (or more) nuclei. These they called molecular orbitals and for the hydrogen molecule they suggested the following form:

$$\psi_{mo} = N'\{\phi_A + \phi_B\} \qquad (6.5.1)$$

There are important points to be made about this wave function. The factor N' is again a normalising constant (Box 6.1). The atomic orbital functions are the normalised 1s functions of nuclei A and B as before; but note that there is, as yet, no mention of the electrons. It is for this reason that we have a sum of the two functions rather than a product. At this stage we are concerned merely to establish a form for the molecular orbital which we shall later populate with two spin-paired electrons. As with the Heitler–London function, the Hund–Mulliken function (Equation (6.5.1)) is an approximation. The justification for its form is solely that we know that the functions ϕ_A and ϕ_B are the exact atomic orbitals an electron occupies when it is in the region of a single hydrogen nucleus, remote from all other atoms. Hund and Mulliken's function simply assumes that this remains the case when two such atoms approach to within bonding distance. The calculation will show just how valid this assumption is. We must now introduce the electrons into our molecular orbital. Both can be accommodated in the one orbital if they have opposite spins and we write:

$$\Psi_{mo} = \psi_{mo}(1) \cdot \psi_{mo}(2) \qquad (6.5.2)$$

The calculation with the Hund and Mulliken function gives a binding energy of 2.68 eV and a bond length of 85.0 pm (Table 6.1). This result is markedly inferior to that of Heitler and London, but the molecular orbital method of obtaining a wave function has proved to be one of the most fruitful concepts in the application of quantum mechanics to chemical problems, as we shall see.

6.6 IMPROVING THE WAVE FUNCTIONS

The results of the early calculations were very encouraging, but it was clear that improvements were required and directions which these should take were rapidly recognised. The first improvements suggested for both were the same and they are important not only because they give better numerical results for the calculation of the H–H binding energy, but also for the physical insight which they bring to the problem. But before we embark upon a brief description of that work we need to address an important problem: what is the criterion by which we shall judge the improvement, or lack of it?

The object of the calculation is to determine the energy of the hydrogen molecule which lies markedly lower than the simple calculations predict, i.e. the calculated value of D_e is smaller than the experimental value. It seems natural, therefore, to suggest that the lower our calculated energy the better the wave function. But is there not a danger that we might choose a wave function that predicts an energy lower than the experimental value, and what would such a result imply? Fortunately, this dilemma will not trouble us. A theorem known as the variation theorem, which is proved and applied to another problem in Appendix 2, shows that no legitimate wave function for any quantum-mechanical problem can give an energy which is less than the true lowest energy eigenvalue of the system which it describes. Therefore, we can modify our MO and VB wave functions in any reasonable way, confident that the closer we get to the experimental energy the better is our calculation and the wave function used.

6.6.1 The value of Z

In the isolated hydrogen atom the electron moves under the electrostatic attraction of a single nucleus, but in the molecule it moves in the field of two nuclei. Though it may not feel twice the attractive force which it does in the atom, on account of the screening effect of the other electron and the fact that the two nuclei are not coincident, it is, nevertheless, to be expected that an electron will be drawn in more closely to the two nuclei than it is in the atom. How can this be allowed for in the wave function?

The algebraic expression for the normalised 1s atomic orbital (AO) of the hydrogen atom is (Appendix 5):

$$\phi = \frac{1}{\sqrt{\pi}} \left(\frac{Z}{a_0} \right)^{\frac{3}{2}} \cdot \exp \left(\frac{-Zr}{a_0} \right) \qquad (6.6.1)$$

Here, a_0 is the Bohr radius, r is the distance between the electron and the proton and Z represents the charge on the nucleus in units of the charge on the proton. In the isolated hydrogen atom $Z = 1$. It was suggested by S.C. Wang that the increase in the attraction to the nuclei felt by electrons in the hydrogen molecule as opposed to the hydrogen atom implied an increase in the value of Z, which would have the effect of drawing the AOs ϕ_A and ϕ_B a little closer around their respective nuclei. In 1928 detailed calculations showed that increasing Z did indeed produce improved values of D_e and R_e (Table 6.1) and the best values of Z were found to be 1.166 and 1.197 for the VB and MO wave functions respectively.

6.6.2 Polarisation

In 1931, N. Rosen noted that the presence of another nucleus at the side of each 1s AO would be expected to distort that AO from the pure spherical shape it would have in

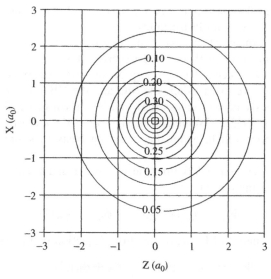

Figure 6.4   Polarisation of the hydrogen 1s AO by addition of $2p_z$

an isolated atom. He proposed that this effect could be allowed for by building a VB or MO wave function from a composite atomic wave function centred on each nucleus. The major contributor to the function would be the 1s AO as before, but there would be a small addition of a 2p AO pointing in the direction of the other nucleus. Neglecting normalising factors and choosing the z-direction as the internuclear axis, the polarised atomic function, ϕ', would be of the form:

$$\phi' = [\phi_{1s} + \gamma \phi_{2p_z}] \tag{6.6.2}$$

γ is simply a number which, in accordance with the variation theorem, is to be varied until the lowest energy (greatest binding energy) is found. Because orbitals have a phase, the positive combination of a 1s and a $2p_z$ orbital produces a function which has a larger electron probability in the positive z-direction than in the negative. Thus, the electron density is no longer spherically symmetrical with respect to the nucleus but is polarised in the positive z-direction which is just what we require. The distortion is illustrated in Figure 6.4 for $\gamma = 0.15$. The other H atom can be polarised in the negative z-direction by forming the negative combination of 1s and $2p_z$. Rosen carried out a calculation based on the VB wave function plus polarisation and found the improved values of D_0 and R_e shown in Table 6.1. The best value of γ was 0.123 (1.5 % of ϕ_{2p_z}) for Z = 1.17.

6.7   UNIFICATION: IONIC STRUCTURES AND CONFIGURATION INTERACTION

The improvements in the wave function suggested by Wang and Rosen have a very clear physical basis in the mutual electrostatic attraction of positive and negative charges. They show us that, even in the realms of quantum mechanics, simple electrostatic reasoning can still be applied with very beneficial results. In fact, as Hellmann and Feynman independently proved, we require Schrödinger's equation to determine the electron distribution in a quantum-mechanical system but once we have that distribution we can calculate the potential energy using classical electrostatics.

However, in spite of the great success of the theory thus far, we are still some way from the level of agreement between theory and experiment which we require to prove that quantum mechanics is *the* theory which interprets the chemical bond quantitatively. And more importantly, we have two models, the MO and the VB. The two different methods cannot be simultaneously correct unless they can be shown to be different formulations of the same theory. In 1933 S. Weinbaum brought the MO and VB approaches together in the following manner which, again, has a strong basis in simple physical-chemical reasoning.

We first write out the simple VB function of Heitler and London (Equation (6.4.3)), neglecting the normalising constants:

$$\Psi_{vb} = \phi_A(1) \cdot \phi_B(2) + \phi_B(1) \cdot \phi_A(2) \tag{6.7.1}$$

The simple, un-normalised MO wave function (Equation (6.14)) is:

$$\Psi_{mo} = [\phi_A + \phi_B](1) \cdot [[\phi_A + \phi_B](2)$$
$$\equiv [\phi_A(1) + \phi_B(1)] \cdot [\phi_A(2) + \phi_B(2)] \tag{6.7.2}$$

It can be multiplied out in exactly the same way as the product of two brackets in algebra giving four terms:

$$\Psi_{mo} = \phi_A(1) \cdot \phi_A(2) + \phi_B(1) \cdot \phi_B(2) + \phi_A(1) \cdot \phi_B(2) + \phi_B(1) \cdot \phi_A(2) \tag{6.7.3}$$

We note that the last two terms of this function are exactly the same as those of the VB function with the electrons in different AOs, while the first two terms have either both electrons in the AO ϕ_A or both in ϕ_B. The physical meaning of the various terms is clear. Those contributors to the wave function in which both AOs are occupied with one electron each distribute the two electrons over the molecule as one would expect in a covalent bond. The other two terms place both electrons in the region of one nucleus so that the molecule is negatively charged at one end and positively charged at the other. This is an *ionic* structure, H^+–H^-, and we see that the MO wave function consists of ionic and covalent structures in equal proportions (Equation (6.7.4)):

$$\Psi_{mo} = \chi(\text{covalent}) + \chi(\text{ionic}) \tag{6.7.4}$$

whereas the VB wave function has only covalent structures. From our knowledge of chemistry, we do not expect ionic bonding to play a very important role in the structure of H–H, though it may well do so in the case of a molecule such as H–F, where there is a large difference in the electronegativities of the two atoms involved. It therefore seems likely that the MO theory over-emphasises the importance of ionic contributions to the bonding and it might well improve the wave function if a way could be found of reducing the importance of these ionic terms. In the same vein, although we expect the contributions of the ionic terms to the bonding to be small, it would improve the VB function if we could add a small proportion of ionic structures to the purely covalent description which we currently have. Weinbaum did this by writing the VB wave function as a sum of covalent and ionic terms in the form:

$$\Psi_{vb} = \chi(\text{covalent}) + \lambda\chi(\text{ionic}) \tag{6.7.5}$$

and varying λ to find the maximum D_e. The value of λ found was in the region of 0.25, the exact figure being determined by the values of the other variable parameters involved, i.e. Z and γ. It should be noted that this is a very heavy calculation indeed, in that the effects of screening and polarisation have also to be considered since the values obtained from earlier calculations cannot be taken over unchanged into a different wave function. We saw this earlier when two different values of Z were found for the MO and VB wave functions (Section 6.6.1).

At this point it is natural to ask what change might be made to the MO wave function to improve it by reducing the contribution of the ionic terms. The answer lies in the out-of-phase combination of the hydrogen 1s atomic orbitals which, as we shall see in Section 6.9, represents a higher energy state of the hydrogen molecule. We place the two electrons in this higher-energy MO with the two spin-paired electrons exactly as we did in Equation (6.7.2) and write the simple, un-normalised MO wave function, Ψ^*_{mo}, as:

$$\Psi^*_{mo} = [\phi_A - \phi_B](1) \cdot [[\phi_A - \phi_B](2)$$
$$\equiv [\phi_A(1) - \phi_B(1)] \cdot [\phi_A(2) - \phi_B(2)] \tag{6.7.6}$$

It can be multiplied out in the same way as before giving:

$$\Psi^*_{mo} = \phi_A(1) \cdot \phi_A(2) + \phi_B(1) \cdot \phi_B(2) - \phi_A(1) \cdot \phi_B(2) - \phi_B(1) \cdot \phi_A(2) \tag{6.7.7}$$

We see immediately that the covalent terms now carry a negative sign so that in a combined MO wave function, Φ, of the form:

$$\Phi = \Psi_{mo} - \mu\Psi^*_{mo} \tag{6.7.8}$$

The value of μ can be used to decrease the contribution of the ionic terms just as λ was used in the VB wave function (Equation (6.7.5)) to increase it. The procedure whereby a wave function is improved by the addition of another arrangement of electrons in orbitals, i.e. another configuration, is known as *configuration interaction*. The VB wave function with the addition of ionic terms and the MO wave function with configuration interaction can be made identical with suitably chosen values of λ and μ. With Weinbaum's work unification was complete and the two approaches to the problem had converged, as they must do if they are describing the same truth. But the calculated value of D_e was still only 85 % of the experimental value and this was not good enough to prove that quantum mechanics provides a quantitative description of the covalent bond. Something more was required and it was provided by H.M. James and A.S. Coolidge who recognised the problem of electron correlation.

6.8   ELECTRON CORRELATION

In refining the wave functions used to describe the hydrogen molecule, Rosen realised that the presence of nucleus A would polarise (distort) the spherically symmetrical 1s electron distribution around nucleus B and vice versa (Figure 6.4). This is essentially a static distortion, but there is another, dynamic distortion process that is caused by the mutual repulsion of the two electrons, which therefore move in such a way as to reduce this repulsion. They avoid each other. This effect upon their motion is known as electron correlation. James and Coolidge showed that this effect must be explicitly allowed for, not just in the Hamiltonian operator as in Equation (6.3.1), but also in the wave function itself.

In a formidable calculation, without the aid of a computer and using a wave function that was the sum of 13 terms, they obtained $D_e = 4.72$ eV and $R_e = 74.0$ pm. This value of D_e is 99.4 % of the experimental value and this impressive result left little room for doubt concerning the applicability of quantum mechanics to the problem of the chemical bond.

In 1960 Kolos and Roothaan repeated the James and Coolidge calculation with a 50-term function and obtained a result that agrees with experiment to within experimental accuracy (Table 6.1). In fact, the result is better than it should have been because the Born–Oppenheimer approximation, though good, is not that good. The exceptional agreement of theory and experiment is attributed to a fortunate cancellation of the small errors due to the Born–Oppenheimer approximation in this particular problem. In very recent years calculations which do not invoke the Born–Oppenheimer approximation at all have been carried out. The results completely confirm the positive conclusions above and today we do not doubt that quantum mechanics provides a quantitative theoretical model of the chemical bond in all its forms.

But as Wigner and Seitz so clearly recognised, a wave function composed of 50 terms is not a very visual thing and we need a more pictorial description of the bond. In Figure 6.5 we attempt to provide this picture. The horizontal axis of the figure runs through the two hydrogen nuclei which are shown as two dots at ± 0.70 a_0. The vertical axis gives the number of electrons that can be found within a disc of infinite radius and 1 a_0 thick at right-angles to the internuclear axis. The continuous lines are the graphs for the AOs ϕ_A and ϕ_B centred on the two protons as they would be if there were no interaction between the two hydrogen atoms. The area under each graph is equivalent to one electron. The dotted line is the graph corresponding to the most simple MO wave function (Equation (6.1.14)) but multiplied by two so that the area beneath it is equivalent to two electrons and is therefore directly comparable with the sum of the areas under the two AO plots. We see that the electron density in the bonding MO is greater in the internuclear region than the sum of the two AO densities. This is a direct result of the fact that the electron density is given by the square of the wave function so that $(\phi_A + \phi_B)^2$ is greater than $(\phi_A)^2 + (\phi_B)^2$ because of the presence of the cross-term, $2\phi_A\phi_B$, in the former. The increased electron density in the internuclear region draws the nuclei together and is the primary cause of the

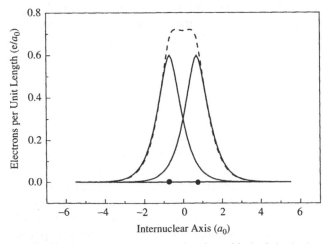

Figure 6.5 Electron distribution in the bonding molecular orbital of the hydrogen molecule

H–H bond. Here again we see that, once quantum mechanics has been used to determine electron distribution, we can understand the forces acting between the various particles forming a molecule using classical electrostatics. Recall at this point that the data we have plotted are drawn from the most simple MO wave function, which gives a rather poor binding energy. Allowing for the polarisation of the basic 1s AOs and a reduction in importance of the ionic terms, as described above, increases the difference between the MO electron density and the sum of the two AO densities in the internuclear region and outside it.

6.9 BONDING AND ANTIBONDING MOs

In the above interpretation of the origin of the covalent hydrogen bond we noted that, as a result of the cross-term $+2\phi_A\phi_B$, which arises for the MO wave function (Equation (6.5.1)), there was a build-up of electronic charge between the two protons. But if we had chosen to form a molecular orbital by taking the AO combination $\phi_A - \phi_B$, we would have found a negative cross-term, $-2\phi_A\phi_B$, and a consequent diminution of charge in the internuclear region. Let us call these two possibilities ψ_+ and ψ_- i.e.:

$$\psi_+ = N_+\{\phi_A + \phi_B\} \tag{6.5.1a}$$

$$\psi_- = N_-\{\phi_A - \phi_B\} \tag{6.9.1}$$

But why should we even consider the possibility of combining the two AOs with opposite signs, i.e. of taking the out-of-phase combination? There is more than one answer to this question, but one of the most important is that such an MO is known from experiment to exist. When the hydrogen molecule is irradiated with light of a wavelength in the region of 110 nm (9×10^8 cm^{-1}) it absorbs that light and is excited to a higher energy state that has been shown by spectroscopic studies to be a state in which the MO ψ_+ contains one electron and the MO ψ_- one electron.

A graph of the MO ψ_- of the same form as Figure 6.5 is shown in Figure 6.6. We now have a situation in which the electron density in the internuclear region is much reduced

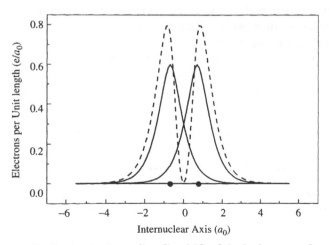

Figure 6.6 Electron distribution in the antibonding MO of the hydrogen molecule

(because of the negative cross-term) compared with the sum of the two AO densities. Indeed, since the AOs are combined with opposite phases, the electron density must be exactly zero mid-way between the nuclei. MOs of this type are called antibonding MOs since electrons occupying them cause the atoms to separate rather than to form a bond. This means that molecules with electrons in antibonding orbitals are unstable and reactive and they frequently decompose or enter into chemical reactions. The absorption of light is the most common cause of electrons entering antibonding MOs and the chemical reactions which then ensue are called photochemical reactions. The most important photochemical reaction is the process of photosynthesis which takes place in green plants. Another important, but less beneficial, one is the primary reaction in the sequence which generates photochemical smog:

$$NO_2(gas) + h\nu \longrightarrow NO(gas) + O(gas)$$

6.10   WHY IS THERE NO HE–HE BOND?

When we recognise that antibonding as well as bonding MOs exist, an explanation of the fact that there is no helium (He–He) molecule is very clear. Consider the energy level scheme (Figure 6.7(b)), which is qualitatively the same as that for H–H (Figure 6.7(a)), apart from the fact that we now have four electrons to be placed in the available MOs. Both bonding and antibonding MOs are now full and bonding and antibonding forces are quite evenly balanced. In fact, more sophisticated calculations show that the antibonding exceeds the bonding and no stable helium molecule can be formed.

6.11   ATOMIC ORBITAL OVERLAP

In the last section, the sign and magnitude of the cross-term generated when a MO wave function is squared was seen to be a very important criterion in the description of a covalent bond. A large positive term places a large electron density between the nuclei drawing them towards it to form a bond, while a large negative term removes electron density from between the nuclei which then have nothing to pull them together. Rather, the mutual repulsion of their positive charges, lacking the screening of an intermediate electron cloud, drives the nuclei apart.

The nature of the cross-term is such that it only has a value in those regions of space where both ϕ_A and ϕ_B each have a significant value, i.e. in the region where the two AOs *overlap*. For this reason we call $\phi_A\phi_B$ an orbital overlap and it plays a central role

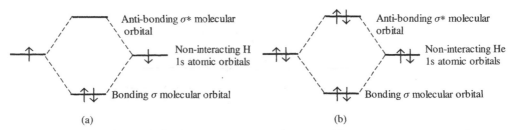

Figure 6.7   Molecular orbital energy-level schemes for H_2 and He_2

in qualitative discussions of bonding in both diatomic and polyatomic molecules. Apart from its sign, the overlap is characterised by its symmetry with respect to rotation about the internuclear axis in the following way.

6.11.1 σ (sigma) overlap

σ overlap is cylindrically symmetrical with respect to rotation about the internuclear axis (Figure 6.8(a)). That is, if we view the bond from one end along the internuclear axis then a rotation of the molecule causes no apparent change in what we see. Both the positive and the negative $\phi_A\phi_B$ overlap in the hydrogen molecule are of this type and we distinguish them by using a * to indicate the antibonding overlap. We also add information about the atomic orbitals involved in the overlap and speak of $\sigma(1s)$ and $\sigma(1s)^*$ overlap or $\sigma(1s)$ bonds and $\sigma(1s)^*$ antibonds.

Cylindrically symmetrical overlap can also arise when p-AOs overlap with s-AOs or with other p-AOs, as illustrated in Figure 6.8(a). Typical descriptions of this type of overlap are $\sigma(1s\text{-}2p)$ or $\sigma(2p)^*$. Generally, since it is strongly directed along the internuclear axis, σ overlap produces the strongest bonds and the strongest antibonds. That is, the decrease in energy when a σ bond is formed is larger than when other types of bonds are formed and the increase in energy when a σ^* antibond is formed is correspondingly greater.

6.11.2 π (pi) overlap

When two AOs overlap in such a way that rotation about the internuclear axis produces a change of sign every 180° then the overlap is termed π overlap (Figure 6.8(b)). A

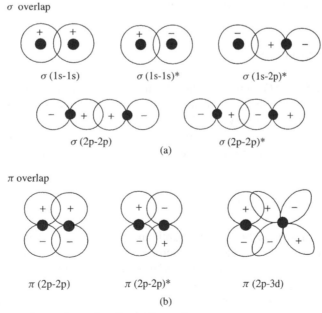

Figure 6.8 $\sigma-, \pi - \delta-$ and non-bonding AO overlap

δ overlap

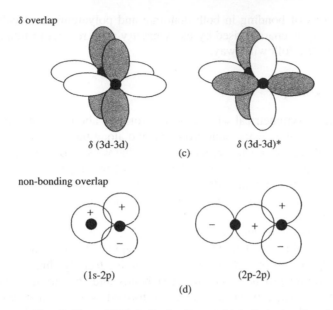

δ (3d-3d)   δ (3d-3d)*

(c)

non-bonding overlap

(1s-2p)   (2p-2p)

(d)

Note: In Figure 6.8(c) shading has been used to indicate phase in order to make the relative positions and orientations of the two $d_{x^2-y^2}$ orbitals clear.

Figure 6.8   (*continued*)

good example is the overlap of two 2p-AOs which are orientated at right-angles to the internuclear axis, as in the carbon-carbon π bond in ethene. The * to denote antibonding is also used and in ethene we have $\pi(2p)$ bonding and $\pi(2p)^*$ antibonding interactions. $\pi(2p\text{-}3d)$ is a rather special case of overlap between p and d AOs which is important in organometallic chemistry and in the phosonitrilic compounds.

6.11.3   δ (delta) overlap

If two nd_{xy} or $nd_{x^2-y^2}$ AOs with their nuclei on the z axis are brought together, face-to-face so to speak, then they overlap in such a way that rotation about the internuclear axis produces a change of sign every 90°. Such overlap is termed δ overlap (Figure 6.8(c)) and we can have bonding δ(3d) and antibonding δ(3d)* interactions. This form of overlap plays an important role in some metal-metal bonds, usually with $n > 3$.

6.11.4   Non-bonding overlap

Apart from the examples of bonding and antibonding overlap we have described above, it is also possible for AOs to overlap in such a way that there is no interaction between them. An example of this non-bonding overlap can be seen in Figure 6.8(d) where the overlap of a 2s AO of atom A with a 2p AO of atom B is illustrated. The 2p AO is orientated at right-angles to the internuclear axis and there are equal and opposite regions of overlap between the two AOs. The bonding and antibonding interactions between them

Table 6.2   Atomic 1s, 2s and 2p orbital overlap and bonding

The overlapping atomic orbitals	Type of bond[a]	Bonding or antibonding
$1s + 1s$	σ	Bonding
$1s - 1s$	σ^*	Antibonding
$2s + 2s$	σ	Bonding
$2s - 2s$	σ^*	Antibonding
$2s + 2p_z$	σ	Bonding
$2s - 2p_z$	σ^*	Antibonding
$2p_z + 2p_z$	σ	Bonding
$2p_z - 2p_z$	σ^*	Antibonding
$2p_x + 2p_x$	π	Bonding
$2p_x - 2p_x$	π^*	Antibonding
$2p_y + 2p_y$	π	Bonding
$2p_y - 2p_y$	π^*	Antibonding
$2s + 2p_x$		non-bonding[b]
$2s + 2p_y$		non-bonding[b]
$2p_x + 2p_y$		non-bonding[b]
$2p_x + 2p_z$		non-bonding[b]
$2p_y + 2p_z$		Non-bonding[b]

[a] The internuclear axis is the z-axis.
[b] Non-bonding overlap does not depend upon the sign of the orbital combination. In-phase (+) and out-of-phase (−) combinations are both non-bonding.

cancel exactly and the total interaction between the two AOs is zero. It is important to recognise that not all the AOs of an atom can necessarily take part in the formation of chemical bonds or antibonds, even when their spatial distributions overlap. The possible overlaps and bonding between 1s, 2s and 2p AOs are summarised in Table 6.2.

6.12   THE HOMONUCLEAR DIATOMIC MOLECULES FROM LITHIUM TO FLUORINE

It is much easier to construct simple, qualitative energy-level schemes for molecules with the MO model than with the VB. In constructing such schemes to describe the bonding in the above molecules we must take account not only of the relative magnitudes and signs of the overlap, but also of the relative energies of the 1s, 2s and 2p atomic orbitals of the elements in question. The 1s always lies well below the 2s in energy and the 2s and 2p have equal energies in the case of hydrogen and move further apart as we go towards fluorine. The situation is further complicated by the fact that, as we have seen above, 2s and 2p AOs can interact if the latter are orientated along the internuclear axis. The most important effect of these changes in the relative energies of the 2s and 2p AOs and the 2s-2p σ interaction is that the order of $E(\pi 1)$ and $E(\sigma 3)$ changes (Figure 6.9) and we require two energy-level diagrams, one for the molecules lithium to nitrogen and another for oxygen and fluorine. Coulson[3] and Murrell, Kettle and Tedder[6] have good discussions of these points.

But once we have constructed suitable energy level schemes, it is very easy to determine the electronic configurations of the molecules in question. We simply fill the MOs with the

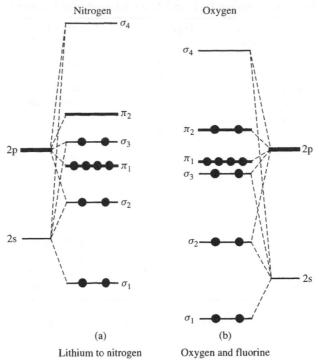

Figure 6.9 Molecular orbital energy-level schemes for homonuclear diatomic molecules of the first row of the periodic table

available electrons starting with the lowest (aufbau or building-up principle) and placing two, but not more than two, electrons (Pauli principle) in each MO. This is illustrated for nitrogen in Figure 6.9(a) and for oxygen in Figure 6.9(b). The concept of bonding and antibonding overlap allows us to draw conclusions about the strength of the bond joining the two atoms. In the case of nitrogen, for example, the 1s-1s σ bonding is very small because the two 1s AOs are very close to their respective nuclei and their overlap is almost zero. Thus, no bonding (or antibonding) results from these four *core* electrons. The $\sigma(2s)$ bonding of MO σ_1, is effectively cancelled by the $\sigma(2s)^*$ antibonding of MO σ_2, so no bond results from the four electrons which we place in these two MOs. The $2p_x$-$2p_x$ and $2p_y$-$2p_y$ bonding interactions produce two bonding π MOs of equal energy (π_1 degenerate) and they can accommodate four electrons. Each doubly-occupied bonding MO constitutes one bond so we have two bonds. Finally, the last two electrons can be placed in the bonding σ MO formed by the $2p_z$-$2p_z$ overlap, σ_3. They provide one more bond for the molecule with the result that we have a very strong N≡N triple bond, in agreement with the usual description of the molecule. σ_2 contains a little $2p_z$ and σ_3 a little 2s, but the s/p mixing is small because of the large E_{2s}/E_{2p} energy gap.

For the oxygen molecule we proceed in the same way until we come to the point where only two electrons remain to be placed in MOs. The MOs next in energy are a pair of degenerate π^* MOs (π_2) and, following Hund's rule, we must place one electron in each of these MOs and the spins of the two electrons must be the same. The result of this parallel orientation of two electron spins is that theory predicts that the oxygen molecule is paramagnetic (Section 12.2) and this is very unusual for a simple, stable molecule. But the oxygen molecule is indeed paramagnetic and the simple and clear way in which this result is obtained in the MO description is a very strong point in favour of the applicability that theory. A count of bonding electron pairs minus antibonding electron pairs (the two electrons in the two π^* MOs each contribute half a bond) shows that the oxygen molecule has an excess of two bonds over antibonds in agreement with the usual formulation of the molecule as O=O.

6.13   HETERONUCLEAR DIATOMIC MOLECULES

When we seek to apply the MO theory to heteronuclear diatomic molecules we have to consider a problem which did not arise in the homonuclear case. There, there was never any question of the proportions of AOs involved in an MO since, by symmetry, each atom had to contribute equally. This is not the case if the atoms are not the same as, for example, in the hydroxyl radical, OH, which is observed spectroscopically when hydrogen burns in oxygen. It is also formed by the photolysis of water at high altitudes and is important in the chemistry of troposphere, i.e. the atmosphere below 10 km. The 1s AO of the hydrogen atom is able to form a bonding overlap with both the 2s and the $2p_z$ AOs of the oxygen atom (z is the internuclear axis) and we may formulate MO wave functions such as:

$$\psi = C_1 \phi_{H1s} + C_2 \phi_{O2s} + C_3 \phi_{O2p_z} \qquad (6.13.1)$$

But we do not know the relative magnitudes of the three coefficients, C_1, C_2 and C_3, i.e. we do not know to what degree the two atoms contribute their AOs to the MO. The method by which the values of C_1, C_2 and C_3 could be found was known even before the advent of quantum mechanics since similar types of problem occur in classical mechanics. But apart from a few small molecules and the special case of the conjugated hydrocarbons (see Section 12.1), it was scarcely possible to solve the necessary equations before the advent of the digital computer in the 1950s.

Nowadays, numerous computer programs for solving the problem at different levels of approximation are available. The method employed is to use the variation method to find the best values of the C_is by setting up the secular equations in the manner described in detail in Appendix 2. In this particular case the measureable property, P, of Appendix 2 is the energy, E, and the associated operator is the Hamiltonian, $\hat{\mathcal{H}}$.

An energy-level diagram for O–H calculated with such a program for the experimentally observed O–H bond length of 97.0 pm is shown in Figure 6.10. The positive lobe of the oxygen $2p_z$ orbital is directed towards the hydrogen atom. On the right of the diagram the 1s AO of hydrogen is shown at an energy of -13.60 eV. On the left of the diagram the 2s and 2p AOs of the oxygen atom are plotted at -32.30 eV and -14.80 eV respectively. The oxygen 1s AO is not involved in the bonding and does not feature in the diagram. In the centre of the figure the calculated energy levels of O–H are shown at -33.39 eV, -15.85 eV, -14.80 eV (doubly degenerate) and $+6.62$ eV. The dots indicate the filling of the available MOs with electrons. On the far right of the figure the coefficients of the AOs, i.e. C_1, C_2 and C_3, for each σ MO are shown. They tell us how each AO contributes to each MO and the signs also show the bonding or antibonding nature of the MO. Immediately beneath each coefficient its square is entered because it is the square of the wave function which gives the electron density and the second row

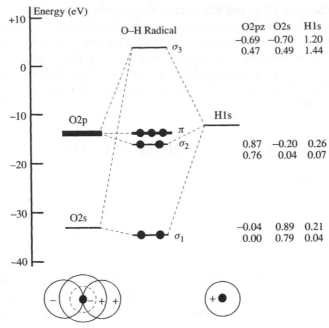

Figure 6.10   A molecular orbital energy-level scheme for the OH radical

of figures therefore reveals how the electrons are distributed in that particular MO; if it contains any.

Let us first consider the MO of lowest energy; σ_1, $E = -33.39$ eV. The major contributor with a coefficient of $+0.889$ is the oxygen 2s AO and hydrogen 1s also plays a part with a coefficient of $+0.212$. The contribution of oxygen $2p_z$ is too small to concern us. Note that the orbital is bonding, because the two AO contributors overlap in phase, and of σ type. We place two of our seven electrons in this MO.

The MO next in energy (σ_2, $E = -15.85$ eV) involves the same three AOs but the oxygen $2p_z$ now plays by far the most important role. Note that the oxygen 2s-hydrogen 1s overlap is antibonding while the oxygen $2p_z$-hydrogen 1s overlap is bonding. The last is by far the greater and dominates so that the MO is bonding in total. But it is important to observe that, where many AOs are involved, both bonding and antibonding interactions can be found within the same MO.

The next two MOs, the degenerate pair of π symmetry at $E = -14.80$ eV, are non-bonding MOs. The π-type oxygen $2p_x$ and oxygen $2p_y$ AOs have zero overlap with the σ-type hydrogen 1s AO and there is no interaction between them. Consequently these two oxygen 2p AOs are found in the molecule with an unchanged energy of -14.80 eV. Their AO coefficients are each 1.0. Since seven electrons are available to fill the MOs of O–H (the oxygen 1s electrons being omitted from the diagram), the last three electrons must be placed in these two orbitals, of which one will be only half-full. The unpaired electron makes the O–H radical paramagnetic and very reactive.

The highest MO (σ_3) is very high in energy ($E = +6.62$ eV) and strongly antibonding in all its overlaps. If electrons find their way into this MO, by absorbing light for example, the molecule becomes even more reactive.

In order to indicate, in a simple manner, the way in which the various AOs contribute to the MOs lines connecting AOs and MOs are frequently drawn, as in Figure 6.10. Occasionally lines of differing thickness or dotted lines are used to indicate greater or smaller AO contributions.

6.14 CHARGE DISTRIBUTION

A further piece of interesting information can be obtained from the results of an MO calculation. From the atomic orbital coefficients for each occupied MO and the calculated overlap integrals the distribution of the electrons throughout the molecule may be calculated. In the present example of the O–H radical it is found that the seven valence electrons are distributed such that 6.48 of them are on the oxygen atom and 0.52 on the hydrogen.[‡] Thus, the hydrogen carries a partial positive charge of approximately $+e/2$ and the oxygen a partial negative charge of $-e/2$, where e is the charge on the proton. This type of analysis of the distribution of the electrons within the molecule is usually termed Mulliken population analysis, after R.S. Mulliken who first introduced the idea.

6.15 HYBRIDISATION AND RESONANCE

The fact that a description of the electronic structure of a molecule is usually easier in terms of the MO rather than the VB model stems from the fact that in the former we leave all consideration of the electrons and their spins until the final step–the filling of the MOs with electrons. In the VB model, on the other hand, we form our trial wave function from atoms complete with their electrons, which have to be unpaired in their atomic orbitals so that they can pair with unpaired electrons on the other reacting atom(s). In Sections 6.15 and 6.16 we introduce the concepts of *hybridisation* and *resonance*, which arose from this requirement for complete atoms with unpaired electrons and form an important part of the VB model.

6.15.1 Hybridisation: Pauling 1931

Linus Carl Pauling (1901–1994) was the greatest chemist of the 20th century. During a long career he made momentous contributions to chemistry and was also active in world affairs. He was awarded the Nobel Prize for Chemistry in 1954 and the Nobel Peace Prize in 1963. The concept of hybridisation, which he put forward in 1931, is more important in the VB theory than in the MO, as we shall see in Section 6.15.2. But we can readily appreciate the need for such an idea, even within the MO theory. The construction of MOs for H–H is very simple since we must always have equal contributions from the two hydrogen atoms and there are only two hydrogen 1s AOs to be considered. When we seek to form MOs for molecules which have 2s and 2p valence AOs many new problems present themselves, even when just two identical atoms are to be combined, as we have seen in Section 6.12. If the atoms to be combined are different, and especially

---

[‡] This statement means that each electron spends 6.48/7 or $\approx$93 % of its time in the vicinity of the oxygen atom and 0.52/7 or $\approx$7 % of its time in the vicinty of the hydrogen atom.

if there are more than two, the problem of choosing suitable combinations of AOs can be very complicated indeed. Consider the three MOs of O–H. Two AOs of oxygen, the 2s and the $2p_z$, are always involved, albeit sometimes in a rather unequal measure. But they both play a very significant role in the MO of highest energy with very nearly equal contributions from each. This observation leads to a potential simplification in the construction of MOs. Rather than consider the oxygen 2s and oxygen $2p_z$ AOs separately, we first make combinations of them of the forms:

$$\phi_+ = \sqrt{\tfrac{1}{2}}(\phi_{2s} + \phi_{2p_z}) \qquad (6.15.1a)$$

and

$$\phi_- = \sqrt{\tfrac{1}{2}}(\phi_{2s} - \phi_{2p_z}) \qquad (6.15.1b)$$

These orbitals are called hybrid atomic orbitals and a contour diagram of ϕ_+ is shown in Figure 6.11. ϕ_- has exactly the same form but the larger lobe is directed along $-z$. Each hybrid orbital is much enhanced in one direction and is therefore particularly suited to bonding interaction with the AOs of a second atom lying in that direction. In this view, the lowest σ MO (Ψ_σ) will be the combination of ϕ_+ with hydrogen 1s, i.e.:

$$\Psi_\sigma = \sqrt{\tfrac{1}{2}}(\phi_+ + \phi_{1s}) = \tfrac{1}{2}(\phi_{2s} + \phi_{2p_z}) + \sqrt{\tfrac{1}{2}}\phi_{1s} \qquad (6.15.2a)$$

There will also be a strong antibonding interaction between ϕ_+ and H1s corresponding to the wave function $\Psi_\sigma{}^*$:

$$\Psi_\sigma{}^* = \sqrt{\tfrac{1}{2}}(\phi_+ - \phi_{1s}) = \tfrac{1}{2}(\phi_{2s} + \phi_{2p_z}) - \sqrt{\tfrac{1}{2}}\phi_{1s} \qquad (6.15.2b)$$

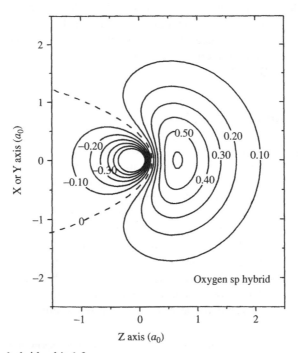

Figure 6.11   An sp hybrid orbital for oxygen

The other hybrid (ϕ_-) points away from the hydrogen atom and will therefore play little part in the bonding to that atom, i.e. it will be a non-bonding orbital. Thus, we have simplified our thoughts about the bonding in O–H by reducing the problem to one of the interaction of a single hybrid orbital (ϕ_+) rather than two atomic orbitals, ϕ_{2s} and ϕ_{2p_x}. An energy level scheme for O–H constructed with hybrid oxygen AOs is shown in Figure 6.12. This scheme is purely qualitative and no attempt has been made to obtain numerical values for the energy levels, for reasons which will be made clear in Section 6.15.5.

The correspondence between the calculated MO scheme in Figure 6.10 and the one constructed with hybrid AOs is not particularly good in this case, since the assumption that the 2s and $2p_z$ AOs of the oxygen atom contribute equally to the two hybrids is a very poor one for the two lower MOs. Close inspection of the AO coefficients in Figure 6.10, or their squares, shows that the oxygen atom contribution to the bonding MO (σ_1) is entirely oxygen 2s while the contribution to the non-bonding MO (σ_2) is almost exclusively oxygen 2p. This illustrates a weakness of the hybrid orbital approach to chemical bonding, but it has the great advantage that we can construct an energy-level scheme, quite literally, '*on the back of an envelope*'. Furthermore, the use of hybrid AOs leads to very specific predictions concerning molecular geometry, as we shall see in more detail in Section 6.15.3. For the moment we note that the formation of a hybrid AO usually leads to an AO that points strongly in a particular direction in space and is therefore very well suited to overlap with the AOs of a second atom which lies in that direction, while the overlap with the AOs of an atom in any other position is much diminished.

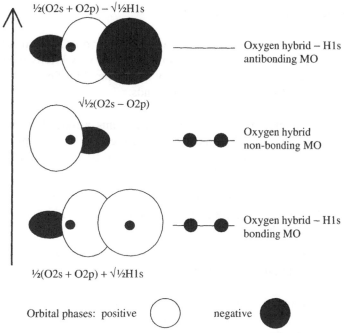

Figure 6.12 Molecular orbital energy-level scheme for the σ MOs of O–H using sp hybrid oxygen orbitals

6.15.2   Hybridisation and the valence bond theory

Recall the essential difference between the ways in which the simplest trial wave function for the hydrogen molecule was constructed using the MO and the VB theories. In the MO theory, the form of a molecular orbital was first determined (as the sum of two hydrogen 1s AOs) and the two electrons were then placed in that MO with their spins paired. In the VB theory, however, the two hydrogen atoms, complete with their electrons, were brought together in such a way that their spins paired. The hydrogen molecule is exceptional in that a fair approximation to the molecular wave function can be obtained, by either the MO or the VB method, using only two AOs. In almost all other molecules more than two AOs are involved in the bonding of two or more atoms. It is quite easy to extend the MO theory to embrace a large number of AOs, but this is not true of the VB theory where we need to find a single electron in an AO on one combining atom which can be paired with another unpaired electron occupying an AO on another combining atom. (We require a singly occupied AO because we must pair electrons to form bonds.) Furthermore, we normally require a hybrid orbital on at least one of the combining atoms because we need to direct the orbital very specifically towards the other atom if we are to form a strong bond. But in atoms the electrons, following the aufbau (building-up) and Pauli principles, occupy the orbitals in pairs with opposed spins. Thus, the first step in a VB description of the chemical bonding in a polyatomic molecule is the formulation of a set of singly occupied hybrid AOs for the atoms involved. The classic and most important example of this is the description of the bonding between carbon and hydrogen and between carbon and carbon in the hydrocarbons.

6.15.3   Hybridisation of carbon AOs

In Section 6.15.1 we noted briefly that the formation of hybrid AOs has many implications for molecular geometry and it is useful, therefore, to review very briefly the major features of the geometry of the hydrocarbons, which any theory should reflect, before we proceed to discuss the AO hybridisation in these compounds.

Type 1:   The carbon atom is bonded to four other atoms arranged at the corners of a tetrahedron with the carbon atom at the centre. The bond angles at the carbon atom are very close to the tetrahedral angle of $109.5°$. Examples are methane and the higher saturated hydrocarbons.

Type 2:   The carbon atom is bonded to three other atoms and all four atoms lie in a plane with bond angles at the carbon atom of approximately $120°$. Examples are ethene and the aromatic hydrocarbons.

Type 3:   The carbon atom is bonded to two other atoms and the bond angle at the carbon atom is $180°$. Ethyne is the best example of this type of bonding.

The carbon atom in its state of lowest electronic energy has the configuration $(1s)^2(2s)^2(2p)^2$ where, in accordance with Hund's rule, the two 2p electrons occupy different 2p AOs with parallel spins. The carbon atom has only two unpaired electrons and we might therefore expect that it would form just two bonds. Furthermore, since any two p AOs are orientated at right-angles to each other we might expect that these two bonds would enclose an angle at the carbon atom of approximately 90°. These predictions are totally at variance with the major features of hydrocarbon structure outlined above. If we are to form four bonds we must, in a VB description, first generate four unpaired electrons. There is one obvious way of doing this and that is to promote one of the 2s electrons to the vacant 2p AO. This promotion requires energy. But if it leads to more and stronger bonds then the energy of the molecule finally formed will be that much lower, i.e. the molecule will be more stable, and the investment of energy will have been worth while. The experimental observation of compounds of Type 1, and theoretical calculations prove that the overall energetics of the process are favourable.

But the promotion of a 2s electron to a 2p AO is not of itself sufficient to explain the tetrahedral geometry of Type 1 compounds. In fact, in the case of methane for example, the carbon atom configuration $(1s)^2(2s)^1(2p_x)^1(2p_y)^1(2p_z)^1$ might be expected to produce three identical C–H bonds directed along the x, y and z axes and one different bond with no particular directional characteristics. But, experiment shows that methane has four identical C–H bonds with H–C–H angles of 109.5°. Clearly, something more is required.

Since the geometry of methane is tetrahedral, it is clear that the strongest C–H bonds are formed from hybrids with enhanced electron densities along four tetrahedral directions. Figure 6.13 shows a cube in which the three cartesian axes pass through the centre point of the faces and we place a carbon atom at the centre of the cube, which is also the origin of the coordinate system. If we now place four hydrogen atoms at alternate corners of the cube we have a model of the methane molecule. The hydrogen atom 1 (H_1) at $+x$, $+y$, $+z$, for example, is identically placed with respect to the three carbon 2p orbitals and a combination of all three 2p AOs in equal proportions $(2p_x + 2p_y + 2p_z)$ must, by symmetry, produce a hybrid orbital directed along the line C–H_1. A hybrid AO directed towards H2 can be formed by the combination $-2p_x - 2p_y + 2p_z$. And similarly for H3 and H4. Since the 2s AO is spherically symmetrical it has no influence upon the directional properties of the hybrid AOs, but we must use it since we need four hybrid AOs in total and these can only be constructed from an equal number of unhybridised AOs. We therefore have to fulfil the following requirements for our hybrid AOs:

- The four hybrid AOs must 'consume' all the four 2s and 2p AOs of the carbon atom; no more and no less.

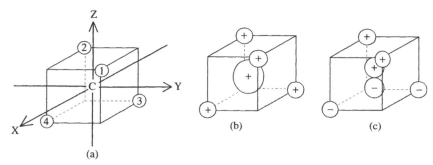

Figure 6.13 The methane molecule with the four hydrogen atoms (1 − 4) arranged at the corners of a cube centred at the carbon atom.

- The resulting hybrid AOs must point to the corners of a tetrahedron, i.e. they must contain the 2p AOs with the appropriate signs as indicated for H1 and H2 above.

- Apart from their direction in space, all the hybrid AOs must be exactly the same, i.e. they must each contain the same proportions of 2s and 2p AOs.

- Each hybrid AO must be normalised to one.

The above requirements are satisfied by the hybrids:

$$\psi_1 = \tfrac{1}{2}\{\phi_{2s} + \phi_{2p_x} + \phi_{2p_y} + \phi_{2p_z}\} \qquad (6.15.3a)$$

$$\psi_2 = \tfrac{1}{2}\{\phi_{2s} - \phi_{2p_x} - \phi_{2p_y} + \phi_{2p_z}\} \qquad (6.15.3b)$$

$$\psi_3 = \tfrac{1}{2}\{\phi_{2s} - \phi_{2p_x} + \phi_{2p_y} - \phi_{2p_z}\} \qquad (6.15.3c)$$

$$\psi_4 = \tfrac{1}{2}\{\phi_{2s} + \phi_{2p_x} - \phi_{2py} - \phi_{2p_z}\} \qquad (6.15.3d)$$

Hybrid AOs of this form are usually known as sp^3 hybrids since they are composed of one quarter s and three quarters p AOs. The proof that they are orthogonal and normalised is given in Box 6.2 and a contour plot is shown in Figure 6.14(a). The hybrid AOs are clearly highly directional and in a VB calculation of the methane molecule one electron would be placed in each hybrid AO and combined with the singly occupied 1s AO of the hydrogen atom to which it is directed. The interaction with hydrogens at the other corners of the tetrahedron would be neglected as a first approximation.

The bonding in hydrocarbons of Type 2 geometry can be interpreted in terms of hybridisation in a similar way. Consider the ethene molecule (Figure 6.15). All six atoms lie in the xy-plane so the carbon $2p_z$ AO plays a role different from that of the $2p_x$ and $2p_y$.

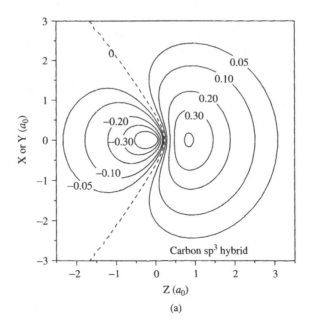

(a)

Figure 6.14

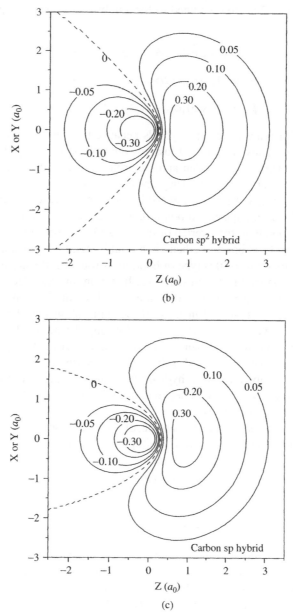

(b)

(c)

Figure 6.14   Carbon sp^3, sp^2 and sp hybrid orbitals

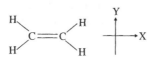

Figure 6.15   Ethene

The carbon 2s, $2p_x$ and $2p_y$ are combined to form hybrid AOs according to the following equations:

$$\psi_1 = (1/\sqrt{6})\{\sqrt{2}\phi_{2s} - \phi_{2p_x} + \sqrt{3}\phi_{2p_y}\} \tag{6.15.4a}$$

$$\psi_2 = (1/\sqrt{6})\{\sqrt{2}\phi_{2s} - \phi_{2p_x} - \sqrt{3}\phi_{2p_y}\} \tag{6.15.4b}$$

$$\psi_3 = (1/\sqrt{3})\{\phi_{2s} + \sqrt{2}\phi_{2p_x}\} \tag{6.15.4c}$$

Hybrids of this form are called sp^2 hybrids because they are composed of one third s and two thirds p AOs. They are directed towards the corners of an equilateral triangle and are therefore ideally suited for overlap and electron pairing with atoms lying in a plane and subtending bond angles of 120°. They are also orthogonal and normalised. The $2p_z$ AO of the carbon atom is not involved in the hybrids and therefore remains available for bonding with $2p_z$ AOs on adjacent carbon atoms. Whereas the overlap of the sp^2 hybrid AOs gives rise to σ bonding, the overlap of the $2p_z$ AOs on adjacent carbon atoms results in π bonding and the C–C bond in ethene is a double bond with σ and π contributions. The most important example of this type of bonding is found in the aromatic hydrocarbons of which benzene is the simplest representative. In π bonding the electrons are not tightly confined in the internuclear region as they are in σ bonding. The nature of π bonding is such that the electrons involved in it lie above and below the plane of the molecule and are able to move over all the carbon atoms forming the molecule. These *delocalised* electrons are responsible for the many properties of the aromatic hydrocarbons which distinguish them from the sp^3-bonded, saturated hydrocarbons. For example, almost all organic dyes are based upon aromatic hydrocarbons and the fact that they absorb light in the visible region of the spectrum, rather than at higher energy in the ultra-violet, is due to the presence of π-electron MOs.

Ethyne (Figure 6.16) is a typical example of hydrocarbons of Type 3 in which the geometry at the carbon atom is linear. The simple sp hybrid AOs:

$$\psi_1 = \sqrt{\tfrac{1}{2}}\{\phi_{2s} - \phi_{2p_x}\} \tag{6.15.5a}$$

and

$$\psi_2 = \sqrt{\tfrac{1}{2}}\{\phi_{2s} + \phi_{2p_x}\} \tag{6.15.5b}$$

produce bonds to the hydrogen and carbon atoms with an H–C–C bond angle of 180°. The remaining, unhybridised $2p_y$ and $2p_z$ AOs on the carbon atoms form two π bonds giving a total of three bonds in accord with our usual formulation of the C≡C triple bond in ethyne. Since they are formed from 2p AOs at 90°, the two π bonds are also at right angles to each other. As in the case of Type 2 compounds, this π bonding provides a way in which the electrons involved can be delocalised over the whole carbon-atom system if the triple bonded carbons are linked to others as in the polyethynes. Some values of C–H and C–C overlap integrals calculated with Slater-type AOs (Box 6.3) and hybrid AOs are given in Tables 6.3 and 6.4.

H—C≡≡≡C—H

Figure 6.16   Ethyne

Table 6.3 Carbon–hydrogen atomic orbital overlap for Slater-type carbon 2p and 2s and AOs[a] with $Z' = 3.25$ and a hydrogen 1s AO with $Z' = 1.2$

$r(C–H) = 109.0$ pm	$r(C–H) = 109.0$ pm
$\langle H1s|C2s\rangle = 0.5851$	$\langle H1s|Csp^3\rangle = 0.7696$
$\langle H1s|C2p_z\rangle = 0.5509$	$\langle H1s|Csp^2\rangle = 0.7876$
	$\langle H1s|Csp\rangle = 0.8033$

[a]Box 6.3 Note how the overlap of the hydrogen 1s with the carbon hybrid AOs exceeds the overlap with unhybridised carbon AOs.

Table 6.4 Carbon–carbon atomic orbital overlap for Slater-type AOs[a] with $Z' = 3.25$

Bond	C–C	C=C	C≡C
r/pm	154.3	135.3	120.8[b]
$\langle 2s|2s\rangle$	0.3391	0.4301	0.5070
$\langle 2p|2p\rangle^\sigma$	0.3288	0.3266	0.2931[c]
$\langle 2s|2p\rangle$	0.3640	0.4285	0.4704
$\langle 2p|2p\rangle^\pi$	0.1909	0.2642	0.3334[d]
$\langle sp^3|sp^3\rangle$	0.6466	0.7236	0.7539
$\langle sp^2|sp^2\rangle$	0.6754	0.7651	0.8078
$\langle sp|sp\rangle$	0.6979	0.8069	0.8704

[a]Box 6.3.
[b]The bond lengths correspond to typical single, double and triple carbon–carbon bonds.
[c]The $\langle 2p|2p\rangle^\sigma$ overlap decreases as the bond length decreases due to the overlap of positive and negative portions of the 2p AOs.
[d]The $\langle 2p|2p\rangle^\pi$ overlap increases as the bond length decreases.

6.15.4 The choice of hybrid orbitals

The reader may regard the preceding discussion of hybridisation in hydrocarbons as a rationalisation of the experimentally observed geometries, rather than a description of a procedure by means of which the experimental energies and geometries may be calculated. However, to take methane as our example, the choice of tetrahedral hybrids is simply a rational first step, given that a set of singly occupied orbitals on the carbon atom is an important simplification in the application of the VB theory to methane. If we choose to use the 2s and three 2p AOs of carbon as our starting point, we would obtain exactly the same result in the end, and it would be possible to express the molecular wave function obtained in exactly the same form as that determined from an initial assumption of sp^3 hybrids, but the calculation would be considerably more complicated. The final results, ground state energy and wave function, are not determined by the starting point.

An interesting and detailed analysis of this problem, from a view-point which appears to have been taken by few other authors, has been given by J. W. Linnett.[7,8] Linnett explored the consequences of the Pauli exlusion principle on the valence electron distribution in atoms. An outline of his approach is given in Box 6.4. The fact that electrons of the same spin cannot occupy the same region of space because of the Pauli principle, together with their mutual electrostatic repulsion, means that electrons of different spins tend to keep together while those of the same spin keep apart. Thus, for neon, the electron distribution of highest probability, as far as spin correlation effects are concerned, is an outer shell consisting of four electron pairs at the corners of a regular tetrahedron.

This would also be true of the C^{4-} ion, if it existed. Thus, when C^{4-} binds to four protons the strongest bonds will be formed if the protons are arranged at the corners of a regular tetrahedron. The formation of electron-pair bonds reinforces the pairing tendency of electrons having opposed spins. In this view, the choice of four tetrahedral sp^3 hybrids for carbon would be a natural choice, even if we did not know the geometry of methane. These ideas do not appear to have been widely explored and this may be because a theory of molecular geometry, apparently based purely upon electron repulsion, which leads to essentially the same conclusions in a manner which is easier to apply, has found wider acceptance. This approach to molecular geometry is described in detail in Section 6.17, but it should be noted here that the concepts outlined immediately above provide the theoretical foundation of the simpler method (Box 6.4).

6.15.5   The properties of hybrid-orbital bonds

The fact that the obvious choice of hybrid orbitals reflects so precisely the salient features of the molecular geometry of hydrocarbons and emphasises the two-electron nature of each identical C–H bond are the strong points of the VB theory and go a long way to explain why concepts derived from it, resonance (see Section 6.16) for example, are so prevalent in organic chemistry. Indeed, these points are so appealing to the chemist seeking a quantum-mechanically based understanding of molecular structure that the hybrid orbitals themselves can easily be elevated to a significance above their true status. This problem is especially clear when we consider energies; and we again take methane as our example.

Because all four C–H bonds are known experimentally to be identical, and this fact is echoed in the VB description of the molecule using four sp^3 hybrids, it is tempting to conclude that there is a four-fold degeneracy in the electronic energy levels of methane. But if we measure the energies of the methane valence electrons using photo-electron spectroscopy, we find a set of three degenerate energy levels at about -13 eV and a single energy level at about -23 eV. Calculations show that the triply degenerate levels arise from the bonding of the four hydrogen 1s AOs to the carbon 2p AOs and the single level from the bonding between the hydrogen 1s orbitals and the carbon 2s AO. That this should be so is immediately confirmed by a study of Figure 6.13, which shows that, because of the high cubic symmetry, each of the three carbon 2p AOs has exactly the same bonding interaction with the appropriate phase combination of the four hydrogen 1s AOs while that of the carbon 2s AO is clearly different. (Figures 6.13(b) and 6.13(c) show the orbital phases for bonding overlap between the four hydrogen 1s AOs and the carbon 2s and carbon $2p_z$ AOs respectively. For clarity the AOs are not shown overlapping.) The price that we pay for the simple description of bonds and their orientation provided by hybridisation is the loss of information on energy which use of the unhybridised carbon 2s and 2p AOs makes obvious. Hybrid orbitals, unlike atomic and molecular orbitals, are not eigenfunctions of an energy operator and it is important to grasp this fact. But quantum-mechanical wave functions do not have to be eigenfunctions of some operator in order to be of value to us.

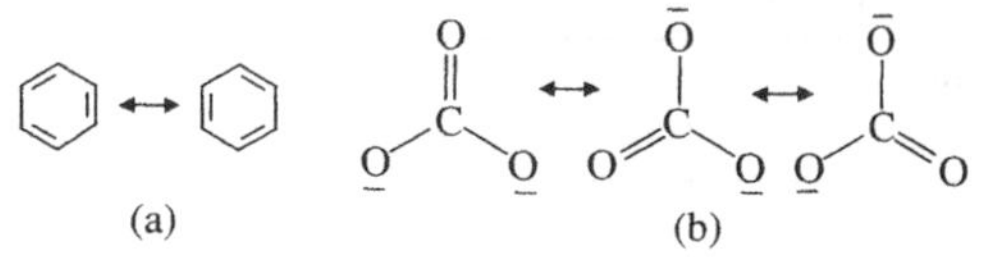

Figure 6.17 Resonance structures of benzene (a) and the carbonate anion (b)

6.16 RESONANCE AND THE VALENCE BOND THEORY

The VB theory is also the origin of and provides the theoretical justification for the concept of resonance which is very widely used in chemistry; in spite of being declared incompatible with dialectical materialism in Stalinist Russia. Consider the familiar structural formulae for benzene (Figure 6.17(a)). The three lines that have different positions in the two structures represent the different possible pairings of the electrons in the carbon $2p_z$ AOs. Both of these pairing schemes are required to describe the bonding in benzene in a VB calculation where the electrons, with specified spin functions, are assigned to AOs in the first step of the calculation. The two structures are known as resonance forms of benzene and we must be careful to recognise that they have no independent existence. Each contributes equally to our VB description (wave function) of the bonding in benzene. Both are essential, be it in a representation of the bonding on paper or in the mathematical formulation of the VB wave function. (The formulation of benzene as a ring inside a regular hexagon is more akin to an MO representation of the electronic structure.) Resonance structures occur extremely frequently in discussions of molecular electronic structure. In the case of the carbonate anion (Figure 6.17(b)), for example, three structures are required to express the fact that all C–O bonds are exactly the same. This cannot be shown with one classical valence (electron pairing) structure.

6.17 MOLECULAR GEOMETRY

Any self-respecting theory of molecular electronic structure must have the ability to predict molecular geometry. If calculations of sufficient accuracy can be performed, then it is only necessary to repeat the calculation of the total electronic energy of the molecule for a variety of geometries and to determine the geometry which gives the lowest (most negative) energy. This will be the equilibrium experimental geometry of the molecule. But there are aspects of this apparently straightforward approach which are unsatisfactory if we seek a widely applicable solution to the problem.

The most important difficulty arises from the required accuracy of the calculation. The change in the total energy of a molecule which results from a small change of bond angle will normally be less than 0.1 % of the total energy. Clearly, a very accurate and therefore tedious calculation is required to achieve this level of accuracy for a small molecule, and it may well be impossible for a large molecule. The practising chemist needs a theory that can be applied more easily, and here we again recall the maxim of Wigner and Seitz (Section 6.2). What we require is a pictorial view of a molecule from which we can draw conclusions about its geometry and particularly its bond angles. The VB theory with its emphasis upon the overlapping and interaction of two singly occupied AOs, one from each of the atoms forming the bond, is better suited to this purpose than is the MO theory.

Table 6.5 Bond angles in simple hydrides

Molecule XH$_3$	H–X–H angle (degrees)	X–H (pm)	Molecule XH$_2$	H–X–H angle (degrees)	X–H (pm)
X			X		
N	107.8	101.7	O	104.5	95.7
P	93.6	141.9	S	92.1	133.6
As	91.8	151.9	Se	91	146
Sb	91.3	170.7	Te	90	170.7

Consider the water molecule (H_2O). Neglecting the 1s electrons, the electron configuration of the oxygen atom is: $(2s)^2(2p_x)^2(2p_y)^1(2p_z)^1$. Four of the six valence electrons occupy the oxygen 2s and one of the oxygen 2p AOs (we have arbitrarily chosen the $2p_x$) with their spins paired and the remaining two electrons occupy the other two oxygen 2p AOs with parallel spins and are available to form bonds. Since the two singly-occupied oxygen AOs have their maximum electron densities directed along the y- and z-axes, the greatest overlap and hence the strongest bond should be formed when the two hydrogen atoms lie on those axes and form an H–O–H bond angle of 90°. In fact, the experimentally determined bond angle of the water molecule is 104.5°. If the principal quantum number 2 is replaced by 3, then the argument just given also applies to the H–S–H bond angle in hydrogen sulfide, the experimental value of which is 92.1°. It should also apply to the hydrides of Group 15 (5A) elements. Table 6.5 gives some examples. These results are remarkable for such simple considerations.

6.17.1   The valence-shell electron-pair repulsion (VSEPR) model

An even simpler theory which, in its application, makes minimal appeal to the concept of atomic orbitals was first suggested by N.V. Sidgwick and H.M. Powell in 1940 and elaborated by R.J. Gillespie and R.S. Nyholm in 1957. It is known as the Sidgwick-Powell-Gillespie-Nyholm or as the Valence Shell Electron-Pair Repulsion (VSEPR) model. The valence shell electrons are those in the outermost shell of the atom and they are the only electrons that play a significant role in the formation of chemical bonds. Sidgwick and Powell, recognising that the valence shell of a chemically-bonded atom contained bond pairs and lone pairs of electrons, suggested that the orientation in space of the electron pairs, and hence of the bonds around a central atom, would be that which minimises the inter-electronic repulsion between them. The concept which the VSEPR theory embodies, and which was new in 1940, is the idea that lone pairs, i.e. electrons in the valence shell but not involved in bonding, also have a very important part to play in the determination of molecular geometry.

Sidgwick and Powell assumed that the bond pairs and lone pairs were of equal importance, but among many refinements of the theory Gillespie and Nyholm proposed that the electrostatic repulsion between electron pairs diminishes in the sequence: lone pair-lone pair > bond pair-lone pair > bond pair-bond pair. The following interpretation of the situation is commonly given. Since a bond pair is drawn away from the central atom towards the atom to which it is bonded, the region of electron density associated with such a pair lies further away from the central atom than does the corresponding region of a lone pair and consequently the repulsion exerted by it is less. However, recent theoretical work has thrown doubt upon this explanation and it appears that the effect is more subtle. But the order given above certainly appears to apply and it provides a useful guide to molecular geometry. We must hope that further work will clarify the basis of this valuable idea which is very easy to use.

We can illustrate the VSEPR approach to molecular geometry by again considering the shape of water. We focus our attention upon the region of space around the central oxygen atom. The valence shell of oxygen, i.e. the shell with principal quantum number equal to 2, contributes six electrons to this region and each hydrogen contributes its single valence electron. Therefore, there are eight electrons, i.e. four pairs, of valence electrons surrounding the oxygen atom. If the four pairs of electrons were all exactly the same, their mutual repulsion would be at a minimum if they were located in four regions directed towards the corners of an imaginary tetrahedron centred on the oxygen atom. The H–O–H bond angle would then be $109.5°$. But the electron pairs are not all identical; we have two bond pairs and two lone pairs and the unequal repulsion between these pairs will result in an increase of the angle between the two lone pairs, which we cannot measure, and a decrease in the H–O–H angle in accord with observation. Since the sulfur atom also has six valence electrons ($n = 3$) exactly the same reasoning can be applied to the hydrogen sulfide molecule so that both water and hydrogen sulfide are predicted to be bent with H–X–H angles less than the tetrahedral angle (Table 6.5). In ammonia (NH_3), with three bond pairs and one lone pair, we predict that the lone pair-bond pair repulsions will reduce the H–N–H bond angle below the tetrahedral angle, as indeed we observe (H–N–H in ammonia $= 107.8°$).

A number of effects may contribute to the larger decrease of the H–X–H angle from $109.5°$ when X is a second-row element. If we compare water and hydrogen sulfide, the two bonding electron pairs are further apart in hydrogen sulfide than in water because of the greater size of the sulfur atom and because the smaller electronegativity of the sulfur atom allows the electrons to move away towards the hydrogen atoms; note the longer S–H bond length in Table 6.5. This permits a reduction in the H–S–H angle relative to the H–O–H angle. If we refine our argument by introducing the concept of the sort of orbitals which the electron pairs might occupy then, since our starting point is a tetrahedral arrangement, we naturally think of sp^3 hybrid orbitals. In this light we might then argue that the tendency to hybridisation of the s and p orbitals is less in sulfur than in oxygen so that the S–H bonds are formed with sulfur 3p AOs giving H–S–H angles of $90°$. This is equivalent to saying that the promotional energy required to hybridise the 3s and 3p AOs of sulfur cannot be recovered in the formation of two S–H bonds.

Finally, we should repeat the warning above. Although these explanations of the trends in bond angle appear plausible we are dealing with very subtle effects and our views may require modification in the light of further research on this subject. Linnett's ideas, which focus more attention on electron spin (Section 6.15.4 and Box 6.4), may play an important role in a full understanding of the VSEPR model.

6.17.2   The VSEPR model and multiple bonds

The VSEPR theory treats multiple bonds in much the same way as it does single bonds regarding the two or three electron pairs of a double or triple bond as a single super-pair.

As an example we may compare the geometries of carbon dioxide and sulfur dioxide. In the case of carbon dioxide (CO_2), the carbon atom contributes four valence electrons and we must assume that the oxygen atoms each contribute two since oxygen is invariably divalent. (Speaking of the bonding in the usual way, we would say that the carbon atom forms one σ bond and one π bond with each oxygen atom.) Thus, the valence electron region surrounding the carbon atom contains eight electrons, which we group into two super-pairs. Their mutual repulsion is minimised by an O–C–O angle of $180°$ and carbon dioxide is predicted to be linear. In the case of sulfur dioxide (SO_2), we have ten electrons, which we group into two super-pairs (SPs) plus a lone pair. The arrangement of lowest energy for three groups of electron pairs is basically one in which they are disposed at

Table 6.6 VSEPR predictions of molecular shapes and examples

Electron pairs	Basic shape	BPs	LPs		Example
2	linear	2	0	two BP's at $180°$; linear	BeF_2
3	trigonal	3	0	three BP's at $120°$; trigonal	BF_3
3	,,	2	1	two BP's at $<120°$; bent	$SnCl_2{}^a$
4	tetrahedral	4	0	four BP's at $109.5°$; tetrahedral	CH_4
4		3	1	three BP's at $<109.5°$; trigonal bipyramidal	NH_3
4	,,	2	2	two BP's at $<109.5°$; bent	OH_2
5	trigonal	5	0	trigonal bypyramid of BP's	PCl_5
5	bipyramidal	4	1	LP in equatorial position	SF_4
5		3	2	two equatorial LP's; "T"-shaped	ClF_3
5		2	3	three equatorial LP's; linear	XeF_2
6	octahedral	6	0	six BP's at $90°$; octahedral	SF_6
6		5	1	square pyramid	IF_5
6	,,	4	2	two axial LP's; planar	XeF_4
6	,,	3	3	implies a central atom with 9 electrons	
6	,,	2	4	implies a central atom with 10 electrons	

Linear
ABA = $180°$

Trigonal
ABA = $120°$

Tetrahedral
ABA = $109.5°$

A——B——A

Trigonal-bipyramidal
ABA = $90°$; EBE = $120°$

Octahedral
ABA = $90°$

^{a}In the vapor phase, not in the solid state.

Table 6.7 VSEPR interpretation of molecular shapes

Molecule AB_n	Valence electrons		Charge	Total electrons	Number of pairs			Molecular shape
	A	nB			BPs	SPs	LPs	
CH_4	4	4	0	8	4	0	0	tetrahedral
$[NO_2]^+$	5	4	+1	8	0	2	0	linear
$[ICl_2]^+$	7	2	+1	8	2	0	2	bent[a]
$[NO_2]^-$	5	4	−1	10	0	2	1	bent[b]
$[ICl_2]^-$	7	2	−1	10	2	0	3	linear[c]
O_3	6	4	0	10	0	2	1	bent[b]
PCl_5	5	5	0	10	5	0	0	trigonal bipyramidal
$[NO_3]^-$	5	6	−1	12	0	3	0	trigonal
$[CO_3]^{2-}$	4	6	−2	12	0	3	0	trigonal
XeF_4	8	4	0	12	4	0	2	square planar[d]
$[ClO_4]^-$	7	8	−1	16	0	4	0	tetrahedral
$[SO_4]^{2-}$	6	8	−2	16	0	4	0	tetrahedral
$[PO_4]^{3-}$	5	8	−3	16	0	4	0	tetrahedral

[a] A tetrahedral disposition of four electron pairs with two lone pairs.
[b] A trigonal disposition of three electron pairs with one lone pair.
[c] The three lone pairs occupy the axial positions of the trigonal bipyramidal array of five electron pairs.
[d] The two lone pairs occupy the axial positions of the octahedral array of six electron pairs.

$120°$ to each other. The observed O–S–O angle of $119.5°$ suggests that the bond pair-bond pair and lone pair-bond pair repulsions are very similar in this molecule. This might be rationalised by observing that the repulsion due to the bond pairs has increased because they are now superpairs of four electrons each. Be that as it may, there is a very positive prediction of a linear carbon dioxide molecule and an angled sulfur dioxide, and this is the major feature that we have to explain. Tables 6.6 and 6.7 summarise the application of the VSEPR model.

6.18 COMPUTATIONAL DEVELOPMENTS

The advent of digital computers and their ever-increasing power has had an enormous impact on the calculation of molecular properties by quantum-mechanical methods. At first, apart from special problems such as H_2, calculations on molecules of chemical interest could only be carried out if all the electron-repulsion integrals and many others were treated as parameters which were not calculated but determined by comparison of theory and experiment and inserted into the calculation at the outset. Notable advances were made with these *semi-empirical* MO methods and they are only now decreasing in importance as the *ab-initio* MO methods, in which all integrals are calculated from scratch, are becoming increasingly feasible. The corresponding developments in VB theory did not take place since it proved to be much more difficult to implement the VB theory computationally, largely on account of the requirement that the electron spins, correctly paired, be introduced at the start of the calculation rather than at the end; compare the two treatments of the hydrogen molecule (Sections 6.4 to 6.7). But notable advances in the VB theory and coding it were made in the last two decades of the 20th century and *ab-initio* VB calculations for chemically interesting molecules are now reported on a regular basis. Concurrently with the rise of *ab-initio* VB calculations a new kid appeared on the block, *density functional theory* (DFT), which takes a very different approach to the problem of calculating the physical properties of molecules, but appears to hold out great promise in terms of quality of results and computational efficiency.

The bibliography, to which I add some comments here, contains a few leading references to this extensive and still rapidly advancing field of chemical research. Levine[9] has a long chapter which provides an excellent introduction to MO methods, both *semi-empirical* and *ab-initio*, and to DFT. The book by Hinchliffe[10] is a compact introduction to *ab-initio* MO methods. Two volumes in the Elsevier series, *Theoretical and Computational Chemistry*, comprise advanced treatments of modern DFT[11] and VB[12] work by multiple authors. Gerratt, a pioneer of the modern VB methodology, and some colleagues have written a chemistry-orientated review[13] of that theory.

This is also an appropriate place to note an exciting and very recent experimental advance[14] because the new technique has provided, for the first time, an image of a molecular orbital which can be directly compared with theoretical results. Molecular orbitals are essentially a mathematical construct (Section 6.5), but the fact that so many experimental measurements can be quantitatively interpreted by MO methods has given us reason to believe that they represent much more than that. Now, experiments using laser pulses on the femto-second (10^{-15} s) time scale have provided an image of the highest occupied MO (HOMO) of nitrogen (N_2) (σ_3 in Figure 6.9) which compares remarkably well, quantitatively, with a similar image produced by an *ab-initio* MO calculation. It should be emphasised that the image is of the MO itself, not of the square or electron density, and we see the expected pattern of phases. The authors believe that it will be possible to obtain such images on a time scale which will allow us to follow the change in electron distribution during a chemical reaction – an insight into the very essence of chemistry.

6.19 BIBLIOGRAPHY AND FURTHER READING

1. W.G. Palmer, *A History of the Concept of Valency to 1930*, Cambridge University Press, 1965.
2. C.A. Russell, *The History of Valency*, Leicester University Press, 1971.
3. C.A. Coulson, *Valence*, 2 nd edn, Oxford University Press, 1961. R McWeeney, *Coulson's Valence*, 3 rd edn, Oxford University Press, 1979.
4. C.S.G. Phillips and R.J.P. Williams, *Inorganic Chemistry*, Oxford University Press, 1965.
5. E.P. Wigner and F. Seitz, *Solid State Physics*, **1**, 97 (1955).
6. J.N. Murrell, S.F.A. Kettle and J.M. Tedder, *Valence Theory*, 2nd edn, Wiley, London, 1970.
7. J.W. Linnett, *Wave Mechanics and Valency*, Methuen, London, 1960.
8. J.W. Linnett, *The Electronic Structure of Molecules*, Methuen, London, 1964.
9. I.N. Levine, *Quantum Chemistry*, 5th edn, Prentice-Hall Inc., New Jersey, 2000.
10. A. Hinchliffe, *Computational Quantum Chemistry*, Wiley, Chichester, 1988.
11. J.M. Seminario and P. Politzer (Eds), *Modern Density Functional Theory*, Elsevier, Amsterdam, 1995.
12. D.L. Cooper (Ed), *Valence Bond Theory*, Elsevier, Amsterdam, 2002.
13. J. Gerratt, D.L. Cooper and P.B. Karadakov, *Chem. Soc. Revs.*, **26**, 87 (1997).
14. J. Itani, J. Levesque, D. Zeidler, H. Nikura, H. Pépin, J.C. Kieffer, P.B. Corkum and D.M. Villeneuve, *Nature*, **432**, 867 (2004).

The following three books illustrate well the application of the quantum-mechanical theory of chemical bonding to a very wide range of real chemical problems, in ways which emphasise the pictorial rather than the mathematical aspects of the theory.

15. T.A. Albright, J.K. Burdett and M.H. Whangbo, *Orbital Interactions in Chemistry*, Wiley, New York, 1985.
16. N.W. Alcock, *Bonding and Structure, Structural Principles in Inorganic and Organic Chemistry*, Ellis Horwood Ltd, Chichester, 1990.
17. I. Fleming, *Frontier Orbitals and Organic Chemical Reactions*, Wiley, Chichester, 1976.

BOX 6.1   Normalisation of the valence bond (VB) and molecular orbital (MO) wave functions for the hydrogen molecule

The VB wave function to be normalised is:

$$\Psi = N[\phi_A(1)\phi_B(2) + \phi_B(1)\phi_A(2)]$$

If the wave function is normalised then the value of the normalising constant N must be such that:

$$\iint \Psi^2 \, dv_1 \, dv_2 = N^2 \iint [\phi_A(1)\phi_B(2) + \phi_B(1)\phi_A(2)]^2 \, dv_1 \, dv_2 = 1.0$$

$$\Rightarrow N^2 \iint \{ [\phi_A(1)\phi_B(2)]^2 + [\phi_B(1)\phi_A(2)]^2 + 2[\phi_A(1)\phi_B(2)]$$

$$\cdot [\phi_B(1)\phi_A(2)] \} \, dv_1 \, dv_2 = 1.0$$

$$\Rightarrow N^2 \left\{ \int \phi_A^2(1) \, dv_1 \int \phi_B^2(2) \, dv_2 + \int \phi_B^2(1) \, dv_1 \int \phi_A^2(2) \, dv_2 \right.$$

$$\left. + 2 \int \phi_A(1)\phi_B(1) \, dv_1 \int \phi_B(2)\phi_A(2) \, dv_2 \right\} = 1.0$$

But, by definition, ϕ_A and ϕ_B are normalised 1s atomic orbitals of the hydrogen atom, i.e.:

$$\int \phi_A^2(1) \, dv_1 = \int \phi_B^2(1) \, dv_1 = 1.0$$

Therefore, the first two terms of the expression above are each equal to 1.0 and we have:

$$N^2 \{ 2 + 2 \int \phi_A(1)\phi_B(1) \, dv_1 \int \phi_B(2)\phi_A(2) \, dv_2 \} = 1.0$$

Since all electrons are identical, the two integrals in the above equation are equal and they can only have a value in the region of space where the atomic wave functions, ϕ_A and ϕ_B, both have non-zero values. If we draw boundary surfaces for the two AO's separated by the H–H bond length (Figure B6.1.1) then we see that this condition is only fulfilled where the two AOs overlap. For this reason, integrals of this kind are called *overlap integrals*.

The overlap integral is usually denoted by S:

$$S = \iiint \phi_A(1)\phi_B(1) \, dv_1$$

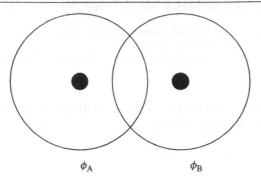

Figure B6.1.1   The overlap of two 1s atomic orbitals

Table B6.1.1   The values of $S \equiv \langle 1s|1s \rangle$ as a function of the orbital exponent (Z) and the inter-nuclear distance (R) in atomic units

Z	R (a_0)						
	0.8	1.0	1.2	1.4	1.6	1.8	2.0
0.9	0.921	0.882	0.838	0.791	0.742	0.692	0.641
1.0	0.905	0.858	0.807	0.753	0.697	0.641	0.586
1.1	0.887	0.833	0.775	0.714	0.652	0.592	0.533
1.2	0.868	0.807	0.742	0.675	0.608	0.544	0.483
1.3	0.848	0.780	0.708	0.636	0.565	0.498	0.435

where the triple integral sign reminds us that this is an integration over all three spatial co-ordinates of electron 1. Some values of the integral are given in Table B6.1.1. We therefore have:

$$N^2(2 + 2S^2) = 1.0 \Rightarrow N = \{2(1 + S^2)\}^{-\frac{1}{2}}$$

The MO wave function to be normalised is:

$$\Psi = N'\{(\phi_A + \phi_B)(1)\}$$

If we call the normalisation constant N', then for a normalised MO we must have:

$$\int \Psi^2 \, dv_1 = N'^2 \int \{(\phi_A + \phi_B)(1)\}^2 \, dv_1 = 1.0$$

$$\Rightarrow N'^2 \int \{\phi_A^2(1) + \phi_B^2(1) + 2\phi_A(1)\phi_B(1)\} \, dv_1 = 1.0$$

$$\Rightarrow N'^2\{2 + 2S\} = 1.0$$

Therefore:

$$N' = \{2(1 + S)\}^{-\frac{1}{2}}$$

BOX 6.2   Proof that the four sp³ hybrid AOs of Equation (6.15.3) are normalised and orthogonal

NORMALISATION

$$\Phi_1 = \tfrac{1}{2}\{\phi_{2s} + \phi_{2p_x} + \phi_{2p_y} + \phi_{2p_z}\}$$

$$\therefore \int \Phi_1^2 \, dv = \tfrac{1}{4} \int \{\phi_{2s} + \phi_{2p_x} + \phi_{2p_y} + \phi_{2p_z}\}^2 \, dv$$

(For the sake of simplicity we omit the electron in these equations.)
When the above wave function is squared we find terms of two types:

a) squared functions such as ϕ_{2s}^2 and $\phi_{2p_x}^2$

b) cross-terms such as $\phi_{2s} \cdot \phi_{2p_x}$ and $\phi_{2p_x} \cdot \phi_{2p_y}$

Since AOs of the same atom, i.e. $\phi_{2s}, \phi_{2p_x}, \phi_{2p_y}$ and ϕ_{2p_z}, are orthogonal and normalised, the four squared terms each give 1.0 and the 12 cross-terms each contribute zero:

$$\int \phi_{2s}^2 \, dv = 1.0 \text{ and } \int \phi_{2s} \cdot \phi_{2p_x} = 0.0$$

Therefore

$$\int \Phi_1^2 \, dv = \tfrac{1}{4} \cdot 4 = 1.0$$

QED.

ORTHOGONALITY

$$\int \Phi_1 \Phi_2 \, dv = \tfrac{1}{4} \int \{[\phi_{2s} + \phi_{2p_x} + \phi_{2p_y} + \phi_{2p_z}] \cdot [\phi_{2s} - \phi_{2p_x} - \phi_{2p_y} + \phi_{2p_z}]\} \, dv$$

As before, all the cross-terms give zero. For the four squared terms we have:

$$\tfrac{1}{4} \int \{\phi_{2s}^2 - \phi_{2p_x}^2 - \phi_{2p_y}^2 + \phi_{2p_z}^2\} \, dv = \tfrac{1}{4}\{1 - 1 - 1 + 1\} = 0$$

QED.

BOX 6.3   Slater-Type Orbitals (STOs)

The vast majority of the calculations of molecular wave functions, VB and MO, proceed from a basis set of atomic orbitals. As we have seen (Section 5.7), the radial functions of accurate self-consistent-field orbitals for an atom with more than one electron cannot be expressed in analytical form and would therefore be totally impracticable in general use. Furthermore, since the electron cloud surrounding each atom must undergo considerable changes when bonds are formed, approximate atomic orbital functions may provide as suitable a starting point as more accurate ones. One of the earliest proposals for the construction of suitable analytical atomic radial functions for calculations on molecules was made by Slater in 1930 (J.C. Slater, *Phys. Rev.*, **36**, 57 (1930)) and these functions have been extensively used, both in their original forms and with some modifications. The original forms have been adopted for the calculation of atomic orbital overlap in this chapter.

Slater first made a radical simplification of the radial function by proposing a functional form which had no nodes as opposed to the $n - 1$ radial nodes found in the exact wave functions (see Chapter 5, Section 2.1 and Box 5.3). He proposed the normalised functions:

$$\Psi_{nlm} = \left\{ \frac{(2Z'/n)^{2n+1}}{(2n)!} \right\}^{1/2} r^{n-1} \exp(-Z'r/n) \ Y_{lm}(\theta, \phi)$$

where Z' is an *effective* nuclear charge for a single electron in a central field resulting from the nuclear charge and the repulsion of the remaining electrons in the atom. The other symbols have the meanings explained in Chapter 5.

From an analysis of atomic spectral data, Slater obtained a set of rules for calculating the screening, S, of the electrons which is then subtracted from the true nuclear charge, Z, to give Z' (in units of $+e$). S is calculated according to the following rules:

1. The electrons are divided into groups. Each electron in a group has the same shielding constant which differs from that of the electrons in other groups. The groups are 1s; 2s + 2p; 3s + 3p; 3d; 4s + 4p; 4d; 4f; 5s + 5p; etc.

2. The screening constant, S, is then given as a sum of the individual contributions, s, which are calculated as follows:
   (a) For any electron outside the shell[†] under consideration s = 0.0.

   (b) For each other electron in the group considered s = 0.35, except for the 1s where s = 0.30.

   (c) If the group considered is an s + p group then s = 0.85 for each electron in the next inner shell[†] and s = 1.00 for all electrons further in.

   (d) If the group is a d or f group then s = 1.00 for every electron inside it.

The following examples illustrate the calculation of Z' for the occupied orbitals of the iron atom, $(1s)^2 (2s)^2 (2p)^6 (3s)^2 (3p)^6 (3d)^6 (4s)^2$.

First form the required electron groups: $[1s]^2$; $[2s + 2p]^8$; $[3s + 3p]^8$; $[3d]^6$; $[4s]^2$.

$Z' (1s) = 26 - (1 \times 0.30) = 25.70$

$Z' (2s) = 26 - (7 \times 0.35) - (2 \times 0.85) = 21.85 = Z' (2p)$

$Z' (3s) = 26 - (7 \times 0.35) - (8 \times 0.85) - (2 \times 1.00) = 14.75 = Z' (3p)$

[†]A shell is the set of atomic orbitals having the same principal quantum number, n.

$$Z'\,(3d) = 26 - (5 \times 0.35) - (18 \times 1.00) = 6.25$$

$$Z'\,(4s) = 26 - (1 \times 0.35) - (14 \times 0.85) - (10 \times 1.00) = 3.75$$

Because the functions have no radial nodes, STOs provide a very poor description of the inner regions of the atom. However, they are a much better representation of the outer atomic regions, and since these are the regions of most significance for chemical binding STOs have been widely used in theoretical chemistry, though their use is now in decline. The Slater rules also require that the higher principal quantum numbers, n, of 4, 5 and 6 be replaced by $3.7^{\ddagger}$, 4.0 and $4.2^{\ddagger}$ respectively and the quality of the STOs deteriorates significantly as n rises. One way of improving on STOs is to use a combination of two of them, each with a different value of Z', to represent one atomic orbital. Orbitals of this type are known as double-zeta functions.

‡The normalisation factor in the STO formula given above does not apply for non-integral values of n.

BOX 6.4   Electron spin correlation: An example following J.W. Linnett[1]

Our purpose here is to demonstrate in more detail the way in which the Pauli principle, by allowing only wave functions which are antisymmetric with respect to electron exchange, exerts an important influence upon the energies of states having different multiplicities and also upon the orientation of electron distributions in space. Suppose that we have an atom with a closed 1s electron shell, which we ignore, and one electron in each of the 2s and $2p_z$ orbitals, i.e. configuration $1s^2 2s^1 2p_z^1$. In Section 4.9 it is shown that if the two electrons have opposed spins the resulting state is a singlet state and if their spins are parallel we have a triplet state and that the spin wave functions of the states with $S_z = 0$ are:

$$\Phi_{\text{spin}}(\text{singlet}) = (1/\sqrt{2})\{\alpha(1)\beta(2) - \beta(1)\alpha(2)\}$$

and

$$\Phi_{\text{spin}}(\text{triplet}) = (1/\sqrt{2})\{\alpha(1)\beta(2) + \beta(1)\alpha(2)\}.$$

When the electrons are exchanged the sign of the singlet function changes while that of the triplet remains the same so that the singlet spin function is *antisymmetric* with respect to electron exchange while the triplet is *symmetric*. But the Pauli principle requires that the total electronic wave function be antisymmetric with respect to electron exchange, so to fulfil that requirement the symmetric spin function must be combined with an antisymmetric space function, and vice versa. This is how spin functions determine space functions and hence electron repulsion energies and the subject is explored in more detail in Section 11.5.

The appropriate space functions are of very similar form and we choose the symmetric function for the singlet and the antisymmetric function for the triplet:

$$\Phi_{\text{space}}(\text{singlet}) = (1/\sqrt{2})\{2s(1)2p_z(2) + 2p_z(1)2s(2)\}$$

and

$$\Phi_{\text{space}}(\text{triplet}) = (1/\sqrt{2})\{2s(1)2p_z(2) - 2p_z(1)2s(2)\}$$

We must now give the individual 2s and $2p_z$ functions a more specific form. The 2s is purely a function of r since it is spherically symmetrical and the $2p_z$ is a function of r multiplied by an angular part of $\cos\theta$ (Appendix 5). Therefore, we write $2s = f_s(r)$ and $2p_z = f_p(r) \cdot \cos\theta$ and assume that the functions are normalised. It is not necessary to make the functions more exact, they already contain the essential elements as we shall see. Since we are concerned with electron repulsion, we only need to study the space functions which can now be written:

$$\Phi_{\text{space}}(\text{singlet}) = (1/\sqrt{2})\{f_s(\text{r1}) \cdot f_p(\text{r2}) \cdot \cos\theta_2 + f_p(\text{r1}) \cdot \cos\theta_1 \cdot f_s(\text{r2})\}$$

and

$$\Phi_{\text{space}}(\text{triplet}) = (1/\sqrt{2})\{f_s(\text{r1}) \cdot f_p(\text{r2}) \cdot \cos\theta_2 - f_p(\text{r1}) \cdot \cos\theta_1 \cdot f_s(\text{r2})\}$$

When the two electrons are at the same radius, r' say, the expressions above become:

$$\Phi_{\text{space}}(\text{singlet}) = (1/\sqrt{2})f_s(\text{r'}) \cdot f_p(\text{r'})\{\cos\theta_2 + \cos\theta_1\}$$

and

$$\Phi_{\text{space}}(\text{triplet}) = (1/\sqrt{2})f_s(\text{r'}) \cdot f_p(\text{r'})\{\cos\theta_2 - \cos\theta_1\}$$

The electron probability is proportional to the square of the wave function which is at a maximum for the singlet function when $\theta_1 = \theta_2 = 0$ or when $\theta_1 = \theta_2 = \pi$.

Conversely, the maximum of the triplet function occurs when $\theta_1 = 0$ and $\theta_2 = \pi$ or vice versa. Thus, if the electrons in the 2s and $2p_z$ orbitals have a high probability of being at the same distance from the nucleus then, for the singlet state, the spin correlation places the electrons on the same side of the nucleus whereas, for the triplet state, the most likely position for the electrons is on opposite sides of the nucleus. The hydrogen atom radial electron density functions illustrated in Figure 5.3 show a very marked overlap of the 2s and 2p functions so that we may expect spin correlation to play an important role in the determination of molecular geometry. Furthermore, it is clear that the interelectronic repulsion of the electrons is lower in the triplet state than in the singlet. This illustrates the underlying reason for Hund's rule and invites us to enquire further into the theoretical foundations of the VSEPR method (Section 6.17.1).

In Section 11.5 we show that the electron-repulsion energy of the singlet and triplet states defined above are given by:

$$E_{\text{triplet}} = J_{2s,2p} - K_{2s,2p} \quad \text{and} \quad E_{\text{singlet}} = J_{2s,2p} + K_{2s,2p}$$

where

$$J_{2s,2p} = \langle 2s(1)2p(2)|1/r_{12}|2s(1)2p(2)\rangle$$

and

$$K_{2s,2p} = \langle 2s(1)2p(2)|1/r_{12}|2p(1)2s(2)\rangle.$$

The electron repulsion $J_{2s,2p}$ is called the Coulombic repulsion and represents the classical repulsion of two electrons, one in the 2s orbital and the other in the 2p.

The repulsion $K_{2s,2p}$ is called the exchange repulsion and makes a contribution to the interelectronic repulsion which has no classical counterpart. Though $K_{2s,2p}$ itself is a positive quantity, the fact that it appears with a negative sign in the expression for $E_{triplet}$ should alert us to its non-classical nature. It represents the repulsion of the two electrons in the overlap regions, i.e. electron density 2s(1)2p(1) repelling density 2s(2)2p(2). Such a term can only arise in quantum mechanics where electron probability is given by the *square* of a function, not by the function itself. The presence of such a term in the expression for repulsion energy immediately confronts us with a very fundamental question. Are we justified in talking about electron-pair repulsion in molecules in purely classical terms?

Calculations by J. Lennard-Jones and J.A. Pople[2] throw some light upon that question. They used Slater-type orbitals (Box 6.3) to calculate the values of $J_{2s,2p}$ and $K_{2s,2p}$, in atomic units, for the beryllium atom and found the values shown in Table B6.4.1. Z' is the effective nuclear charge.

The data in the first row of the table show that the non-classical term is about 22 % of the classical term which provides some justification for our neglect of it when we apply the VSEPR theory. The values in the second row are also relevant to this discussion, but they first require further explanation. From the 2s and 2p orbitals we can form normalised hybrid orbitals (called *equivalent* orbitals by Lennard-Jones and Pople) of the forms:

$$h_+ = (1/\sqrt{2})\{2s + 2p_z\} \qquad h_- = (1/\sqrt{2})\{2s - 2p_z\}$$

and if we calculate J and K with these hybrids we obtain the results in the second row of the table. We first note that the total electron repulsion energy, $J - K$, is constant, independent of the change to hybrid orbitals, which is quite generally true as Lennard-Jones and Pople[2] prove. The formulation of the electron distribution in terms of hybrids reduces the non-classical term from 22 % to 2.4 % of the classical. Clearly, in this case, a classical view of electron repulsion is more justified when hybrids rather than the 2s and 2p orbitals themselves are used. The same is true in other situations, e.g. four sp^3 hybrids as opposed to a 2s and three 2p orbitals, though the difference is less marked. The reason for the decrease in the magnitude of K when hybrid orbitals are used is the reduction of the overlap electron densities due to the localised and directional character of the hybrids. It is fortunate, therefore, that the localised groups of bonding or lone-pair electrons which we envisage when we apply the VSEPR method are just those electron distributions in space which we also associate with hybrid orbitals. Though it is not essential to invoke hybridisation when applying the VSEPR theory, it does appear that hybrid orbitals make a classical approach to electron repulsion more acceptable from a strictly quantum-mechanical point of view.

Table B6.4.1 Electron repulsion integrals for Be: $1s^2 2s^1 2p^1$

	$J_{a,b}/E_H$	$K_{a,b}/E_H$	$(J_{a,b} - K_{a,b})/E_H$
a = 2s, b = 2p	0.1816 Z'	0.0401 Z'	0.1415 Z'
a = h_+, b = h_-	0.1450 Z'	0.0035 Z'	0.1415 Z'

REFERENCES

1. J.W. Linnett, *Wave Mechanics and Valency*, Methuen, London, 1960.
2. J. Lennard-Jones and J.A. Pople, *Proc. Roy. Soc.*, **A202**, 166 (1950).

PROBLEMS FOR CHAPTER 6

1. Use the following data to draw an energy-level diagram, similar to Figure 6.1, for the $^7Li_2$ molecule. Calculate D_e.

$$Li \longrightarrow Li^+ + e \qquad IE = 5.39 \text{ eV} \approx 43,473 \text{ cm}^{-1}$$

$$Li_2 \longrightarrow Li + Li \qquad D_0 = 1.03 \text{ eV} \approx 8,307 \text{ cm}^{-1}$$

Zero-point energy $= \frac{1}{2}h\nu = 0.022$ eV ≈ 175.7 cm^{-1}

2. P. Morse proposed that, for a diatomic molecule, the variation of the energy (E) with internuclear distance (r) could be written:

$$E = D_e[1 - \exp\{\beta(R_e - r)\}]^2$$

For Li_2 R_e, the equilibrium internuclear distance $= 267.3$ pm and β is a constant related to the vibrational frequency (ν) of the molecule and the reduced mass (μ):

$$\mu = \frac{m_1 m_2}{(m_1 + m_2)} = 3.509 \text{ a.m.u.} \qquad \beta = 4.0617 \times 10^{-4} \cdot \nu \cdot \left(\frac{\mu}{D_e}\right)^{\frac{1}{2}}$$

(The last expression gives β in cm^{-1} when μ is entered in a.m.u., ν in s^{-1} and D_e in cm^{-1}.)

Plot the Morse curve for Li_2 for $r = 5$–160 nm using the data above. (This is a problem where a spread sheet with a graph-plotting facility, e.g. EXCEL, is very useful.)

3. The function ϕ is a normalised 1s atomic orbital.

$$\phi = \frac{1}{\sqrt{\pi}} \left(\frac{Z}{a_0}\right)^{3/2} \cdot \exp(-Zr/a_0) \qquad a_0 = \text{Bohr radius} = 52.92 \text{ pm}$$

Illustrate the extension of the orbital into the space surrounding the atom by plotting the function ϕ versus r for $Z = 0.5, 1.0, 1.5$. (Here, again, a spread sheet can be very useful.)

4. Show that the sp^2 hybrid orbitals (Equation (6.15.4)) are normalised, orthogonal and orientated at $120°$ to each other. (Vector algebra, regarding a normalised p-orbital as a unit vector along the appropriate axis, or trigonometry can be used for the orientation problem.)

5. This problem is essentially a repeat of Heitler and London's first VB calculation of the binding energy of the hydrogen molecule. We work in atomic units which are much easier to use in calculations. The energies which we determine can be easily converted to SI energy units, if that is required.

In a.u. the Hamiltonian $(\hat{\mathcal{H}})$ of Equation (6.3.1) becomes:

$$\hat{\mathcal{H}} = -\tfrac{1}{2}\nabla_1^2 - \tfrac{1}{2}\nabla_2^2 - \frac{1}{r_{A1}} - \frac{1}{r_{B1}} - \frac{1}{r_{A2}} - \frac{1}{r_{B2}} + \frac{1}{r_{12}} + \frac{1}{R_{AB}}$$

You will find it useful to note that a pair of terms such as $-\tfrac{1}{2}\nabla_1^2 - \frac{1}{r_{A1}}$ or $-\tfrac{1}{2}\nabla_1^2 - \frac{1}{r_{B1}}$ constitute the Hamiltonian operator for a hydrogen atom. Therefore, if we denote such a pair by $\hat{\mathcal{H}}_{1s}$ and a hydrogen 1s orbital by ϕ_A then:

$$\hat{\mathcal{H}}_{1s}\phi_A(1) = E_{1s}\phi_A(1)$$

The VB wave function (Equation (6.4.3)) is:

$$\Psi = N\{\phi_A(1) \cdot \phi_B(2) + \phi_B(1) \cdot \phi_A(2)\}$$

where the normalising constant (N) is determined in Box 6.1.

Show that the expectation value of the energy $(\overline{E})$ is given by:

$$\overline{E} = \frac{\langle \psi | \hat{\mathcal{H}} | \psi \rangle}{\langle \psi | \psi \rangle} = \frac{J' + K'}{1 + S^2}$$

Where:

$$J' = 2E_{1s} + 2Q + J + 1/R_{AB}, \quad K' = 2S^2 E_{1s} + 2SQ' + K + S^2/R_{AB}$$

$$Q = \langle \phi_A(1) | -1/r_{B1} | \phi_A(1) \rangle, \quad Q' = \langle \phi_A(1) | -1/r_{B1} | \phi_B(1) \rangle, \quad S = \langle \phi_A(1) | \phi_B(1) \rangle$$

$$J = \langle \phi_A(1) \cdot \phi_B(2) | 1/r_{12} | \phi_A(1) \cdot \phi_B(2) \rangle \text{ and}$$

$$K = \langle \phi_A(1) \cdot \phi_B(2) | 1/r_{12} | \phi_B(1) \cdot \phi_A(2) \rangle$$

Show that if we set $2E_{1s} = 0.0$ then the difference in energy (ΔE) between the molecule and two hydrogen atoms is:

$$\Delta E = \frac{2Q + 2SQ' + J + K}{1 + S^2} + \frac{1}{R_{AB}}$$

Show that the trial wave function, $\psi' = \phi_A(1) \cdot \phi_B(2)$, is normalised and gives an expectation value for the energy of:

$$\overline{E}' = \frac{\langle \psi' | \hat{\mathcal{H}} | \psi' \rangle}{\langle \psi' | \psi' \rangle} = J'$$

Using the data in the following table, plot graphs of $\overline{E}$ and $\overline{E}'$ against internuclear distance, R_{AB}. The result shows that the binding energy obtained without allowing for electron 'exchange' is about one-sixth of that obtained with the full wave function. But the wave function ψ' is quite unacceptable since it contradicts the fundamental theoretical requirement that both terms, $\phi_A(1) \cdot \phi_B(2)$ and $\phi_A(2) \cdot \phi_B(1)$, must be included because electrons are indistinguishable.

R_{AB}/a_0	S	Q/E_H	Q'/E_H	J/E_H	K/E_H
0.25	0.98973	−0.96735	−0.97350	0.61986	0.60974
0.75	0.91521	−0.81270	−0.82664	0.58259	0.50684
1.25	0.79386	−0.65225	−0.64464	0.52313	0.36479
1.50	0.72517	−0.58369	−0.55783	0.49034	0.29684
1.65	0.68322	−0.54682	−0.50893	0.47061	0.25940
1.75	0.65527	−0.52398	−0.47788	0.45760	0.23612
2.00	0.58645	−0.47253	−0.40601	0.42597	0.18416
2.50	0.45831	−0.39057	−0.28730	0.36839	0.10662
3.00	0.34851	−0.33003	−0.19915	0.31980	0.05851
3.50	0.25919	−0.28454	−0.13589	0.27994	0.03076
4.00	0.18926	−0.24958	−0.09158	0.24755	0.01560
4.50	0.13609	−0.22207	−0.06110	0.22119	0.00760
5.00	0.09658	−0.19995	−0.04043	0.19957	0.00318

Find the expectation energy, $\overline{E}''$, for the wave function ψ'', which represents an unstable state of H_2:

$$\psi'' = N''\{\phi_A(1) \cdot \phi_B(2) - \phi_A(2) \cdot \phi_B(1)\}$$

Plot $\overline{E}''$ against R and compare the curve with your results for ψ and ψ'.

6. Consider the molecule shown in the figure:

Hydrogen atoms 2 and 3 of the H_3 molecule are equally bonded to atom 1 but the bond between H1 and H2 is not necessarily of the same strength and this is reflected in the values of the interaction elements of the energy (Hamiltonian) matrix:

$$\langle\phi_1|\hat{\mathcal{H}}|\phi_2\rangle = \langle\phi_1|\hat{\mathcal{H}}|\phi_3\rangle = \beta, \quad \text{but } \langle\phi_2|\hat{\mathcal{H}}|\phi_3\rangle = \gamma$$

where ϕ_n is a hydrogen 1s AO on atom n. Note that β and γ are negative quantities because they represent bonding interactions which reduce the energy of the molecule *vis-à-vis* the isolated atoms. If we arbitrarily fix the zero of the energy scale by setting the on-diagonal elements of the matrix to α we have:

$\hat{\mathcal{H}}$	ϕ_1	ϕ_2	ϕ_3
ϕ_1	α	β	β
ϕ_2	β	α	γ
ϕ_3	β	γ	α

Find the eigenvalues of this matrix (Appendix 3). (If you form rows and columns of ϕ_1, $(\phi_2 + \phi_3)/\sqrt{2}$ and $(\phi_2 - \phi_3)/\sqrt{2}$ you will find that the matrix blocks out into a 1×1 and a 2×2 matrix.)

Two extreme forms of the molecule may be envisaged; one (the linear molecule) in which $\gamma \approx 0$ and the other in which $\gamma = \beta$ (the equilateral triangle). Determine the eigenvalues for these two species.

Use a spread sheet to plot a correlation diagram relating the molecular orbital energies over the whole range of relative values of β and γ. (It is convenient to set $\alpha = 0.0$, $\beta = -10.0$ and $\gamma = 0.0$ to -10.0 in steps of 1.0.)

If the energies of the molecular orbitals of an H_3 species were measured and found to be -4.2 eV, $+0.8$ eV and $+3.4$ eV:

(a) Where on the correlation diagram should the molecule be placed?
(b) What would the relative values of β and γ be?
(c) What would the absolute values of β and γ be?

[Answers: (a) $\gamma = -3$; (b) $\gamma/\beta = 0.3$; (c) $\beta = -2.67$, $\gamma = -0.8$]

CHAPTER 7

Bonding, Spectroscopy and Magnetism in Transition-Metal Complexes

7.0   Introduction   181
7.1   Historical development   182
7.2   The crystal field theory   182
7.3   The electronic energy levels of transition-metal complexes   187
      7.3.1   The weak-field scheme for d^2 (example of ^3F in an octahedral field) .   189
      7.3.2   The weak-field scheme for d^2 (inclusion of ^3P)   190
      7.3.3   The d^2 energy levels for weak, strong and intermediate
              octahedral fields   191
      7.3.4   The strong-field scheme for d^2 in an octahedral field   193
      7.3.5   Spin-orbit coupling   195
7.4   The electronic spectroscopy of transition-metal complexes   196
7.5   Pairing energies; low-spin and high-spin complexes   197
7.6   The magnetism of transition-metal complexes   197
7.7   Covalency and the ligand field theory   199
7.8   Bibliography and further reading   203
      Problems for Chapter 7   212

7.0   INTRODUCTION

The colours and magnetic properties of the complex compounds of the transition-metal ions presented a great challenge to the theory of the chemical bond and, although the problem is now well understood in principle, quantitative interpretation of the many subtle effects continues to engage the theoretician, not least on account of their significance in biology and technological applications. One thinks, for example, of transition-metal catalysts, the biological functions of metallo-proteins, magnetic storage devices and up-converters. Historically, the interpretation of these phenomena has been by means of ligand field (LF) theory, but density functional theory and $X\alpha$ scattered-wave methods are now seeing increasing application. The words 'ligand field theory' convey to the practitioners a range of methods, varying widely in their sophistication, of applying effectively the same theory. The essentials of the theory are its origins in crystal field (CF) theory,

The Quantum in Chemistry   R. Grinter
© 2005 John Wiley & Sons, Ltd

extensive and sometimes advanced use of group theory and the treatment of the interaction between metal ion and ligand as covalent as well as electrostatic. A brief outline of the developments which have led to the present state of the art will make these points clearer; but first a comment on the use of group theory and matrix diagonalisation.

The theory of the transition-metal ions relies heavily upon group theory and many excellent descriptions of the subject which incorporate the required group theory are available.[1] It therefore appears superfluous to include group theory here and little or no appeal to it will be made in the following chapter. The use of group-theoretical symbols to identify electronic states should present no problems; the reader who has no knowledge of the subject may regard these symbols simply as labels.

In what follows we shall also make frequent appeal to the concept of determining energies by setting up and diagonalising the Hamiltonian matrix for a problem. Readers unfamiliar with this idea should read Appendix 3 before going beyond Section 7.1.

7.1 HISTORICAL DEVELOPMENT

In 1929, J. Becquerel proposed that the central metal ion in a transition-metal complex was subject to an electrostatic field originating from the surrounding ligands. In the same year, Hans Albrecht Bethe (1906–) used symmetry and group theory to place Becquerel's idea on the firm theoretical foundation we now call CF theory. Just three years later, in 1932, John Hasbrouck Van Vleck (1899–1980) demonstrated the power of the new theory when he interpreted the paramagnetism of the first-row transition-metal complexes and the rare earths with good quantitative accuracy. Van Vleck and his co-workers made many other seminal contributions to the development of the theory and its applications over the following decade.

A valence bond approach to the problem was introduced in the 1940s by Linus Pauling, but it proved less successful than the CF theory in the interpretation of electronic spectra and we shall not pursue it further here. It is described in detail in Pauling's masterpiece, *The Nature of the Chemical Bond*.[2]

With the growth of the molecular orbital (MO) theory in the second half of the 20th century, the time appeared ripe to tackle the most obvious shortcoming of the CF theory. The interaction between ligand and central metal ion is clearly more than a purely electrostatic one; there is also a significant element of covalency which should be amenable to a MO treatment. The range of theories which grew from this seed carry the collective name of ligand field theory.

7.2 THE CRYSTAL FIELD THEORY

Many of the complexes of the transition metal ions are very symmetrical. Bethe showed how the symmetry could be elegantly exploited by means of group theory to determine the way in which a purely electrostatic field from the surrounding ligands could remove the five-fold degeneracy of the partially occupied metal d orbitals. As noted above, there are many excellent accounts of this subject.[1] As an illustration of the quantitative effect of a crystal field on five d orbitals we shall calculate the matrix elements of an octahedral field.

An electron with charge $-e$ and polar coordinates r, θ and ϕ and a point charge $-q$ at r_q, θ_q and ϕ_q is shown in Figure 7.1. The atomic nucleus is at 0,0,0 and the distance

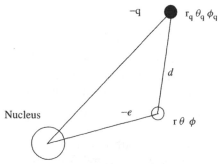

Figure 7.1   An electron in the field of a negative electrostatic charge

between the point charge and the electron is d. We wish to calculate the energy of each of the five d-electron distributions in the electrostatic field of an array of six such charges, $-q$, located at the vertices of a regular octahedron. We shall arrange our results in the form of a matrix, i.e. we shall calculate the matrix elements of the crystal-field Hamiltonian, $\hat{\mathcal{H}}_{cf}$. For the single charge in Figure 7.1 we require:

$$\langle \psi_{n'l'm'} | qe/d | \psi_{nlm} \rangle = qe \langle \psi_{n'l'm'} | 1/d | \psi_{nlm} \rangle$$

for all possible combinations of two d-orbitals. The brackets, $\langle |$ and $| \rangle$, imply integration over the electronic coordinates r, θ and ϕ. We use m rather than m_l to reduce the complexity of the notation where there is no possibility of confusion.

Our first step is to express the inverse distance $(1/d)$ in terms of the spherical harmonics (Appendix 5). $Y_{k\alpha}(\theta, \phi)$ describes the position of the electron, for which the co-ordinates θ and ϕ can vary over their whole ranges, and $Y_{k\alpha}{}^*(\theta_q, \phi_q)$ that of the charge which is stationary at the point r_q, θ_q, ϕ_q. The result is:[3]

$$\frac{1}{d} = \sum_{k=0}^{\infty} \frac{4\pi}{2k+1} \cdot \frac{r^k}{r_q^{k+1}} \cdot \sum_{\alpha=-k}^{\alpha=+k} Y_{k\alpha}(\theta, \phi) \cdot Y_{k\alpha}^*(\theta_q, \phi_q) \qquad (7.2.1)$$

where we have assumed that the charge is further from the nucleus than the electron at all points of the latter's distribution, i.e. r_q is always greater than r. This is not essential but it simplifies the present problem while retaining the essentials of the method. Inserting the expression for $1/d$, our matrix element becomes:

$$\left\langle \psi_{n'l'm'} \left| \frac{qe}{d} \right| \psi_{nlm} \right\rangle = qe \sum_{k=0}^{\infty} \frac{4\pi}{2k+1} \cdot \frac{1}{r_q^{k+1}} \cdot \sum_{\alpha=-k}^{\alpha=+k} \langle \psi_{n'l'm'} | r^k Y_{k\alpha}(\theta, \phi) \cdot Y_{k\alpha}^*(\theta_q, \phi_q) | \psi_{nlm} \rangle$$

$$(7.2.3)$$

Because the brackets, $\langle |$ and $| \rangle$, imply integration over the electronic coordinates r, θ and ϕ, we can take $Y_{k\alpha}{}^*(\theta_q, \phi_q)$ outside the integration and write:

$$\left\langle \psi_{n'l'm'} \left| \frac{qe}{d} \right| \psi_{nlm} \right\rangle = qe \sum_{k=0}^{\infty} \frac{4\pi}{2k+1} \cdot \frac{1}{r_q^{k+1}} \cdot \sum_{\alpha=-k}^{\alpha=+k} Y_{k\alpha}^*(\theta_q, \phi_q) \langle \psi_{n'l'm'} | r^k Y_{k\alpha}(\theta, \phi) | \psi_{nlm} \rangle$$

$$= qe \sum_{k=0}^{\infty} \cdot \sum_{\alpha=-k}^{\alpha=+k} A_{k\alpha} \langle \psi_{n'l'm'} | r^k Y_{k\alpha}(\theta, \phi) | \psi_{nlm} \rangle \qquad (7.2.4)$$

where

$$A_{k\alpha} \equiv \frac{4\pi}{2k+1} \cdot \frac{1}{r_q^{k+1}} \cdot Y_{k\alpha}^*(\theta_q, \phi_q) \qquad (7.2.5)$$

But each wave function, ψ_{nlm}, can be written (Section 5.4) as a product of a spherical harmonic $(Y_{lm}(\theta, \phi))$ and a radial part $(R_{nl}(r))$ and we can separate the radial and angular integrations. We then have, simplifying the notation by dropping the (θ, ϕ) from $Y_{lm}(\theta, \phi)$:

$$\langle \psi_{n'l'm'} | r^k Y_{k\alpha} | \psi_{nlm} \rangle = \langle R_{n'l'} | r^k | R_{nl} \rangle \langle Y_{l'm'} | Y_{k\alpha} | Y_{lm} \rangle \equiv \overline{r^k} \cdot \Gamma_{k\alpha} \qquad (7.2.6)$$

where $\overline{r^k}$ is the expectation value of the kth power of the distance of the electron from the nucleus and $\Gamma_{k\alpha}$ stands for the angular integral. We now examine the angular integral in detail in order to make use of symmetry and the properties of the spherical harmonics to reduce the sums over α and k.

For d orbitals $l' = l = 2$ and, as we know from Chapter 3 (Equation (3.6.1)), the combination (coupling) of $Y_{2m}(\theta, \phi)$ with $Y_{2m'}(\theta, \phi)$ gives the functions $Y_{l''m''}(\theta, \phi)$, where $(2-2) \le l'' \le (2+2)$, i.e. l'' can only take the integer values 0 to 4. The angular integral then reduces to a sum of terms of the form $\langle Y_{l''m''} | Y_{k\alpha} \rangle$ and since the spherical harmonics are a set of orthogonal functions, it will be non-zero only for values of k of 0, 1, 2, 3 and 4.

A further property, the parity, of the integral $\langle Y_{2m'} | Y_{k\alpha} | Y_{2m} \rangle$ can now be used to limit the possible values of k even further. For an integral to be non-zero the integrand, $Y_{2m'} \cdot Y_{k\alpha} \cdot Y_{2m}$ in our case, must be even. The parity of a spherical harmonic Y_{lm} is odd or even depending on whether its lobes change sign or retain their sign on inversion in the origin of coordinates so that the function is odd when l is odd and even when l is even or zero. The d orbitals $(l = 2)$ are even (see Figure 5.6). It follows therefore, since odd $\times$ odd = even $\times$ even = even while odd $\times$ even = odd, that when k is odd, $\Gamma_{k\alpha}$ is zero and values of $k = 1$ or 3 result in an integral of zero. These values of k can therefore be neglected.

Thus our d-electron matrix element can be expressed as a sum of fifteen terms:

$$\left\langle \psi_{n'2m'} \left| \frac{qe}{d} \right| \psi_{n2m} \right\rangle = qe \left\{ A_{00} \overline{r^0} \Gamma_{00} + \sum_{\alpha=-2}^{\alpha=+2} A_{2\alpha} \overline{r^2} \Gamma_{2\alpha} + \sum_{\alpha=-4}^{\alpha=+4} A_{4\alpha} \overline{r^4} \Gamma_{4\alpha} \right\} \qquad (7.2.7)$$

This expression can be used to calculate the d-electron matrix elements for any arrangement of charges surrounding a metal ion. Each of the surrounding charges will give a contribution of the above form to each matrix element. The contribution from each charge will be characterised by the different values of the $A_{k\alpha}$ corresponding to the particular values of r_q, θ_q and ϕ_q for that charge. The values of the $\Gamma_{k\alpha}$ will be unchanged.

But where the disposition of the surrounding charges is highly symmetrical the number of terms in the summation can be reduced further by placing restrictions on the possible values of α, as will now be demonstrated for the case of an octahedral field. An octahedral complex with identical charges of $-q$ at the ends of the three Cartesian axes is shown in Figure 7.2. The relationship between the polar and Cartesian coordinate systems is shown in Appendix 7.

Since the six charges are identical and the electron density of a d-electron wave function (in its complex form) is cylindrically symmetrical with respect to rotation about the Z axis, the contribution to a matrix element of the charge at $+X$ (polar coordinates r_q, $\pi/2$, 0) will be equal to that of the charge at $+Y$ $(r_q, \pi/2, \pi/2)$. Thus, the values of the $A_{k\alpha}$

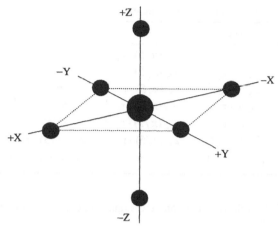

Figure 7.2 An atom surrounded by an octahedral array of negative electrostatic charges

at these two positions must be equal. Since the only difference between the two positions is in the value of ϕ_q, we need examine only that part of the expression for $A_{k\alpha}$ which is $\Phi(\phi_q) = \exp(-i\alpha\phi_q) = \cos\alpha\phi_q + i\sin\alpha\phi_q$. We have:

at $+X$ $\qquad\qquad\qquad A_{k\alpha} \propto \Phi(0) = \cos 0 + i\sin 0$

and at $+Y$ $\qquad\qquad A_{k\alpha} \propto \Phi(\alpha\pi/2) = \cos\alpha\pi/2 + i\sin\alpha\pi/2$

The two values of $A_{k\alpha}$ can only be equal when $\alpha\pi/2$ is equal to zero or an integer multiple of 2π. Therefore, $\alpha = 0, \pm 4, \pm 8\ldots$, and since the upper limit of $|\alpha|$ is 4 we have now restricted the possible values of α to just three, 0 and ± 4, giving:

$$\left\langle \psi_{n'2m'} \left| \frac{qe}{d} \right| \psi_{n2m} \right\rangle = qe\left\{ A_{00}\overline{r^0}\Gamma_{00} + A_{20}\overline{r^2}\Gamma_{20} + A_{40}\overline{r^4}\Gamma_{40} + A_{4-4}\overline{r^4}\Gamma_{4-4} + A_{44}\overline{r^4}\Gamma_{44} \right\}$$

Furthermore, simple evaluation shows that $A_{44} = A_{4-4}$ for each of the four possible values of $\phi_q(0, \pm\pi/2, \pi)$ so that:

$$\left\langle \psi_{n'2m'} \left| \frac{qe}{d} \right| \psi_{n2m} \right\rangle = qe\left\{ A_{00}\overline{r^0}\Gamma_{00} + A_{20}\overline{r^2}\Gamma_{20} + A_{40}\overline{r^4}\Gamma_{40} + A_{44}\overline{r^4}(\Gamma_{4-4} + \Gamma_{44}) \right\}$$

We are now in a position to calculate and tabulate (Table 7.1) the values of $A_{k\alpha}$ and $\Gamma_{k\alpha}$, which we need to evaluate the required matrix elements. The values of $A_{k\alpha}$ can be readily obtained by substituting appropriate values of θ_q and ϕ_q into Equation (7.2.5). It is convenient to sum the columns of $A_{k\alpha}$ values in the table because, for any particular matrix element, the angular integral, $\Gamma_{k\alpha}$, is independent of the position of the charge.

In evaluating and tabulating the angular integrals, $\langle Y_{2m'}|Y_{k\alpha}|Y_{2m}\rangle$, it is useful to recall that the integration over ϕ is zero unless $m + \alpha - m' = 0$ (see Box 7.1) and, therefore, that we have only diagonal matrix elements, except where $\alpha = \pm 4$. By writing out the spherical harmonics in their full algebraic forms (Appendix 5) and a rather tedious, but straight-forward, integration (Box 7.1) we obtain Table 7.2.

It now remains to combine the results of Tables 7.1 and 7.2 to obtain the required matrix elements. We find the following, where $\Xi \equiv 6qer_q^{-1}$ and $\Omega \equiv qe\overline{r^4}r_q^{-5}$.

Table 7.1 Values of $A_{k\alpha}$ for an octahedral complex

	θ_q	ϕ_q	A_{00}	A_{20}	A_{40}	A_{44}
+X	$\pi/2$	0	$2\sqrt{\pi}\cdot r_q^{-1}$	$-\sqrt{(\pi/5)}\cdot r_q^{-3}$	$(\sqrt{\pi}/4)\cdot r_q^{-5}$	$\sqrt{(35\pi/288)}\cdot r_q^{-5}$
+Y	$\pi/2$	$\pi/2$	$2\sqrt{\pi}\cdot r_q^{-1}$	$-\sqrt{(\pi/5)}\cdot r_q^{-3}$	$(\sqrt{\pi}/4)\cdot r_q^{-5}$	$\sqrt{(35\pi/288)}\cdot r_q^{-5}$
+Z	0	0	$2\sqrt{\pi}\cdot r_q^{-1}$	$2\sqrt{(\pi/5)}\cdot r_q^{-3}$	$2\sqrt{(\pi/3)}\cdot r_q^{-5}$	0
−X	$\pi/2$	0	$2\sqrt{\pi}\cdot r_q^{-1}$	$-\sqrt{(\pi/5)}\cdot r_q^{-3}$	$(\sqrt{\pi}/4)\cdot r_q^{-5}$	$\sqrt{(35\pi/288)}\cdot r_q^{-5}$
−Y	$\pi/2$	$-\pi/2$	$2\sqrt{\pi}\cdot r_q^{-1}$	$-\sqrt{(\pi/5)}\cdot r_q^{-3}$	$(\sqrt{\pi}/4)\cdot r_q^{-5}$	$\sqrt{(35\pi/288)}\cdot r_q^{-5}$
−Z	π	0	$2\sqrt{\pi}\cdot r_q^{-1}$	$2\sqrt{(\pi/5)}\cdot r_q^{-3}$	$2(\sqrt{\pi/3})\cdot r_q^{-5}$	0
Sum			$12\sqrt{\pi}\cdot r_q^{-1}$	0	$7(\sqrt{\pi/3})\cdot r_q^{-5}$	$\sqrt{(35\pi/18)}\cdot r_q^{-5}$

Table 7.2 Values of the non-zero angular integrals for an octahedral complex

	$k = 0$	$k = 2$	$k = 4$
$\langle Y_{20}\lvert Y_{k0}\rvert Y_{20}\rangle$	$1/2\sqrt{\pi}$	$\sqrt{(5/\pi)}/7$	$3/7\sqrt{\pi}$
$\langle Y_{2\pm1}\lvert Y_{k0}\rvert Y_{2\pm1}\rangle$	$1/2\sqrt{\pi}$	$\sqrt{(5/\pi)}/14$	$-2/7\sqrt{\pi}$
$\langle Y_{2\pm2}\lvert Y_{k0}\rvert Y_{2\pm2}\rangle$	$1/2\sqrt{\pi}$	$-\sqrt{(5/\pi)}/7$	$1/14\sqrt{\pi}$
$\langle Y_{2\pm2}\lvert Y_{k\pm4}\rvert Y_{2\pm2}\rangle$	0	0	$\sqrt{(5/14\pi)}$

The Hamiltonian matrix for the five d orbitals in an octahedral crystal field.

$\hat{\mathcal{H}}_{cf}$	$\lvert\psi_{20}\rangle$	$\lvert\psi_{2-1}\rangle$	$\lvert\psi_{2+1}\rangle$	$\lvert\psi_{2-2}\rangle$	$\lvert\psi_{2+2}\rangle$
$\langle\psi_{20}\rvert$	$\Xi + \Omega$	0	0	0	0
$\langle\psi_{2-1}\rvert$	0	$\Xi - 2\Omega/3$	0	0	0
$\langle\psi_{2+1}\rvert$	0	0	$\Xi - 2\Omega/3$	0	0
$\langle\psi_{2-2}\rvert$	0	0	0	$\Xi + \Omega/6$	$5\Omega/6$
$\langle\psi_{2+2}\rvert$	0	0	0	$5\Omega/6$	$\Xi + \Omega/6$

On examining the above matrix we see that the term Ξ occurs in every diagonal element. It represents a uniform increase in energy which all five d orbitals experience as a result of the surrounding six negative charges. If we set this common term aside for a moment, we find that the diagonal elements have the values Ω (ψ_{20}), $-2\Omega/3$ ($\psi_{2\pm1}$) and $\Omega/6$ ($\psi_{2\pm2}$). The off-diagonal matrix element which connects the orbitals with $m = \pm2$ is $5\Omega/6$ and when we diagonalise the 2×2 matrix we find eigenvalues of Ω and $-2\Omega/3$.

The following significant results of the foregoing calculation may now be noted:

- The effect of the octahedral field is to raise the energy of all the d orbitals to a new centre of energetic gravity at Ξ.

- More importantly, it splits the orbital energies into two groups, one twofold degenerate and one threefold. For the doubly degenerate (group theoretical symbol; e_g), $E = \Xi + \Omega$ and for the triply degenerate (group theoretical symbol; t_{2g}) $E = \Xi - 2\Omega/3$. The energy-level splitting of $5\Omega/3$ is usually given the symbol Δ and, since it is such common practice, we shall express energies in terms of Δ rather than Ω here. The new, raised centre of gravity is not changed by the splitting.

- Since there is an off-diagonal matrix element between ψ_{2-2} and ψ_{2+2} these two functions are mixed in equal proportions by the crystal field to give the orbitals which we call ψ_{xy} and $\psi_{x^2-y^2}$. These are the one-electron eigenfunctions of the Hamiltonian operator when it includes the octahedral crystal field.

- There is no off-diagonal matrix element connecting the two functions ψ_{2-1} and ψ_{2+1} so these functions are not mixed by an octahedral field. However, we generally find it convenient to think of them in terms of the two mixed eigenfunctions ψ_{xz} and ψ_{yz} which have the same energy as ψ_{2-1} and ψ_{2+1}.

- The e_g orbital set with energy $\Xi + 3\Delta/5$ is:

$$\psi_{z^2} = \psi_{20} \text{ and } \psi_{x^2-y^2} = (1/\sqrt{2}) \cdot (\psi_{2-2} + \psi_{2+2})$$

- The t_{2g} orbital set with energy $\Xi - 2\Delta/5$ is:

$$\psi_{xy} = (i/\sqrt{2}) \cdot (\psi_{2-2} - \psi_{2+2}), \ \psi_{xz} = (1/\sqrt{2}) \cdot (\psi_{2-1} - \psi_{2+1}) \text{ and}$$

$$\psi_{yz} = (i/\sqrt{2}) \cdot (\psi_{2-1} + \psi_{2+1})$$

- If we imagine the d orbitals in their real form (Figure 5.6) then it is easy to see, qualitatively, why those for which the lobes point directly towards the point charges, ψ_{z^2} and $\psi_{x^2-y^2}$, are raised more in energy that those where the lobes point between the charges, ψ_{xy}, ψ_{xz} and ψ_{yz}. It is far from obvious why the energies of the ψ_{z^2} and $\psi_{x^2-y^2}$, are raised equally, but our quantitative calculation shows that this is indeed the case.

- In principle, an attempt might be made to evaluate the matrix elements exactly. But this is very difficult on account of our poor knowledge of the electron distribution which is required to evaluate the terms in $\overline{r^4}$. Furthermore, the point charge model is a very approximate one and it is unlikely that the quantitative results obtained would justify the effort of calculating them. The CF model is therefore used in a qualitative manner as examples of its applications below will show.

Similar calculations can be carried out for any number and disposition of charges around a central metal ion. The results of such calculations for equal charges placed at the vertices of a cube, an octahedron and a tetrahedron are illustrated in Figure 7.3. Note that the increase in the centre of gravity is directly proportional to the number of charges so that Ξ cube : Ξ octahedron : Ξ tetrahedron $= 4:3:2$. The splittings of the levels are not simply related to the number of charges; Δ cube : Δ octahedron : Δ tetrahedron, are in the ratio $8:9:4$. The displacement of the triply degenerate levels (t) from the centre of gravity is always $2\Delta/5$ and that of the doubly degenerate levels (e) is always $3\Delta/5$, so the centre of gravity is maintained. Note however, that in the cube and the tetrahedron $E(t) > E(e)$ whereas in the octahedron $E(t) < E(e)$.

7.3 THE ELECTRONIC ENERGY LEVELS OF TRANSITION-METAL COMPLEXES

The CF theory, as we have described it above, is a one-electron theory which immediately provides a qualitative interpretation of the fact that most transition-metal complexes are

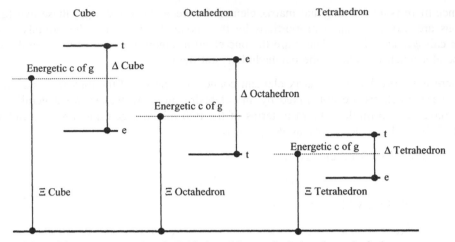

Figure 7.3   d-Electron energy levels fields in cubic, octahedral and tetrahedral symmetry

coloured, i.e. they absorb radiation in the visible region which lies at the low-energy end of the electronic spectral range. For example, an aqueous solution of the Ti^{3+} ion (configuration: Ar $3d^1$) is red and we ascribe the colour to the absorption of green light by a transition of the single d-electron from the t_{2g} levels to the e_g in the octahedral complex $[Ti(H_2O)_6]^{3+}$. This idea can be extended to ions with more electrons but, as we have seen in Chapter 5, we cannot describe the electronic spectra of atoms without considering the effects of inter-electronic repulsion and spin-orbit coupling, and this is equally true of ions. Furthermore, we have here the additional complication of the crystal field. That is, the Hamiltonian operator for the problem is of the form:

$$\hat{\mathcal{H}} = \hat{\mathcal{H}}_0 + \hat{\mathcal{H}}_{er} + \hat{\mathcal{H}}_{SO} + \hat{\mathcal{H}}_{cf}$$

in which $\hat{\mathcal{H}}_0$ represents the kinetic energy of the electrons on the central ion and their attraction to the nucleus of the ion, $\hat{\mathcal{H}}_{er}$ is the inter-electronic repulsion, $\hat{\mathcal{H}}_{SO}$ the spin-orbit coupling and $\hat{\mathcal{H}}_{cf}$ the crystal field. As in the case of the free atom where there was no $\hat{\mathcal{H}}_{cf}$, we have to decide the order in which we apply terms 2, 3 and 4 to correct $\hat{\mathcal{H}}_0$. It turns out that, of the several possibilities, only three are important in practice.

In the *rare earth coupling scheme*, where $\hat{\mathcal{H}}_{SO}$ is large, we diagonalise the matrices of $\hat{\mathcal{H}}_{er}$ and $\hat{\mathcal{H}}_{SO}$ before using $\hat{\mathcal{H}}_{cf}$. Thus, in this case we determine the energy levels of the free ion as exactly as possible, before considering the perturbation of those levels by the crystal field. Problems of this type are few in chemistry and we shall not consider them further here since the other two possibilities are much more important.

In the *weak-field coupling scheme*, $\hat{\mathcal{H}}_{cf}$ is applied as a perturbation to the terms of the free-ion which result from $\hat{\mathcal{H}}_0 + \hat{\mathcal{H}}_{er}$ and the spin-orbit coupling, $\hat{\mathcal{H}}_{SO}$, is added last.

In the *strong-field coupling scheme* the one-electron orbitals of the central ion are first combined under the influence of $\hat{\mathcal{H}}_{cf}$, and then $\hat{\mathcal{H}}_{er}$ and $\hat{\mathcal{H}}_{SO}$ are introduced, in that order. Thus, in the two schemes which are of most interest to chemistry, the effect of spin-orbit coupling is introduced last. This is reflected in our treatment here in that in Sections 7.3.1 to 7.3.4 we first study the effects of $\hat{\mathcal{H}}_{cf}$, and $\hat{\mathcal{H}}_{er}$ leaving the introduction of $\hat{\mathcal{H}}_{SO}$ until Section 7.3.5.

7.3.1   The weak-field scheme for d^2 (example of ^3F in an octahedral field)

In the weak-field approach we first determine the terms of the free ion using the methods in Chapter 5. We then calculate the effect of the crystal field upon these terms and we are usually most interested in the term of highest multiplicity because, according to Hund, this will be the ground term with the lowest energy. In the case of the ^3F arising from d^2 we can work with the $M_S = 1$ component of the triplet for which the required wave functions are (Appendix 10):

$$|3,3\rangle = |+2^+, +1^+\rangle \equiv (1/\sqrt{2}) \cdot \{|\psi_{2,+2}(1) \cdot \psi_{2,+1}(2) - \psi_{2,+1}(1) \cdot \psi_{2,+2}(2)\}$$

$$|3,2\rangle = |+2^+, 0^+\rangle$$

$$|3,1\rangle = \sqrt{(2/5)} \cdot |+1^+, 0^+\rangle + \sqrt{(3/5)} \cdot |+2^+, -1^+\rangle$$

$$|3,0\rangle = (2/\sqrt{5}) \cdot |+1^+, -1^+\rangle + (1/\sqrt{5}) \cdot |+2^+, -2^+\rangle$$

$$|3,-1\rangle = \sqrt{(2/5)} \cdot |0^+, -1^+\rangle + \sqrt{(3/5)} \cdot |+1^+, -2^+\rangle$$

$$|3,-2\rangle = |0^+, -2^+\rangle$$

$$|3,-3\rangle = |-1^+, -2^+\rangle$$

Note that all *kets* on the right-hand sides of the above equations are two-electron Slater determinants (Appendix 6) characterised by the m values of the occupied d-orbitals. A superscript $+$ sign indicates that the electron has $m_s = +\frac{1}{2}$. Each determinant can be expanded as a difference of two orbital products as has been illustrated above for the case of $|+2^+, +1^+\rangle$. The subscripts on ψ are the l and m values which characterise that orbital.

The calculation of the matrix elements of $\hat{\mathcal{H}}_{cf}$ is illustrated with the example of the element, $\langle 3,3|\hat{\mathcal{H}}_{cf}|3,3\rangle$ in Box 7.2.

The complete matrix is found to be:

$\hat{\mathcal{H}}_{cf}$	$|3,+3\rangle$	$|3,-1\rangle$	$|3,-3\rangle$	$|3,+1\rangle$	$|3,+2\rangle$	$|3,-2\rangle$	$|3,0\rangle$
$\langle 3,+3|$	$-3\Delta/10$	$-\sqrt{15}\Delta/10$	0	0	0	0	0
$\langle 3,-1|$	$-\sqrt{15}\Delta/10$	$-\Delta/10$	0	0	0	0	0
$\langle 3,-3|$	0	0	$-3\Delta/10$	$-\sqrt{15}\Delta/10$	0	0	0
$\langle 3,+1|$	0	0	$-\sqrt{15}\Delta/10$	$-\Delta/10$	0	0	0
$\langle 3,+2|$	0	0	0	0	$7\Delta/10$	$\Delta/2$	0
$\langle 3,-2|$	0	0	0	0	$\Delta/2$	$7\Delta/10$	0
$\langle 3,0|$	0	0	0	0	0	0	$-3\Delta/5$

The eigenvalues of this matrix are:

$-3\Delta/5$, threefold degenerate: group theoretical designation, $^3T_1$.
$+\Delta/5$, threefold degenerate: group theoretical designation, $^3T_2$.
$+6\Delta/5$, singly degenerate: group theoretical designation, $^3A_2$.

It is useful to note here that the presence of three off-diagonal elements in the above matrix means that the three pairs of ^3F basis states $|3,+3\rangle$ and $|3,-1\rangle$, $|3,+2\rangle$ and $|3,-2\rangle$, $|3,+1\rangle$ and $|3,-3\rangle$ are mixed by the field in proportions which depend upon the value of Δ. One basis state, the $|3,0\rangle$ which is a component of the $^3T_1$, is not mixed by the field and we shall find this useful when we take account, as we do now, of the presence of another state of $^3T_1$ symmetry formed by the d orbitals, the ^3P.

7.3.2   The weak-field scheme for d² (inclusion of ³P)

The electron configuration d^2 gives rise to two sets of triplet states, ^3P and ^3F, and in accordance with Hund's rules, $E(^3P)$ is greater than $E(^3F)$. However, the energy gap between the two is not, in general, so large that we can ignore the ^3P levels; especially in the interpretation of spectra. Therefore, we must now examine the effect of the octahedral crystal field on the ^3P state. Using the same notation as above, the three $|L, M_L\rangle$ components of the state having $M_S = 1.0$ may be written:

$$|1, 1\rangle = \sqrt{(2/5)} \cdot |2^+, -1^+\rangle - \sqrt{(3/5)} \cdot |1^+, 0^+\rangle$$

$$|1, 0\rangle = (2/\sqrt{5}) \cdot |2^+, -2^+\rangle - (1/\sqrt{5}) \cdot |1^+, -1^+\rangle$$

$$|1, -1\rangle = \sqrt{(2/5)} \cdot |1^+, -2^+\rangle - \sqrt{(3/5)} \cdot |0^+, -1^+\rangle.$$

The 3×3 matrix of $\hat{\mathcal{H}}_{cf}$ in this basis can be constructed in exactly the same manner as that used above for ^3F. However, we find only three equal diagonal elements of Ξ, which we neglect, and no elements containing Δ appear. This should come as no surprise. A P state has the same electron distribution as a p orbital and if we think of the three real p orbitals (Chapter 5) it is clear that each one points in an identical manner to two of the six charges forming the octahedral field. The triple degeneracy of p orbitals or a P state is not lifted by an octahedral field, as the symmetry designation $^3T_1$ confirms. But there are matrix elements of $\hat{\mathcal{H}}_{cf}$ between components of the ^3P and ^3F states and so our 7×7 matrix for the latter should be extended to 10×10 in order to include the former. Fortunately, this is unnecessary. If we require *only the energies* of the mixed ^3P and ^3F states, we can make use of the facts that the field does not mix the $|3, 0\rangle$ component of the ^3F with any other component of that state and the $|1,0(^3P)\rangle$ interacts solely with the $|3, 0\rangle$ component of the ^3F state. Therefore, we need to evaluate only the matrix element $\langle 3,0(^3F)|\hat{\mathcal{H}}_{cf}|1,0\ (^3P)\rangle$.

Using the well-tried methods we find:

$\hat{\mathcal{H}}_{cf}$	$|3, 0(^3F)\rangle$	$|1, 0(^3P)\rangle$
$\langle 3, 0(^3F)|$	$-3\Delta/5$	$+2\Delta/5$
$\langle 1, 0(^3P)|$	$+2\Delta/5$	$0 + \Lambda$

where Λ is the combination of electron repulsion integrals which are responsible for the difference in energy of the ^3P and ^3F states (see Box 7.3). It is interesting to determine the eigenvalues of this matrix at the weak-field extreme, $\Delta = 0$, and the strong-field extreme, $\Delta \gg \Lambda$. When $\Delta = 0$ the eigenvalues are 0 and Λ, i.e. the separation of the two states is, as we would expect, the same as it would be in the free gaseous ion. When $\Delta \gg \Lambda$ we can neglect Λ and find that the eigenvalues are $-4\Delta/5$ and $+\Delta/5$. We can confirm that these are the expected strong-field eigenvalues by noting that, in an octahedral crystal field, the energies of the five d orbitals split into two groups, e_g at $\Xi + 3\Delta/5$ and t_{2g} at $\Xi - 2\Delta/5$. Neglecting the common contribution to the energy, Ξ, we see that the lowest possible energy of our d^2 system in a strong crystal field occurs when both electrons occupy a

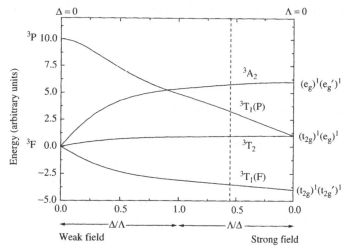

Figure 7.4   Correlation diagram of d^2 in an octahedral field

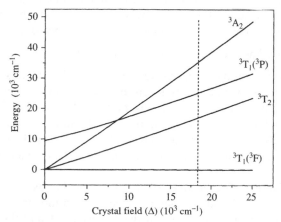

Figure 7.5   An Orgel or Tanabe-Sugano diagram for d^2

different t_{2g} orbital with parallel spins, giving $E\{(t_{2g})^1(t'_{2g})^1\} = -4\Delta/5$. If we have one electron in a t_{2g} and one in an e_g the energy is $E\{(t_{2g})^1(e_g)^1\} = -2\Delta/5 + 3\Delta/5 = +\Delta/5$. We should also note that $E\{(e_g)^1(e'_g)^1\} = +6\Delta/5$. We now have the theoretical analysis required to determine the triplet energy levels of a d^2 ion in an octahedral crystal field of any magnitude and for any value of the $^3P - ^3F$ splitting. The results are illustrated in Figures 7.4 and 7.5 which show, in different ways, how the energies depend on Λ and Δ.

7.3.3   The d^2 energy levels for weak, strong and intermediate octahedral fields

First, we summarise our expressions for the energies.

In the intermediate region, the energies of the ^3F states which do not mix with ^3P are:

$$E(^3A_2 : ^3F) = +6\Delta/5 \text{ and } E(^3T_2 : ^3F) = +\Delta/5$$

The energies of the interacting ^3P and ^3F states, obtained by diagonalising the interaction matrix above, are:

$$E(^3T_1 : {}^3P) = +\tfrac{1}{2}(\Lambda - 3\Delta/5) + \{\tfrac{1}{4}([3\Delta/5] - \Lambda)^2 + 3\Lambda\Delta/5 + (2\Delta/5)^2\}^{\frac{1}{2}}$$

$$E(^3T_1 : {}^3F) = +\tfrac{1}{2}(\Lambda - 3\Delta/5) - \{\tfrac{1}{4}([3\Delta/5] - \Lambda)^2 + 3\Lambda\Delta/5 + (2\Delta/5)^2\}^{\frac{1}{2}}$$

So that in the strong- and weak-field limits we have:

Weak-field limit	Strong-field limit
$\Lambda \neq 0, \Delta = 0$	$\Lambda = 0, \Delta \neq 0$
$E(^3P) = \Lambda$	$E\{(e_g)^1 (e_g')^1\} = +6\Delta/5$
	$E\{(e_g)^1 (t_{2g})^1\} = +\Delta/5$
$E(^3F) = 0$	$E\{(t_{2g})^1 (t_{2g}')^1\} = -4\Delta/5$

In Figure 7.4 the energies of the four d^2 states, calculated using the expressions above, are plotted against Δ/Λ in the left-hand half of the diagram and against Λ/Δ in the right-hand half. The far left is the weak-field limit where $\Delta = 0$, while the far right corresponds to the strong-field limit where $\Lambda = 0$, and the figure represents all possible relative values of the two parameters. The energies have been divided by $0.1\Lambda\{1.0 + \chi^2\}^{\frac{1}{2}}$ on the left and by $0.2\Delta\{1.0 + (1/\chi)^2\}^{\frac{1}{2}}$ on the right, where $\chi = 2\Delta/\Lambda$. The factors 0.1Λ and 0.2Δ are chosen to make the total span of energy levels at both extremes of the diagram the same. The factors in parentheses ensure that levels meet at $\Delta = \Lambda$. Low[5] has measured the electronic spectrum of V^{3+} ions (configuration: Ar $3d^2$) in an aluminium oxide host and finds prominent bands at 17 400, 25 200 and 34 500 cm^{-1}. This is a d^2 ion octahedrally coordinated by oxide ions and represents experimental data which may be interpreted using Figure 7.4.

We proceed as follows. The energies of the three bands observed by Low are in the approximate ratios $1 : 1.5 : 2$ and we therefore seek a position on the horizontal axis of Figure 7.4 where the energy gaps between the lowest level and the three above it are in these ratios. We find it on the strong-field side of the diagram where $\Lambda/\Delta \approx 0.55$. This assigns the spectrum and gives us the ratio of Δ to Λ. By noting that the energy gap between the $^3A_2$ and $^3T_2$ states is Δ, we find their approximate absolute values to be $\Delta \approx 17\,100$ and $\Lambda \approx 9400$ cm^{-1}. The word *approximate* is used for three reasons. Firstly, since we have three energy differences and two unknowns, the problem is over-determined and slightly different values of the parameters would be obtained by using the data in a different way. Secondly, the observed bands are broadened by vibrational effects and the position of the pure electronic transition cannot be determined to better than ±500 cm^{-1}. Thirdly, and most importantly, this simple CF model does not justify more exact analysis; the agreement of theory with experiment is already quite remarkable and the major aim of assigning the spectrum has been achieved.

A second diagrammatic way of analysing the spectrum of a d^2 ion in an octahedral environment is shown in Figure 7.5. Diagrams of this type were first derived by Tanabe and Sugano and by Orgel in 1954/5. Figure 7.5 is a reproduction of part of the Tanabe and Sugano diagram for d^2. The effect upon the triplet state energies of increasing Δ at a constant value of Λ (9500 cm^{-1} in this figure) is shown in wave number units. The energy of the $^3T_1(^3F)$ ground state is subtracted from each of the four energy levels. It therefore appears in the diagram as a horizontal line of zero energy so that the energy gaps to the other three states are quantitatively represented by their vertical heights above this ground state line. A good approximation to the experimental data, and therefore an assignment of the spectrum, is obtained where $\Delta \approx 18\,300$ cm^{-1}. The difference in the value of Δ found from the two diagrams should not be thought significant. Since both are based on exactly the same theoretical analysis, the same values of the two parameters, Δ and Λ, could be deduced from either; but any attempt to bring the results closer together would be quite unjustified by the simple model and the uncertainties inherent in the assignment of precise energies to broad experimental bands.

Tanabe and Sugano reported diagrams of the above type for d^2 to d^8 ions and reproductions of them can be found in numerous books. Both axes are usually graduated in units of Racah's B parameter, a measure of the electron repulsion in the ion. For the d^2 case, for example, $B = \Lambda/15$. In this way a single diagram can be made valid for all ions having the same number of d electrons, in contrast to Figure 7.5 which applies only to a d^2 ion with $\Lambda = 9500$ cm^{-1}.

A few remarks concerning the relative merits of Figures 7.4 and 7.5 are in order. At first glance, the energy-level plots of Figure 7.5 appear to be straight lines, and the fact that the energies of $^3A_2$ and $^3T_2$ depend only on Δ suggests that they should indeed be linear. The energies of the two $^3T_1$ states depend on Λ as well as Δ and are therefore expected to be curves. However, because we have made $^3T_1(F)$ a base line its curvature has been superimposed upon the plots of the other three energies. But the curvature is slight. More important is the fact that Figure 7.5 shows energies as a function of Δ but with a fixed, pre-selected value of Λ and therefore applies for that value of Λ only. This objection does not apply to Tanabe-Sugano diagrams in general, nor to Figure 7.4 in which all possible relative values of Δ and Λ are shown in the one figure. The drawback is that only *relative* values are represented and the diagram conceals the linear dependence of some states on Δ.

7.3.4   The strong-field scheme for d^2 in an octahedral field

The power and ready availability of digital computers have made it easy to calculate the energy levels of a d^2 ion in an octahedral field for any combination of parameters, even when the singlet states are included. It was not always so. Furthermore, when understanding as well as a quantitative analysis is sought, a thoughtful, algebraic approach to a problem has much to recommend it and so we now approach the problem from the strong-field end. In the strong-field approximation to the d^2 problem we first diagonalise $\hat{\mathcal{H}}_{cf}$ and obtain the real d orbitals with energies $\Xi + 3\Delta/5$ and $\Xi - 2\Delta/5$, as in Section 7.2. If we then focus our attention on the states of highest multiplicity, i.e. states in which the two d electrons occupy different orbitals with their spins parallel, we see that we can form the following configurations:

$$(t_{2g})^1(t_{2g}')^1 \quad E = -2\Delta/5 - 2\Delta/5 = -4\Delta/5 \equiv E(^3T_1 : t_{2g}{}^2)$$

$$(t_{2g})^1(e_g)^1 \quad E = -2\Delta/5 + 3\Delta/5 = +\Delta/5 \equiv E(^3T_1 : t_{2g}e_g) \text{ and } E(^3T_2)$$

$$(e_g)^1(e_g')^1 \quad E = 3\Delta/5 + 3\Delta/5 = +6\Delta/5 \equiv E(^3A_2)$$

The uniform rise in energy of 2Ξ has been neglected and the two states of T_1 symmetry are distinguished by the electron configurations from which they originate. Note that two distinct states, $^3T_1$ and $^3T_2$, arise from the configuration $(t_{2g})^1(e_g)^1$. These states also differ in energy when electron repulsion is included due to the different spatial distributions of the z^2 and $x^2 - y^2$ d-orbital wave functions (see Box 7.3).

We now have to introduce the perturbation of the inter-electronic repulsion represented by $\hat{\mathcal{H}}_{er}$. This will change the energies of the electronic states and we can calculate this change by expressing each configuration as a Slater determinant in which the real d orbitals are expressed in terms of the $\psi_{l,m}$. The calculation is quite straightforward but tedious and we will not do it here. It is well described by Ballhausen.[1] We find that the electron repulsion raises the energy of each state as shown in the matrix below. Since we are only interested in energy differences, a constant term (Racah's A) which arises in each diagonal has been omitted. Interelectronic repulsion may also cause states to mix, a phenomenon known as configuration interaction. However, group theoretical principles tell us that states of different symmetries cannot be mixed by an operator like $\hat{\mathcal{H}}_{er}$, which has the full symmetry of the system, and we therefore expect that there will be mixing only between the two states of T_1 symmetry. The complete energy matrix, including the crystal field but with $2\Xi + A$ omitted from each diagonal element, is:

$\hat{\mathcal{H}}_{er}$	$E(^3T_1 : t_{2g}{}^2)$	$E(^3T_1 : t_{2g}e_g)$	$E(^3T_2)$	$E(^3A_2)$
$E(^3T_1 : t_{2g}{}^2)$	$-4\Delta/5 - \Lambda/3$	$-2\Lambda/5$	0	0
$E(^3T_1 : t_{2g}e_g)$	$-2\Lambda/5$	$+\Delta/5 + 4\Lambda/15$	0	0
$E(^3T_2)$	0	0	$+\Delta/5 - 8\Lambda/15$	0
$E(^3A_2)$	0	0	0	$+6\Delta/5 - 8\Lambda/15$

The eigenvalues of the 2×2 $^3T_1$ matrix are $-3\Delta/10 - \Lambda/30 \pm \frac{1}{2}\{\Delta^2 + \Lambda^2 + 6\Lambda\Delta/5\}^{\frac{1}{2}}$. We can check this result by noting that at zero electron repulsion ($\Lambda = 0$) the $^3T_1$ eigenvalues are $-4\Delta/5$ and $+\Delta/5$, while at zero field ($\Delta = 0$) their energy separation is Λ. These are the correct limiting results. The square root in the above energies makes their use in assigning spectra and determining the values of the parameters Λ and Δ a little difficult. But in the strong field regime where $\Delta \gg \Lambda$ terms in the energy which involve $(\Lambda/\Delta)^2$ or higher powers may be neglected and we can approximate the square root as follows:

$$\{\Delta^2 + \Lambda^2 + 6\Lambda\Delta/5\}^{\frac{1}{2}} = \Delta\{1 + (\Lambda/\Delta)^2 + 6\Lambda/5\Delta\}^{\frac{1}{2}}$$

$$\approx \Delta\{1 + 6\Lambda/5\Delta\}^{\frac{1}{2}} \approx \Delta\{1 + \tfrac{1}{2} \cdot 6\Lambda/5\Delta\} = \Delta + 3\Lambda/5$$

With this approximation $E(^3T_1 : t_{2g}{}^2) = -4\Delta/5 + \Lambda/5$ and the energy differences, in ascending order, which might be compared with experimental data, are:

$$E(^3T_2) - E(^3T_1 : t_{2g}{}^2) = \Delta - \Lambda/5$$

$$E(^3T_1 : t_{2g}e_g) - E(^3T_1 : t_{2g}{}^2) = \Delta + 3\Lambda/5$$

$$E(^3A_2) - E(^3T_1 : t_{2g}{}^2) = 2\Delta - \Lambda/5.$$

Alexander and Gray[6] have reported bands at $22\,700$ cm^{-1} and $27\,200$ cm^{-1} in the spectrum of the complex, 3d^2 ion $[V(CN)_6]^{3-}$. If we assign the first of these bands to the

$E(^3T_1 : t_{2g}{}^2) \rightarrow E(^3T_2)$ transition and the second to the $E(^3T_1 : t_{2g}{}^2) \rightarrow E(^3T_1 : t_{2g}e_g)$ we readily find $\Delta \approx 23\,800$ cm^{-1} and $\Lambda \approx 5600$ cm^{-1}. The large value of Δ/Λ shows that we really are in the strong-field regime and justifies the above neglect of $(\Lambda/\Delta)^2$ and higher powers. With these values of Δ and Λ we predict that the $E(^3T_1 : t_{2g}{}^2) \rightarrow E(^3A_2)$ band should lie at about $46\,500$ cm^{-1} where it is hidden under much stronger bands due to other electronic transitions.

7.3.5   Spin-orbit coupling

The final influence upon energy levels is that of spin-orbit coupling which, for the transition metals of the first series, is much smaller than the crystal field or the electron repulsion. It becomes increasingly important as we descend the periodic table and especially when we encounter the rare earths. As an illustration we consider the effect of increasing spin-orbit coupling on the configuration d^2 in a strong crystal field (Figure 7.6). On the left the four triplet energy levels are plotted. The matrix of all 45 possible d^2 states has been diagonalised with $\Lambda = 5610$, $\Delta = 23\,800$ and the spin-orbit coupling constant, ζ_d, $= 0 - 2000$ cm^{-1}. These values of the parameters were chosen so as to reproduce the spectrum of $[\mathrm{V(CN)_6}]^{3-}$ discussed above at low values of ζ_d. The energy of $^3T_1(^3F)$ when ζ_d, $= 0$ has been set to zero and, in order to include all 13 distinct levels in one diagram, $19\,000$, $20\,000$ and $37\,000$ cm^{-1} have been subtracted from the energies of the $^3T_2$, $^3T_1$ and $^3A_2$ states respectively. All ^3T levels behave qualitatively as if they were ^3P, that is they give rise to states having $J\,(= L + S)$ values of 2, 1 and 0 for non-zero ζ_d, and the 5-fold degeneracy of the first is lifted further by the crystal field. The development of the levels as ζ_d is increased from zero to 2000 cm^{-1} is shown with the J values, and degeneracies $(2J + 1)$ on the right. Note that at low values of ζ_d the five-fold degeneracy of the states with $J = 2$ is scarcely lifted and in this range the levels in each group obey an approximate Landé interval rule, i.e. $E(J = 2) - E(J = 1) \approx 2\{E(J = 1) - E(J = 0)\}$.

In order to assess the effect of spin-orbit coupling on electronic spectra we first note that the ranges of the experimentally determined values of ζ_{nd} are 50–850, 200–1900 and 300–5000 cm^{-1} for $n = 3$, 4 and 5 respectively[4] whilst experimental band widths

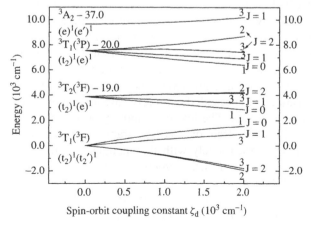

Figure 7.6   The effect of spin-orbit coupling on the strong-field triplet states of d^2

typically lie between 1000 and 4000 cm^{-1}. Thus, the splittings in Figure 7.6 attain the order of magnitude of typical bandwidths only on the right-hand side of the diagram where $\zeta_d = 2000$ cm^{-1}. Such values of ζ_d are not found until the end of the second transition metal series and in the third. The result is that it is rare to see evidence of spin-orbit coupling in the form of band splittings in electronic spectra. By mixing states of different multiplicity, spin-orbit coupling also causes breakdown of the strict $\Delta S = 0$ selection rule (see Section 5.11) but this effect also requires that ζ_d have a substantial value and spectral bands made allowed in this manner are rarely observed with light transition metals. However, Alexander and Gray[6] observed two such bands, $^3T_{1g} \rightarrow {}^1T_{2g}$ at 14 600 cm^{-1} and $^3T_{1g} \rightarrow {}^1T_{1g}$ at 19 200 cm^{-1}, in their spectrum of $K_3V(CN)_6$. For V^{3+} $\zeta_d \approx 860$ cm^{-1}.

Thus far in our account of the theory of the electronic structure and bonding in transition-metal complexes we have given a fairly detailed analysis of d^2 systems. In a non-specialised volume such as this, space does not permit a similar treatment of d^3 to d^{10} for which we must refer the reader to the bibliography.[1,4] We have been drawn into a consideration of electronic spectra because spectral data provide the most direct way of determining Λ and Δ, the essential parameters of the CF model. No account of the theory of the transition-metal ions would be complete without some reference to their electronic spectroscopy. But this is a vast and complicated subject which is well treated in a number of specialist volumes[4,7] and here we can only draw attention to a few significant points.

7.4 THE ELECTRONIC SPECTROSCOPY OF TRANSITION-METAL COMPLEXES

Thus far we have compared the positions of calculated and observed electronic transitions without asking whether or not those transitions were sufficiently intense to be seen. Since we have been concerned exclusively with d electrons, the states involved must all be of even parity and the transitions, all d $\rightarrow$ d, between states of the same parity. Spectroscopic selection rules are discussed in detail in Sections 5.11, 8.4 and 8.5, so it suffices here to say that transitions between states of the same parity are strictly forbidden; the Laporte rule. The fact that these d $\rightarrow$ d bands, though rather weak, are indeed observed is due to the loss of parity which can arise in either a static or a dynamic manner. If the ligand environment is not strictly octahedral, e.g. because of a slight static distortion, then the centre of symmetry of the complex as a whole may be lost and with it the definitive even parity of the d orbitals and states. A d $\rightarrow$ d transition may then become weakly allowed, frequently because the lowered symmetry allows p and d orbitals to mix. This is particularly true of tetrahedral complexes. The tetrahedron has no centre of symmetry and the d $\rightarrow$ d transitions of tetrahedral complexes are invariably more intense than those of their octahedral counterparts. The loss of symmetry need not be permanent and static; vibrations can have the same effect of making a transition partially allowed.

The CF model is concerned solely with the energies of the d-electron states but the intense colours of many transition-metal complexes, e.g. the purple colour of $[MnO_4]^-$ and the blood red of $[Fe(CNS)_6]^{3-}$, arise not from d $\rightarrow$ d transitions, which as we have seen above are normally very weak, but from *charge-transfer* transitions in which electrons are transferred from the ligands to the metal (LMCT). $[Fe(2,2'-bipyridine)_3]^{2+}$ shows a strong absorption at 522 nm due to the reverse process, a metal-to-ligand charge transfer

(MLCT). Transitions of this type are further evidence of covalency and require a molecular orbital treatment for their quantitative understanding.

Apart from in Section 7.3.5, we have referred only to transitions between states of the same multiplicity, triplet states in the case of d^2. $d \rightarrow d$ transitions which involve a change of multiplicity, triplet to singlet in the case of d^2, are doubly forbidden and are very weak indeed. But as spin-orbit coupling grows in the 4d and 5d series of metals, and especially in the rare earths, nominally spin-forbidden transitions gain in intensity because of the loss of a clearly defined multiplicity for each state. The multiplicity of the ground state of a transition-metal complex is not only important for the electronic spectroscopy of the compound, there are other interesting consequences, some of which we explore in the next two sections.

7.5   PAIRING ENERGIES; LOW-SPIN AND HIGH-SPIN COMPLEXES

Consider the aufbau principle as applied to a d^4 complex in a strong octahedral field. Following Hund's rules, the first three electrons will be placed in the triply degenerate t_{2g} orbitals with their spins parallel. In placing the fourth electron we have a choice between double occupation of a t_{2g} orbital, giving a triplet state, or the use of an empty e_g orbital giving a quintet. The actual orbital occupation will be that of lower energy which depends upon the relative magnitudes of the interelectronic repulsion in a doubly occupied t_{2g} orbital and the energy gap, Δ, between the t_{2g} and e_g orbitals. Calculation shows that:

$$e_g \qquad -\uparrow- \quad \underline{\qquad} \qquad\qquad \underline{\qquad} \quad \underline{\qquad}$$
$$t_{2g} \qquad -\uparrow- \quad -\uparrow- \quad -\uparrow- \qquad -\uparrow\downarrow- \quad -\uparrow- \quad -\uparrow-$$
$$\text{Energy } {}^5E = 6A - 21B + \Delta \qquad \text{Energy } {}^3T_1 = 6A - 15B + 5C$$

In these equations the electron repulsion energy is written in terms of Racah parameters, A, B and C[4] (Box 7.3) and we do not require to delve into the exact form of these integrals to see that when $\Delta = 6B + 5C$ the energies of the two states are equal. Similar equalities arise in the d^5, d^6 and d^7 cases so that a large crystal-field splitting causes electron spins to pair whilst small values of Δ lead to the maximum number of unpaired spins. The corresponding complexes are referred to as *low-spin* and *high-spin* respectively. Though the consequences of the two possibilities for electron pairing can be detected in the spectra of the complex and in many of its other physical and chemical properties, the effect is most marked in its magnetic behaviour which is largely determined by the number of unpaired electrons.

7.6   THE MAGNETISM OF TRANSITION-METAL COMPLEXES

The following formula for the magnetic moment (μ) of a molecule in Bohr magnetons is discussed in Section 12.2:

$$\mu_{\text{total}} = -[\{L(L+1)\} + 4\{S(S+1)\}]^{\frac{1}{2}} \mu_B$$

It is appropriate for situations in which the multiplet splittings that result from spin-orbit coupling are small compared with kT and should therefore be applicable to the transition-metal ions from Ti^{3+} to Cu^{2+} for which the spin-orbit coupling constant, ζ_d, is small. L and S are the total orbital and spin angular momenta respectively. In Table 7.3 the above formula is used to calculate the magnetic moments of the free transition-metal ions of the first transition series and the results are compared with the measured moments for the octahedrally co-ordinated ions in their high-spin states. The μ_S (spin-only) values were calculated by setting $L = 0$. The configurations marked † are those for which there can be an orbital contribution to the magnetic moment (see below) n = number of d electrons, n′ = number unpaired.

Comparison of the theoretical and experimental data in Table 7.3 reveals that, in many cases, the experimental magnetic moment is quite close to the spin-only value, i.e. the electronic orbital angular momentum appears to contribute very little to the total magnetic moment of a $3d^n$ transition-metal ion. This has the important, and particularly useful, consequence that $\mu_{exp.}$ gives a very direct indication of the number of unpaired electrons. The reason for the small contribution from the orbital motion of the electrons is that this has been *quenched* by the interaction of the metal d-electrons with the ligands. In our model this interaction is purely electrostatic, but the fact that the quenching is in many cases so complete strongly suggests that there is covalent bonding between ligand and metal. The latter provides an even better quenching mechanism since it prevents the electrons from circulating the metal nucleus by locking them into localised chemical bonds.

Though the lack of an orbital contribution to the magnetic moment is readily qualitatively understandable, it is of interest to enquire whether we can place the concept of quenching on a firmer theoretical basis in order to understand why, in some cases at least, significant orbital contributions to the magnetic moment remain. In pursuit of this objective we first recall the matrix of the five d orbitals in an octahedral crystal field

Table 7.3 Calculated magnetic moments, in Bohr magnetons, of the ground states of some free transition-metal ions and the experimental magnetic moments for the octahedrally co-ordinated ions in their high-spin configurations

Ion	S	L	n	n′	μ_S	μ_{total}	μ_{exp}
Ti^{3+}	$\frac{1}{2}$	2	1	1	1.73	3.00	$1.7 - 1.8^\dagger$
V^{4+}	$\frac{1}{2}$	2	1	1	1.73	3.00	$1.7 - 1.8^\dagger$
V^{3+}	1	3	2	2	2.83	4.47	$2.6 - 2.8^\dagger$
Cr^{3+}	$\frac{3}{2}$	3	3	3	3.87	5.20	$3.7 - 3.9$
Cr^{2+}	2	2	4	4	4.90	5.48	$4.7 - 4.9$
Mn^{3+}	2	2	4	4	4.90	5.48	$4.9 - 5.0$
Mn^{2+}	$\frac{5}{2}$	0	5	5	5.92	5.92	$5.6 - 6.1$
Fe^{3+}	$\frac{5}{2}$	0	5	5	5.92	5.92	$5.7 - 6.0$
Fe^{2+}	2	2	6	4	4.90	5.48	$5.1 - 5.5^\dagger$
Co^{3+}	2	2	6	4	4.90	5.48	ca. $5.4^\dagger$
Co^{2+}	$\frac{3}{2}$	3	7	3	3.87	5.20	$4.1 - 5.2^\dagger$
Ni^{2+}	1	3	8	2	2.83	4.47	$2.8 - 3.5$
Cu^{2+}	$\frac{1}{2}$	2	9	1	1.73	3.00	$1.7 - 2.2$

†The configurations marked † are those for which there can be an orbital contribution to the magnetic moment, see text. n = number of d electrons, n′ = number unpaired.
H.L. Schläfer and G. Gliemann, *Basic Principles of Ligand Field Theory*. © 1969, John Wiley & Sons Ltd; reprinted with permission.

(Section 7.2) and the remarks following it. The two d orbitals, having m values of $+2$ and -2, were found to have exactly the same diagonal matrix element ($\Xi + \Omega/6$) and to be connected, and hence mixed, by an off-diagonal matrix element of $5\Omega/6$. Since the diagonal elements are equal, the resulting mixed wave functions, d_{xy} and $d_{x^2-y^2}$, will each contain equal parts of $m = +2$ and $m = -2$ and will therefore have no z-component of orbital angular momentum. Since there are no other off-diagonal matrix elements, the remaining three d orbitals will retain their z-components of angular momentum of $\pm 1h/2\pi$ and 0. In an octahedral field, the description of these orbitals as d_{xz}, d_{yz} and d_{z^2} rather than as d_{+1}, d_{-1} and d_0 is a matter of choice, not of necessity. Therefore, when a magnetic field is applied along the z-direction the d_{-1} will be slightly lower in energy than the d_{+1}, because of the field produced by the orbiting electron, and if the configuration is such that it is possible to place more electrons in d_{-1} than in d_{+1} an orbital contribution to the total magnetic moment of the material may result. But note that this potential orbital moment can still be quenched by forces not considered in our simple electrostatic interaction matrix. Identifying the two groups of degenerate d orbitals in the octahedral field by their group-theoretical symbols, the configurations which can manifest an orbital contribution to the magnetic moment are, for high spin, i.e. minimum possible electron pairing: $(t_{2g})^1$, $(t_{2g})^2$, $(t_{2g})^4(e_g)^2$ and $(t_{2g})^5(e_g)^2$. These configurations are marked † in Table 7.3. In the low spin configurations with maximum spin pairing the configurations $(t_{2g})^4$ and $(t_{2g})^5$ may contribute orbital moments. When we compare the high-spin predictions with the above table we see that the ions with six and seven d electrons are indeed those where the highest experimental moments match μ_{total} most closely. The minimal orbital contributions to the ions with one or two d electrons are thought to be due to deviations of the complexes from exact octahedral geometry. This may be a result of forces in the crystal lattice (all the measurements in Table 7.3 were made on solids) or they may be due to a Jahn-Teller effect, the origin of which space does not allow us to pursue further here. In either case, a small distortion links the d_{+1} and d_{-1} orbitals with a small off-diagonal matrix element and causes them to mix forming d_{xz} and d_{yz}, whereby the z-component of their orbital angular momentum is lost.

As a last word on the complex subject of orbital contributions to magnetic moment we might note that, if spin-orbit coupling is significant, in configurations having less than five electrons the spin and orbital angular momenta are antiparallel in the state of lowest energy so the orbital contribution has to be subtracted from the spin contribution. The reverse is the case where the number of d electrons is greater than five.

7.7 COVALENCY AND THE LIGAND FIELD THEORY

As early as 1935, Van Vleck recognised that any widely applicable theory of transition-metal complexes would have to include the effects of covalent interaction between metal and ligand, even though compelling experimental evidence for it was unavailable at that time. In the 1950s, however, firstly electron paramagnetic resonance (EPR) and then nuclear magnetic resonance (NMR) measurements left no room for doubt and provided quantitative evidence. The essential nature of the evidence is easy to explain. In the case of EPR spectroscopy,[8] we measure the absorption of radiation (in the microwave region) by one or more unpaired electrons in a magnetic field where the energies of electrons having $m_s = +\frac{1}{2}$ and $m_s = -\frac{1}{2}$ are different. Apart from the magnetic field imposed by the spectrometer, the electron sees other fields, in particular the fields due to the magnetic nuclei (Chapter 10), which it visits as it makes its peregrinations through the molecule. These fields cause splittings, known as *fine structure*, in the spectra and the magnitude of a particular splitting is proportional to the product of the nuclear magnetic moment and the probability that the unpaired electron will be found at that particular nucleus.[‡] In this way the interaction of unpaired electrons from the Mn^{2+} ion with F^- ions in MnF_2 was observed and measured; to mention just one of many similar experiments. Clearly, such a hyperfine interaction is only possible if there is mixing of metal and ligand atomic orbitals to form molecular orbitals, i.e. covalency. The NMR evidence is similarly direct and will not be discussed here. Thus, covalency cannot be denied, but the great success of the simple CF theory also leads us to expect that there will be important similarities between the results of an MO theory of transition-metal complexes and those of the crystal field, and indeed there are.

We can illustrate this point by applying the MO theory, in linear combination of atomic orbitals (LCAO) form, to a simple, hypothetical, octahedral complex, MH_6, in which the d-orbitals of the central metal ion form covalent bonds with the 1s orbitals of six surrounding hydrogen atoms placed equally distant from the metal atom at $\pm x$, $\pm y$ and $\pm z$ (Figure 7.7).

Our calculation will use a very simple MO theory, quite like the Hückel procedure (Section 12.1) but, because the interactions between the metal 3d orbitals and the surrounding hydrogen 1s orbitals, ϕ_i, are not all equal, we shall find it necessary to use the $\langle 3d \mid 1s \rangle$ overlap to estimate the bonding between them. We take the d orbitals in their real forms and we note immediately that those for which the lobes lie *between* the Cartesian axes, i.e. d_{xz}, d_{yz} and d_{xy} have zero overlap, and hence zero interaction with the hydrogen atoms. This is because each hydrogen 1s orbital has an equal and opposite overlap with the lobes of the above d-orbitals which lie on either side of it (Figure 7.8). By contrast, the $d_{x^2-y^2}$ and d_{z^2} orbitals, which are directed *along* the Cartesian axes, overlap strongly with their neighbouring hydrogen 1s orbitals. We can already see here the basis of a distinction

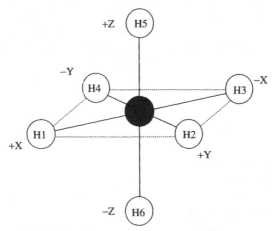

Figure 7.7 Octahedral MH_6 molecule

[‡] The phrase '*at the nucleus*' may appear a little strange, but we are talking here not of the dipole-dipole but the Fermi-contact interaction for which the electron has to have a finite probability of actually being at the nucleus of the atom in question. Carrington and McLachlan[8] give a nice treatment of the theory involved. See also Section 4.9.

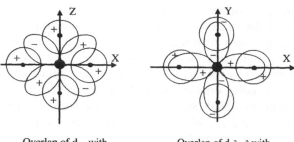

Overlap of d_{xz} with
H1, H3, H5 and H6

Overlap of $d_{x^2-y^2}$ with
H1, H2, H3 and H4

Figure 7.8 Atomic orbital overlap in MH_6

between these two sets of d orbitals which is central to the success of the CF theory. The overlaps which we require may be expressed in terms of an arbitrary parameter, σ, as follows:

$$\langle d_{z^2}|\phi_1\rangle = \langle d_{z^2}|\phi_2\rangle = \langle d_{z^2}|\phi_3\rangle = \langle d_{z^2}|\phi_4\rangle = -\sigma$$

$$\langle d_{z^2}|\phi_5\rangle = \langle d_{z^2}|\phi_6\rangle = 2\sigma$$

$$\langle d_{x^2-y^2}|\phi_1\rangle = -\langle d_{x^2-y^2}|\phi_2\rangle = \langle d_{x^2-y^2}|\phi_3\rangle = -\langle d_{x^2-y^2}|\phi_4\rangle = \sqrt{3}\sigma$$

And we assume that the interactions, which we express in the form of a Hamiltonian energy matrix (Appendix 3), are proportional to these overlaps:

$\hat{\mathcal{H}}$	$d_{x^2-y^2}$	d_{z^2}	ϕ_1	ϕ_2	ϕ_3	ϕ_4	ϕ_5	ϕ_6
$d_{x^2-y^2}$	Ed	0	$\sqrt{3}\beta$	$-\sqrt{3}\beta$	$\sqrt{3}\beta$	$-\sqrt{3}\beta$	0	0
d_{z^2}	0	Ed	$-\beta$	$-\beta$	$-\beta$	$-\beta$	2β	2β
ϕ_1	$\sqrt{3}\beta$	$\sqrt{3}\beta$	Es	0	0	0	0	0
ϕ_2	$-\sqrt{3}\beta$	$-\beta$	0	Es	0	0	0	0
ϕ_3	$\sqrt{3}\beta$	$-\beta$	0	0	Es	0	0	0
ϕ_4	$-\sqrt{3}\beta$	$-\beta$	0	0	0	Es	0	0
ϕ_5	0	2β	0	0	0	0	Es	0
ϕ_6	0	2β	0	0	0	0	0	Es

In this matrix, Ed and Es are the energies of the metal d orbitals and hydrogen 1s orbitals respectively. The off-diagonal elements are the interaction energies expressed in terms of a parameter β, proportional to σ, which, for our present purposes, we do not need to quantify further. When the matrix is diagonalised (Box 7.4) we find four eigenvalues of Es. The eigenfunctions that correspond to these eigenvalues are combinations of hydrogen 1s orbitals which do not have the correct symmetry, e_g or t_{2g}, to combine with any of the metal d orbitals. We also find two eigenvalues of:

$$E_b = \frac{Ed + Es - \sqrt{(Ed - Es)^2 + 48\beta^2}}{2}$$

and two of:

$$E_a = \frac{Ed + Es + \sqrt{(Ed - Es)^2 + 48\beta^2}}{2}$$

These energies are those of the bonding, E_b, and antibonding, E_a, molecular orbitals formed by interaction of the metal $d_{x^2-y^2}$ and d_{z^2} orbitals with combinations of the hydrogen 1s orbitals of the hydrogen atom ligands which have the same symmetry. Since the d orbitals are both of e_g symmetry the molecular orbitals must also be of e_g symmetry. The symmetry is confirmed by the fact that we have a degenerate pair of bonding and a degenerate pair of anti-bonding orbitals. As we have seen above, the hydrogen atoms of our MH_6 molecule offer no orbitals suitable for combination with metal d orbitals of t_{2g} symmetry, so the d_{xz}, d_{yz} and d_{xy} orbitals of the metal take no part in the bonding and remain triply degenerate with energy Ed, which lies between the bonding and antibonding e_g molecular orbitals. However, where the ligands have valence p orbitals interaction with the d_{xz}, d_{yz} and d_{xy} orbitals of the metal is possible and triply degenerate bonding and antibonding molecular orbitals result. Thus the MO model reproduces all the essential feature of the CF. Our results for the MH_6 complex are illustrated in Figure 7.9.

Thus far, the energy-level scheme above reflects the CF scheme quite well but a problem arises when we ask which energy levels are occupied and which unoccupied. Six hydrogen atoms provide six electrons, four of which would fill the lowest pair of e_g levels leaving two for the three t_{1u} and the a_{1g} levels. Therefore, at least seven d electrons from the metal are required before any enter the t_{2g} and e_g levels, the occupation of which is central to the CF model. Our simple MO model of MH_6 appears to be at odds with the CF description. But we should recall that in the majority of simple transition-metal complexes the ligands each donate two electrons to the structure. These two electrons are usually a lone pair, as we see very clearly in the coordination complexes of water and ammonia. If we assume that the hydrogen 1s orbitals we have used are, in effect, lone pair orbitals containing two electrons, all the essentials, including the symmetry, of the energy-level scheme (Figure 7.9) remain unchanged, but we now have 12 ligand electrons at our disposal; just enough to fill the six lowest levels, i.e. e_g, t_{1u} and a_{1g}. Any metal d electrons must then occupy the t_{2g} and e_g levels, in excellent correspondence with the CF model.

The pairs of bonding and antibonding wave functions for the energy levels of e_g symmetry bear closer scrutiny. They arise from two identical 2×2 matrices (see Box 7.4) and therefore have the same form in both cases. Taking, as our example, the combination

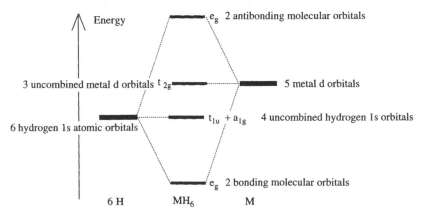

Figure 7.9 Molecular orbital energy-level scheme for MH_6

of ligand orbitals, $\Psi_{eg(z^2)}$, which bonds to the d_{z^2} metal orbital, we can write the bonding and antibonding molecular orbitals in the forms:

$$\Psi_{bonding} = \rho d_{z^2} + \tau \Psi_{eg(z^2)} \quad \text{and} \quad \Psi_{antibonding} = \tau d_{z^2} - \rho \Psi_{eg(z^2)}$$

where $\rho^2 + \tau^2 = 1.0$ if orbital overlap between the d orbitals and the hydrogen 1s orbitals is neglected in the normalisation. At one extreme where $\tau = 1.0$ and $\rho = 0.0$ we have the CF theory. The lower-energy MO, $\Psi_{bonding}$, is composed solely of ligand orbitals, $\Psi_{eg(z^2)}$, and the higher-energy MO, $\Psi_{antibonding}$, is a pure d_{z^2} orbital of the metal. The three uncombined metal d orbitals of t_{2g} symmetry lie immediately below the latter. When $\tau \approx \rho \approx 1/\sqrt{2}$ (τ and ρ will only be exactly equal to $1/\sqrt{2}$ when $Ed = Es$) we have Pauling's covalent valence bond description[2] with metal d-orbitals and ligand orbitals contributing approximately equally to the two molecular orbitals. In principle, a high-quality MO calculation will determine just where within this range of possibilities a particular complex lies. But even today MO or DFT calculations with such a large number of electrons present considerable problems. Comprehensive packages by means of which these theories can be applied to transition metal complexes are available but their *modi operandi* require careful study if they are to be used with confidence and the limitations on their results understood.

In an important semi-empirical development, Roald Hoffman proposed his 'extended Hückel theory'. Matrix elements analogous to Es, Ed and β above are represented by parameters the values of which are obtained by fitting the results of trial calculations to experiment. The inter-orbital bonding interactions are determined with reference to the corresponding overlaps as we have done for β above, but in a more sophisticated manner. Electron repulsion is not considered explicitly. Though purists have sometimes been rather scathing about the theory, it has found widespread application and acceptance in the semi-quantitative interpretation of some of the vast quantity of data which organo-metallic chemists in particular have generated in the last 50 years.

7.8 BIBLIOGRAPHY AND FURTHER READING

1. H.L. Schläfer and G. Gliemann, *Basic Principles of Ligand Field Theory*, Wiley-Interscience, London, 1969.
   F.A. Cotton, *Chemical Applications of Group Theory*, 2nd edn, Wiley-Interscience, New York, 1971.
   C.J. Ballhausen, *Introduction to Ligand Field Theory*, McGraw-Hill, New York, 1962.
   S.B. Piepho and P.N. Schatz, *Group Theory in Spectroscopy*, Wiley-Interscience, New York, 1983.
   [Helpful introductions to group theory may be found in:
   S.F.A. Kettle, *Symmetry and Structure*, 2nd edn, Wiley, New York, 1995.
   G. Davidson, *Group Theory for Chemists*, MacMillan, London, 1991.]
2. L. Pauling, *The Nature of the Chemical Bond*, 2nd edn, Cornell University Press, Ithaca, 1940.
3. H. Eyring, J. Walter and G.E. Kimball, *Quantum Chemistry*, Wiley, New York, 1944.
4. J.S. Griffith, *The Theory of Transition-Metal Ions*, Cambridge University Press, 1961.
5. W. Low, *Zeit. physik. Chem.*, **13**, 107 (1957).
6. J.J. Alexander and H.B. Gray, *J. Amer. Chem. Soc.*, **90**, 4260 (1968).
7. A.B.P. Lever, *Inorganic Electronic Spectroscopy*, 2nd edn, Elsevier, Oxford, 1984.
8. A. Carrington and A.D. McLachlan, *Introduction to Magnetic Resonance*, Harper & Row, London, 1967.

BOX 7.1　Calculation of an angular integral

Example of the evaluation of an angular integral; $\langle Y_{2-2}|Y_{4-4}|Y_{2+2}\rangle \equiv I$.

Since $\langle Y|$ requires that we take the complex conjugate of Y we have:

$$\langle Y_{2-2}| = |Y_{2+2}\rangle = \Theta_{2+2}(\theta) \cdot \Phi_{+2}(\phi) = \frac{\sqrt{15}}{4} \sin^2 \theta \cdot \frac{1}{\sqrt{2\pi}} \exp(2i\phi)$$

and

$$Y_{4-4} = \Theta_{4-4}(\theta) \cdot \Phi_{-4}(\phi) = \frac{3\sqrt{35}}{16} \sin^4 \theta \cdot \frac{1}{\sqrt{2\pi}} \exp(-4i\phi)$$

Therefore the integral I is:

$$I = \frac{45\sqrt{35}}{256} \cdot \frac{1}{2\pi} \cdot \frac{1}{\sqrt{2\pi}} \int_0^\pi \sin^8 \theta \cdot \sin \theta \, d\theta \int_0^{2\pi} \exp(0) \, d\phi$$

The integration over ϕ gives a multiplicative factor of 2π.

It is perhaps worth noting here that, had the argument of the exponential been anything other than zero, $in\phi$ say, then the result of the integration would be zero because $\sin(n\phi)$ is zero at 0 and 2π while $\cos(n\phi) = 1$ at both limits and the two values cancel. Hence the requirement that the m values of the three spherical harmonics, allowing for a change of sign of the first, sum to zero.

Now, setting $-\sin^2 \theta = \cos^2 \theta - 1$ and using the substitution $x = \cos\theta$, $dx = -\sin\theta \, d\theta$, we have:

$$I = \frac{45\sqrt{35}}{256} \cdot \frac{1}{\sqrt{2\pi}} \int_0^\pi \sin^8 \theta \cdot \sin \theta \, d\theta = -\frac{45\sqrt{35}}{256} \cdot \frac{1}{\sqrt{2\pi}} \int_1^{-1} (x^2 - 1)^4 \, dx$$

Expansion of the bracket and integration of the resulting polynomial in x gives:

$$I = \frac{45\sqrt{35}}{256} \cdot \frac{1}{\sqrt{2\pi}} \cdot 2 \left[\frac{1}{9} - \frac{4}{7} + \frac{6}{5} - \frac{4}{3} + 1 \right] = \frac{45\sqrt{35}}{256} \cdot \frac{1}{\sqrt{2\pi}} \cdot \frac{256}{315} = \sqrt{\frac{5}{14\pi}}$$

BOX 7.2　Calculation of the matrix elements of $\mathcal{H}_{cf}$ for the ^3F state of d^2 in an octahedral field

The calculation of the matrix elements of $\mathcal{H}_{cf}$ will be illustrated with the example of the element, $\langle 3, 3|\hat{\mathcal{H}}_{cf}|3, 3\rangle$.

Expanding each Slater determinant, and noting that we must evaluate $\mathcal{H}_{cf}$ for each electron and sum their contributions, we have:

$$\langle 3, 3|\hat{\mathcal{H}}_{cf}|3, 3\rangle$$
$$= (\tfrac{1}{2}) \cdot \{\langle \psi_{+2}(1), \psi_{+1}(2) - \psi_{+1}(1), \psi_{+2}(2)|\hat{\mathcal{H}}_{cf}(1)|\psi_{+2}(1), \psi_{+1}(2)$$
$$- \psi_{+1}(1), \psi_{+2}(2)\rangle\}$$

$$+ (\tfrac{1}{2}) \cdot \{\langle \psi_{+2}(1), \psi_{+1}(2) - \psi_{+1}(1), \psi_{+2}(2) | \hat{\mathcal{H}}_{cf}(2) | \psi_{+2}(1), \psi_{+1}(2)$$
$$- \psi_{+1}(1), \psi_{+2}(2)\rangle\}$$

where the l subscript, which is always 2 for d orbitals, has been omitted to simplify the notation.

But, in quantum mechanics all electrons are indistinguishable, so each of the above two terms must give the same result and we may therefore write:

$$\langle 3, 3 | \hat{\mathcal{H}}_{cf} | 3, 3\rangle$$
$$= \langle \psi_{+2}(1), \psi_{+1}(2) - \psi_{+1}(1), \psi_{+2}(2) | \hat{\mathcal{H}}_{cf}(1) |$$
$$\psi_{+2}(1), \psi_{+1}(2) - \psi_{+1}(1), \psi_{+2}(2)\rangle$$

Each of the four combinations of bra with ket gives rise to a product of two integrals, one over each electron:

$$\langle \psi_{+2}(1), \psi_{+1}(2) | \hat{\mathcal{H}}_{cf}(1) | \psi_{+2}(1), \psi_{+1}(2)\rangle$$
$$= \langle \psi_{+2}(1) | \hat{\mathcal{H}}_{cf}(1) | \psi_{+2}(1)\rangle \cdot \langle \psi_{+1}(2) | \psi_{+1}(2)\rangle$$

The first of these integrals expresses the interaction of an electron in a d_{+2} AO with the crystal field and its value has been shown in Section 7.2 to be $\Omega/6$ (neglecting Ξ as before). The second is 1.0 because the d orbitals are normalised. Thus, neglecting the rise in the d-orbital centre of gravity, we have:

$$\langle \psi_{+2}(1), \psi_{+1}(2) | \hat{\mathcal{H}}_{cf}(1) | \psi_{+2}(1), \psi_{+1}(2)\rangle = +\Omega/6$$

Similarly:

$$\langle \psi_{+1}(1), \psi_{+2}(2) | \hat{\mathcal{H}}_{cf}(1) | \psi_{+1}(1), \psi_{+2}(2)\rangle = \langle \psi_{+1}(1) | \hat{\mathcal{H}}_{cf}(1) | \psi_{+1}(1)\rangle$$
$$\cdot \langle \psi_{+2}(2) | \psi_{+2}(2)\rangle = -2\Omega/3$$

but:

$$\langle \psi_{+2}(1), \psi_{+1}(2) | \hat{\mathcal{H}}_{cf}(1) | \psi_{+1}(1), \psi_{+2}(2)\rangle \text{ and}$$
$$\langle \psi_{+1}(1), \psi_{+2}(2) | \hat{\mathcal{H}}_{cf}(1) | \psi_{+2}(1), \psi_{+1}(2)\rangle$$

are both zero because $\langle \psi_{+1}(2) | \psi_{+2}(2)\rangle = \langle \psi_{+2}(2) | \psi_{+1}(2)\rangle = 0$.

As it happens, $\langle \psi_{+2}(1) | \hat{\mathcal{H}}_{cf}(1) | \psi_{+1}(1)\rangle$ is also zero.

In total therefore:

$$\langle 3, 3 | \hat{\mathcal{H}}_{cf} | 3, 3\rangle = \{\Omega/6 - 2\Omega/3\} = -\Omega/2.$$

BOX 7.3   Calculation of the inter-electronic repulsion for some states of d^2

The general two-electron repulsion integral takes the form:

$$\iint \Psi_a^*(1)\Psi_b^*(2)\ e^2/r_{12}\Psi_c(1)\Psi_d(2)\ d\tau_1\ d\tau_2 \equiv \langle a(1)b(2) | e^2/r_{12} | c(1)d(2)\rangle$$

There are two special forms of the integral.

The Coulomb integral, $J(a,b)$, in which $a = c$ and $b = d$, represents the mutual repulsion of an electron having the distribution $\Psi_a^*\Psi_a$ and an electron with the distribution $\Psi_b^*\Psi_b$.

The exchange integral, $K(a,b)$, in which $a = d$ and $b = c$, represents the mutual repulsion of the two overlap distributions $\Psi_a^*\Psi_b$.

But integrals which belong to neither of these special cases are frequently encountered and all non-zero two-electron repulsion integrals between d-orbital distributions are listed in the two tables below.

The angular parts of the integrals are the same whatever the principal quantum number, n. But the radial functions, R_i, and hence the radial parts of the integrals vary with n and from metal ion to metal ion, depending upon its environment etc. This dependence upon radial function is contained within the radial integrals, F_0, F_2 and F_4 which, for d-orbital functions, are defined as follows:

$$F_0 = \iint R_a(1)R_b(2)\frac{1}{r_>} R_c(1)R_d(2)r_1^2 r_2^2 \, dr_1 \, dr_2$$

$$F_2 = \frac{1}{49} \iint R_a(1)R_b(2)\frac{r_<^2}{r_>^3} R_c(1)R_d(2)r_1^2 r_2^2 \, dr_1 \, dr_2$$

$$F_4 = \frac{1}{441} \iint R_a(1)R_b(2)\frac{r_<^4}{r_>^5} R_c(1)R_d(2)r_1^2 r_2^2 \, dr_1 \, dr_2$$

in which $r_>$ is the greater and $r_<$ the lesser of r_1 and r_2.

The F_0, F_2 and F_4 given in Tables B7.3.1 and B7.3.2 include the factors of 1/49 and 1/441 in F_2 and F_4 respectively.

In ligand field theory, the F_k are usually treated as adjustable parameters and they are frequently taken in the convenient combinations first suggested by Racah who defined:

$$A = F_0 - 49F_4, \quad B = F_2 - 5F_4 \quad \text{and} \quad C = 35F_4$$

Electron repulsion in the T_1 and T_2 states of the configuration $(t_{2g})^1(e_g)^1$

A $(t_{2g})^1(e_g)^1$ configuration of T_2 symmetry is $(xy)^1(z^2)^1$ which when written as an expanded Slater determinant is:

$$(1/\sqrt{2})\{xy(1)\, z^2(2) - z^2(1)xy(2)\}$$

The mutual repulsion of the two electrons is:

$$\frac{1}{2}\langle xy(1)z^2(2) - z^2(1)\, xy(2)|\, e^2/r_{12}|xy(1)\, z^2(2) - z^2(1)\, xy(2)\rangle$$

$$= \frac{1}{2}\{\langle xy(1)\, z^2(2)|\, e^2/r_{12}|xy(1)\, z^2(2)\rangle + \langle z^2(1)\, xy(2)|\, e^2/r_{12}|z^2(1)\, xy(2)\rangle$$

$$- \langle xy(1)\, z^2(2)|\, e^2/r_{12}|z^2(1)\, xy(2)\rangle - \langle z^2(1)xy(2)|\, e^2/r_{12}|xy(1)\, z^2(2)\rangle\}$$

$$= \frac{1}{2}\{J(xy, z^2) + J(xy, z^2) - K(xy, z^2) - K(xy, z^2)\} = J(xy, z^2) - K(xy, z^2)$$

Which, using Table B7.3.1:

$$= F_0 - 4F_2 + 6F_4 - (4F_2 + 15F_4) = F_0 - 8F_2 - 9F_4$$

Or in terms of Racah parameters:

$$= A - 8B = A - 8\Lambda/15$$

where Λ is the energy difference between the ^3P and ^3F states of d^2 as defined in Section 7.3.2 and calculated below.

Similarly, a $(t_{2g})^1(e_g)^1$ configuration of T_1 symmetry is $(xy)^1(x^2 - y^2)^1$ which when written as an expanded Slater determinant is:

$$(1/\sqrt{2})\{xy(1)x^2 - y^2(2) - x^2 - y^2(1)xy(2)\}$$

For which the mutual repulsion is:

$$J(xy, x^2 - y^2) - K(xy, x^2 - y^2) = F_0 + 4F_2 - 34F_4 - (35F_4) = F_0 + 4F_2 - 69F_4$$

Or in terms of Racah parameters: $= A + 4B = A + 4\Lambda/15$

Determination of Λ

To calculate Λ we express the ^3P and ^3F states of d^2 as in Section 7.3.1 and Appendix 10:

$$^3F = | + 2^+(1), +1^+(2)\rangle = (\sqrt{\tfrac{1}{2}})\{(2, 1) - (1, 2)\}$$

$$^3P = \sqrt{(\tfrac{2}{5})}| + 2^+(1), -1^+(2)\rangle - \sqrt{(\tfrac{3}{5})}| + 1^+(1), 0^+(2)\rangle$$

$$= \sqrt{(\tfrac{2}{10})}\{(2, -1) - (-1, 2)\} - \sqrt{(\tfrac{3}{10})}\{(1, 0) - (0, 1)\}$$

Since all components of a term have the same energy, even when inter-electronic repulsion is included, we can take any component to calculate the electron repulsion. In each case we take the component of the term which has the maximum M_S and M_L values since these have the simplest wave functions. Each of the three microstates is a 2×2 Slater determinant which has been expanded above with a simplification of the notation which assumes that the electrons in the state symbols are always in the order 1, 2. Now, for ^3F we have:

$$\tfrac{1}{2}\langle(2, 1) - (1, 2)| \ e^2/r_{12}|(2, 1) - (1, 2)\rangle$$

$$= \tfrac{1}{2}\langle(2, 1)| \ e^2/r_{12}|(2, 1)\rangle + \langle(1, 2)| \ e^2/r_{12}|(1, 2)\rangle$$

$$- \langle(2, 1)| \ e^2/r_{12}|(1, 2)\rangle - \langle(1, 2)| \ e^2/r_{12}|(2, 1)\rangle$$

Inspection of the above four integrals shows that the first two are both Coulomb integrals representing the repulsion of an electron in a d_{+2} orbital with one in a d_{+1}. The second pair are exchange integrals representing the mutual repulsion of two electron-density distributions which are described by the product of d_{+2} and d_{+1}. Therefore:

$$\tfrac{1}{2}\langle(2, 1) - (1, 2)| \ e^2/r_{12}|(2, 1) - (1, 2)\rangle$$

$$= \langle(2, 1)| \ e^2/r_{12}|(2, 1)\rangle - \langle(2, 1)| \ e^2/r_{12}|(1, 2)\rangle = J(1, 2) - K(1, 2)$$

We can now look up the electron repulsion integrals in Table B7.3.2, which is the same as Table B7.3.1 but based upon the complex rather that the real d orbitals. We find:

$$\langle(2, 1)|\ e^2/r_{12}|(2, 1)\rangle = F_0 - 2F_2 - 4F_4$$

and:

$$\langle(2, 1)|\ e^2/r_{12}|(1, 2)\rangle = +6F_2 + 5F_4$$

so that the electron repulsion in the ^3F state is: $F_0 - 8F_2 - 9F_4$.

For the ^3P state we find three contributions to the energy:

$$(\tfrac{1}{5})\{2\langle(2, -1)|\ e^2/r_{12}|(2, -1)\rangle - 2\langle(2, -1)|\ e^2/r_{12}|(-1, 2)\rangle\}$$

$$= \tfrac{2}{5}J(2, -1) - \tfrac{2}{5}K(2, -1)$$

$$= (\tfrac{2}{5})\{F_0 - 2F_2 - 4F_4 - 35F_4\} = (\tfrac{2}{5})\{F_0 - 2F_2 - 39F_4\}$$

$$(\tfrac{3}{10})\{2\langle(1, 0)|\ e^2/r_{12}|(1, 0)\rangle - 2\langle(1, 0)|\ e^2/r_{12}|(0, 1)\rangle\} = \tfrac{3}{5}J(1, 0) - \tfrac{3}{5}K(1, 0)$$

$$= (\tfrac{3}{5})\{F_0 + 2F_2 - 24F_4 - F_2 - 30F_4\} = (\tfrac{3}{5})\{F_0 + F_2 - 54F_4\}$$

and a cross-term composed of integrals which are of neither J nor K type:

$$-\left(\frac{2\sqrt{6}}{10}\right)\{2\langle(2, -1)|\ e^2/r_{12}|(1, 0)\rangle - 2\langle(2, -1)|\ e^2/r_{12}|(0, 1)\rangle\}$$

$$= -\left(\frac{2\sqrt{6}}{5}\right)\{-\sqrt{6}F_2 + 5\sqrt{6}F_4 - 2\sqrt{6}F_2 + 10\sqrt{6}F_4\} = -(\tfrac{12}{5})\{-3F_2 + 15F_4\}$$

The total electron repulsion in the ^3P state is therefore, $F_0 + 7F_2 - 84F_4$, so that:

$$\Lambda = E(^3P) - E(^3F) = F_0 + 7F_2 - 84F_4 - (F_0 - 8F_2 - 9F_4) = 15F_2 - 75F_4$$

Note that in using Tables B7.3.1 and B7.3.2 the integral $\langle a(1), b(2)|e^2/r_{12}|c(1), d(2)\rangle$:

(a) Will always be zero if the spin function of orbitals a and c (electron 1) and of b and d (electron 2) are not the same. But the spins of the two electrons need not be the same.

(b) Expresses the mutual repulsion of two electron densities $\langle a(1)|c(1)\rangle$ and $\langle b(2)|d(2)\rangle$, and electrons are identical.

Therefore:

$$\langle a(1), b(2)|\ e^2/r_{12}|c(1), d(2)\rangle = \langle c(1), b(2)|\ e^2/r_{12}|a(1), d(2)\rangle$$

$$= \langle a(1), d(2)|\ e^2/r_{12}|c(1), b(2)\rangle = \langle c(2), d(1)|\ e^2/r_{12}|a(2), b(1)\rangle \text{ etc.}$$

LITERATURE

The monographs by Ballhausen[1] and Griffith[4], noted in the bibliography of Chapter 7, are important sources for readers who wish to take this subject further.

Table B7.3.1 Two-electron repulsion integrals between d orbitals in their real forms

a	b	c	d	F_0	F_2	F_4	Integral type
z^2	z^2	z^2	z^2	1	+4	+36	J
xz	xz	xz	xz	1	+4	+36	J
yz	yz	yz	yz	1	+4	+36	J
xy	xy	xy	xy	1	+4	+36	J
x^2-y^2	x^2-y^2	x^2-y^2	x^2-y^2	1	+4	+36	J
x^2-y^2	xz	x^2-y^2	xz	1	−2	−4	J
x^2-y^2	yz	x^2-y^2	yz	1	−2	−4	J
xy	xz	xy	xz	1	−2	−4	J
xy	yz	xy	yz	1	−2	−4	J
xz	yz	xz	yz	1	−2	−4	J
z^2	xz	z^2	xz	1	+2	−24	J
z^2	yz	z^2	yz	1	+2	−24	J
z^2	xy	z^2	xy	1	−4	+6	J
z^2	x^2-y^2	z^2	x^2-y^2	1	−4	+6	J
x^2-y^2	xy	x^2-y^2	xy	1	+4	−34	J
xy	yz	yz	xy	0	+3	+20	K
xy	xz	xz	xy	0	+3	+20	K
xz	yz	yz	xz	0	+3	+20	K
x^2-y^2	xz	xz	x^2-y^2	0	+3	+20	K
x^2-y^2	yz	yz	x^2-y^2	0	+3	+20	K
z^2	x^2-y^2	x^2-y^2	z^2	0	+4	+15	K
z^2	xy	xy	z^2	0	+4	+15	K
z^2	xz	xz	z^2	0	+1	+30	K
z^2	yz	yz	z^2	0	+1	+30	K
x^2-y^2	xy	xy	x^2-y^2	0	0	+35	K
xz	z^2	xz	x^2-y^2	0	$-2\sqrt{3}$	$+10\sqrt{3}$	Other
yz	z^2	yz	x^2-y^2	0	$+2\sqrt{3}$	$-10\sqrt{3}$	Other
xz	xz	z^2	x^2-y^2	0	$+\sqrt{3}$	$-5\sqrt{3}$	Other
yz	yz	z^2	x^2-y^2	0	$-\sqrt{3}$	$+5\sqrt{3}$	Other
z^2	xy	xz	yz	0	$+\sqrt{3}$	$-5\sqrt{3}$	Other
z^2	xy	yz	xz	0	$+\sqrt{3}$	$-5\sqrt{3}$	Other
z^2	xz	xy	yz	0	$+2\sqrt{3}$	$-10\sqrt{3}$	Other
x^2-y^2	xy	xz	yz	0	+3	−15	Other
x^2-y^2	xy	yz	xz	0	−3	+15	Other

Table B7.3.2 Two-electron repulsion integrals between d orbitals in their complex forms

m_a	m_b	m_c	m_d	F_0	F_2	F_4	Integral type
+2	+2	+2	+2	1	+4	+1	J
+2	+1	+2	+1	1	−2	−4	J
+2	+1	+1	+2	0	+6	+5	K
+2	0	+2	0	1	−4	+6	J
+2	0	+1	+1	0	$+\sqrt{6}$	$-5\sqrt{6}$	Other
+2	0	0	+2	0	+4	+15	K
+2	−1	+2	−1	1	−2	−4	J
+2	−1	+1	0	0	$-\sqrt{6}$	$+5\sqrt{6}$	Other
+2	−1	0	+1	0	$+2\sqrt{6}$	$-10\sqrt{6}$	Other
+2	−1	−1	+2	0	0	+35	K
+2	−2	+2	−2	1	+4	+1	J
+2	−2	+1	−1	0	−6	−5	Other
+2	−2	0	0	0	+4	+15	Other
+2	−2	−1	+1	0	0	−35	Other
+2	−2	−2	+2	0	0	+70	K
+1	+2	+1	+2	1	−2	−4	J
+1	+1	+1	+1	1	+1	+16	J
+1	+1	0	+2	0	$+\sqrt{6}$	$-5\sqrt{6}$	Other
+1	0	+1	0	1	+2	−24	J
+1	0	0	+1	0	+1	+30	K
+1	0	−1	+2	0	$+2\sqrt{6}$	$-10\sqrt{6}$	Other
+1	−1	+1	−1	1	+1	+16	J
+1	−1	0	0	0	−1	−30	Other
+1	−1	−1	+1	0	+6	+40	K
+1	−1	−2	+2	0	0	−35	Other
0	+2	0	+2	1	−4	+6	J
0	+1	0	+1	1	+2	−24	J
0	+1	−1	+2	0	$-\sqrt{6}$	$+5\sqrt{6}$	Other
0	0	0	0	1	+4	+36	J
0	0	−1	+1	0	−1	−30	Other
0	0	−2	+2	0	+4	+15	Other
−1	+2	−1	+2	1	−2	−4	J
−1	+1	−1	+1	1	+1	+16	J
−1	+1	−2	+2	0	−6	−5	Other
−2	+2	−2	+2	1	+4	+1	J

Integrals not listed explicitly may be obtained by replacing m_i by $-m_i$ across a row or by interchanging m_a with m_d and m_b with m_c. Also see notes in last paragraph of text in box.

BOX 7.4   Diagonalising the MH_6 matrix

The easiest way to obtain the eigenvalues of the Hamiltonian matrix for MH_6 is by means of a little group theory.

The molecule belongs to the symmetry point group O_h and it is not difficult to show that the symmetry of the molecule is reflected in the following combinations of the hydrogen 1s atomic orbitals. The hydrogen atoms are numbered as in Figure 7.7 and the orbital combinations have been normalised with neglect of ligand orbital overlap, i.e. $\langle \phi_i | \phi_j \rangle = \delta_{ij}$.

$$\Psi_{a1g} = (1/\sqrt{6})\{\phi_1 + \phi_2 + \phi_3 + \phi_4 + \phi_5 + \phi_6\}$$

$$\Psi_{eg(x^2-y^2)} = (\tfrac{1}{2})\{\phi_1 - \phi_2 + \phi_3 - \phi_4\}$$

$$\Psi_{eg(z^2)} = (1/\sqrt{12})\{-\phi_1 - \phi_2 - \phi_3 - \phi_4 + 2\phi_5 + 2\phi_6\}$$

$$\Psi_{t1ux} = (1/\sqrt{2})\{\phi_1 - \phi_3\} \quad \Psi_{t1uy} = (1/\sqrt{2})\{\phi_2 - \phi_4\} \quad \Psi_{t1uz} = (1/\sqrt{2})\{\phi_5 - \phi_6\}$$

The combinations of hydrogen 1s orbitals have been designated by the symmetry species to which they belong, a_{1g}, e_g, t_{1u}, and the metal d (x^2-y^2, z^2) or p (x, y, z) orbitals which have the same symmetry and with which they will therefore interact. Since we are not considering metal p orbitals here, the t_{1u} combinations of hydrogen 1s atomic orbitals will play no part in the bonding of MH_6. If we now reform our Hamiltonian matrix, replacing the rows and columns which were formerly headed by the individual hydrogen 1s orbitals by rows and columns representing the above combinations of the hydrogen 1s orbitals, we find the following matrix. To do this we first re-calculate the columns. The column headed Ψ_{a1g}, for example, is simply the sum of the six columns headed ϕ_1 to ϕ_6 divided by $\sqrt{6}$. Having completed the operation for the columns, we repeat it for the rows and obtain:

$\hat{\mathcal{H}}$	$d_{x^2-y^2}$	$\Psi_{eg(x^2-y^2)}$	$d_z{}^2$	$\Psi_{eg(z^2)}$	Ψ_{a1g}	Ψ_{t1ux}	Ψ_{t1uy}	Ψ_{t1uz}
$d_{x^2-y^2}$	Ed	$2\sqrt{3}\beta$	0	0	0	0	0	0
$\Psi_{eg(x^2-y^2)}$	$2\sqrt{3}\beta$	Es	0	0	0	0	0	0
$d_z{}^2$	0	0	Ed	$2\sqrt{3}\beta$	0	0	0	0
$\Psi_{eg(z^2)}$	0	0	$2\sqrt{3}\beta$	Es	0	0	0	0
Ψ_{a1g}	0	0	0	0	Es	0	0	0
Ψ_{t1ux}	0	0	0	0	0	Es	0	0
Ψ_{t1uy}	0	0	0	0	0	0	Es	0
Ψ_{t1uz}	0	0	0	0	0	0	0	Es

The use of symmetrised sets of ligand orbitals as basis functions results in a matrix which is blocked out; compare the matrix based on the unsymmetrised ligand orbitals in Section 7.7. The new matrix is actually six separate matrices four of which are simply 1×1 matrices with eigenvalues of Es. The other two are identical 2×2 matrices:

$\hat{\mathcal{H}}$	$d_z{}^2$	$\Psi_{eg(z^2)}$
$d_z{}^2$	Ed	$2\sqrt{3}\beta$
$\Psi_{eg(z^2)}$	$2\sqrt{3}\beta$	Es

$\hat{\mathcal{H}}$	$d_{x^2-y^2}$	$\Psi_{eg(x^2-y^2)}$
$d_{x^2-y^2}$	Ed	$2\sqrt{3}\beta$
$\Psi_{eg(x^2-y^2)}$	$2\sqrt{3}\beta$	Es

and their eigenvalues, λ, can be found by solving the characteristic equation (Appendix 3):

$$(E\text{d} - \lambda)(E\text{s} - \lambda) - 12\beta^2 = 0$$

the roots of which are:

$$\lambda = \frac{E\text{d} + E\text{s} \pm \sqrt{(E\text{d} - E\text{s})^2 + 48\beta^2}}{2}$$

Each identical matrix gives two values of λ one of which corresponds to ligand-metal bonding (the negative sign before the square root) and one to antibonding. In Section 7.7 these pairs of degenerate levels are designated E_b and E_a respectively.

PROBLEMS FOR CHAPTER 7

1. Show that $A_{44} = A_{4-4}$ for $\phi_\text{q} = 0, \pm\pi/2$ and π. ($A_{k\alpha}$ is defined by Equation (7.2.5)).

2. Follow the method in Section 7.3.1 and Box 7.2 to calculate the off-diagonal matrix element $\langle 3, +2|\hat{\mathcal{H}}_\text{cf}|3, -2\rangle$ for the ^3F state of d^2 in an octahedral crystal field.

3. Follow the method in Section 7.3.2 and Box 7.3 to calculate the inter-electronic repulsion in the $|1, 0\rangle$ component of the ^3P state of d^2 in an octahedral crystal field.

$$|1, 0\rangle = (2/\sqrt{5})|2^+, -2^+\rangle - (1/\sqrt{5})|1^+, -1^+\rangle$$

4. Consider two d^2 configurations of an octahedral transition metal complex, a) in which the electrons occupy the d_{xy} and d_{xz} orbitals with paired spins and b) in which the electrons occupy the d_{xy} orbital only.

   Use the information in Box 7.2 to calculate the electron-repulsion energy of these two configurations. You should find:
   Electron repulsion for configuration a $= F_0 + F_2 + 16F_4$.
   Electron repulsion for configuration b $= F_0 + 4F_2 + 36F_4$.
   Note: The wave function for configuration a is best obtained by recognising that it must be the product of a spin function and a space function of the form described in Section 11.5.

5. Use the operator $\zeta\hat{\boldsymbol{l}} \cdot \hat{\mathbf{s}} = \frac{1}{2}\zeta(\hat{l}_+\hat{s}_- + \hat{l}_-\hat{s}_+) + \zeta\hat{l}_z\hat{s}_z$ to draw up the 10×10 matrix for the spin orbit coupling in the configuration d^1. Show that the matrix has six eigenvalues of $+\zeta$ and 4 of $-3\zeta/2$, as one would expect for the states $^2D_{5/2}$ and $^2D_{3/2}$.

6. Imagine a tetrahedral, transition-metal complex ion, $[MH_4]^{4-}$, in which the four H^- ions are arranged around the central metal atom as shown in the figure:

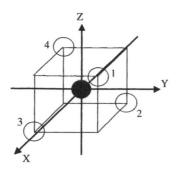

With the aid of this figure you should be able to convince yourself that the overlaps of each of the four H^- 1s AOs with the d_{xy}, d_{xz} and d_{yz} AOs of the metal are equal in magnitude but depend for their sign upon the sign of the nearest lobe of the d-orbital. Equally, it should be clear that the overlap of the $d_{x^2-y^2}$ AO with each of the H^- 1s AOs is zero. You can see that this is also the case for the d_{z^2} AO when you recall that $d_{z^2} = (1/\sqrt{3})(d_{z^2-x^2} + d_{z^2-y^2})$.

Assuming that there is zero overlap of the H^- 1s AOs with each other, draw up an energy matrix for the MH_4 complex like the one for MH_6 in Section 7.7. Diagonalise the matrix by reforming the columns and rows using the following combinations of the H^- 1s AOs:

$$\Psi_{xy} = \tfrac{1}{2}(s_1 - s_2 - s_3 + s_4) \qquad \Psi_{xz} = \tfrac{1}{2}(s_1 + s_2 - s_3 - s_4)$$

$$\Psi_{yz} = \tfrac{1}{2}(s_1 - s_2 + s_3 - s_4) \qquad \Psi_s = \tfrac{1}{2}(s_1 + s_2 + s_3 + s_4)$$

You should find two sets of triply degenerate levels, one pair of doubly degenerate levels and one singly degenerate level.

Assuming that:

a) the initial energy of the five d-orbitals is -50×10^3 cm^{-1}
b) the initial energy of the 1s orbitals of the four H^- ions is -110×10^3 cm^{-1}
c) where the overlap is positive (negative), the interaction energy of an interacting d-orbital with an H^- 1s orbital is $-(+)15 \times 10^3$ cm^{-1}.

show that we expect the d-d transition of a d^1 complex with this structure to be found in the region of 12.5×10^3 cm^{-1}.

CHAPTER 8

Spectroscopy

8.0	The interaction of radiation with matter	216
8.1	Electromagnetic radiation	216
	8.1.1 The electric field	217
	8.1.2 The magnetic field	219
8.2	Polarised light	219
	8.2.1 Linearly polarised light	220
	8.2.2 Circularly polarised light	221
8.3	The electromagnetic spectrum	222
	8.3.1 Three forms of electromagnetic radiation	222
8.4	Photons and their properties	223
	8.4.1 Velocity	223
	8.4.2 Energy	224
	8.4.3 Mass	224
	8.4.4 Linear momentum	224
	8.4.5 Angular momentum	225
	8.4.6 Parity	225
8.5	Selection rules	225
	8.5.1 The Bohr–Einstein condition	226
8.6	The quantum mechanics of transition probability	227
8.7	The nature of the time-independent interaction $\langle \phi_f | \hat{V}(x, y, z) | \phi_i \rangle$	233
	8.7.1 The transition dipole moment	234
	8.7.2 The relative intensities of UV–VIS, IR and NMR transitions	238
	8.7.3 The particle and wave views of spectroscopic transitions	242
8.8	Spectroscopic time scales	245
8.9	Quantum electrodynamics	247
8.10	Spectroscopic units and notation	247
	8.10.1 The energy/frequency/wavelength axis	248
	8.10.2 The intensity/absorbance axis	249
8.11	The Einstein coefficients	250
8.12	Bibliography and further reading	250
	Problems for Chapter 8	259

8.0 THE INTERACTION OF RADIATION WITH MATTER

Spectroscopy may be defined as the study of the interaction of electromagnetic radiation with matter. Its application to chemistry, essentially in the second half of the 20th century, has been primarily responsible for the detailed knowledge of the structure of molecules which we now have. Chemistry, as we know it today, is inextricably entwined with the methods of molecular spectroscopy and totally dependent upon them. It is the purpose of this chapter to outline the principles of atomic and molecular spectroscopy from a quantum mechanical point of view. This will help us to understand, in general terms, how electromagnetic radiation interacts with matter and how information about the structure of molecules can be obtained by studying that interaction. The specific branches of spectroscopy which are of particular importance to chemistry–electronic, infrared and nuclear magnetic resonance–will be discussed in the following chapters. Our first task is to set down a suitable description of electromagnetic radiation.

8.1 ELECTROMAGNETIC RADIATION

There were few developments in the science of optics during the 18th century and, for the most part, Newton's corpuscular view of light held sway.[1] In the present context, it is interesting to note that the generalisation of Newtonian mechanics by Lagrange and Laplace, which was also believed to be applicable to particles of light, was at least partially responsible for the general confidence in the corpuscular description of light at the end of that century. Nevertheless, in 1801 Thomas Young (1773–1829) revived the wave theory of Christian Huygens (1629–1695) because he had come to believe that colours might be associated with vibrations of light as notes are associated with sound vibrations. Later in the century, two French amateur scientists, Armand Fizeau (1819–1896) and Jean Foucault (1819–1868), independently improved the accuracy with which the velocity of light could be measured and Foucault showed that light travels more slowly in water than in air. This result is in agreement with the prediction of the wave theory but in contradiction to the corpuscular theory of light. In the years 1864 to 1873, which immediately followed Foucault's researches, James Clerk Maxwell (1831–1879) developed a mathematical theory of light in the form of four differential equations which describe light as oscillating electric and magnetic fields. The new theory of electromagnetic radiation was a triumph. Not only did it show how electricity, magnetism and light are connected, a fact first demonstrated experimentally by Michael Faraday (1791–1867) in 1846, but it made the relationships quantitative. In Maxwell's time, electricity was measured in one of two units. It could be measured either in terms of charge in electrostatic units (e.s.u.) or in terms of its magnetic effect in electromagnetic units (e.m.u.). Maxwell derived the simple relationship between these two units in which electricity might be measured; he found that:

$$\text{value in e.s.u.} = \text{velocity of light} \times \text{value in e.m.u.} \tag{8.1.1}$$

in agreement with the experimental discovery of this result by Kirchhoff in 1857.

But Maxwell's description of visible light as electromagnetic radiation went much further. Though knowledge of the visible spectrum is clearly as old as man himself, only the spectral regions immediately adjacent to visible light were known in Maxwell's day. The near infrared, which adjoins the visible on the long-wavelength side, was discovered by Sir William Herschel (1738–1822) in 1800 and the ultraviolet, on the other side of the visible, was found by Johann Wilhelm Ritter (1776–1810) in 1801. But Maxwell's equations predicted that there was an infinite spectrum of electromagnetic radiation, just waiting to be discovered, outside the then-known range. This prediction was brilliantly confirmed for long wavelengths by Heinrich Rudolf Hertz (1857–1894) and in modern spectroscopy we make use of a spectral range of frequency (or wavelength) which covers approximately 12 powers of ten (Table 8.1). Maxwell's equations passed through the great upheavals which overtook theoretical physics at the beginning of the 20th century quite unchanged. They provide us with a description of electromagnetic radiation which is very valuable when we wish to consider how that radiation interacts with matter, the process which is the basis of spectroscopy. We therefore now examine this description.

8.1.1   The electric field

Imagine (Figure 8.1(a)) an electric field represented by a vector E_0 which is rotating in the yz-plane with an angular velocity of ω radians per second. For this velocity the time, t', required by the vector E_0 to execute one complete revolution is:

$$t' = 2\pi/\omega \qquad (8.1.2)$$

Table 8.1   The regions of the electromagnetic spectrum

Branch of spectroscopy	Nominal boundary[a]	λ/m	$\bar{\nu}$/cm^{-1}	$\bar{\nu}$/m^{-1}	ν/s^{-1}
X-ray					
		10^{-11}	10^9	10^{11}	3×10^{19}
Auger					
		10^{-9}	10^7	10^9	3×10^{17}
Vacuum ultraviolet		10^{-7}	10^5	10^7	3×10^{15}
- - - - - - - - - -	- - - - - - -	2×10^{-7}			
Ultraviolet					
- - - - - - - - - -	- - - - - - -	3.8×10^{-7}			
Visible					
- - - - - - - - - -	- - - - - - -	7.8×10^{-7}			
Near infrared		10^{-6}	10^4	10^6	3×10^{14}
- - - - - - - - - -	- - - - - - -	3×10^{-6}			
Infrared		10^{-5}	10^3	10^5	3×10^{13}
- - - - - - - - - -	- - - - - - -	3×10^{-5}			
Far infrared		10^{-4}	10^2	10^4	3×10^{12}
- - - - - - - - - -	- - - - - - -	3×10^{-4}			
Microwave					
- - - - - - - - - -		10^{-3}	10	10^3	3×10^{11}
Electron paramagnetic resonance					
- - - - - - - - - -		10^{-2}	1	10^2	3×10^{10}
Nuclear magnetic resonance					
- - - - - - - - - -		10	10^{-3}	10^{-1}	3×10^7

[a]Set for the range 10^{-8} to 3×10^{-4} m in Report No. 6 of the Joint Committee on Nomenclature in Applied Spectroscopy, *Analytical Chemistry*, **24**, 1349 (1952).

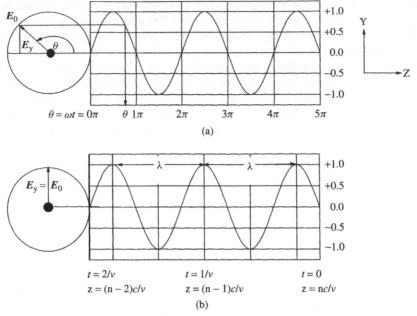

Figure 8.1 Graphs of E_0 versus θ (a) and t (b)

The inverse of this time is the frequency of revolution, ν, in s^{-1}, therefore:

$$\nu = \omega/2\pi \qquad (8.1.3)$$

If the tail of the vector is fixed, then its head describes a circle in the yz-plane and the angle, θ, between the z-axis and the vector is given by Equation (8.1.4):

$$\theta = \omega t \qquad (8.1.4)$$

and the component of E_0 in the y-direction, E_y, is:

$$E_y = E_0 \sin(\theta) = E_0 \sin(\omega t) \qquad (8.1.5)$$

Thus, the electric field in the y-direction, E_y, varies with time as a sine function and oscillates as illustrated in Figure 8.1(a).

In Maxwell's description, this oscillating electric field propagates through space in a direction, usually chosen to be z, at $90°$ to the oscillation direction, with the velocity of light, c. Therefore, a particular field vector, $E_y(0)$ generated at time $t = 0$, moves away from the light source at velocity c. In the time, $t' = 1/\nu$, required by the rotating vector E_0 to complete one cycle, the propagating E_y vector will have moved along the z-direction to a position $z = ct' = c/\nu$ and after n cycles when $\theta = 2n\pi$, E_y will have reached $z = nc/\nu$. Therefore, eliminating n, $\theta = 2\pi z\nu/c$. Thus, the wave reproduces itself every interval of c/ν along the z-direction (Figure 8.1(b)) and this repeat distance is known as the wavelength, λ, so that:

$$\lambda = c/\nu \quad \text{or} \quad c = \lambda\nu \qquad (8.1.6)$$

and

$$E_y = E_0 \sin(2\pi z\nu/c) = E_0 \sin(\omega z/c) \qquad (8.1.7)$$

The dependence of E_y upon time (Equation (8.1.5)) and distance (Equation (8.1.7)) may be combined to give:

$$E_y = E_0 \sin[\omega(t - z/c)] \qquad (8.1.8)$$

where the two parts of the argument of the sine function are combined with the negative sign because (Figure 8.1(b)) if we move along the wave in the positive z-direction we are going backwards in time. If this idea presents a problem, consider the light arriving on the earth today from a distant star. The light which we now see is a record of events that took place on the star at the time that the light left it, many years ago. But if we could journey towards the star, i.e. in the negative z-direction, we would see ever more recent light until, upon arriving at the star's surface, we would see events as they are currently happening. The reverse is true if we move away from the star in the direction of propagation of the light.

8.1.2 The magnetic field

Apart from the oscillating electric field discussed above, Maxwell's description of light also includes a magnetic field, H, which oscillates in phase with the electric field but is always directed at 90° to it, i.e. along x in Figure 8.1. Thus, the full description is one of an electric and magnetic field oscillating in phase and orientated at right angles to each other and to the propagation direction. These oscillating fields, which propagate through space (vacuum) with a speed of 2.997925×10^8 ms^{-1}, constitute what we call light in the region where we can detect it with our eyes and, in general, electromagnetic (e-m) radiation. It is characterised particularly by its wavelength and frequency, which are related by Equation (8.1.6). In much of the following we use the word 'light' rather than 'electromagnetic radiation' on account of its brevity, but the latter is always implied.

8.2 POLARISED LIGHT

The electric and magnetic fields of a light beam oscillate in directions perpendicular to the direction of propagation of the light. Thus, if we choose z as the direction of propagation, E and H must lie in the xy-plane. (Clearly, we could choose any direction for the direction of propagation, but unless it is stated otherwise we will always choose z.) The vectors E and H are frequently said to vibrate, rather than oscillate, in the xy-plane. Light from the usual sources, e.g. the sun, incandescent metals, electrical discharges has electric and magnetic vectors vibrating in every direction in the xy-plane (Figure 8.2(a)). But by using optical devices known as polarisers, it is possible to obtain light which vibrates in only one direction (Figure 8.2(b)). Such light is said to be linearly or plane polarised in the direction in which the E vector vibrates. Polarised light is very important in spectroscopy because the direction of polarisation of the light can have a decisive effect upon whether or not the light is absorbed by a particular sample. In this connection it is useful to distinguish two kinds of polarised light; *linearly* and *circularly* polarised. In what follows we shall mention only the E vector; but the H vector is always present following exactly the same pattern of behaviour as the E vector, but always at right-angles to it. Both E and H are important, but usually in different forms of spectroscopy.

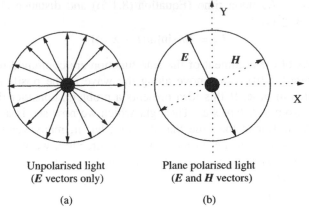

Figure 8.2   Unpolarised light (a) and linearly (plane) polarised light (b)

8.2.1   Linearly polarised light

The electric vector of a polarised light beam propagating along z may be resolved onto any two mutually perpendicular directions in the xy-plane. It is of particular interest in connection with the relationship between linearly and circularly polarised light to resolve the E vector onto two such axes each of which lies at 45° to E (Figure 8.3). We then find that the two components of the E vector, E_x and E_y say, form two beams of linearly polarised light with mutually perpendicular planes of polarisation which vibrate in phase and have equal amplitude; i.e. $E_x = E_y = E / \sqrt{2}$. Thus, a beam of linearly polarised light can always be described in terms of two beams of linearly polarised light with mutually perpendicular planes of polarisation, equal E_0 values and vibrating in phase.

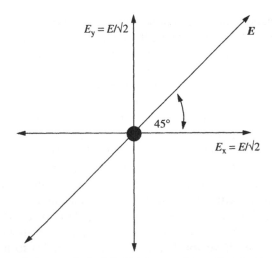

Figure 8.3   Resolution of plane polarised light into x and y components

8.2.2 Circularly polarised light

Light propagating through some crystalline materials, e.g. gypsum, mica and quartz, travels at different velocities depending upon the direction in which it is polarised. By passing plane polarised light through a plate of such a material, known as a quarter-wave plate, it is possible to create a situation whereby two perpendicular and equal E vectors vibrate 90° out of phase so that one is at a maximum when the other is a minimum, and vice-versa, though only at one specific wavelength. The resultant is an E vector which describes a circle and if we also think of the propagation of the light wave along z, then the E vector describes a spiral, like a staircase, and returns to the same orientation every time the wave advances by a wavelength (Figure 8.4). Left- and right-handed spirals can be formed by suitable choices of phase (Figure 8.5). Light having these properties is known as circularly polarised light, LCP and RCP for the left- and right-handed versions respectively. To ensure that there is no confusion, we must define more exactly what we mean by right- and left-handed spirals. The convention is that if the head of the E vector moves clockwise when viewed by an observer looking at the light source then the light is RCP. Counter-clockwise rotation is LCP. The physical significance of these two forms of polarised light lies in the angular momentum properties of the photons with which they are associated. This will be made clear in Section 8.4.5 below.

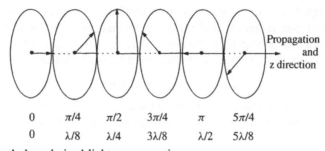

Figure 8.4   Circularly polarised light – propagation

$$E_R = (E/\sqrt{2})(E_x \cos 2\pi\nu t - E_y \sin 2\pi\nu t)$$

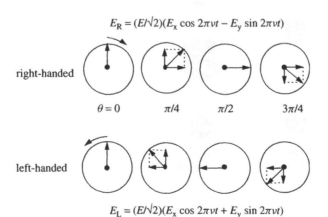

$$E_L = (E/\sqrt{2})(E_x \cos 2\pi\nu t + E_y \sin 2\pi\nu t)$$

Figure 8.5   Circularly polarised light – sense of rotation

8.3   THE ELECTROMAGNETIC SPECTRUM

There is, in principle, no limit to the wavelength and frequency of e-m radiation. However, as far as chemical spectroscopy is concerned, the range of interest runs from $\lambda = 10^{-11}$ m ($\nu = 3 \times 10^{19}$ s^{-1}) to $\lambda = 10$ m ($\nu = 3 \times 10^7$ s^{-1}). This is an extremely large range and it is divided into a number of smaller regions (Table 8.1). There is, of course, no sharp division between any two ranges, but the experimental techniques used for spectroscopy in the different regions and the type of information obtained differ widely. It is therefore convenient to discuss spectroscopy in the different regions of the e-m spectrum under different headings, e.g. ultraviolet and visible spectroscopy, infrared spectroscopy etc., as we shall presently see.

8.3.1   Three forms of electromagnetic radiation

The e-m radiation generated by the rotating electric field vector, $\boldsymbol{E}_0$, as described in Sections 8.1 and 8.2 is called *electric dipole radiation* because it has exactly the same form as the radiation emitted by an oscillating dipole composed of two equal and opposite charges, q_+ and q_-, separated by a distance l (Figure 8.6(a)) and each varying with time according to the equation:

$$q_- = -q_+ = q_0 \cos \omega t$$

However, though this is the most common form of e-m radiation it is not the only one. Two other types of e-m radiation will make occasional appearances in this book. Two identical electric dipoles arranged co-linearly end to end (Figure 8.6(b)) with the central charge and those at the ends varying in time according to the equation:

$$2q_- = -q_+ = q_0 \cos \omega t$$

emit *electric quadrupole radiation*. Finally, an oscillating electric current flowing in a loop (Figure 8.6(c)) constitutes an oscillating magnetic dipole and emits *magnetic dipole radiation*.

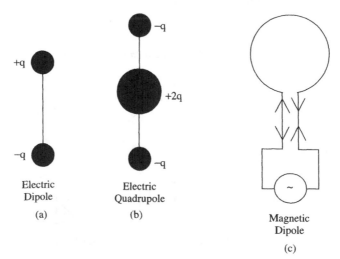

Electric Dipole
(a)

Electric Quadrupole
(b)

Magnetic Dipole
(c)

Figure 8.6   Three forms of electromagnetic radiation

8.4 PHOTONS AND THEIR PROPERTIES

Before we embark upon the study of the interaction of radiation with matter in particular regions of the spectrum we must recall (Section 2.6) that, at the end of the 19th century, the wave theory of light was found to be difficult to reconcile with the results obtained from a study of the photoelectric effect; the ejection of electrons from a metal surface when that surface is irradiated with visible or ultraviolet light. This was just one of the problems prevalent at that time which, in 1905, led Einstein to propose that processes where light is absorbed or emitted by a substance might be better understood in terms of light quanta or photons of energy ($h\nu$). Einstein did not himself coin the name 'photon'; he used the term light-energy quantum throughout his 1905 paper. The word photon for the corpuscle of light was proposed by Gilbert Newton Lewis (1875–1946) in 1926 and the suggestion appears to have received immediate acceptance. Einstein's corpuscular concept of light was successfully applied by Bohr in his 1913 interpretation of the hydrogen atom spectrum. In 1916 it was fully substantiated, with respect to the photoelectric effect, by Millikan's comprehensive experimental studies (Section 2.6), and since that time this view of the energetics of the interaction of light with matter has never been challenged. It is therefore appropriate that we now consider some properties of the photon.

Photons are the quanta of the electromagnetic field. Like other particles, they have energy, angular momentum and parity; properties which must be conserved when photons interact with other quantum-mechanical entities such as atoms and molecules. Our interest in these conservation requirements arises because they form the basis of the spectroscopic selection rules. In particular, when a photon is absorbed (emitted) by an atom or molecule the energy, angular momentum and parity of the system, photon + atom or molecule, must be the same before and after the absorption (emission). This bald statement requires further elucidation in the case of the conservation of angular momentum. In quantum mechanics, when two angular momenta, j_1 and j_2, are added together $2j_< + 1$ values of the resultant sum, J, are possible, where $j_<$ is the smaller of j_1 and j_2 (Section 4.6.1). These values form the Clebsch–Gordan series (Equation (4.6.1)):

$$J = j_1 + j_2, \ j_1 + j_2 - 1, \ j_1 + j_2 - 2 \ldots |j_1 - j_2|$$

Values of J, j_1 and j_2 which satisfy any of the above $2j_< + 1$ equations are said to satisfy a triangular condition which is written $\Delta(j_1, j_2, J)$ and is not equal to zero only if J is one of the Clebsch–Gordan series formed from j_1 and j_2. Thus, if j_i, j_f and j_p are the angular momenta of the initial state, final state and photon respectively, then these three angular momenta must satisfy the triangular condition $\Delta(j_i, j_p, j_f)$ if angular momentum is to be conserved in the absorption/emission process. The fact that $2j_< + 1$ different outcomes can all satisfy the condition for the conservation of angular momentum seems, at first sight, a little surprising. However, this is necessary in order to take account of both absorption and emission processes as we shall see when we consider electric dipole transitions below.

8.4.1 Velocity

The velocity of the photon is the velocity of light, which is equal to the product of the wavelength (λ in m) and the frequency (ν in s^{-1}):

$$c = \lambda \times \nu = 2.997925 \times 10^8 \ \mathrm{m \, s^{-1}} \tag{8.4.1}$$

The above figure is the velocity of light in a vacuum. In any other medium the velocity is less by a factor of $1/n$, where n is the refractive index of the medium. Furthermore, in any medium other than a vacuum the velocity of light depends upon its wavelength. The dependence of velocity or refractive index upon wavelength is responsible for the fact that light is dispersed into a spectrum when it passes through a transparent prism. As a simple illustration of Equation (8.4.1), we take one of a pair of lines known as the sodium D lines; two closely spaced atomic spectral lines which are responsible for the strong yellow colour of the sodium flame or street lamp and which derive from the transitions $^2P_{\frac{3}{2}} \Leftrightarrow {}^2S_{\frac{1}{2}}$ and $^2P_{\frac{1}{2}} \Leftrightarrow {}^2S_{\frac{1}{2}}$ of the sodium atom (Chapter 5). The first of these lines has $\lambda = 5.892 \times 10^{-7}$ m and $\nu = 5.088 \times 10^{14}$ s^{-1}. It is clear that these data satisfy Equation (8.4.1).

8.4.2 Energy

The energy of a photon depends upon its frequency according to the equation proposed by Einstein in 1905:

$$E = h\nu \tag{8.4.2}$$

h is Planck's constant, for which we shall use the truncated value of 6.626×10^{-34} J s in this section. Thus, for the above D-line photon we find:

$$E = 6.626 \times 5.088 \times 10^{-20} = 3.371 \times 10^{-19} \text{ J} \tag{8.4.3}$$

8.4.3 Mass

A relationship also deduced by Einstein, but in connection with the special theory of relativity, may be used to determine the mass of a photon:

$$E = mc^2 \quad \text{or} \quad m = E/c^2 \tag{8.4.4}$$

For our particular example we find:

$$m = 3.371 \times 10^{-19}/(2.998 \times 10^8)^2 = 3.751 \times 10^{-36} \text{ kg}$$

Note that, because the mass of the photon depends upon its energy, its mass also depends upon its wavelength or frequency. If we also recall that the mass of a particle moving with a velocity, v, comparable with that of light is related to its mass when at rest, m_0, by yet another equation due to Einstein:

$$m_0 = m\sqrt{1 - (v/c)^2} \tag{8.4.5}$$

then we see that, because the photon moves with the velocity of light, $v/c = 1$ and the rest mass of the photon is zero.

8.4.4 Linear momentum

The linear momentum, p, of a particle of mass m moving with velocity v is given by $p = mv$ and since for the photon $v = c$, in a vacuum our D-line photon has a linear momentum of:

$$p = mc = 3.751 \times 2.998 \times 10^{-28} = 1.125 \times 10^{-27} \text{ kg m s}^{-1} \tag{8.4.6}$$

Compton's interpretation of the scattering of X-rays by electrons (see Box 8.1) provides one of the most convincing pieces of evidence for the photon as a discrete particle which has mass and, hence, linear momentum.

8.4.5 Angular momentum

Experiment shows that the electric dipole photon has a spin angular momentum characterised by a spin quantum number of 1, i.e. a total spin angular momentum of $\sqrt{2}h/2\pi$ (Chapter 4). But, unlike a particle with a finite rest mass, which would have z-components of $+1$, 0 and -1 in units of $h/2\pi$ (Chapter 4), the photon has only two components of $\pm h/2\pi$ directed parallel ($+$) and antiparallel ($-$) to its direction of propagation. These two components of angular momentum are related to the property of light which is known as polarisation. As we have seen in Section 8.2.2, two types of circularly polarised light exist. LCP light has a z component of angular momentum of $+h/2\pi$ and RCP light has a z component of angular momentum of $-h/2\pi$. The photons in a beam of linearly polarised light are not in states which are eigenfunctions of the operator for the z component of angular momentum. The expectation value for a measurement of the z component of their angular momentum is zero, since a large number of measurements would record equal numbers of photons having angular momenta of $+h/2\pi$ and $-h/2\pi$. In SI units the angular momentum of a photon is:

$$l = \sqrt{2}h/2\pi = \sqrt{2} \times 6.626 \times 10^{-34}/2\pi = 1.491 \times 10^{-34} \text{ J s} \qquad (8.4.7)$$

The value of l has been determined experimentally in a very ingenious way by Beth.[2]

Note that l is independent of the wavelength or frequency of the photon and is therefore the same for all photons.

8.4.6 Parity

Like atomic wave functions (Section 5.10.2), an odd (u) or even (g) character may be assigned to photons and, as with atoms, the parity of the photon depends upon its state. Parity and angular momentum together lie at the heart of the stringent conditions which determine the way in which photons interact with matter. These conditions are called *selection rules*.

8.5 SELECTION RULES

Three combinations of photonic angular momentum and parity play a particular role in the selection rules of chemical spectroscopy. Each corresponds to a particular form of e-m radiation.

The photons of electric dipole radiation are of odd (u) parity and carry one unit of angular momentum. For an allowed electric dipole transition, therefore, the parities of the initial and final states must differ, i.e. u $\rightarrow$ g or g $\rightarrow$ u, and the triangular condition $\Delta(j_i, 1, j_f)$ must be obeyed. Since the total angular momentum quantum number, j, is always positive, one might naively think that the absorption of a photon must be accompanied by an increase in the angular momentum of the absorbing species, and emission by a decrease. But the fact that angular momentum is conserved if the resulting j_f satisfies any of the $2j_< + 1$ Clebsch–Gordan relationships removes this apparent restriction. For example, if an atom in a P state ($j = 1$) absorbs a photon, states having increased ($j = 2$ [D]), unchanged ($j = 1$ [P]) or decreased ($j = 0$ [S]) angular momentum may result. And in all three cases angular momentum is conserved. However, in the case of the P $\rightarrow$ P transition there is no change of parity, so it would not be electric-dipole allowed. Electric dipole transitions are responsible for the vast majority of the absorption (emission) bands observed in the electronic spectra of atoms and molecules and in all infrared absorption and emission processes. They are much stronger (by a factor $\approx 10^4$) than all other transitions which are forbidden by these selection rules.

The photons of electric quadrupole radiation carry two units of angular momentum and are of even (g) parity. The extra angular momentum is usually called orbital angular momentum, but the use of the terms 'spin' and 'orbital' in the case of photons is to a large extent a matter of convenience because in most cases the two cannot be separated. For an electric quadrupole transition, therefore, the triangular condition $\Delta(j_i, 2, j_f)$ must be obeyed and the parities of the initial and final states must be the same. The very weak d $\rightarrow$ d (g $\rightarrow$ g) and f $\rightarrow$ f (u $\rightarrow$ u) transitions of gas-phase metal atoms and ions with open d or f shells are of this type.

The photons of magnetic dipole radiation carry one unit of angular momentum and are of even (g) parity. For an allowed magnetic dipole transition, therefore, the triangular condition $\Delta(j_i, 1, j_f)$ must be obeyed and the parities of the initial and final states must be the same. These transitions are weak and rarely seen in electronic spectroscopy and never in the infrared; but their significance is great because they are the transitions which are observed in electron and nuclear magnetic resonance spectroscopies.

The selection rules that derive directly from the requirements of conservation, especially conservation of energy, angular momentum and parity, impose many restrictions upon the process of absorption or emission of a photon by a molecule. The fact that the selection rules are so restrictive should not, in general, be regarded as a disadvantage. This selectivity is the basic cause of the highly characteristic spectrum of each chemical compound which is, in turn, the reason for the value of spectroscopy as a means of identifying atoms and molecules. We shall return to the subject of selection rules with some examples in Section 8.7.3.

We might note here that, since energy and mass are related by Equation (8.4.4), the conservation of these two quantities should be considered together. But this fact plays no discernable role in the spectroscopy of interest to chemists and we confine our attention here to conservation of energy which forms the basis of the most fundamental of all the selection rules: the Bohr–Einstein condition.

8.5.1   The Bohr–Einstein condition

Strange as it may seem to us today, Einstein's concept of the light particle or photon received little attention in the years immediately following its publication in 1905. But in 1913 the idea played a central role in Bohr's pre-quantum-mechanical theory of the hydrogen atom (Sections 2.7.1 and 2.7.2). Having quantised the orbital angular momentum, and hence the energy, of his planetary electron, Bohr postulated that when an electron changed orbit it emitted or absorbed the energy difference between the energy of the initial, E_i, and the final, E_f, states according to the equation:

$$\Delta E = |E_f - E_i| = h\nu. \tag{8.5.1}$$

Thus, if the energy of the photon is $h\nu$, as Einstein had proposed, Equation (8.5.1) is the equation of energy conservation and determines the region of the e-m spectrum in which each type of atomic and molecular process is observed spectroscopically. To appreciate the conservation of other quantities, especially the most important of them angular momentum, we must examine the interaction of e-m radiation and matter in more detail, and we do this in Section 8.6.

In addition to the fundamental concepts associated with conservation of mass, energy and angular momentum noted above, at least one other general view of the selection rules can be discerned. If we again consider e-m radiation as a wave rather than as a stream of photons, we may enquire as to how the radiation actually interacts with a molecule. What is the mechanism by means of which an e-m radiation wave can change the electron distribution in a molecule or induce a higher energy mode of vibration in it? The 'hands' with which the e-m radiation 'grasps' the molecule are the oscillating electric and magnetic fields. The electric field, for example, can interact with a permanent dipole moment, if one exists in a molecule; the magnetic field can interact with the magnetic moment of an electron, or of a nucleus if that nucleus has a magnetic moment. These are the concepts with which we shall seek to understand the nature of spectroscopic selection rules which apply in the individual spectral regions. In the next section of this chapter we set the scene by giving a semi-classical description of the interaction of e-m radiation with matter. In Section 8.9 attention will be drawn to a purely quantum-mechanical theory of the interaction, *quantum electrodynamics*.

8.6 THE QUANTUM MECHANICS OF TRANSITION PROBABILITY

This section will be one of the more mathematical parts of the book and readers who would prefer to leave it aside at present may do so if they note carefully the following points, which will also serve as a guide for those who propose to work through the material.

1. We first set up a wave function that describes a system which is changing in time; i.e. one which is making a transition from an initial state having a wave function Ψ_i and an energy E_i to a final state with wave function Ψ_f and energy E_f.

2. We recognise that such a wave function has time-dependent and time-independent parts and we focus our attention upon the former.

3. We discover that the combination of the time-dependent part of the wave function and the time-dependency of the radiation ($\sin \omega t$), acting over a finite period of time, T, results in a non-zero interaction only when the Bohr–Einstein condition of energy conservation is satisfied. It must be emphasised that the Bohr–Einstein condition is not introduced into the analysis; it arises quite naturally.

4. This time-dependent part of the problem is common to all branches of spectroscopy and the Bohr–Einstein condition therefore represents a very fundamental and widely applicable selection rule.

5. The time-independent part of the problem, however, varies with the type of spectroscopy involved and does not take the same form in all branches of the subject.

We now attempt to answer the questions: How does an electromagnetic wave interact with an atom or molecule and cause it to change its energy state by absorbing energy from the radiation? What is the mechanism of this process and how does it give rise to selection rules?

With problems such as the above in mind, Erwin Schrödinger formulated two quantum-mechanical equations; the time-independent and the time-dependent Schrödinger equations. We have already discussed and used the first of these equations in Chapters 2 and 3. It takes the form:

$$\hat{\mathcal{H}}\phi_a = E_a\phi_a \tag{8.6.1}$$

in which ϕ_a is an eigenfunction of the energy operator, $\hat{\mathcal{H}}$, with eigenvalue E_a. $\hat{\mathcal{H}}$ and ϕ_a are functions of the spatial co-ordinates x, y and z only.

The time-dependent equation takes the form:

$$\hat{\mathcal{H}}\Psi = \frac{ih}{2\pi} \cdot \frac{\partial\Psi}{\partial t} \tag{8.6.2}$$

$\hat{\mathcal{H}}$ is the same as in Equation (8.6.1), i is $\sqrt{(-1)}$, h is Planck's constant and the simplest form of Ψ is a product of the ϕ_a from Equation (8.6.1) and a time-dependent exponential term:

$$\Psi = \phi_a \exp(-i2\pi E_a t / h) \tag{8.6.3}$$

We see that it is a function of the three spatial co-ordinates (from ϕ_a) and of the time, t. It is convenient to simplify the above expression for Ψ by noting that:

$$E = h\nu = h\omega/2\pi \quad \text{or} \quad \omega = 2\pi E/h \tag{8.6.4}$$

where ω is an angular frequency measured in radians per second.

Using Equation (8.6.4), Ψ can be written:

$$\Psi = \phi_a \exp(-i\omega_a t) \tag{8.6.5}$$

In this simple form Ψ is an eigenfunction of the time-dependent Schrödinger equation with the eigenvalue E_a. But the most general form of Ψ is a linear combination of functions of the form:

$$\Psi = c_1\phi_1 \exp(-i\omega_1 t) + c_2\phi_2 \exp(-i\omega_2 t) + c_3\phi_3 \exp(-i\omega_3 t) + \cdots$$

i.e.:

$$\Psi = \Sigma_a c_a\phi_a \exp(-i\omega_a t) \tag{8.6.6}$$

where c_a are coefficients which tell us how much of each function $\phi_a \exp(-i\omega_a t)$ there is in Ψ. The energy of a system described by such a general wave function as Ψ is not constant. It changes with time (Box 8.2), which is just what we require for a description of a spectroscopic transition.

In absorption spectroscopy we are normally particularly interested in a transition between just two states, an initial state, Ψ_i, and a final state, Ψ_f, which have energies of E_i and E_f with $E_i < E_f$. Then the wave function which describes all stages of the process whereby the system absorbs energy and goes from Ψ_i to Ψ_f during the time interval $t = 0$ to $t = T$ is:

$$\Psi = c_i\phi_i \exp(-i\omega_i t) + c_f\phi_f \exp(-i\omega_f t) \qquad (8.6.7)$$

and

at $t = 0$: $\Psi = \Psi_i = \phi_i \exp(-i\omega_i t)$ with precise energy E_i and $c_i = 1$, $c_f = 0$

at $t = T$: $\Psi = \Psi_f = \phi_f \exp(-i\omega_f t)$ with precise energy E_f and $c_i = 0$, $c_f = 1$

We see that the coefficients, c, are also functions of time and we shall use the change in the value of $c_f{}^*c_f$ to track the progress of the transition from Ψ_i to Ψ_f. It is clear that for this process to take place there must be some property of the electromagnetic radiation which links the two states Ψ_i and Ψ_f. Let us represent that property by an operator, $\hat{V}$, which we write is a product of two parts, one a function of the spatial co-ordinates only and the other a function of the time:

$$\hat{V} = \hat{V}(x,y,z) \times \hat{V}(t) \qquad (8.6.8)$$

It will also be convenient to define the symbol V_{fi} as:

$$V_{fi} \equiv \langle\phi_f|\hat{V}(x,y,z)|\phi_i\rangle \cdot \langle\exp(-i\omega_f t)|\hat{V}(t)|\exp(-i\omega_i t)\rangle \qquad (8.6.9)$$

which will have a finite value for an allowed transition.

We say that the effect of the electromagnetic radiation is to perturb the system; the operator $\hat{V}$ represents that perturbation. This is expressed in quantum mechanics (see Appendix 4) by writing a new Hamiltonian operator, $\hat{\mathcal{H}}$, for the system as a sum of the original Hamiltonian, $\hat{\mathcal{H}}^0$, and the perturbation:

$$\hat{\mathcal{H}} = \hat{\mathcal{H}}^0 + \hat{V} \qquad (8.6.10)$$

The Schrödinger equation which describes the system under the influence of the perturbation is therefore:

$$(\hat{\mathcal{H}}^0 + \hat{V})\Psi = \frac{ih}{2\pi} \cdot \frac{\partial\Psi}{\partial t} \qquad (8.6.11)$$

If Ψ describes our simple two-state system (Equation (8.6.7)) then substitution into Equation (8.6.11) gives:

$$c_i\hat{\mathcal{H}}^0\phi_i \exp(-i\omega_i t) + c_f\hat{\mathcal{H}}^0\phi_f \exp(-i\omega_f t) + c_i\hat{V}\phi_i \exp(-i\omega_i t) + c_f\hat{V}\phi_f \exp(-i\omega_f t)$$

$$= \frac{ih}{2\pi}\left\{ \frac{\partial c_i}{\partial t} \cdot \phi_i \exp(-i\omega_i t) + c_i\frac{\partial\phi_i}{\partial t} \cdot \exp(-i\omega_i t) \right.$$

$$\left. + \frac{\partial c_f}{\partial t} \cdot \phi_f \exp(-i\omega_f t) + c_f\frac{\partial\phi_f}{\partial t} \cdot \exp(-i\omega_f t)\right\} \qquad (8.6.12)$$

Note that in forming $\partial\Psi/\partial t$ on the right-hand side of the above equation the functions to be differentiated have been treated as products because the coefficients, c, are also functions of time as we saw above.

Because of Equations (8.6.2) and (8.6.3), the first two terms on the left of Equation (8.6.12) cancel the second and the fourth on the right and the equation reduces to:

$$c_i\hat{V}\phi_i \exp(-i\omega_i t) + c_f\hat{V}\phi_f \exp(-i\omega_f t) = \frac{ih}{2\pi}\left\{ \frac{\partial c_i}{\partial t} \cdot \phi_i \exp(-i\omega_i t) + \frac{\partial c_f}{\partial t} \cdot \phi_f \exp(-i\omega_f t)\right\}$$

$$(8.6.13)$$

If we multiply each term of Equation (8.6.13), on the left, by $\langle \phi_f \exp(-i\omega_f t)|$ and integrate over all spatial co-ordinates and time we have:

$$c_i \langle \phi_f \exp(-i\omega_f t)| \hat{V} | \phi_i \exp(-i\omega_i t) \rangle + c_f \langle \phi_f \exp(-i\omega_f t)| \hat{V} | \phi_f \exp(-i\omega_f t) \rangle$$

$$= \frac{ih}{2\pi} \left\{ \frac{\partial c_i}{\partial t} \langle \phi_f \exp(-i\omega_f t)| \phi_i \exp(-i\omega_i t) \rangle + \frac{\partial c_f}{\partial t} \langle \phi_f \exp(-i\omega_f t)| \phi_f \exp(-i\omega_f t) \rangle \right\}$$

$$(8.6.14)$$

In this *bra-ket* notation, which we have used before in Chapter 3, the $<>$ means integration over all the co-ordinates in the operator, i.e. x, y, z and t. Note that the exponential term on the left-hand side of each bracket is required by quantum mechanics to be the complex conjugate ($-i \rightarrow +i$; Appendix 8) of $\phi_f \exp(-i\omega_f t)$. However, this is not written explicitly, rather it is *implied* by the *bra* ($\langle|$). The first integration on the right-hand side of Equation (8.6.14) gives zero because the functions $\phi_i \exp(-i\omega_i t)$ and $\phi_f \exp(-i\omega_f t)$ are orthogonal to each other. The integration of the second term gives one because $\phi_f \exp(-i\omega_f t)$ is normalised. Simplifying the notation using Equation (8.6.9) and rearranging we obtain Equation (8.6.14) in the compact form:

$$\frac{\partial c_f}{\partial t} = \frac{2\pi}{ih} \cdot \{c_i V_{fi} + c_f V_{ff}\} \tag{8.6.14a}$$

At the onset of the perturbation $c_i = 1$ and $c_f = 0$ so that:

$$\frac{\partial c_f}{\partial t} = \frac{2\pi}{ih} V_{fi} \tag{8.6.15}$$

This equation is an expression for the change of c_f with time. By monitoring its value we can follow the change in the system with time under the influence of the perturbation. Suppose that the perturbation lasts for a time T. The value of c_f at the end of that time, c_f^T, can be found by evaluating the integral:

$$c_f^T = \int_0^T \partial c_f = \frac{2\pi}{ih} \int_0^T V_{fi} \partial t \tag{8.6.16}$$

If we recall that the definition of V_{fi} (Equation (8.6.9)) involves integration over the spatial co-ordinates x, y and z, then Equation (8.6.16) requires integration over all spatial and time variables and it is convenient to separate the integrations over space and time. This is why we have written V_{fi} as a product of a space-dependent and a time-dependent part in Equation (8.6.9), which we repeat here for convenience:

$$V_{fi} \equiv \langle \phi_f | \hat{V}(x,y,z) | \phi_i \rangle \cdot \langle \exp(-i\omega_f t)| \hat{V}(t) | \exp(-i\omega_i t) \rangle \tag{8.6.9}$$

The first term, the time-independent term, is quite specific to the type of spectroscopy and is not the same, for example, for infrared and NMR spectroscopy. We therefore leave the time-independent term aside for the moment and consider it again in Section 8.7. The time-dependent term, on the other hand, is exactly the same in all cases and it will therefore be considered here in the general discussion. Combining the two exponential functions (remember that the *bra* implies the complex conjugate of the function written inside it) the integral can be written:

$$\int_0^T \exp[i(\omega_f - \omega_i)t]\hat{V}(t)\,dt \equiv \int_0^T \exp[i\,\Delta\omega t]\hat{V}(t)\,dt \qquad (8.6.17)$$

where $\Delta\omega = \omega_f - \omega_i$.

Since $\exp(i\,\Delta\omega t) = \cos(\Delta\omega t) + i\sin(\Delta\omega t)$ (Box 3.1 and Appendix 8), we see that the exponential function is a function which oscillates in time and the integral of such a function will normally contain a large number of equal and opposite contributions which will sum to zero (Figure 8.7(a)). We therefore require that $\hat{V}(t)$ should also be an oscillating function which, if suitably chosen, cancels the oscillations of the exponential term giving a non-zero value for the integral. The electric field of a beam of light seems a good choice because it has the required oscillatory character and the presence of the field will change the potential energy of the charged particles of which atoms and molecules are formed. We therefore investigate the function:

$$\hat{V}(t) = E_0 \sin \omega t = E_0\{\exp(i\omega t) - \exp(-i\omega t)\}/2i \qquad (8.6.18)$$

because it is of the form which we have used earlier to describe the waves of electromagnetic radiation. Substituting Equation (8.6.18) into Equation (8.6.17) our integral over time becomes:

$$E_0 \int_0^T \exp[i\,\Delta\omega t] \cdot \sin \omega t\,dt = \frac{E_0}{2i} \int_0^T \exp[i\,\Delta\omega t] \cdot \{\exp(i\omega t) - \exp(-i\omega t)\}\,dt$$

$$= \frac{E_0}{2i}\left[\int_0^T \exp\{i(\Delta\omega + \omega)t\}\,dt - \int_0^T \exp\{i(\Delta\omega - \omega)t\}\,dt\right] \qquad (8.6.19)$$

since $E_f > E_i$, $\omega_f > \omega_i$ and $\Delta\omega > 0$. Therefore, the argument of the first exponential term is always large giving a rapidly oscillating function which integrates to a very small quantity (Figure 8.7(a)). But the argument of the second exponential can be small and

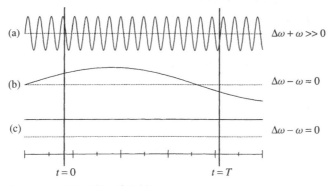

Figure 8.7 The integral in Equation (8.6.19)

is, in fact, zero when $\omega = \Delta\omega$. Therefore, the second term in the integral is slowly varying in the frequency range $\omega \approx \Delta\omega$ and it can contribute to a significant value of the integral (Figure 8.7(b)). The maximum contribution will be obtained when $\omega = \Delta\omega$ since the exponential function is then $\exp(0) = 1$ which shows no fluctuation with time, (Figure 8.7(c)). Thus the maximum interaction of the molecule with the radiation arises when:

$$\omega = \Delta\omega = \omega_f - \omega_i$$

or (using Equation (8.6.4)) in terms of frequency and energy, when:

$$2\pi\nu = (2\pi/h) \cdot (E_f - E_i) \quad \text{or} \quad E_f - E_i = h\nu$$

We have obtained the Bohr–Einstein or energy conservation rule without any mention of energy conservation. This result may be taken as confirmation, if such is needed, of Einstein's proposal that the energy of an electromagnetic wave could be expressed in terms of photons of energy $h\nu$. Note that the argument would have been made in exactly the same way, and with exactly the same result, had we been considering stimulated emission for which $E_f < E_i$, $\omega_f < \omega_i$ and $\Delta\omega < 0$. Then it would have been the first term of Equation (8.6.19) which was the important one.

Now we can complete our analysis by carrying out the integration, which is quite simple. Neglecting the first term on the right in Equation (8.6.19) we have:

$$-\frac{E_0}{2i} \int_0^T \exp\{i(\Delta\omega - \omega)t\}\,dt = -\frac{E_0}{2i} \cdot \left[\frac{\exp\{i(\Delta\omega - \omega)t\}}{i(\Delta\omega - \omega)}\right]_0^T$$

$$= \frac{E_0[\exp\{i(\Delta\omega - \omega)T\} - 1]}{2(\Delta\omega - \omega)}$$

We have evaluated the integral in Equation (8.6.16) to obtain $c_f{}^T$, the value at $t = T$ of the wave function coefficient c_f defined in Equation (8.6.7). Since in quantum mechanics, probability is determined by the product of the wave function with its complex conjugate, the time-dependent part of the transition probability, $W(t)$, is proportional to the product of this result and its complex conjugate:

$$W(t) = \frac{E_0[\exp\{i(\Delta\omega - \omega)T\} - 1]}{2(\Delta\omega - \omega)} \cdot \frac{E_0[\exp\{-i(\Delta\omega - \omega)T\} - 1]}{2(\Delta\omega - \omega)}$$

$$= \frac{E_0^2[\exp\{0\} + 1 - \exp\{i(\Delta\omega - \omega)T\} - \exp\{-i(\Delta\omega - \omega)T\}]}{4(\Delta\omega - \omega)^2}$$

$$= \frac{E_0^2[2 - 2\cos\{(\Delta\omega - \omega)T\}]}{4(\Delta\omega - \omega)^2} = \frac{E_0^2[\sin^2\{\frac{1}{2}(\Delta\omega - \omega)T\}]}{(\Delta\omega - \omega)^2}$$

This important function is sometimes written in the form:

$$W(t) = \frac{E_0^2 \sin^2\{\frac{1}{2}(\Delta\omega - \omega)T\}}{4\{\frac{1}{2}(\Delta\omega - \omega)\}^2} \tag{8.6.20}$$

$W(t)$ is plotted against $(\Delta\omega - \omega)$ from $-3\pi/T$ to $+3\pi/T$ with $E_0 = 1$ and $T = 2$ in Figure 8.8. It is interesting to note that this function was first obtained theoretically

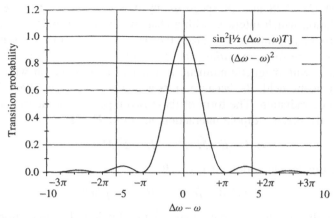

Figure 8.8 Graph of the function of Equation (8.6.20)

in the early 1930s, but no experimental conformation of its peculiar form was available until 1972 when it was observed directly in a molecular beam experiment.[3]

When the Function (8.6.20) is multiplied by the squared, time-independent, part of the interaction $|\langle \phi_f | \hat{V}(x,y,z) | \phi_i \rangle|^2$ and the factor $(2\pi/ih) \cdot (-2\pi/ih)$ from Equation (8.6.16), we have as our final expression for the transition probability between the two states Ψ_i and Ψ_f:

$$W(x,y,z,t) = E_0^2 \cdot \frac{4\pi^2}{h^2} |\langle \phi_f | \hat{V}(x,y,z) | \phi_i \rangle|^2 \cdot \frac{\sin^2\{\frac{1}{2}(\Delta\omega - \omega)T\}}{4\{\frac{1}{2}(\Delta\omega - \omega)\}^2} \qquad (8.6.21)$$

The time-independent function is the origin of all spectroscopic selection rules, apart from the Born–Einstein condition, and consequently it is responsible for much of the detail in the spectroscopic information we obtain from molecules. In the next section we examine its particular form for three major branches of chemical spectroscopy – electronic, infrared and nuclear magnetic resonance.

8.7 THE NATURE OF THE TIME-INDEPENDENT INTERACTION $\langle \phi_f | \hat{V}(x, y, z) | \phi_i \rangle$[‡]

To proceed further with our investigation of the interaction of radiation with matter, we must consider the explicit form which the time-independent part of the interaction, characterised by the operator $\hat{V}(x,y,z)$, takes in the case of the individual spectroscopies. Though all branches of spectroscopy have made important contributions to chemical knowledge, electronic (ultraviolet-visible, UV–VIS), infrared (IR) and nuclear magnetic resonance (NMR) spectroscopy have played by far the greatest role in this context. Furthermore, these three branches of the subject are used much more extensively every day in every chemical research laboratory and accordingly we here confine our attention to these three only.

[‡] Comment on notation: In this section we need to distinguish between wave functions which relate to different aspects of a molecular energy state; in particular we wish to distinguish between electronic and vibrational wave functions. To make this possible we now write i and f as superscripts leaving the subscript position vacant for later use.

In the case of UV–VIS and IR spectroscopy it is the *electric field*, E, of the electromagnetic radiation which interacts with a changing electric dipole moment, M, in the molecule and through this interaction causes the electronic or vibrational state (or both) of the molecule to change from an *initial state* ϕ^i to a *final state* ϕ^f. In NMR spectroscopy a magnetic nucleus with magnetic moment, μ, changes its orientation with respect to the applied, static magnetic field, B_0, because of its interaction with the *magnetic field*, B_1, of the electromagnetic radiation. The form of these two types of interaction is very similar:

UV–VIS and IR: electric dipole interaction energy $= E \cdot M$[†]

$$\Rightarrow \langle \phi^f | E \cdot \hat{M} | \phi^i \rangle = E \cdot \langle \phi^f | \hat{M} | \phi^i \rangle \qquad (8.7.1)$$

NMR: magnetic dipole interaction energy $= B_1 \cdot \mu$

$$\Rightarrow \langle \phi^f | B_1 \cdot \hat{\mu} | \phi^i \rangle = B_1 \cdot \langle \phi^f | \hat{\mu} | \phi^i \rangle \qquad (8.7.2)$$

These are simply the classical equations for the dipole-field interaction energy. The 'dot' $\cdot$ indicates the scalar product of two vectors which is defined by the equation:

$$\mathbf{A} \cdot \mathbf{B} = A_x B_x + A_y B_y + A_z B_z$$

so that:

$$E \cdot \langle \phi^f | \hat{M} | \phi^i \rangle = E_x \langle \phi^f | \hat{M}_x | \phi^i \rangle + E_y \langle \phi^f | \hat{M}_y | \phi^i \rangle + E_z \langle \phi^f | \hat{M}_z | \phi^i \rangle$$

We see that in order for there to be an interaction between the electric (magnetic) field of the radiation there must be a matrix element of the electric (magnetic) dipole moment operator in the same direction. It is this which gives spectroscopy with polarised light the ability to provide information about directions within a molecule.

To calculate the intensity of a spectral line, we must evaluate the above matrix elements of $\hat{M}$ and $\hat{\mu}$ for the spectroscopies in question. In the case of UV–VIS and IR spectroscopy we require the electric dipole moment, $\langle \phi^f | \hat{M} | \phi^i \rangle$, associated with the transition and in the case of NMR spectroscopy the corresponding magnetic moment. We first consider the UV–VIS and IR case.

8.7.1 The transition dipole moment

In Appendix 9 we show that the transition dipole moment, $M_{i,f}$, for a transition from an initial state, ϕ^i_{total}, to a final state, ϕ^f_{total}, is given by the equation:

$$M_{i,f} = \langle \phi^f_{total} | \hat{M} | \phi^i_{total} \rangle = \langle \psi^f_{elec} \psi^f_{vib} | \hat{M} | \psi^i_{elec} \psi^i_{vib} \rangle \qquad (8.7.3)$$

$M_{i,f}$ is the transitory dipole moment generated when the molecule changes its state and, in a wave-based view of the interaction of radiation and matter, it is the 'handle' by means of which the oscillating electric field of the radiation is able to couple with the molecule.

In deriving Equation (8.7.3) the Born–Oppenheimer approximation (Section 6.3) has been used to express the total wave function as a product of an electronic and a vibrational function:

$$\phi_{total}(x,X) = \psi_{elec}(x,X) \cdot \psi_{vib}(X) \qquad (8.7.4)$$

[†] Note that this expression for the interaction assumes that the wavelength of the radiation is much greater than the dimensions of the atom or molecule. For everyday chemical spectroscopy this assumption is very well justified.

where the presence of both electronic (x) and nuclear (X) co-ordinates following ψ_{elec} reminds us of the fact that the electronic wave function depends upon the positions of the nuclei as well as upon those of the electrons. For a discussion which includes the rotation of the molecule see Section 10.5. [In Equation (8.7.3) the (X) and (x,X) have been dropped to simplify the notation.]

Equation (8.7.3) may be written in the form:

$$M_{i,f} = \langle \psi_{\text{vib}}^f | \; \langle \psi_{\text{elec}}^f | \hat{M} | \psi_{\text{elec}}^i \rangle | \; \psi_{\text{vib}}^i \rangle = \langle \psi_{\text{vib}}^f | M_X(e^f e^i) | \; \psi_{\text{vib}}^i \rangle \qquad (8.7.5)$$

in which $M_X(e^f e^i)$ is the result of evaluating the integral $\langle \psi_{\text{elec}}^f | \hat{M} | \psi_{\text{elec}}^i \rangle$ over the electronic co-ordinates for a particular nuclear configuration represented by X. The presence of the symbols X, e^f and e^i in $M_X(e^f e^i)$ again reminds us that the result of the integration depends upon the two electronic wave functions involved, ψ_{elec}^f and ψ_{elec}^i, and the particular nuclear co-ordinates chosen, X.

But we must allow for the dependence of $M_{i,f}$ upon the nuclear configuration and therefore to proceed further we express $M_X(e^f e^i)$ as a sum of two terms:

$$M_X(e^f e^i) = M_0(e^f e^i) + \sum_{k=1}^{3N-6} Q_k M_k'(e^f e^i) \qquad (8.7.6)$$

The first term is the value of $M_X(e^f e^i)$ at the equilibrium nuclear configuration. In the second term, $M_k'(e^f e^i) \equiv \partial M_0(e^f e^i)/\partial Q_k$, the derivative of $M_X(e^f e^i)$ with respect to Q_k at the equilibrium configuration. Q_k is the kth normal co-ordinate. The motion of the atoms of a molecule along a normal co-ordinate is a vibration which is an eigenfunction of the vibrational Hamiltonian of the molecule (see Section 10.7.1).

Thus, $Q_k M_k'(e^f e^i)$ is the change in the value of $M_X(e^f e^i)$ when the nuclear configuration differs from its equilibrium value by a displacement along the normal co-ordinate Q_k. The actual value of that displacement, $\langle \psi_{\text{vib}}^f | Q_k | \psi_{\text{vib}}^i \rangle$, will enter the calculation when the integration over the vibrational wave function is performed. The basic assumption of this expansion is that the first term is very much larger than the second which, in turn, is very much larger than any further terms, e.g. $Q_j Q_k M_j' M_k'$ $(e^f e^i)$ or $Q_k^2 M''_k(e^f e^i)$. In Section 8.7.2 we shall find that this assumption is well justified. Equation (8.7.5) can now be rewritten in the form:

$$M_{i,f} = \langle \psi_{\text{vib}}^f | \; M_0(e^f e^i) | \; \psi_{\text{vib}}^i \rangle + \left\langle \psi_{\text{vib}}^f \left| \sum_{k=1}^{3N-6} Q_k M_k'(e^f e^i) \right| \psi_{\text{vib}}^i \right\rangle$$

$$= \langle \psi_{\text{vib}}^f | \psi_{\text{vib}}^i \rangle \cdot M_0(e^f e^i) + \sum_{k=1}^{3N-6} \langle \psi_{\text{vib}}^f | \; Q_k | \; \psi_{\text{vib}}^i \rangle \cdot M_k'(e^f e^i) \qquad (8.7.7)$$

We can take $M_0(e^f e^i)$ and $M_k'(e^f e^i)$ outside the integration over the vibrational functions implied by $\langle \rangle$ because they are both constants; though they do, of course, have units. Equation (8.7.7) is a very general expression which applies to any molecule, and we can recognise in it the following three important special cases.

Case 1: $\quad \psi_{\text{elec}}^f = \psi_{\text{elec}}^i$ but $\psi_{\text{vib}}^f \neq \psi_{\text{vib}}^i$

In this case there is a change of vibrational state but no change of electronic state, i.e. we have an infrared transition. The first term on the right in Equation (8.7.7) is zero because the two different vibrational wave functions belong to the same electronic state and are therefore orthogonal. Thus, the transition dipole moment is given by the second term. The vibrational states, ψ_{vib}^f and ψ_{vib}^i, can be written as products of normal vibrational modes (Section 10.7) but because $\psi_{vib}^f \neq \psi_{vib}^i$ the two products will not be identical. Let us examine the possibility that a particular vibrational mode, Q_j say, can lose or gain quanta to go from $\psi^i(Q_j'')$ in ψ_{vib}^i to $\psi^f(Q_j')$ in ψ_{vib}^f while the numbers of quanta in all other vibrational modes remain unchanged. The second term in Equation (8.7.7) is then:

$$
\begin{aligned}
\boldsymbol{M}_{i,f} &= \sum_{k=1}^{3N-6} \langle \psi_{vib}^f | Q_k | \, \psi_{vib}^i \rangle \boldsymbol{M}_k'(e^i e^i) \\
&= \sum_{k=1}^{3N-6} \langle \psi^f(Q_1)\psi^f(Q_2)\dots\psi^f(Q_j')\dots\psi^f(Q_{3N-6}) | Q_k | \\
&\qquad\qquad \psi^i(Q_1)\psi^i(Q_2)\dots\psi^i(Q_j'')\dots\psi^i(Q_{3N-6}) \rangle \boldsymbol{M}_k'(e^i e^i) \\
&= \langle \psi^f(Q_1)|\psi^i(Q_1)\rangle \dots \langle \psi^f(Q_j')|Q_j|\psi^i(Q_j'')\rangle \dots \langle \psi^f(Q_{3N-6})|\psi^i(Q_{3N-6})\rangle \boldsymbol{M}_j'(e^i e^i) \\
&= \langle \psi^f(Q_j')|Q_j|\psi^i(Q_j'')\rangle \boldsymbol{M}_j'(e^i e^i) \qquad\qquad (8.7.8)
\end{aligned}
$$

The sum over k reduces to a single term in the penultimate line of Equation (8.7.8) since it is only when $k = j$ that we do not have the integral $\langle \psi^f(Q_j')|\psi^i(Q_j'')\rangle$ which is zero because we have assumed that the same mode, Q_j, has different quantum numbers in $\psi^f(Q_j')$ and $\psi^i(Q_j'')$. The simplification in the last line is a consequence of the fact that all the normal vibrational modes of the same electronic state are orthogonal and normalised:

$$
\langle \psi(Q_m)|\psi(Q_n)\rangle = \delta_{mn} \qquad\qquad (8.7.9)
$$

The result, Equation (8.7.8), shows that if an infrared transition in the normal mode Q_j is to be allowed there must be a change of dipole moment associated with that vibration, i.e. $\boldsymbol{M}_j'(e^i e^i) \neq 0$. The infrared selection rule is embodied in the multiplying factor $\langle \psi^f(Q_j')|Q_j|\psi^i(Q_j'')\rangle$, which is evaluated and discussed in Section 10.5. It will be sufficient here if we say that it is zero unless the vibrational quantum numbers of the wave functions $\psi^f(Q_j')$ and $\psi^i(Q_j'')$ differ by ± 1.

Case 2: $\quad \psi_{elec}^f \neq \psi_{elec}^i$ and $M_0(e^f e^i) \neq 0$; $\psi_{vib}^f \neq \psi_{vib}^i$ or $\psi_{vib}^f = \psi_{vib}^i$
This is the case of an allowed electronic transition $[M_0(e^f e^i) \neq 0]$, which may be accompanied by a simultaneous change of vibrational state, $\psi_{vib}^f \neq \psi_{vib}^i$, or by no change, $\psi_{vib}^f = \psi_{vib}^i$. Both terms in Equation (8.7.7) contribute to the transition dipole moment but, for the reasons given above, the second term is neglected in comparison with the first term which is very much larger. Thus, the transition dipole moment is given by:

$$
\boldsymbol{M}_{i,f} = \langle \psi_{vib}^f | \, \psi_{vib}^i \rangle M_0(e^f e^i) \qquad\qquad (8.7.10)
$$

We see that the transition dipole moment for the vibronic (*vibronic* = product of vibrational and electronic states) transition as a whole is determined by the electronic transition dipole moment calculated at the equilibrium internuclear distance multiplied by the overlap of the vibrational wave functions of the initial and final states. The two vibrational functions are not orthogonal because $\psi_{elec}^f \neq \psi_{elec}^i$. This result was first derived quantum-mechanically by Edward Uhler Condon (1902–1974), but James Franck (1882–1964) had obtained it earlier using classical theory. Accordingly, it is known as the Franck–Condon principle. With its aid even qualitative analysis of the form of the vibrational structure of an electronic absorption band can give us valuable information about the relative shapes of the ground and excited state potential energy curves. This is illustrated in Box 8.3. The concept is also helpful in describing the vibrational structure of forbidden electronic transitions which are allowed because of molecular vibrations, as we shall see when we consider Case 3.

Case 3: $\psi_{elec}^f \neq \psi_{elec}^i$ and $M_0(e^f e^i) = 0$; $\psi_{vib}^f \neq \psi_{vib}^i$ or $\psi_{vib}^f = \psi_{vib}^i$

This is the case of a forbidden electronic transition [$M_0(e^f e^i) = 0$], which may be weakly observed because of the presence of the second term in Equation (8.7.7) which determines whether transitions having $M_0(e^f e^i) = 0$ occur as a result of molecular vibrations. The vibrational states may be expanded in terms of the normal co-ordinates as in Equation (8.7.8) giving:

$$M_{i,f} = \sum_{k=1}^{3N-6} \langle \psi_{vib}^f | Q_k | \psi_{vib}^i \rangle \cdot M_k'(e^f e^i)$$

$$= \sum_{k=1}^{3N-6} \langle \psi^f(Q_1) | \psi^i(Q_1) \rangle \ldots \langle \psi^f(Q_k) | Q_k | \psi^i(Q_k) \rangle \ldots \langle \psi^f(Q_{3N-6}) | \psi^i(Q_{3N-6}) \rangle M_k'(e^f e^i)$$

$$(8.7.11)$$

since the two sets of normal modes belong to two different electronic states $\langle \psi^f(Q_n) | \psi^i(Q_n) \rangle \neq 1$. But such integrals are also unlikely to be zero, unless the two electronic potential energy curves are very different. Nevertheless, the vibrational contribution to the intensity of the forbidden electronic spectral band can be evaluated, though this is not normally done using Equation (8.7.11). The value of M_k' is required and the usual method employs perturbation theory (Appendix 4) to calculate the degree to which the vibrations, by changing the shape of the molecule, mix ψ_{elec}^f and ψ_{elec}^i with other electronic states. In most cases of practical interest ψ_{elec}^i is the ground state, which does not mix significantly with other states since they are so much higher in energy. The mixing is therefore confined to the mixing of ψ_{elec}^f with nearby excited states. The result of the mixing is that the transition is no longer purely $\psi_{elec}^i \rightarrow \psi_{elec}^f$ because there are some states among those newly mixed into ψ_{elec}^f to which transitions from ψ_{elec}^i are allowed. In this way the 'forbidden' transition acquires some intensity, which it is said to have 'borrowed' or 'stolen' from one or more allowed transitions.

If the potential energy curves of the two electronic states are quite similar in form and differ only in energy, which may well be the case for the excitation of one electron in a large molecule, then the vibrational eigenfunctions of the two states will be similar also. Then, to a good approximation, $\langle \psi^f(Q_n) | \psi^i(Q_n) \rangle = 1$ and the major contributions to the transition dipole moment from each allowing vibration will be of the form:

$$\langle \psi^f(Q_k) | Q_k | \psi^i(Q_k) \rangle M_k'(e^f e^i)$$

In Section 10.5 it is shown that the integral $\langle \psi^f(Q_k) | Q_k | \psi^i(Q_k) \rangle$ is zero unless the vibrational quantum numbers associated with $\psi^f(Q_k)$ and $\psi^i(Q_k)$ differ by ± 1. Therefore, where the electronic energy surfaces of the ground and excited electronic states are similar, we anticipate that the vibrational fine structure of a forbidden electronic transition will correspond predominantly to changes of ± 1 in the vibrational quantum numbers of the allowing vibrations. The allowing vibrations are effective because they change the shape of the molecule; lowering its symmetry. It is common to find that the first peak due to an allowing vibration is followed by a series of peaks in which the allowing vibration and an increasing number of quanta of a totally symmetric vibration are simultaneously excited. Such a series of bands is known as a *progression*.

Three further remarks should be made before we leave the long and complicated subject of transition dipole moments.

1. Equation (8.6.12) shows that the intensity of a transition between two states is proportional to the *square* of the transition dipole moment between them or, in the event that the transition dipole moment is a complex quantity, the product of the transition dipole moment and its complex conjugate.

2. A similar analysis can be made with respect to transitions stimulated by the magnetic field of the light, and we shall use this when we come to consider an NMR transition.

3. In the case of the 1s → 2p transition of the hydrogen atom for example, though the quantum mechanical expression is quite clear, it is difficult not to ask how the process of interaction between the light wave and the spherically symmetrical atom 'gets started'. Where is the dipole which provides the first means of interaction for the e-m wave? In answer we may note that, since the nucleus and the electron are particles, the atom is only spherically symmetrical when the positions of the nucleus and the electron are averaged over some finite period of time; at any instant in time the atom is a dipole. A further examination of the consequences of the time-dependency of the hydrogen-atom wave functions can be found in Box 8.2.

   An alternative, 'more classical' view of the problem is the following. When light passes through a transparent medium it does interact with the medium even though it is not absorbed; we know, for example, that its velocity is changed. However, when the frequency of the light fulfils the Born–Einstein condition, the previously limited response of the electrons is much magnified and a condition of resonance is established whereby electrons can absorb the light energy and change their energy states. In terms of the old analogy, a column of soldiers marching over a bridge will certainly cause the bridge to shake but will do nothing more if they are not all in step. However, if they are in step and the frequency of their step matches a natural vibrational frequency of the bridge, the bridge may oscillate much more violently than is safe. It was a phenomenon of this type, caused by normal pedestrians rather than soldiers, which necessitated structural modifications to the Millennium Bridge over the River Thames in London. It appears that there was a remarkable two-way process in which the oscillations of the bridge caused by the crossing pedestrians induced in those same people a synchronised stepping frequency which increased the oscillation of the bridge still further.

8.7.2   The relative intensities of UV–VIS, IR and NMR transitions

Armed with the above results we can proceed to examine the simplest example of an electronic transition: the 1s → 2p transition of the hydrogen atom. If we select as our final state the ϕ_{2pz} then the transition moment which we require is:

$$\boldsymbol{M} = e\langle\phi_{2pz}|\mathbf{r}|\phi_{1s}\rangle = e\langle\phi_{2pz}|\mathsf{x}|\phi_{1s}\rangle + e\langle\phi_{2pz}|\mathsf{y}|\phi_{1s}\rangle + e\langle\phi_{2pz}|\mathsf{z}|\phi_{1s}\rangle \qquad (8.7.12)$$

and evaluation of the three integrals shows that only the last gives a value that is not zero. If we choose ϕ_{2px} or ϕ_{2py} as our 2p orbital then the matrix elements containing the operators x and y respectively become the non-zero quantities in Equation (8.7.12). All three contributions to M are equal since the three axes are equivalent. We shall evaluate the last term, M_z, by taking expressions for the two hydrogen wave functions (in atomic units, Appendix 1) from Appendix 5 and replacing z by $r \cos \theta$ in order to have all our quantities in polar co-ordinates. Recalling (Appendix 7) that the volume element in polar co-ordinates is $r^2 \sin \theta \; dr \; d\theta \; d\phi$, we have:

$$M_Z = \frac{e}{4\sqrt{2\pi}} \int_0^\infty \int_0^\pi \int_0^{2\pi} r \cdot \exp\left(\frac{-r}{2}\right) \cos\theta \cdot r\cos\theta \cdot \exp(-r) r^2 \sin\theta \; dr \; d\theta \; d\phi$$

$$= \frac{e}{4\sqrt{2\pi}} \int_0^\infty r^4 \exp\left(\frac{-3r}{2}\right) dr \int_0^\pi \cos^2\theta \sin\theta \; d\theta \int_0^{2\pi} d\phi$$

$$= \frac{e}{2\sqrt{2}} \int_0^\infty r^4 \exp\left(\frac{-3r}{2}\right) dr \int_0^\pi \cos^2\theta \sin\theta \; d\theta$$

$$= \frac{-e}{3\sqrt{2}} \int_0^\infty r^4 \exp\left(\frac{-3r}{2}\right) dr = \frac{-e}{3\sqrt{2}} \cdot \frac{24}{(\frac{3}{2})^5} = -0.744936 \text{ au}$$

Thus, $M_z^2 = 0.5549$ and $M^2 = 3 \times M_z^2 = 1.6648$ atomic units. Converting from atomic to SI units we find that a dipole moment of 0.744936 au $= 6.3158 \times 10^{-30}$ C m so that the square of the 1s $\rightarrow$ 2p transition dipole moment of hydrogen is 39.889×10^{-60} C^2 m^2, which agrees exactly with the experimental value. This is a very satisfying result in that it confirms that the quantum mechanical method of calculating transition probabilities is correct. But we should not be lulled into a false sense of security. Uniquely in the case of one-electron atoms, we have extremely accurate algebraic wave functions at our disposal for calculations of this type. As soon as we move to multi-electron atoms, and *a fortiori* to molecules, the interelectronic repulsion drastically reduces the quality of the available wave functions (see Chapter 5) and agreement with experiment deteriorates rapidly. For a small molecule such as ethene agreement to within a factor of two or three is often as much as we can hope for.

However, our purpose here is not to obtain accurate theoretical intensities but rather to have some insight into their relative values in the UV–VIS, IR and NMR spectral regions. Therefore, we do not consider possible reasons for discrepancies further but go on to consider a typical IR transition, the fundamental $v = 0$ to $v = 1$ transition of the hydrogen chloride (HCl) molecule.

We have to evaluate the Expression (8.7.8), which is a product of $\langle \psi_{vib}^f | r | \psi_{vib}^i \rangle$ (r is the only normal co-ordinate of a diatomic molecule) and the rate of change of the molecular dipole moment with bond length, $M'(e'e)$, at the equilibrium bond length, r_e. Using data from Box 10.2 we find:

$$\langle \psi_{vib}^f | r | \psi_{vib}^i \rangle = (h/8\pi^2 \mu v)^{\frac{1}{2}} = 7.59 \text{ pm}.$$

If we assume that the charge distribution does not change with bond length for small amplitude vibrations about r_e, then the change in dipole moment with bond length depends only upon the change of bond length, and the rate of change of dipole moment with bond length may be obtained by dividing the former $(3.67 \times 10^{-30}$ C m) by the latter (127.5 pm):

$$M' \equiv \partial M_0/\partial R = 3.67 \times 10^{-30}/127.5 \times 10^{-12} = 2.88 \times 10^{-20} \text{ C}$$

whence:

$$M = 7.59 \times 2.88 \times 10^{-32} = 2.19 \times 10^{-31} \text{ C m} \Rightarrow M^2 = 4.78 \times 10^{-62} \text{ C}^2 \text{ m}^2$$

The best experimental estimate of M is 2.18×10^{-31} C m giving a much closer agreement between experiment and our theoretical value than we have any right to expect!

Note that the value of the square of the transition dipole moment, and hence the intensity, for a typical strong electronic transition is some four orders of magnitude greater than that of a strong infrared transition. This justifies the assumption underlying Equation (8.7.10) and raises the question as to why it should be so. The essential reason is that in an electronic spectral transition the transition dipole moment involves the electrons directly, and they each carry a large charge. In infrared spectroscopy the transition dipole moment depends upon a change of dipole moment, which depends upon the *differences* in the distribution of electronic and nuclear charge in the molecule. The difference in the intensities of UV–VIS and IR bands has an important experimental consequence. Infrared absorption spectra can frequently be measured on samples of pure substance whereas UV–VIS experiments almost always require dilution of the sample in a suitable solvent or inert matrix.

For the NMR case we shall again take a simple example and consider the magnetic resonance of a proton, which can take up two orientations to an applied, static magnetic field. If we assume the conventional experimental set-up with the static magnetic field, B_0, orientated along the z-direction and the oscillating radio frequency field, B_{1x}, along the x-direction, then the orientation of lower energy has $m_I = +\frac{1}{2}$ and the higher energy orientation has $m_I = -\frac{1}{2}$ so that for absorption of radiation we must evaluate (see Chapter 9 for the relation $\hat{\mu}_x = \gamma\hbar\hat{I}_x$):

$$B_{1x}\langle\phi^f|\hat{\mu}_x|\phi^i\rangle \Rightarrow B_{1x}\langle m_I = -\tfrac{1}{2}|\hat{\mu}_x|m_I = +\tfrac{1}{2}\rangle$$

$$= B_{1x}\langle-\tfrac{1}{2}|\gamma\hbar\hat{I}_x|+\tfrac{1}{2}\rangle = B_{1x}\gamma\hbar\langle-\tfrac{1}{2}|\hat{I}_x|+\tfrac{1}{2}\rangle = B_{1x}\gamma\tfrac{1}{2}\hbar \quad (8.7.13)$$

On inserting the value of the magnetogyric ratio of the proton ($\gamma = 26.572 \times 10^7$ rad T^{-1} s^{-1}) we find that the interaction is $B_{1x} \times 14.106 \times 10^{-27}$ rad T^{-1} J.

At this point comparison with IR and UV–VIS experiments becomes rather tenuous because, from the experimental point of view, the intensity of an NMR signal is governed much more by the populations of the upper and lower levels connected by the transition than by any other factor. In Section 9.6 it is shown that in a sample of protons in a field of four Tesla at 300 K, if there are 10^6 spins in the lower energy state then there will be only 28 less in the upper state! Therefore, application of radiation at the resonance frequency soon equalises the populations in the two states and no signal is observed since the number of upward (absorbing) transitions is equalled by the number of downward (emitting).

Furthermore, there is another important difference between the techniques of IR and UV–VIS spectroscopies on the one hand and NMR on the other. As far as the most basic experiments are concerned, and those are the only ones which we can consider here, in the IR and UV–VIS regions the phenomenon measured is the intensity of a light beam before it enters a sample and after it emerges from the sample; i.e. the quantity of light absorbed is measured, usually as the absorbance, A. In the basic NMR experiment however, the magnetisation induced in the sample by the $\boldsymbol{B}_1$ field is measured.

Therefore, it is not possible to make a comparison with experiment which stands on the same footing as the above examples taken from infrared and electronic spectroscopy. But it is possible to obtain a purely theoretical measure of the relative sensitivity of the three spectroscopies by comparing the magnitudes of the transition dipole matrix elements and the magnetic and electric fields of the radiation with which they couple.

With regard to the latter, we have written the perturbing field, $\hat{V}(t)$, in Equation (8.6.18) as:

$$\hat{V}(t) = \boldsymbol{E}_0 \sin \omega t \qquad (8.6.18)$$

If $\hat{V}(t)$ represents the electric field, $E(t)$ in V m^{-1}, of the radiation then we have:

$$\boldsymbol{E}(t) = \boldsymbol{E}_0 \sin \omega t$$

and the magnetic field of the same light beam, $\boldsymbol{B}(t)$ in T (Tesla), is given by:

$$\boldsymbol{B}(t) = (\boldsymbol{E}_0/c) \sin \omega t = \boldsymbol{B}_0 \sin \omega t,$$

where c is the velocity of light.

Using these results and the previously determined transition dipole matrix elements, we can draw up Table 8.2, which shows the relative theoretical intensities for a light beam having an electric field amplitude, $\boldsymbol{E}_0$, of 1.0 V m^{-1} and the corresponding magnetic field value of 3.33×10^{-9} T. [For a 100 W filament tungsten lamp inside a spherical envelope of 60 mm diameter (i.e. a domestic 100 W lamp) the value of $\boldsymbol{E}_0$ at the envelope is approximately 2×10^{-3} V m^{-1}.] The intensity, I, or absorbance, A, is proportional to the square of the interaction energy given in units of J^2 in the last row of Table 8.2 (the units of I and A are not themselves J^2; see Section 8.10.2). In so far as the energy of interaction can be regarded as a measure of the relative absorbance, we see from these data that the primary cause of the low sensitivity of NMR spectroscopy lies not in the value of the transition dipole matrix element which, for protons at least, is markedly larger than the corresponding quantities in our IR and UV–VIS examples. The problem lies with the low value of $\boldsymbol{B}(t)$ compared with $\boldsymbol{E}(t)$ and is exacerbated by the population problem which plays effectively no role in UV–VIS or IR spectroscopy and has not been included in our deliberations above.

Table 8.2 Theoretical relative intensities/absorbances of electronic, infrared and proton NMR transitions (Calculated for a transition dipole moment in one Cartesian direction only)

Electronic	Infrared	Proton NMR
H atom 1s $\rightarrow$ 2π	HCl $\upsilon = 0 \rightarrow \upsilon = 1$	$m_I = +\frac{1}{2} \rightarrow m_I = -\frac{1}{2}$
$\langle \phi^f | \mu | \phi^i \rangle =$	$\langle \phi^f | m | \phi^i \rangle =$	$\langle \phi^f | \mu | \phi^i \rangle =$
63.2×10^{-31} C m	21.9×10^{-32} C m	14.1×10^{-27} J T^{-1}
$E(t) = 1.0$ V m^{-1}	$E(t) = 1.0$ V m^{-1}	$B(t) = 3.33 \times 10^{-9}$ T
I or $A \propto 3.99 \times 10^{-58}$ J^2	I or $A \propto 4.78 \times 10^{-62}$ J^2	I or $A \propto 2.20 \times 10^{-69}$ J^2

8.7.3   The particle and wave views of spectroscopic transitions

The particle and wave descriptions of electromagnetic radiation play a role wherever the subject is discussed. In spectroscopy too, we tend to take the corpuscular or wave approach depending upon which seems to be most helpful in the interpretation of a particular observation. In the case of UV–VIS radiation the choice of description seems obvious. It is a photon which is detected by a spot on a photographic plate or by an electron ejected from a metal surface. The waves appear to be essential to the interpretation of interference phenomena. At the long-wavelength end of the spectrum we speak almost exclusively of waves and it is difficult to envisage a discussion of radio aerials in terms of photons. But the angular momentum changes in NMR and EPR spectroscopy seem to demand a photonic interpretation. In this section these points are illustrated with a discussion of some selected examples of spectroscopic transitions showing how they are used to obtain further information about the emitting/absorbing species. We assume that energy is conserved in all cases and it will not be discussed further.

As an example, drawn from atomic electronic spectroscopy, of an electric dipole transition, we take the transitions from the $^2S_{\frac{1}{2}}$ to $^2P_{\frac{1}{2}}$ and $^2P_{\frac{3}{2}}$ states of sodium. These are the transitions which give rise to the famous D-lines. An energy level diagram is shown in Figure 8.9 in which the different M_J levels are spaced out as they would be if the atom were placed in a strong magnetic field, the Zeeman effect (Section 5.10).

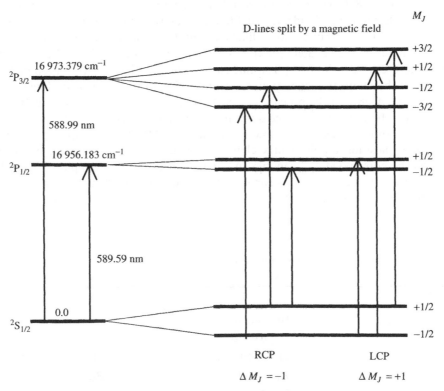

Figure 8.9   The transitions $^2S_{\frac{1}{2}}$ to $^2P_{\frac{1}{2}}$ and $^2P_{\frac{3}{2}}$ of sodium; the D lines

Note that the diagram is not drawn to scale; the spacing between the M_J levels is very small ($<0.1 \, \mathrm{cm}^{-1}$) compared with the separation between $^2P_{\frac{1}{2}}$ and $^2P_{\frac{3}{2}}$ ($\sim 17 \, \mathrm{cm}^{-1}$) and the excitation energy $^2S_{\frac{1}{2}}$ to $^2P_{\frac{1}{2}}$ ($\sim 16\,956 \, \mathrm{cm}^{-1}$). The levels between which transitions are allowed are connected with arrows. The total angular momentum of the system is conserved for all six allowed transitions since $1 + \frac{1}{2} \Rightarrow 1/2$ ($^2P_{\frac{1}{2}}$) or $3/2$ ($^2P_{\frac{3}{2}}$). In order that the z-component of the angular momentum be also conserved the three transitions on the left of the diagram can only take place when an RCP photon is absorbed while the three on the right require an LCP photon. To excite these transitions we require light propagating parallel to the magnetic field. The four transitions in which M_J is unchanged require a photon which has no z-component of angular momentum, i.e. a photon linearly polarised in the xy-plane. These excitations require light propagating at right-angles to the field. Parity is conserved. The parity of the S state is $+1$ and that of the photon -1. Therefore the parity of their combination is $(+1) \times (-1) = -1$, which is the correct parity for a P state.

As an example of the conservation of angular momentum in an electric quadrupole transition we take a line observed at ~ 436.3 nm in the emission spectra of certain cosmic nebulae. This and other similar lines puzzled spectroscopists for many years but it is now known to be due to the $^1S_0 \to {}^1D_2$ emission of O^{2+} atoms. Since this is an $S \to D$ transition, it would appear that a photon having two units of angular momentum must be involved so that the triangular condition is $\Delta(j_i, 2, j_f)$ (Section 8.4). However, when we express the states in terms of the atomic orbitals involved we find that the transition is due to an electron jumping from one 2p orbital to another. But $\Delta(1, 2, 1)$ is also quite acceptable as far as the conservation of angular momentum is concerned and we note further that since the parity of the product of two 2p orbitals is g the photon must also be g, which rules out an electric dipole photon but not an electric quadrupole photon. More detailed study reveals that at the orbital level the transition is $p_{+1} \to p_{-1}$ or $p_{-1} \to p_{+1}$ so that the emission does not change the total angular momentum of the atom but changes its z-component by $\pm 2\hbar$, which only a photon carrying at least two units of angular momentum can do.

The value of the wave-based concept of the direction of linear polarisation can be illustrated with an example drawn from the electronic spectroscopy of molecules; e.g. ethene (C_2H_4). The ethene molecule shows a band in the region of 160 nm that is due to a transition from the π bonding to π^* antibonding orbital (Section 7.1). The Hückel orbitals (Section 12.1) are:

$$\pi = \sqrt{(\tfrac{1}{2})}\{\phi_1 + \phi_2\} \quad \text{and} \quad \pi^* = \sqrt{(\tfrac{1}{2})}\{\phi_1 - \phi_2\} \tag{8.7.14}$$

where the ϕ_1 and ϕ_2 are the 2pz atomic orbitals of the two carbon atoms. Therefore, neglecting the vibrational factor, the transition dipole moment is $M_0(e^f e^i)$ (Equation (8.7.10)), and is given by:

$$M_{(\pi \to \pi^*)} = \langle \pi^* | \mathbf{er} | \pi \rangle = \tfrac{1}{2}\langle (\phi_1 - \phi_2) | \mathbf{er} | (\phi_1 + \phi_2) \rangle = \tfrac{1}{2}\langle \phi_1 | \mathbf{er} | \phi_1 \rangle - \tfrac{1}{2}\langle \phi_2 | \mathbf{er} | \phi_2 \rangle \tag{8.7.15}$$

since the orbital overlap contributions cancel. The electron density $\langle \phi_1 | \phi_1 \rangle$ is centred at carbon nucleus 1 and $\langle \phi_2 | \phi_2 \rangle$ at nucleus 2 so the only contribution to $M_{\pi \to \pi^*}$ is obtained from the component of r which lies along the C=C bond. Therefore the transition dipole moment lies along the C=C bond. This being the case, it is easy to see that if we had a sample of ethene in which all the molecules were orientated in the same direction then only light polarised, i.e. with its electric field vibrating, in that direction would be capable of interacting with the molecules and hence of being absorbed. The orientated sample would be completely transparent to light polarised at right angles to the C=C bond.

The last example leads us into a consideration of the selection rules in infrared spectroscopy which have much in common with it. Consider the C=O stretching vibrations of carbon dioxide. There are two such vibrations; the symmetric stretching where both C=O bonds grow longer and shorter at the same time and the asymmetric stretching where one grows longer as the other grows shorter, and vice versa. The corresponding transitions occur at $\sim$1330 cm^{-1} and $\sim$2349 cm^{-1} respectively but only the latter is observed in the infrared. The reason is, of course, that the change of the two C=O bond dipoles as the bonds change in length cancel each other exactly in the case of the symmetric stretching, so that there is no change of dipole moment for the molecule as a whole and hence no means for it to couple with the oscillating electric field of the light, no matter in what direction it is polarised. In the case of the asymmetric stretching there is a change in dipole moment and infrared radiation can be absorbed, provided that it is polarised along the O=C=O axis of the molecule. This link between the direction of polarisation and molecular geometry is extremely valuable in the study of orientated molecules, e.g. crystals and stretched polymers, by IR and UV–VIS methods.

When the sensitivity of Fourier-transform infrared spectroscopy made it possible to measure monolayers of molecules adsorbed on metal surfaces, extension of the above ideas led to the *metal-surface selection rule* from which the orientations of molecules adsorbed on the metal surface could be determined. When a molecule absorbed on a metal surface vibrates the valence electrons in the metal are sufficiently free (Section 12.2) to follow the oscillating dipole, and they do so in such a way as to produce an image of the dipole (Figure 8.10). Thus a vibration producing an oscillating dipole parallel to the metal surface generates an image dipole in the metal surface which cancels it out (Figure 8.10(a)) so that there is no interaction with radiation. If the molecule generates a changing dipole perpendicular to the surface then the image dipole reinforces the molecular dipole (Figure 8.10(b)) and absorption of infrared radiation is possible. Thus, when a particular vibration of a surface-absorbed species is

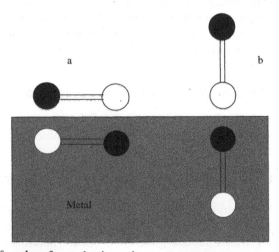

Figure 8.10   The infrared surface selection rule

identified, usually by its wave number, then the oscillating dipole moment generated must be perpendicular to the surface and this enables conclusions to be drawn concerning the orientation of the adsorbed molecule on the surface.

Conservation of linear momentum plays no role in the interpretation of infrared spectra, though angular momentum can be important in some specialised cases (Section 10.8.2). But parity conservation is very significant. The $v = 0$ level of the harmonic oscillator is even while the $v = 1$ level is odd and the parity of the higher levels is determined similarly by the odd/even character of v. Thus, the overall parity, photon plus molecule, is conserved for transitions between adjacent levels.

Since magnetism is so intimately related to angular momentum in quantum mechanics (Section 12.2) we may say that NMR is all about angular momentum. In the fundamental NMR transition of a proton we have with a RCP photon:

$$H(I = \tfrac{1}{2}, m_I = +\tfrac{1}{2}) + \text{Photon } (l = 1, m_l = -1) \rightarrow H^*(I = \tfrac{1}{2}, m_I = -\tfrac{1}{2}).$$

Note that although the z-component of the angular momentum is conserved this does not, at first sight, appear to be the case for the total angular momentum for which we have $\tfrac{1}{2} + 1 \rightarrow \tfrac{1}{2}$. But we must remember that the law for the addition of angular momentum in quantum mechanics (Section 3.6.1) is:

$$J_1 + J_2 = J_1 + J_2, J_1 + J_2 - 1, J_1 + J_2 - 2, \ldots |J_1 - J_2|,$$

so that final total angular momentum quantum numbers of $\tfrac{3}{2}(1 + \tfrac{1}{2})$ or $\tfrac{1}{2}(1 - \tfrac{1}{2})$ are in accord with the requirement for the conservation of angular momentum.

8.8 SPECTROSCOPIC TIME SCALES

There is a further aspect of the fundamentals of spectroscopic transitions that is of importance in applications of spectroscopy to systems which are changing with time. As fast as they are, the processes whereby an atom or molecule changes its energy state as a consequence of absorbing (or emitting) radiation are not instantaneous. It follows, therefore, that if a molecule changes its structure appreciably during the time in which a photon is absorbed we have to ask how the spectrum is affected by this other process. To be precise, let us consider the equilibrium between the keto (I) and enol (II) forms of 2,4-pentane dione (acetyl acetone):

If the proton NMR, IR and UV–VIS spectra of each of the two species involved in the above equilibrium are measured all show clear differences on account of the considerable change in the structure. But under suitable conditions, e.g. in acid aqueous solution, there is a rapid exchange and, intuitively, we envisage two extreme possibilities. If the exchange is very slow compared with the time taken for the particular spectroscopic process which we are using then we expect to see the spectrum of a mixture of I and II in proportions reflecting the particular position of the equilibrium. We could measure the equilibrium constant in this way. On the other hand, if the exchange is very much faster than the spectroscopic transition then we expect the spectrum to appear like the spectrum of one substance of a structure intermediate between I and II; i.e. the C=O stretching frequencies in the IR would lie somewhere between that expected for a free >C=O group and the lower frequency of the hydrogen-bonded species, C–O···H···O–C. Where the rate of exchange and the spectroscopic transition are of comparable magnitude the situation is more complicated. In favourable circumstances, where the rate of exchange can be controlled (e.g. by changing the temperature, pH etc.), the spectra of an exchanging system can be measured at the two extremes and in the complex intermediate region. NMR experiments of this type have been widely reported.

Our purpose here is to attempt to answer the important question, 'How fast is a spectroscopic transition'? or to be more explicit, 'How long, for example, does it take for an electron to absorb a photon and jump from the 1s to the 2p atomic orbital of a hydrogen atom'? There appears to be no precise way of answering this question, but a useful estimate of the required figures may be made using the time–energy uncertainty principle (Section 4.10) in the following manner. When a molecule or atom is making a transition from an initial to a final energy state having energies E_i and E_f, respectively, then its energy is uncertain to the extent $E_f - E_i = \Delta E = h\nu$. Now, according to Heisenberg:

$$\Delta E \times \Delta t \geq h/2\pi \Rightarrow h\nu \times \Delta t \geq h/2\pi \Rightarrow \Delta t \approx 0.16/\nu$$

where Δt represents the lower limit of the time taken to complete the spectroscopic transition. It is convenient to express the Δt in terms of the frequency of the transition, ν. The times for electronic, vibrational and NMR transitions are listed in Table 8.3 for the specific examples we have used throughout this chapter to provide representative numerical data for the three branches of spectroscopy.

For an IR transition, where we have a simple classical concept of a vibrating molecule, we might also argue as follows. It seems intuitively reasonable to assume that we could not observe that the HCl molecule had changed its vibrational state unless we studied it over a period of at least one complete vibration, i.e. for $1/\nu \approx 1.1 \times 10^{-14}$ s. This is approximately six times longer than our earlier estimate of Δt. We get the same result if we make an estimate of the NMR time scale in terms of the Larmor frequency (Equation (9.4.1)), which is 85.15×10^6 Hz for protons in a magnetic field of 2 T. In the case of the UV–VIS region, we might take the time required for an electron to complete a cycle of the lowest Bohr orbit which is 1.52×10^{-16} s; again a somewhat longer time than the value of Δt_{min} deduced from the uncertainty principle. But we know that quantum mechanics tells us we cannot view electrons orbiting a nucleus in the same way as we regard planets circling the sun. This should be a warning to us that we are treading on very insecure ground and

Table 8.3 Estimates of the minimum time taken, Δt_{min}, to complete a spectroscopic transition in the UV–VIS, IR and NMR regions of the spectrum

UV–VIS	IR	Proton NMR
H atom 1s $\rightarrow$ 2p	HCl $\nu = 0 \rightarrow \nu = 1$	$m_I = +\frac{1}{2} \rightarrow m_I = -\frac{1}{2}$
$\lambda = 121.6$ nm	$\bar{\nu} = 298974$ m−1	$B = 2$ Tesla
$\nu = 24.67 \times 10^{14}$ Hz	$\nu = 89.63 \times 10^{12}$ Hz	$\nu = 85.15 \times 10^6$ Hz
$\Delta t_{min} \approx 6.4 \times 10^{-17}$ s	$\Delta t_{min} \approx 1.8 \times 10^{-15}$ s	$\Delta t_{min} \approx 1.9 \times 10^{-9}$ s

are pushing the limits of our classical view of the world too far into those regions where we know that classical mechanics must be replaced by quantum mechanics.

Be that as it may, these crude theoretical estimates of spectroscopic time scales suggest clearly that as the processes which we wish to study become faster we must move to higher spectral energies if we wish to use spectroscopy as our tool. This finding is borne out in the laboratory where NMR spectroscopy is frequently used to study the comparatively slow isomerisation processes in which atoms move, whereas the very fast processes which involve solely electron transfer must be studied in the UV–VIS spectral region. An example of the former is the study of the 2,4-pentane dione keto $\leftrightarrow$ enol equilibrium mentioned previously. The latter might be illustrated by the investigations of the photo-synthetic processes involved when the chlorophyll of green plants absorbs sunlight where events on the femto-second (10^{-15} s) time scale are being studied. Very recently a study of the photoelectron ionisation of neon in which the authors claim to be able to distinguish events separated in time by 10^{-16} s has been reported.[4] Central to these experiments is the generation of pulses of light in the extreme ultraviolet ($\lambda \approx 1 \times 10^{-8}$ m) lasting only 250×10^{-18} s.

8.9 QUANTUM ELECTRODYNAMICS[5]

It was clear in the discussion of spectroscopic time scales that we have been describing a theory which is a mixture of classical and quantum mechanical concepts. Throughout our discussion of the interaction of radiation with matter (Section 8.6) we have treated the atoms and molecules as having quantised energy levels, but in our calculations of intensities we described the electromagnetic radiation in terms of electric and magnetic fields, i.e. a classical viewpoint. A consistent theory must quantise both the energy levels of the atoms and molecules and the fields. The first step in this direction was made by Einstein in 1905 when he suggested that the interactions between electromagnetic radiation and matter known at that time might be better explained in terms of light quanta than in terms of oscillating electric and magnetic fields. The light quanta, which we now call photons, are the quanta of the electromagnetic field. A systematic attack upon the problem of a complete quantum description of the interaction of radiation with matter was begun by Dirac quite early in the history of quantum mechanics.

By the 1970s a sophisticated and detailed theory was available. The modern form of this theory is known as quantum electrodynamics (QED) and it has proved capable of providing extremely accurate values of crucial quantities in the theory of spectroscopy. Indeed, though spectroscopic measurements are among the most accurate in the whole of science, the theories used to analyse them are in many cases equal to the challenge of interpreting these elegant data quantitatively. QED is especially important in the interpretation of experiments involving the adsorption and/or emission of several photons within a very short time interval which are of growing importance in chemistry. However, this is not the place to pursue this topic further.

8.10 SPECTROSCOPIC UNITS AND NOTATION

The most frequently used spectroscopic units and notation are the subjects of the penultimate section of this chapter. Spectroscopists have been somewhat reluctant to adopt SI units in their entirety and non-SI usage is frequently found, even in modern publications (this book included!). Naturally, it also occurs in earlier work and it seems appropriate here to summarise the more commonly used units and notation. It is helpful to divide this section into two parts and to discuss the quantities appearing on the two axes of a spectrum under different headings.

8.10.1 The energy/frequency/wavelength axis

This is normally the horizontal axis of the spectrum and the quantities which may be shown on it are listed in Table 8.4. They relate directly to the position of the absorption or emission band in the electromagnetic spectrum, which implies an energy difference given by the Bohr–Einstein condition, $\Delta E = h\nu$. In SI units this energy would be given in Joules, but spectroscopists have traditionally used other units, some not even strictly energy units, in which to report ΔE. The electron volt (eV), which is the kinetic energy acquired by an electron when it is accelerated in an electric field of one volt, is a true energy unit and is widely used in electronic and photoelectron spectroscopy (Chapter 11). The wave number, $\bar{\nu}$, is not an energy unit but it is directly proportional to energy ($E = hc\bar{\nu}$). Infrared and Raman spectra are almost always discussed in terms of wave numbers. In NMR spectroscopy, the spin-spin coupling energy is always expressed as a frequency, i.e. in Hz, but for both technical and data-transferability reasons, the position of a resonance is invariably reported in terms of its position with respect to the resonance of a standard. (For more information see Chapter 9.)

The relative number of molecules, N_a and N_b, in two energy levels separated by an energy difference of ΔE_{ab} ($= E_a - E_b$) is determined by Boltzmann's law as:

$$N_a/N_b = \exp(-\Delta E_{ab}/kT) \tag{8.10.1}$$

This relationship plays an extremely important role in spectroscopies, notably NMR, where ΔE_{ab} is very small so that $N_a \approx N_b$. The units of k are energy T^{-1} and an absolute temperature of 1 K can be seen as equivalent to an energy of $k\mathcal{E}$, where $\mathcal{E}$ is the energy unit in which k is measured. This provides a way of relating energy to the absolute scale of temperature, which is very useful for discussing spectroscopic phenomena that depend upon temperature, such as the population of energy levels.

Where the energies of single molecules are to be related to practical chemical quantities, the energy per mole in kJ mol^{-1} is the appropriate SI unit.

Table 8.4 Some quantities which may be found on the energy/frequency/wavelength axis of a spectrum; symbols and units

Quantity	Symbol	Definition	SI unit	Other units or symbols
Wavelength	λ		m	nm, Å (Ångstrom) = 10^{-10} m
Frequency	ν	$\nu = c/\lambda$	Hz	s^{-1}; cycles per second
Angular frequency	ω	$\omega = 2\pi\nu$	s^{-1}	rad s^{-1} [a]
Wave number in vacuum	$\bar{\nu}$	$\bar{\nu} = \nu/c_0 = 1/n\lambda$ [c]	m^{-1}	cm^{-1} = 10^{-2} m^{-1} [b]
Wave number in a medium	σ	$\sigma = 1/\lambda$	m^{-1}	cm^{-1} = 10^{-2} m^{-1} [b]
Chemical shift	δ	see Chapter 9		parts per million (ppm)

[a] rad s^{-1} = radians per second (the radian is dimensionless).
[b] The kilokayser (kK) has been proposed as the unit for 1000 cm^{-1}, but this suggestion has not been widely adopted.
[c] n = refractive index.

Table 8.5 Energy-conversion factors

	Wave number $\bar{\nu}$(cm^{-1})	Frequency ν(MHz)	Energy E(J $\times$ 10^{-18})	Energy E(eV)	Molar energy E_m(kJ mol^{-1})	Temperature T(K)
$\bar{\nu}$: 1 cm$^{-1} \approx$	1	2.997925 $\times$ 10^4	1.986447 $\times$ 10^{-5}	4.556335 $\times$ 10^{-6}	11.96266 $\times$ 10^{-3}	1.438769
ν: 1 MHz $\approx$	3.33564 $\times$ 10^{-5}	1	6.626076 $\times$ 10^{-10}	4.135669 $\times$ 10^{-9}	3.990313 $\times$ 10^{-7}	4.79922 $\times$ 10^{-5}
E: 1atto J $\approx$	50341.1	1.509189 $\times$ 10^9	1	6.241506	602.2137	7.24292 $\times$ 10^4
E: 1 eV $\approx$	8065.54	2.417988 $\times$ 10^8	0.1602177	1	2625.500	3.15773 $\times$ 10^5
E_m: 1 kJ mol$^{-1} \approx$	83.5935	2.506069 $\times$ 10^6	1.660540 $\times$ 10^{-3}	1.036427 $\times$ 10^{-2}	1	120.272
T: 1 K $\approx$	0.695039	2.08367 $\times$ 10^4	1.380658 $\times$ 10^{-5}	8.61738 $\times$ 10^{-5}	8.31451 $\times$ 10^{-3}	1

Table 8.5 gives the conversion factors between the various 'energy' units mentioned above. The inverted commas around the word energy serve as a final reminder that some of these units are not energy units and, for example, 1 eV may be thought of as equivalent to, *but not equal to*, 8065.54 cm^{-1}.

8.10.2 The intensity/absorbance axis

The second axis of the spectrum is the one which records just how much radiation has been absorbed or emitted. This axis is especially important when a spectroscopic measurement is being used in quantitative analysis. Such procedures are almost always based upon relative, rather than absolute, measurements since only in this way is the required accuracy obtainable. If the problem is to determine the concentration of copper in a sample of wine,[‡] then the spectroscopic instrument used, an atomic absorption spectrometer in this example, would first be calibrated using standard copper solutions and the copper concentration in the unknown sample determined by comparison with the calibration graph. The measurement of the quantity of light absorbed or emitted is recorded as a change in the output signal of the detector and absolute measurement of the intensity of the radiation is not required.

In UV–VIS and IR spectroscopy the most common measure of the quantity of radiation absorbed is the absorbance, A_λ, which is defined by the equation:

$$A_\lambda = \log_{10}(I_o/I_{tr})_\lambda \qquad (8.10.2)$$

where I_o is the intensity of the radiation falling upon the sample and I_{tr} is the intensity of the transmitted radiation at a particular wavelength, λ. The value of A_λ in Equation (8.10.2) is calculated directly by modern spectrometers and the determination of the absorbance is a particularly important aspect of quantitative analysis in the ultraviolet, visible and infrared regions of the spectrum. A_λ is related to the concentration of light-absorbing substance, C, by the Beer–Lambert–Bouguer law (Box 8.4):

$$A_\lambda = \mathcal{E}_\lambda CL \qquad (8.10.3)$$

[‡] There is Cu^{2+} in wine because a fungal infection of the vines is treated by spraying them with an aqueous suspension of lime and copper sulfate known as *Bordeaux mixture*.

where L is the path length of the radiation through the sample and $\mathcal{E}_\lambda$ is the molar decadic absorption coefficient, which has a characteristic value for every compound at every wavelength, λ, and is directly related to the probability that a photon of that wavelength striking the molecule will be absorbed. The word 'decadic' arises because A is defined in terms of logarithms to base 10 rather than to base e in Equation (8.10.2).

8.11   THE EINSTEIN COEFFICIENTS

Finally, mention should be made of three parameters characterising transition probabilities deduced by Einstein in 1917 using purely thermodynamic reasoning. They are the Einstein coefficients of induced emission $(B_{m \to n})$, absorption $(B_{n \to m})$ and spontaneous emission $(A_{m \to n})$. Einstein's argument and the relationships between these parameters are described in Box 8.5.

8.12   BIBLIOGRAPHY AND FURTHER READING

1. S.F. Mason, *A History of the Sciences*, Collier Books, New York, 1962; Chapters 37 and 38.
2. R.A. Beth, *Phys. Rev.*, **48**, 471 (1935); **50**, 115 (1936).
3. T.R. Dyke, G.R. Tomasevich, W. Klemperer and W. Falconer, *J. Chem. Phys.*, **57**, 2277 (1972).
4. R. Kienberger *et al.*, *Nature*, **427**, 817 (2004).
5. D.P. Craig and T. Thirunamachandran, *Molecular Quantum Electrodynamics*, Academic Press, New York, 1984, Dover Publications Inc., New York, 1998.

BOX 8.1   The Compton effect

In 1923 Arthur Holly Compton (1892–1962) observed that when X-rays are scattered by electrons the wavelength of the scattered rays is slightly longer than that of the incident rays; i.e. the radiation has lost energy. This observation could not be explained by a wave theory of light and Compton suggested the following interpretation, which provides one of the most convincing pieces of evidence for the photon as a discrete particle that has mass and, therefore, linear momentum. Consider a photon with frequency ν_i moving in the x-direction. It strikes a stationary electron of mass m_o and is scattered at an angle of ϕ to its original path and its frequency is reduced to ν_f. The collision causes the electron to move off with velocity v and momentum $\boldsymbol{m}v$ at an angle of ψ to the x-axis, Figure B8.1.1.

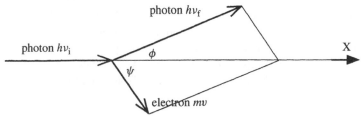

Figure B8.1.1   Collision of a photon with a stationary electron.

Compton analysed the problem using the requirements that energy and linear momentum be conserved. He expressed the energy of the electron in terms of Einstein's equation $E = mc^2$, and obtained the momentum of the photon by combining the two equations for its energy, i.e. $E = h\nu = mc^2 \Rightarrow p = mc = h\nu/c$. The rest mass of the electron, m_o, is related to its mass when in motion, m, by Equation (8.4.5):

$$m_o = m\{1 - (v/c)^2\}^{\frac{1}{2}} \text{ or } m^2(c^2 - v^2) = m_o^2 c^2 \qquad \text{(B8.1.1)}$$

Therefore, before the impact:

Photon:	$E = h\nu_i$	$p = h\nu_i/c$
Electron:	$E = m_o c^2$	$p = 0$

and after the impact:

Photon:	$E = h\nu_f$	$p = h\nu_f/c$
Electron:	$E = mc^2$	$p = mv$

The conservation of momentum requires that along x:

$$h\nu_i/c = (h\nu_f/c)\cos\phi + mv\cos\psi \Rightarrow mvc\cos\psi = h\nu_i - h\nu_f\cos\phi \quad \text{(B8.1.2)}$$

and perpendicular to x:

$$(h\nu_f/c)\sin\phi = mv\sin\psi \Rightarrow mvc\sin\psi = h\nu_f\sin\phi \qquad \text{(B8.1.3)}$$

The conservation of energy requires that:

$$h\nu_i + m_o c^2 = h\nu_f + mc^2 \Rightarrow mc^2 = h(\nu_i - \nu_f) + m_o c^2 \qquad \text{(B8.1.4)}$$

Squaring both sides of Equations (B8.1.2) and (B8.1.3) and adding gives:

$$mv^2 c^2 = h^2\nu_i + h^2\nu_f - 2h^2\nu_i\nu_f\cos\phi = h^2(\nu_i + \nu_f - 2\nu_i\nu_f\cos\phi) \quad \text{(B8.1.5)}$$

Squaring both sides of Equation (B8.1.4) gives:

$$m^2 c^4 = h^2(\nu_i - \nu_f)^2 + m_o^2 c^4 + 2h(\nu_i - \nu_f)m_o c^2$$
$$= h^2(\nu_i + \nu_f - 2\nu_i\nu_f) + m_o^2 c^4 + 2h(\nu_i - \nu_f)m_o c^2 \qquad \text{(B8.1.6)}$$

Subtracting Equation (B8.1.6) from Equation (B8.1.5) we have:

$$mc^2(v^2 - c^2) = -2h^2\nu_i\nu_f(\cos\phi - 1) - m_o^2 c^4 - 2h(\nu_i - \nu_f)m_o c^2$$

and using Equation (B8.1.1):

$$h\nu_i\nu_f(1 - \cos\phi) = (\nu_i - \nu_f)m_o c^2$$

Since $\lambda\nu = c$, this result may be written:

$$1 - \cos\phi = m_o c^2(\nu_i - \nu_f)/h\nu_i\nu_f = \{m_o c^2/h\}\{(1/\nu_f) - (1/\nu_i)\} = \{m_o c/h\}\{\lambda_f - \lambda_i\}$$

This equation is in exact agreement with Compton's observations. The manner of its derivation shows that the photon may be considered to be a particle with momentum $h\nu/c$. The factor $h/m_o c$, which has the units of length, is known as the Compton wavelength and has the numerical value 2.42×10^{-12} m.

BOX 8.2   The transition dipole moment in the hydrogen atom

Readers will have noted that Equation (8.7.8) requires that there is a change of dipole moment associated with the vibration if the transition moment of an infrared transition is to be non-zero, whereas no such condition attaches to the expression (Equation (8.7.10)) for an electronic transition dipole moment. And if, as an example, we consider the strongly allowed $1s \rightarrow 2p$ transition of the hydrogen atom, it is difficult to envisage how the dipole, which is to interact with the electric field of the light, arises. Whereas a classical viewpoint is quite satisfactory for vibrational spectroscopy, and also for NMR where we have a precessing magnet which can interact with an oscillating magnetic field (Chapter 9), the case of an electronic transition demands a quantum-mechanical analysis. We can, however, make the whole process more plausible by considering the time-dependent initial, $1s$, and final, $2p$, wave functions.

According to Equation (8.6.3), the two time-dependent wave functions have the forms:

$$\Psi_{1s} = \phi_{1s} \exp(-i2\pi E_{1s}t/h) \text{ and } \Psi_{2p} = \phi_{2p} \exp(-i2\pi E_{2p}t/h)$$

Each of the states Ψ is an eigenstate of the atomic Hamiltonian operator and we may also form a state which is a superposition of the two:

$$\Psi_{sum} = \Psi_{1s} + \Psi_{2p} = \phi_{1s} \exp(-i2\pi E_{1s}t/h) + \phi_{2p} \exp(-i2\pi E_{2p}t/h)$$

Compare Equation (8.6.6). The above expression for Ψ_{sum} may be written:

$$\Psi_{sum} = \exp(-i2\pi E_{1s}t/h)\{\phi_{1s} + \phi_{2p} \exp(-i2\pi [E_{2p} - E_{1s}]t/h)\}$$

Note that Ψ_{sum} is not an energy eigenstate. At $t = 0$, the probability of finding the hydrogen electron at any point in space is given by the product:

$$\Psi_{sum}^* \Psi_{sum} = \{\phi_{1s} + \phi_{2p}\}^2$$

We now examine how Ψ_{sum} changes with time.

When a time $t = h/2(E_{2p} - E_{1s})$ has elapsed the wave function can be written:

$$\Psi_{sum} = \exp(-i2\pi E_{1s}t/h)\{\phi_{1s} + \phi_{2p} \exp(-i\pi)\} = \exp(-i2\pi E_{1s}t/h)\{\phi_{1s} - \phi_{2p}\}$$

because $\exp(-i\pi) = -1$ (see Box 4.3). We might also have substituted the first t in the above equation, but this is unnecessary since that part of the expression reduces to 1 when we carry out our next step.

The probability of finding the hydrogen electron at any point in space is now:

$$\Psi_{sum}^* \Psi_{sum} = \exp(+i2\pi E_{1s}t/h) \cdot \exp(-i2\pi E_{1s}t/h)\{\phi_{1s} - \phi_{2p}\}^2 = \{\phi_{1s} - \phi_{2p}\}^2$$

After a further period of time equal to $h/2(E_{2p} - E_{1s})$, Ψ_{sum} will have returned to its value at $t = 0$ and the oscillation between $\phi_{1s} + \phi_{2p}$ and $\phi_{1s} - \phi_{2p}$ continues indefinitely.

Because of the differing phases of the two lobes of the 2p function, the centre of the electronic charge distribution moves from one side of the nucleus to the other (Figure B8.2.1), thus creating an incipient oscillating dipole which is always ready

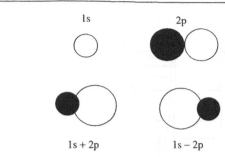

1s   2p

1s + 2p   1s − 2p

Figure B8.2.1 The time-dependence of the 1s/2p electron distribution.

and able to interact with light of the correct frequency, should such light fall on the atom. Indeed, the frequency of the oscillation is $(E_{2p} - E_{1s})/h$ which is exactly the frequency of the 1s → 2p transition, but we must be careful not to draw unwarranted conclusions from this fact. We are in danger of straying into the quantum-mechanical world armed only with classical-mechanical concepts.

BOX 8.3   The Franck–Condon principle

The data in Table 8.4 show that the time required for an vibrational transition is of the order of 100 times that required for an electronic transition. Already in 1925, James Franck (1882–1964) had reasoned that because the masses of nuclei were so much greater than that of electrons the former would move much more slowly and that they might be considered to be essentially stationary during the period required for an electronic transition. The same concept underlies the Born–Oppenheimer approximation (Section 6.3). If we think in terms of the potential-energy curves for the ground and excited states of a diatomic molecule (Figure B8.3.1), then an electronic transition between these states with no motion of the nuclei during that time implies a transition with no change of internuclear distance, i.e. a vertical transition.

In 1929 Edward Uhler Condon (1902–1974) analysed the problem with the new quantum mechanics. He derived Equation (8.7.10), following essentially the method described in Section 8.7.1, Case 2, and came to the same conclusion as Franck. But by introducing the concept of the overlap of the vibrational functions of the ground and excited electronic states, Condon made it possible, in favourable cases, to calculate the relative intensities of vibrational fine structure and provided a method of determining the change of shape which takes place when a molecule is electronically excited. This sort of information is of great value in the study of molecules in their electronically excited states.

And even when a quantitative calculation is not possible, we can draw conclusions about the ground and excited state potential energy curves from the appearance of the vibrational structure. Figure 10.6 shows that, as we move to higher quantum numbers, the lobes of harmonic vibrational functions become disproportionately

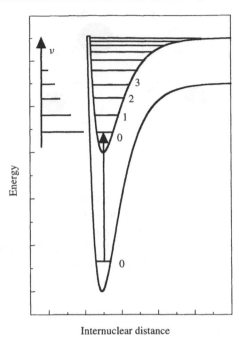

Figure B8.3.1   The Franck–Condon principle with r_e(ground state) $\approx r_e$(excited state)

larger at the turning points of the vibration and this is more marked in the case of anharmonic vibrations. Thus, the maximum overlap occurs where the upper state potential energy curve lies vertically above the equilibrium internuclear configuration of the ground state because, at room temperature, almost all molecules will be in their vibrational ground state. When the equilibrium bond length of the upper state is similar to that of the lower state we have the situation depicted in Figure B8.3.1. The vibronic transition, which has the maximum overlap, is the $0 \rightarrow 0$ transition and so the first band is the most intense and as we go to higher energies the intensity falls off with increasing υ', the vibrational quantum number of the upper state. The resulting spectrum is shown diagrammatically to the left of the potential energy curves. The $^1\Sigma^+(r_e = 149.10$ pm$) \rightarrow {}^1\Pi(r_e = 154.66$ pm$)$ transition of PN, which is found in the region of 40 000 cm^{-1}, is of this type. The relative intensities of the first four $0 \rightarrow \upsilon'$ vibronic bands fall in the sequence 20, 3, 2, 0.

When, as is usually the case, the upper electronic state is more weakly bound than the ground state, the equilibrium bond length of the upper state increases and the particular vibronic transition which has the maximum overlap with the $\upsilon'' = 0$ level of the ground state changes moves to higher values of υ'. Now the vibronic band of maximum intensity has bands of decreasing intensity towards both higher and lower values of υ', i.e. toward both higher and lower energies (Figure B8.3.2). The $^1\Sigma^+(r_e = 112.81$ pm$) \rightarrow {}^1\Pi(r_e = 123.51$ pm$)$ band of CO, found at about 65 000 cm^{-1}, falls into this category.

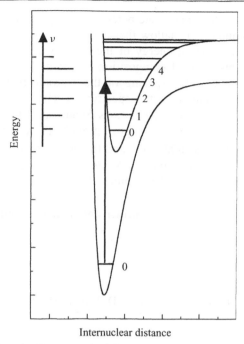

Figure B8.3.2 The Franck–Condon principle with r_e(ground state) $< r_e$(excited state)

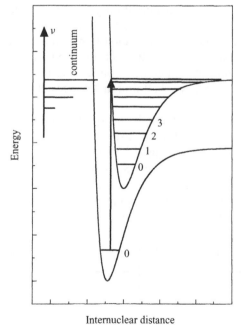

Figure B8.3.3 The Franck–Condon principle when a vertical transition enters the continuum

It is quite common for the upper state potential energy curve to be so far displaced to longer bond length and/or so shallow that maximum vibrational overlap occurs at an energy above the dissociation energy of the molecule, so that absorption of the corresponding photon results in decomposition of the molecule into its constituent atoms. When this is the case the spacing of the vibronic bands decreases until it is unresolvable and the intensity of the spectrum reaches a maximum somewhere in the continuum (Figure B8.3.3). The famous Schumann-Runge bands of oxygen, $^3\Sigma_g^-\,(r_e = 120.74\text{ pm}) \rightarrow {}^3\Sigma_u^-\,(r_e = 160\text{ pm})$, provide an example of this behaviour. The $0 \rightarrow 0$ band is found at $49\,802$ cm^{-1} and the convergence limit at $56\,850$ cm^{-1} at which energy the oxygen molecule dissociates into one ground-state atom O(^{3}P) and one excited atom O(^{1}D). It is little wonder that excitation processes such as this lead to extensive photochemistry!

Herzberg[1] should be consulted for further details and examples of the Franck–Condon principle.

REFERENCE

1. G. Herzberg, *Spectra of Diatomic Molecules*, D. Van Nostrand Inc., New York, 1950.

BOX 8.4 The Beer–Lambert–Bouguer Law[a]

In this box we interpret theoretically the laws of light absorption discovered experimentally by Beer, Lambert and Bouguer during the 19th century.

Suppose (Figure B8.4.1(a)) that a beam of light traverses a path of L m in a solution of an absorbing solute, concentration C mol dm^{-3}, in a transparent solvent contained in a transparent cell or cuvette. Let the intensity of the light of frequency ν be $I_0(\nu)$ on entering the solution and $I(\nu)$ on leaving it. We do not need to specify the units of the intensity since we shall find that our result depends upon an intensity ratio.

As the light makes its way through the solution each succeeding layer of the solution is subject to a diminished intensity of light since some of the light has already been absorbed by the preceding layers. Therefore, in order to calculate the total intensity loss, we consider (Figure B8.4.1(b)) an infinitely thin layer of thickness dl and 1 m^2 cross section at right angles to the direction of propagation of the light. Let the cross-sectional area of the absorbing molecules be a m^2 per molecule, the intensity of the light entering the thin layer of the solution i, the amount of light

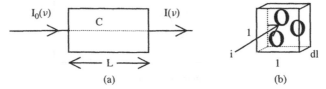

Figure B8.4.1 The passage of a light beam through an absorbing solution.

absorbed in the layer di and the probability that a photon falling upon a molecule is absorbed w. Then the number of molecules per m^3 is $10^3 \cdot CN_A$ where N_A is the Avogadro constant. Therefore, the cross-sectional area of the molecules in the layer is $10^3 \cdot aCN_A dl$. Since the probability of each photon being absorbed is w, the fraction of light absorbed in the layer is $10^3 \cdot waCN_A dl$:

$$-di = 10^3 \cdot waCN_A idl \text{ or } -di = \mathcal{E}' Cidl \text{ where } \mathcal{E}' = 10^3 \cdot waN_A$$

The di has a negative sign because i gets smaller as l increases.

Integrating over the intensity from I_0 to I and path length from 0 to L we have:

$$-\int_{I_0}^{I} \frac{di}{i} = \mathcal{E}'C \int_{0}^{L} dl \Rightarrow \log_e \left[\frac{I_0}{I} \right] = \mathcal{E}'CL$$

or, converting to logarithms to base 10:

$$\log_{10} \left[\frac{I_0}{I} \right] = 0.4343 \mathcal{E}'CL \equiv \mathcal{E}CL$$

The quantity $\log_{10}(I_0/I)$ is called the absorbance, A; it has no dimensions or units. $\mathcal{E}$ is the molar decadic absorption coefficient, its units are $m^2 \text{ mol}^{-1}$, i.e. cross-sectional area per mole, which explains why $\mathcal{E}$ is sometimes spoken of as a cross section in some branches of science and technology, though rarely in chemistry.[b] If we set the cross-sectional area of a molecule to $\approx 10^{-19} m^2$ then for $w = 1.0$ we have the maximum possible value of $\mathcal{E}$ of $\approx 2.5 \times 10^7$. The minimum value is, of course, zero. The quantity T, known as the percentage transmission, is also used as a measure of the light absorption:

$$T = \left[\frac{I}{I_0} \right] \cdot 100\%$$

The equation:

$$A = \mathcal{E}CL$$

or better[c]:

$$A_\lambda = \mathcal{E}_\lambda CL$$

shows that A is proportional to path length (the Bouguer–Lambert law) and also to concentration (the Beer law). All quantitative applications of spectroscopy are based upon Beer's law and its practical significance cannot be overestimated.

Notes:

a Terminology of the type used here (absorbance, absorbance coefficient, path length) is universal in UV–VIS spectroscopy, very much less common in IR spectroscopy and never used in NMR spectroscopy.

b In older literature the *absorption* coefficient is almost always called the *extinction* coefficient and has units of $cm^2 \text{ mol}^{-1}$. The *absorbance* is frequently called the *optical density*.

c It must be remembered that A and $\mathcal{E}$ are functions of the wavelength or frequency of the light. This is the basic fact responsible for what we know as a spectrum.

BOX 8.5 The Einstein coefficients

Although no quantum mechanics is involved, any account of the absorption and emission of radiation would be incomplete without a mention of the coefficients which Einstein introduced, on the basis of thermodynamic arguments, in 1917.

There are three distinct ways in which an atom or molecule can undergo a transition between two quantised energy states, m and n, having energies E_m and E_n. All three are of interest to the chemical spectroscopist. If we assume that $E_m > E_n$, then the transition from n to m will be accompanied by the absorption of a photon of frequency ν_{mn} given by the Bohr–Einstein relationship:

$$\nu_{mn} = (E_m - E_n)/h \tag{B8.5.1}$$

Similarly, the transition from m to n will result in the emission of a photon of frequency ν_{mn}. Einstein proposed that the probability of the transition from n to m, $P_{n \to m}$, is proportional to $\rho(\nu_{mn})$, the energy density of the radiation at the frequency ν_{mn}, and given by the equation:

$$P_{n \to m} = B_{n \to m}\, \rho(\nu_{mn}) \tag{B8.5.2}$$

The constant of proportionality, $B_{n \to m}$, is called the *Einstein coefficient of absorption*. For the transition from the higher to the lower energy state Einstein postulated that there were two processes to be considered. Firstly, the system could simply emit a photon spontaneously and drop to the lower state for no external reason. Secondly, the transition from m to n could be induced by a photon in just the same way as the transition from n to m is induced. The total probability of the transition from m to n is therefore the sum of two processes and:

$$P_{m \to n} = A_{m \to n} + B_{m \to n}\, \rho(\nu_{mn}) \tag{B8.5.3}$$

where $A_{m \to n}$ is Einstein's coefficient of spontaneous emission and $B_{m \to n}$ is Einstein's coefficient of induced emission. The process of spontaneous emission is exactly the same as the spontaneous emission of a γ-ray by a radioactive nucleus. If we now consider a large number of systems at equilibrium with radiation at a temperature, T, then the energy density of the radiation is given according to Planck (Section 2.5) by the equation:

$$\rho(\nu_{mn}) = 8\pi h(\nu_{mn})^3/(c^3[\exp\{h\nu/kT\} - 1]) \tag{B8.5.4}$$

where k is the Boltzmann constant. If the number of systems in state m is N_m and the number in state n is N_n, then the number of systems making the transition from state n to state m in unit time is:

$$N_n B_{n \to m}\, \rho(\nu_{mn}) \tag{B8.5.5}$$

and the number moving in the reverse direction is:

$$N_m\{A_{m \to n} + B_{m \to n} \ \rho(\nu_{mn})\} \tag{B8.5.6}$$

At equilibrium these two numbers must be equal and we have:

$$N_n/N_m = \{A_{m \to n} + B_{m \to n} \ \rho(\nu_{mn})\}/B_{n \to m} \ \rho(\nu_{mn}) \tag{B8.5.7}$$

But according to Boltzmann the ratio N_n/N_m is given by the equation:

$$N_n/N_m = \exp\{(E_m - E_n)/kT\} = \exp\{h\nu_{mn}/kT\} \tag{B8.5.8}$$

Equating the two expressions for N_n/N_m we find the following expression for $\rho(\nu_{mn})$:

$$\rho(\nu_{mn}) = A_{m \to n}/(B_{n \to m} \exp\{h\nu_{mn}/kT\} - B_{m \to n}) \tag{B8.5.9}$$

If this is to be identical with Equation (B8.5.4), the three Einstein coefficients must be related as follows:

$$B_{n \to m} = B_{m \to n} \tag{B8.5.10}$$

and

$$A_{m \to n} = 8\pi h(\nu_{mn}/c)^3 B_{m \to n} \tag{B8.5.11}$$

We find that the coefficients of induced emission and absorption are equal while the coefficient of spontaneous emission differs from them by a factor of $8\pi h(\nu_{mn}/c)^3$.

PROBLEMS FOR CHAPTER 8

1. Combine two plane-polarised waves, E_a^o and E_b^o 90° out of phase and with $E_a^o = 2 \times E_b^o$. What path is traced by the head of the combined E-vector?

2. Calculate $\langle \psi_{2pz}|M_x|\psi_{1s} \rangle$ and $\langle \psi_{2px}|M_x|\psi_{1s} \rangle$ where ψ_{1s} and ψ_{2px} are hydrogen-atom orbitals (Appendix 5)

3. Using the data for $^1H-^{35}Cl$ in Box 10.2 and the vibrational wave functions in Box 10.3, show that:

$$\langle \Psi_{vib}^f|r|\Psi_{vib}^i \rangle = \sqrt{\frac{h}{8\pi^2\mu\upsilon}} = 7.59 \text{ pm}$$

Hint: If f(x) is a function of even powers of x only then:

$$\int_{-\infty}^{+\infty} f(x) \, dx = 2 \times \int_{0}^{+\infty} f(x) \, dx$$

You will also find the following integral useful, it holds for a > 0 and all integer values of n > -1. $\{n! = n \times (n-1) \times (n-2) \ldots \times 1; 0! = 1\}$

$$\int_{0}^{+\infty} x^n e^{-ax} \, dx = \frac{n!}{a^{n+1}}$$

4. Show that the wave function (Equation (8.6.5)):

$$\Psi = \phi_a \exp(-i\omega t)$$

is an eigenfunction of Schrödinger's time-dependent equation (Equation (8.6.2)), but that the wave function (Equation (8.6.6)):

$$\Psi = \Sigma_a c_a \phi_a \exp(-i\omega a t)$$

is not.

5. The energies, E_m, and wave functions, Ψ_m, of the Hückel π-electron molecular orbitals of linear conjugated polyenes may be written in closed algebraic form:

$$E_m = \alpha + 2\cos\left(\frac{m}{N+1}\right)\beta \quad \Psi_m = \sqrt{\frac{2}{N+1}}\sum_r \sin\left(\frac{rm\pi}{N+1}\right)\phi_r$$

where N is the number of conjugated carbon atoms and $r = 1, 2, \ldots N$ numbers the atoms. The definitions of α and β are given in Section 12.1 but this knowledge is not required for this problem.

The highest occupied MO has $m = \frac{1}{2}N$ and the lowest unoccupied MO has $m = \frac{1}{2}N + 1$. The $\pi \to \pi^*$ transition from the lower to the higher of these two MOs is responsible for the strong, long-wavelength absorption band which characterises these compounds and increases markedly in intensity as the chain gets longer. Show that the intensities predicted using the above MOs also increase in the series $N = 4 \to 6 \to 8$. A suggested method for this calculation is the following:

(a) Consider a polyene to be a zig-zag chain of atoms with a uniform bond length of 140 pm and all C–C–C angles 120°. Define an x axis as the line passing through the centre points of all the bonds with $x = 0$ in the centre of the molecule.

(b) Calculate the transition density at each carbon atom assuming, in accord with the Hückel theory, that there is no overlap between the AOs on different carbon atoms. Multiply the transition density by the x co-ordinate of that atom.

(c) Sum the results for each atom to obtain the x component (in e pm) of the transition dipole moment for molecules having $N = 4$, 6 and 8.
(Taking account of symmetry will considerably reduce your work!)
The results for $N = 4$ are tabulated below as a guide.

$r =$	1	2	3	4
Transition density =	+0.362 e	−0.138 e	+0.138 e	−0.362 e
x co-ordinate =	−181.8 pm	−60.6 pm	+60.6 pm	+181.8 pm
Transition dipole moment =	−65.81 e pm	+8.36 e pm	+8.36 e pm	−65.81 e pm

Sum = −114.9 e pm The intensity is proportional to $(\text{sum})^2 = 13{,}202\ e^2\ \text{pm}^2$

CHAPTER 9

Nuclear Magnetic Resonance Spectroscopy

9.0   Introduction 261
9.1   The magnetic properties of atomic nuclei 262
9.2   The frequency region of NMR spectroscopy 264
9.3   The NMR selection rule 264
9.4   The chemical shift 267
      9.4.1   The delta (δ) scale 268
      9.4.2   The shielding constant, σ (sigma) 270
9.5   Nuclear spin–spin coupling 270
9.6   The energy levels of a nuclear spin system 273
      9.6.1   First order spectra 274
      9.6.2   Second order spectra 275
9.7   The intensities of NMR spectral lines 276
9.8   Quantum mechanics and NMR spectroscopy 277
9.9   Bibliography and further reading 278
      Problems for Chapter 9 287

9.0   INTRODUCTION

In Chapter 4 we saw that electrons and many nuclei have an intrinsic angular momentum which we call spin angular momentum. This angular momentum is invariably associated with a magnetic moment as Wolfgang Pauli, originator of the Pauli exclusion principle, proposed in 1924. Pauli's suggestion was confirmed in 1939 when the deflection of atomic beams by magnetic fields was detected. Nuclear magnetic resonance in bulk matter was first reported in 1946. The presence of the magnetic moments associated with the spin of electrons and nuclei causes these particles to interact with each other and with applied magnetic fields and these interactions are very important in several branches of spectroscopy. In nuclear magnetic resonance and electron spin or paramagnetic resonance spectroscopies the interactions of the nuclei and the electrons with applied magnetic fields are studied. The quantum mechanics which underlies the two forms of spectroscopy is very similar, though they are often discussed in very different language. Some important formulae are collected together in Table 9.1 and described in more detail below.

The Quantum in Chemistry   R. Grinter
© 2005 John Wiley & Sons, Ltd

Table 9.1   The Magnetic Properties of the Electron and the Proton

	Electron	Proton
Spin angular momentum in units of $h/2\pi$ characterised by the quantum numbers	$\hbar \equiv h/2\pi = 1.054592 \times 10^{-34}$ Js	
	s and m_s	I and m_I
The magnetic moment in	Bohr magnetons, μ_B $\mu = -g_e\sqrt{[s(s+1)]}$	nuclear magnetons, μ_N $\mu = +g_H\sqrt{[I(I+1)]}$
where	$\mu_B = eh/4\pi m_e$ $= 9.274 \times 10^{-24}$ J T^{-1}	$\mu_N = eh/4\pi M$ $= 5.051 \times 10^{-27}$ J T^{-1}
and the g factors are	$g_e = 2.002319$	$g_H = 5.585691$
In J T^{-1}	$\mu = -g_e\mu_B\sqrt{[s(s+1)]}$	$\mu = +g_H\mu_N\sqrt{[I(I+1)]}$
Alternatively	$\mu = +\gamma_e(h/2\pi)\sqrt{[s(s+1)]}$	$\mu = +\gamma_H(h/2\pi)\sqrt{[I(I+1)]}$
where γ is the magnetogyric ratio	$\gamma_e = -g_e\mu_B/(h/2\pi)$ $= -1.760 \times 10^{11}$ rad T^{-1} s^{-1}	$\gamma_H = g_H\mu_N/(h/2\pi)$ $= +26.752 \times 10^7$ rad T^{-1} s^{-1}
Similarly	$\mu_z = +\gamma_e h m_s/2\pi$	$\mu_z = +\gamma_H h m_I/2\pi$
Since $E = -\boldsymbol{\mu}\cdot\boldsymbol{B} = -\mu_z B_z$ for a magnetic field aligned along the z axis:		
	$E = +g_e\mu_B m_s B$	$E = -g_H\mu_N m_I B$
and $\Delta E(+\frac{1}{2} \leftrightarrow -\frac{1}{2}) =$	$g_e\mu_B B$	$g_H\mu_N B$

The data for the proton also serve as an example for all nuclei. But it must be noted that the values of the nuclear g's and γ's are different for every nucleus, i.e. for every isotope.

9.1   THE MAGNETIC PROPERTIES OF ATOMIC NUCLEI

The spin and magnetic properties of all stable nuclei are now well documented. There are three useful rules which can be used to predict the spin angular momenta of nuclei.

1. Nuclei with an odd mass number have a half-integral spin quantum number, for example: ^{1}H, $I = \frac{1}{2}$; ^{23}Na, $I = \frac{3}{2}$; ^{27}Al, $I = \frac{5}{2}$.

2. Nuclei with an even mass number and an even charge number have zero spin, for example: ^{12}C, ^{16}O, ^{32}S.

3. Nuclei with an even mass number and an odd charge number have an integral spin quantum number, for example: ^{2}H, $I = 1$; ^{10}B, $I = 3$; ^{14}N, $I = 1$.

The spin angular momentum, P, of a nucleus is determined by the quantum number I according to Equation (9.1.1):

$$P = \sqrt{[I(I+1)]}h/2\pi \qquad (9.1.1)$$

and its magnetic moment, μ, is given by the equation:

$$\mu = g_N\mu_N\sqrt{[I(I+1)]} = 2\pi g_N\mu_N P/h \qquad (9.1.2)$$

$\mu_N = eh/4\pi M$ is the nuclear magneton which has the value $5.05095 \times 10^{-27} \, J\,T^{-1}$ and relates the spin angular momentum to the magnetic field produced. (e is the elementary charge, h Planck's constant and M the mass of the proton.) g_N is the nuclear g factor, a quantity which, at the present state of our knowledge, can only be determined by experiment and may be either positive or negative. A negative value indicates that the magnetic moment and the angular momentum vectors point in opposite directions. It is convenient to express μ in terms of a quantity γ_N, the *magnetogyric ratio*, rather than g_N, where:

$$\gamma_N = g_N \mu_N 2\pi / h \tag{9.1.3}$$

and therefore:

$$\mu = \gamma_N P \tag{9.1.4}$$

The magnetogyric ratio is a factor which converts the angular momentum in units of $h/2\pi$ into magnetic moment in units of $J\,T^{-1}$, hence its name. Thus, the nuclear magnetic moment is related to the nuclear spin quantum number, I, by the equation:

$$\mu = g_N \mu_N \sqrt{[I(I+1)]} = \gamma_N \sqrt{[I(I+1)]} h/2\pi \tag{9.1.5}$$

As we know from Chapter 4, for a spin quantum number I there are $2I + 1$ possible orientations of the spin angular momentum in space, each characterised by a different value of the quantum number m_I, which takes all integer or half-integer values from $-I$ to $+I$. Consequently, $2I + 1$ orientations of the nuclear magnet are also possible and the component of the magnetic moment in the z direction, in analogy to Equations (9.1.4) and (9.1.5), is (Box 12.2):

$$\mu_z = \gamma_N P_z = \gamma_N m_I h/2\pi \tag{9.1.6}$$

In general, this will be of no obvious consequence, but in the presence of a magnetic field each orientation will have a different energy, as illustrated in Figure 9.1 for $I = \frac{3}{2}$, e.g. the ^{23}Na nucleus. The energy, E, of a magnet, μ, placed in a magnetic field of strength B is (Box 12.2):

$$E = -\mu B \cos\theta \tag{9.1.7}$$

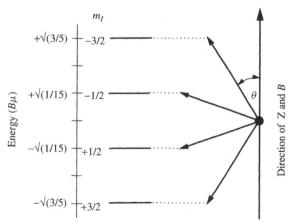

Figure 9.1 The energy levels of the four M_I components of an $I = \frac{3}{2}$ nucleus in a magnetic field

where θ is the angle between the axis of the magnet μ and the direction of B (Figure 9.1). If we orientate the field along the z-direction then Equation (9.1.7) can be written:

$$E = -\mu_z B = -\gamma_N B m_I h / 2\pi \qquad (9.1.8)$$

Thus, if a molecule containing atoms which have nuclear magnetic moments is placed in a magnetic field each of the $2I + 1$ sub-levels will have a different energy and, if the molecule is irradiated with e-m radiation of a suitable frequency, transitions between these energy levels may be induced. Such transitions are known as nuclear magnetic resonance (NMR) and the process forms the basis of NMR spectroscopy.

9.2   THE FREQUENCY REGION OF NMR SPECTROSCOPY

The conservation of energy, as expressed in the Bohr–Einstein condition (Section 8.5.1) determines the region of the e-m spectrum in which NMR spectra will be observed. The condition requires that the energy difference between the two states shall be equal to the energy, $h\nu$, of the photon which is absorbed when it stimulates the transition. Table 9.2 shows the energy gap between the two m_I levels, $m_I = +\frac{1}{2}$ and $m_I = -\frac{1}{2}$, for three nuclei having $I = \frac{1}{2}$ and three magnetic fields. We see that the energy separation, even at high magnetic fields, is very small. The corresponding wavelengths and frequencies are those which we associate with television and radio transmission. Although we may say that NMR is observed in the region of the e-m spectrum between 10 and 1000 MHz, this is a very wide range of frequencies and within it each nucleus is confined to quite a narrow band, well separated from the resonances of other nuclei. Though a modern NMR instrument can be readily adjusted to measure the spectra of a number of different nuclei, e.g. ^{1}H, ^{13}C, ^{19}F, ^{31}P, ..., in chemical applications only the range of resonance frequencies of one particular nuclear species is covered in any one experiment.

9.3   THE NMR SELECTION RULE

The fact that the energy difference between the two energy levels having $m_I = -\frac{1}{2}$ and $+\frac{1}{2}$ satisfies the Bohr–Einstein condition is not in itself sufficient to ensure that transitions between the two levels can be observed when the sample is irradiated with e-m radiation of the above frequencies. There must be a mechanism whereby the photon or e-m wave can interact with the nucleus and cause it to change its energy by absorbing the photon. The mechanism is the interaction between the oscillating *magnetic* field of the radiation

Table 9.2   The energy gap between the nuclear spin levels with $m_I = +\frac{1}{2}$ and $m_I = -\frac{1}{2}$ for three frequently-studied nuclei at three values of the applied magnetic field (The corresponding frequencies and wavelengths are also given)

B(T)	ΔE(J) $\times 10^{27}$	^{1}H ν(MHz)	λ(m)	ΔE(J) $\times 10^{27}$	^{13}C ν(MHz)	λ(m)	ΔE(J) $\times 10^{27}$	^{31}P ν(MHz)	λ(m)
1	28.21	42.58	0.704	7.10	10.71	2.800	11.43	17.25	1.737
4	112.9	170.3	0.176	28.38	42.83	0.700	45.73	69.02	0.434
8	225.7	340.6	0.088	56.76	85.67	0.350	91.46	138.0	0.217

and the magnetic moment of the nucleus. To understand how this happens first recall (Chapter 4) that, even when $m_I = I$, the magnetic moment of the nucleus does not lie exactly along the z-direction.

Now, suppose that we place a bar magnet, in the form of a rod and mounted at its centre on a frictionless bearing, in a magnetic field. If we displace the magnet from alignment with the field there is a torque which tries to return the magnet to the aligned orientation. If we release the magnet it will realign itself with the field, but it will overshoot the position of minimum energy and then oscillate about it. Since the bearing is frictionless, this oscillation will continue indefinitely (Figure 9.2(a)). Suppose further that the magnet is spinning about its axis so that there is angular momentum directed along that axis. The torque which seeks to realign the magnet with the magnetic field is now opposed by the requirement that the angular momentum and its direction in space be conserved. The result of these two conflicting demands is that the magnet precesses about the direction of the field like a gyroscope or top in the gravitational field of the earth. That is, the magnet rotates about the field in such a way that its north and south poles describe circles around the direction of the field as illustrated in Figure 9.2(b). This is exactly the same as the motion of a top when the axis of the top slowly circles around the vertical line drawn through the point at which the top is in contact with the ground. This motion of nuclei in a magnetic field is known as *Larmor precession* and its frequency, f, the Larmor frequency, is given by:

$$f = \gamma_N B / 2\pi \qquad (9.3.1)$$

in which the magnetogyric ratio is again involved.

The precession of μ about the z-axis (Figure 9.3) produces a rotating magnetic field in the xy-plane and this provides the 'handle' by means of which the e-m radiation is able to disturb the precessional motion of the nucleus. If the precessing nucleus is irradiated with circularly polarised radiation then, if the frequency of the radiation is very different from the Larmor frequency, there will be no effect. But if the frequency of the radiation is equal to the Larmor frequency then the rotating magnetic moment and the rotating magnetic field of the radiation will remain in phase with each other and the effect is to tip

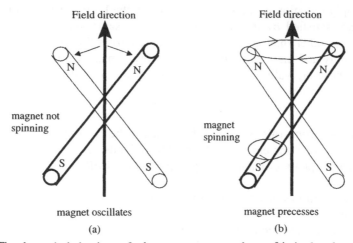

Figure 9.2 The dynamic behaviour of a bar magnet mounted on a frictionless bearing and placed in a magnetic field

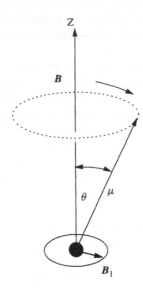

Figure 9.3 Larmor precession

the nucleus over from the $m_I = +\frac{1}{2}$ into the $m_I = -\frac{1}{2}$ orientation. But energy also has to be conserved and we therefore have two criteria which the frequency of the radiation must satisfy:

1. The frequency of the radiation must be equal to the Larmor frequency, i.e. it must satisfy Equation (9.3.1), so that $v = f = \gamma_N B / 2\pi$.

2. The frequency of the radiation must satisfy the Bohr–Einstein condition:

$$h v = \Delta E \Rightarrow v = -\gamma_N B h (-\tfrac{1}{2} - \tfrac{1}{2}) / 2 h \pi = \gamma_N B / 2\pi \qquad (9.3.2)$$

Thus, the energy required for the transition between the two levels $m_I = +\frac{1}{2}$ and $-\frac{1}{2}$ does indeed imply radiation of the same frequency as the Larmor precession and the two requirements are satisfied by the same radiation. The same analysis applies for nuclei with values of m_I other than $\pm\frac{1}{2}$ and, clearly, the criterion that the radiation shall have the required energy and the Larmor frequency is always met by a transition between any pair of levels whose m_I values differ by ± 1, but not for any other transition. The NMR selection rule is therefore:

$$\Delta m_I = \pm 1 \qquad (9.3.3)$$

Note that the selection rule is exactly the same for the absorption of a photon and an increase in energy as for the emission of a photon and a decrease in energy. The fact that both processes are equally probable is very important in NMR spectroscopy and we shall return to it later.

Naturally, quantum mechanics also provides a more precise description of the process whereby interaction with an electromagnetic wave changes the energy of a nucleus in a magnetic field. Recall (Section 8.6) that the link between the initial and final states of a spectroscopic transition brought about by e-m radiation can be written as the product of a time-dependent and a time-independent part. Evaluation of the time-dependent part results in the Bohr–Einstein or energy conservation condition, which is the same in all branches of spectroscopy. The time-independent part, in the case of NMR, takes the form (see Box 9.1 for further justification):

$$(\gamma B_{1x})^2 |\langle \Psi_{final} | \hat{I}_x | \Psi_{initial} \rangle|^2 \tag{9.3.4}$$

where B_{1x} is the magnetic field of the applied radiation that oscillates in the x-direction and $\hat{I}_x$ is the operator for the x-component of the nuclear spin. Note that in NMR it is the oscillating *magnetic* field of the radiation which interacts with the nucleus; in almost all other spectroscopies it is the *electric* field. But this makes no difference in the evaluation of the time-dependent part of the interaction since the electric and magnetic fields of the radiation oscillate in phase and are described by exactly the same cosine or sine functions. Since there are only two spin states for a nucleus having $I = \frac{1}{2}$, the operation of $\hat{I}_x$ on one of them must give the other; in fact, in units of $h/2\pi$ (Box 9.1):

$$\hat{I}_x | + \tfrac{1}{2} \rangle = \tfrac{1}{2}(\hat{I}_+ + \hat{I}_-) | + \tfrac{1}{2} \rangle = \tfrac{1}{2} | - \tfrac{1}{2} \rangle \tag{9.3.5a}$$

and

$$\hat{I}_x | - \tfrac{1}{2} \rangle = \tfrac{1}{2}(\hat{I}_+ + \hat{I}_-) | - \tfrac{1}{2} \rangle = \tfrac{1}{2} | + \tfrac{1}{2} \rangle \tag{9.3.5b}$$

(Note that these are not eigenfunction-eigenvalue equations; the function, $| \pm \frac{1}{2} \rangle$, on either side of the equality sign is not the same.) Therefore, the interaction of an e-m wave with its magnetic field vibrating in the x-direction converts states from $m_I = -\frac{1}{2}$ to $m_I = +\frac{1}{2}$, and vice versa, with equal facility; i.e. both processes are equally probable. Though the case does not arise here we might also note that $\hat{I}_x$ links any two spin states whose m_I values differ by ± 1, e.g. $+\frac{1}{2}$ is linked to $-\frac{1}{2}$ and $+\frac{3}{2}$; $+1$ is linked to 0 and $+2$.

9.4 THE CHEMICAL SHIFT

Examination of the equations above shows that the frequency at which a nucleus resonates depends only upon its magnetogyric ratio and the magnetic field in which it is placed. If this were the whole story NMR spectroscopy would be of little interest to chemists. But in January 1950 the editor of the *Physical Review* received two letters requesting publication of exciting and surprising new observations. W.C. Dickinson wrote: 'Most unexpectedly, it has been found that for ^{19}F the value of the applied magnetic field for nuclear magnetic resonance at a fixed radio frequency depends on the chemical compound containing the fluorine nucleus.' W.G. Proctor and F.C. Yu reported that: 'In the course of measurements on ^{14}N we made the surprising observation that its frequency of resonance, in liquid samples, depended strongly upon the chemical compound in which it was contained.' These observations, so clearly unexpected, were the trigger for an explosive growth of NMR spectroscopy which has completely changed the way in which much of chemistry is practised and shows no sign of diminishing after more than 50 years.

Following the introduction of the Fourier-transform method and the development of NMR into arguably the most important spectroscopic technique for chemical applications, new imaging techniques have made it a vital adjunct to medicine. Pictures of the living human brain and other vital organs are now routinely obtained in all major hospitals where the technique is known as NMR tomography. Those who believe that the only research worth pursuing is research with an obvious application just around the corner should study the history of NMR carefully. The advances in imaging, in particular, arose from experiments in which magnetic nuclei were made to dance a jig to an intricate sequence of pulses of radiation. Who could have dreamed that this esoteric pursuit of knowledge would lead, within about 20 years, to detailed pictures of the living, working human brain?

Returning to chemistry, the cause of the variation in resonance frequency or field with chemical composition was soon clear to those who observed it. The nuclei we meet in chemistry are always surrounded by clouds of electrons and these electrons have an important effect upon the magnetic field experienced by the nuclei. When an atom is placed in a magnetic field the motion of its electrons will be slightly changed and, according to Lenz's law, the change will be such as to oppose the applied field. Therefore, all other things being equal, which is usually but not quite always the case, the nucleus will experience a field slightly less than the applied field and will resonate at a slightly lower frequency than that given by Equation (9.3.2). This difference in resonance frequency, which depends upon the electron density in the immediate neighbourhood of the nucleus, is called the *chemical shift*. It has been found that the resonance frequency of a nucleus is highly characteristic of its chemical environment and can therefore be used to identify molecules and particular groupings of atoms within molecules. The NMR spectrum of a solution of 4-bromo-2,6-dimethylphenol, in which the different resonance frequencies of the three types of ^1H in the molecule can be readily seen, is shown in Figure 9.4. The peaks can even be assigned to the particular ^1H types, since their heights are directly proportional to the number of protons in any particular chemical environment. The two horizontal scales in Figure 9.4 require further explanation and we now turn to the subject of the chemical shift scale.

9.4.1   The delta (δ) scale

The most obvious chemical shift scale might appear to be one in which the resonance frequency of a particular nucleus in a molecule is related to the frequency of that nucleus

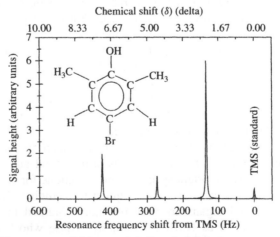

Figure 9.4   The NMR spectrum of 4-bromo-2,6-dimethylphenol

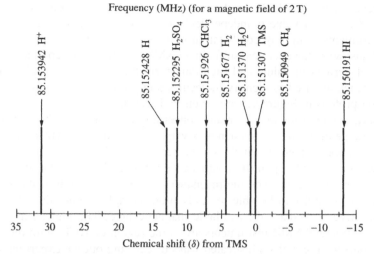

Figure 9.5 Typical proton chemical shifts (δ) from TMS

Table 9.3 Some ^{1}H resonance frequencies (ν) at a field of 2 T and the corresponding shifts (δ) from tetramethyl silane (TMS)

Hydrogen nucleus	ν at 2 T(MHz)	δ
Bare proton, H^+	85.153942	+31.44
Hydrogen atom	85.152428	+13.16
H_2SO_4	85.152295	+11.60
$CHCl_3$	85.151926	+7.27
H_2 molecule	85.151677	+4.34
Water vapour	85.151370	+0.74
TMS	85.151307	0.00
CH_4	85.150949	−4.21
HI	85.150191	−13.11

in the total absence of all its electrons, i.e. the bare nucleus. Figure 9.5 and Table 9.3 present some ^{1}H resonance data in such a form for a magnetic field of 2 T.

However, it is very difficult to measure the exact magnetic field to which a molecule is exposed. This is especially so when it is dissolved in a solvent, contained in a sample tube and surrounded by a thermostat to maintain constant temperature. But this would be necessary if the chemical shifts of the nuclei were to be measured as absolute values against the resonance frequency of the bare nucleus. It has therefore become universal practice to measure the shift relative to the resonance position of a standard substance that has one well defined resonance. For ^{1}H, ^{13}C and ^{29}Si tetramethyl silane [TMS, $Si(CH_3)_4$] is used and the chemical shift scale, known as the δ scale, is defined as:

$$\delta_a = 10^6(\nu_a - \nu_{TMS})/\nu_{TMS} \tag{9.4.1}$$

The factor of 10^6 is included simply to give readily handled numbers and the δ values of the chemical shift scale are said to be in *parts per million* (ppm). This form of chemical shift scale has a further advantage. Since it is a ratio of a frequency difference and a frequency, both of which are directly proportional to the applied magnetic field (Equation (9.3.2)), the δ value is independent of the field and measurements made with different instruments using any combination of field and frequency can be immediately compared. Thus, when the chemical shifts in an NMR spectrum are reported in these units the result is of permanent value, and such data form the basis of the tables of chemical shift values which are used to identify molecules and groupings of atoms in them. The δ values for the protons in Figure 9.5 are given in Table 9.3.

Two important comments may be made concerning these data. In the first instance we note that the range of resonance frequency is very small, some 3.8 kHz as compared with the mean resonance frequency, which is very near to 85 MHz. Secondly, the resonance frequency depends directly upon the magnetic field used. Therefore, the greater the field the greater is the spacing between the resonances, i.e. the resolution of the spectrometer is increased and, in order to be able to study more complex molecules, there has been a continuous drive to increase the available magnetic field. The highest fields currently available in commercial NMR instruments are in the region of 10 T. With the aid of such instruments, plus a variety of sophisticated ways of carrying out the experiment including the use of isotopic labelling with ^{15}N, ^{13}C and ^{2}H, the NMR spectra of proteins with RMR values of the order of 30–40 kDa can now be measured and interpreted.

9.4.2   The shielding constant, σ (sigma)

In Section 9.3 we saw that the frequency, ν, of a nuclear magnetic resonance signal in a field B_0 is given by Equation (9.3.2), $\nu = \gamma_N B_0/2\pi$. We introduce the effect of the chemical shift into this expression by defining a shielding constant, σ_a, for each nucleus, a, in a molecule so that the field, and hence the resonance frequency, is reduced according to the equation:

$$\nu_a = \gamma_N B_0(1 - \sigma_a)/2\pi \qquad (9.4.2)$$

where, in continuing to use γ_N rather than γ_a, it is assumed that only nuclei of the same species are involved in the spectrum.

By substituting Equation (9.4.2) into Equation (9.4.1) we obtain the relation between σ and δ as:

$$\delta_a = 10^6(\sigma_{TMS} - \sigma_a)/(1 - \sigma_{TMS})$$

But since the σ values of nuclei important for chemical NMR spectroscopy are very much less than 1, the form:

$$\delta_a = 10^6(\sigma_{TMS} - \sigma_a) \qquad (9.4.3)$$

is quite sufficient for chemical purposes. Note that increasing δ corresponds to decreasing σ and that spectra are plotted with ν and δ increasing from right to left.

9.5   NUCLEAR SPIN–SPIN COUPLING

In practice, NMR spectra are normally more complex than the above discussion might lead us to expect. For example, the ^1H spectrum of 1,1,2-trichloroethane, (Figure 9.6) shows 5 lines (resonances), although it contains only three hydrogen atoms. A clue as to the assignment of the lines is given by the fact that the sum of the two signals centred

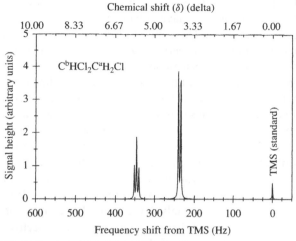

Figure 9.6 The NMR spectrum of 1,1,2-trichloroethane

at 3.95 ppm ($\approx$237 Hz) is twice that of the three signals at 5.77 ppm ($\approx$346 Hz). Since, in NMR, the signal height is usually proportional to the number of nuclei in the same chemical environment, this suggest that the two lines at 3.95 ppm are caused by the protons marked 'a' and the three lines at 5.77 ppm by that marked 'b'. The cause of this splitting of the resonance lines is the nuclear spin–spin coupling, which arises in the following manner.

Each nucleus capable of giving a resonance signal does so because it has a magnetic moment which, in the case of nuclei with $I = \frac{1}{2}$, can be aligned in two directions with respect to the direction of the applied magnetic field, B_0. The magnetic moment of each nucleus is therefore capable of increasing or decreasing the applied field and this change can be detected by other nuclei if they are sufficiently close to it. The process is mutual; if 'a' feels the effect of 'b' then 'b' feels the effect of 'a'. Consider the spectrum of 1,1,2-trichloroethane again. The protons 'a' experience either an increased or a decreased magnetic field depending upon whether the proton 'b' is aligned with the applied field or against it. Therefore, they experience two fields, and resonate at two slightly different frequencies. But this is too simple an explanation of the actual mechanism of spin–spin coupling and we must examine the phenomenon more closely.

Consider first Figure 9.7. The large open circles represent nuclei and the filled circles electrons in a hydrogen molecule. The arrows indicate the orientation of the spin magnetic moments associated with the four particles. On the left-hand side we find that the electron

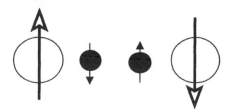

Figure 9.7 Spin–spin coupling in the hydrogen molecule

drawn closest to the nucleus has its magnetic moment in the direction opposite to that of the nucleus, i.e. the two particles have orientated their magnetic moments so as to minimise the energy of interaction. This type of electron nucleus interaction is known as Fermi contact interaction after Enrico Fermi (1901–1954) who first suggested it in 1930 to explain effects seen in atomic spectroscopy.

It is important to note that it is not the same as the dipole-dipole interaction, which is felt when two bar magnets held in the hands are brought close together. The interaction we are concerned with here arises because, as small as they are, the nucleus and the electron are not point charges and this interaction is present when the two are in 'contact'.[‡]

Thus, because of the Fermi contact effect, the two electrons in the H-H bond are not exactly equally distributed in the bond. In the region of the hydrogen nucleus on the left there is a slight excess of the electron with its magnetic moment directed downwards because the magnetic moment of the nucleus is directed upwards. But since the electron spins must be paired in total this must imply that there is also a slight imbalance at the other end of the molecule in the sense that there is an excess of the electron with its magnetic moment directed upwards in the region of the right-hand hydrogen nucleus. The energy of the molecule is therefore slightly lower when the right-hand nucleus has its magnetic moment directed downwards. Thus, the energy of the hydrogen molecule is slightly lower when the nuclear magnetic moments are opposed (antiparallel) than it is when they are parallel; this energy difference is the spin–spin coupling energy. It is extremely small. In the case of the hydrogen molecule its value is $h \times 280$ Hz $= 1.855 \times 10^{-31}$ J, which should be compared with the molecular binding energy of 7.605×10^{-19} J, a factor of approximately 4×10^{12}! Most coupling constants are much smaller than this and it is a tribute to the technology of the NMR experiment that we are able to measure and make use of such tiny energy differences.

Note how the nuclear spin information is transmitted by the electrons. In molecules that do not have a fixed spatial orientation, e.g. in a solution, the dipole-dipole interaction of the nuclear spins exists, but we do not see it because the constantly changing orientation of the molecule with respect to the applied magnetic field averages the dipolar coupling to zero. The spin–spin coupling which we measure is the *electron-mediated* coupling. It should be noted that two other mechanisms, the *spin-dipolar* and *orbital terms*, also contribute to the electron-mediated coupling observed in the fluid phase, but since these mechanisms have a negligible effect on proton-proton coupling we ignore them here.

Figure 9.8 is a similar illustration of the H–C–C–H unit of 1,1,2-trichloroethane and in the first H–C bond the situation is qualitatively exactly the same as that in the H–H

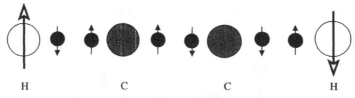

H   C   C   H

Figure 9.8 Spin-spin coupling in a hydrocarbon fragment

[‡] The idea that the electron can be 'in contact with' or 'at' the nucleus is a poor classical description, but the best we have, of a situation which can only be properly described in mathematical terms. But it does make clear why only s-electrons have a Fermi contact interaction with the nucleus; all other atomic orbitals have a node at the nucleus.

bond above and if the carbon atom is a ^{13}C atom with $I = \frac{1}{2}$ then C–H coupling is observed. But the coupling message can be transmitted further along the molecule. The two electrons shown on either side of the carbon atom are involved in different bonds; a H–C bond on the left and a C–C bond on the right. Consequently, these electrons are not paired and, according to Hund's rule (Chapter 5), their energy is lower when their spins, and therefore their magnetic moments, are parallel. This effect transmits the electron spin polarisation into the C–C bond and hence to the second carbon atom, where the same mechanism induces a spin polarisation in the C–H bond. Finally, we arrive at the second hydrogen nucleus, which for lowest energy should have its magnetic moment aligned antiparallel to that of the first. This spin–spin coupling mechanism leads to the result that for coupling over an odd number of bonds the more stable orientation of the nuclear magnetic moments is antiparallel, while for coupling over an even number of bonds the parallel orientation is the more stable. The former coupling is *defined* to be positive and the latter negative; this alternation of the sign of the coupling with the number of intervening bonds is frequently observed in NMR spectroscopy. However, the magnitude of the coupling diminishes quite rapidly as the number of bonds increases.

Though the mechanism of spin–spin coupling by electrons is complex, the quantum mechanical expression of the effect is very simple if we regard the coupling interaction energy between two nuclei, r and s, as a parameter, the coupling constant J_{rs}, to be determined by experiment. The energy can then be written as J_{rs} multiplied by the scalar product of $\hat{\mathbf{I}}_r$ and $\hat{\mathbf{I}}_s$, the nuclear spin operators for the two nuclei:

$$\text{Interaction energy} = E_J = J_{rs}\hat{\mathbf{I}}_r \cdot \hat{\mathbf{I}}_s = J_{rs}[\hat{I}_{rx}\hat{I}_{sx} + \hat{I}_{ry}\hat{I}_{sy} + \hat{I}_{rz}\hat{I}_{sz}]$$

$$= J_{rs}[\hat{I}_{rz}\hat{I}_{sz} + \tfrac{1}{2}(\hat{I}_{r+}\hat{I}_{s-} + \hat{I}_{r-}\hat{I}_{s+})] \tag{9.5.1}$$

and we always divide E_J by Planck's constant and quote the coupling in Hz, the unit in which it is measured experimentally.

9.6   THE ENERGY LEVELS OF A NUCLEAR SPIN SYSTEM

Summarising Sections 9.4 and 9.5: the Hamiltonian operator from which the energy of all the nuclear spins in a molecule can be calculated includes the interaction with the applied field, B_0, allowing for the shielding, σ, and the nuclear spin–spin coupling, J. Using $\hat{I}_z|I, m_I\rangle = m_I|I, m_I\rangle$ to transform Equation (9.1.8) to operator form and combining with Equation (9.5.1) we write the Hamiltonian, in units of frequency, as:

$$h^{-1}\hat{\mathcal{H}} = -(2\pi)^{-1}\sum_r \gamma_r B_0(1 - \sigma_r)\,\hat{I}_{rz} + \sum_{r<s} J_{rs}\hat{\mathbf{I}}_r \cdot \hat{\mathbf{I}}_s$$

or, using Equation (9.4.2), as:

$$h^{-1}\hat{\mathcal{H}} = -\sum_r \nu_r\hat{I}_{rz} + \sum_{r<s} J_{rs}\hat{\mathbf{I}}_r \cdot \hat{\mathbf{I}}_s \tag{9.6.1}$$

As an example we shall consider the simple case of two protons, 'a' and 'b'. Very detailed treatments of much more complex systems may be found elsewhere.[1,2]

There are four possible arrangements of the z components of the spins of two protons (always written in the order a,b), $|+\frac{1}{2} + \frac{1}{2}\rangle$, $|+\frac{1}{2} - \frac{1}{2}\rangle$, $|-\frac{1}{2} + \frac{1}{2}\rangle$, and $|-\frac{1}{2} - \frac{1}{2}\rangle$. These are the basis functions of the problem and we shall follow the approach, described in

Appendix 3, of setting up the matrix of $\hat{\mathcal{H}}$ in this basis and diagonalising it. The first term in Equation (9.6.1) is the chemical shift, which gives a contribution to the on-diagonal elements of our 4×4 matrix. There are no off-diagonal elements since the two nuclei are not linked in any way by $\hat{I}_{rz}$. As an illustration of the operation of this first term on a basis state we have:

$$-\sum_r v_r \hat{I}_{rz}|+\tfrac{1}{2} - \tfrac{1}{2}\rangle = -v_a \hat{I}_{az}|+\tfrac{1}{2} - \tfrac{1}{2}\rangle - v_b \hat{I}_{bz}|+\tfrac{1}{2} - \tfrac{1}{2}\rangle = -\tfrac{1}{2}[v_a - v_b]|+\tfrac{1}{2} - \tfrac{1}{2}\rangle$$

To illustrate the second part of the operator, the scalar coupling, we evaluate:

$$J_{ab}[\hat{I}_{az}\hat{I}_{bz} + \tfrac{1}{2}(\hat{I}_{a-}\hat{I}_{b+} + \hat{I}_{a+}\hat{I}_{b-})]|+\tfrac{1}{2} - \tfrac{1}{2}\rangle = J_{ab}[\tfrac{1}{2} \times -\tfrac{1}{2}]|+\tfrac{1}{2} - \tfrac{1}{2}\rangle + \tfrac{1}{2}J_{ab}|-\tfrac{1}{2} + \tfrac{1}{2}\rangle$$

$$= -\tfrac{1}{4}J_{ab}|+\tfrac{1}{2} - \tfrac{1}{2}\rangle + \tfrac{1}{2}J_{ab}|-\tfrac{1}{2} + \tfrac{1}{2}\rangle$$

The first term gives a diagonal, the second an off-diagonal matrix element. Using the fact that the spin basis functions are orthogonal and normalised the complete matrix is found to be:

	$\|+\tfrac{1}{2} + \tfrac{1}{2}\rangle$	$\|+\tfrac{1}{2} - \tfrac{1}{2}\rangle$	$\|-\tfrac{1}{2} + \tfrac{1}{2}\rangle$	$\|-\tfrac{1}{2} - \tfrac{1}{2}\rangle$
$\langle+\tfrac{1}{2} + \tfrac{1}{2}\|$	$-\tfrac{1}{2}[v_a + v_b] + \tfrac{1}{4}J_{ab}$	0	0	0
$\langle+\tfrac{1}{2} - \tfrac{1}{2}\|$	0	$-\tfrac{1}{2}[v_a - v_b] - \tfrac{1}{4}J_{ab}$	$\tfrac{1}{2}J_{ab}$	0
$\langle-\tfrac{1}{2} + \tfrac{1}{2}\|$	0	$\tfrac{1}{2}J_{ab}$	$\tfrac{1}{2}[v_a - v_b] - \tfrac{1}{4}J_{ab}$	0
$\langle-\tfrac{1}{2} - \tfrac{1}{2}\|$	0	0	0	$\tfrac{1}{2}[v_a + v_b] + \tfrac{1}{4}J_{ab}$

The matrix is blocked out (Appendix 3) and the energies of the spin functions $|+\tfrac{1}{2} + \tfrac{1}{2}\rangle$ and $|-\tfrac{1}{2} - \tfrac{1}{2}\rangle$ can be written down immediately. The spin functions $|+\tfrac{1}{2} - \tfrac{1}{2}\rangle$ and $|-\tfrac{1}{2} + \tfrac{1}{2}\rangle$, however, are mixed by the coupling and the degree of mixing depends upon the separation of the two diagonal elements and the magnitude of their interaction, $\tfrac{1}{2}J_{ab}$. Two extreme cases can be distinguished.

9.6.1   First order spectra

If the chemical shifts of the two protons are very different we have an AX system. Now, the energies of the states $|-\tfrac{1}{2} - \tfrac{1}{2}\rangle$ and $|+\tfrac{1}{2} + \tfrac{1}{2}\rangle$ are as before but the two other states, $|+\tfrac{1}{2} - \tfrac{1}{2}\rangle$ and $|-\tfrac{1}{2} + \tfrac{1}{2}\rangle$ are not mixed equally and their energies must be found by diagonalising the 2×2 matrix. In the extreme where the difference in energy of the on-diagonal elements, $|[v_a - v_x]|$, is very much greater than the off-diagonal element which links them, $\tfrac{1}{2}J_{ax}$, the eigenfunctions of the spin system are effectively the four basis functions and the energy levels are given by the diagonal elements of the above matrix. The NMR spectra from spin systems for which this is true are known as *first order* spectra. Assuming that $v_a > v_x$, we can draw the energy-level diagram (Figure 9.9). The energies of the allowed transitions, i.e. those for which one m_I value changes while the other remains constant are:

X changes:

$$E(+\tfrac{1}{2} - \tfrac{1}{2}) - E(+\tfrac{1}{2} + \tfrac{1}{2}) = v_x - \tfrac{1}{2}J_{ax}$$

$$E(-\tfrac{1}{2} - \tfrac{1}{2}) - E(-\tfrac{1}{2} + \tfrac{1}{2}) = v_x + \tfrac{1}{2}J_{ax}$$

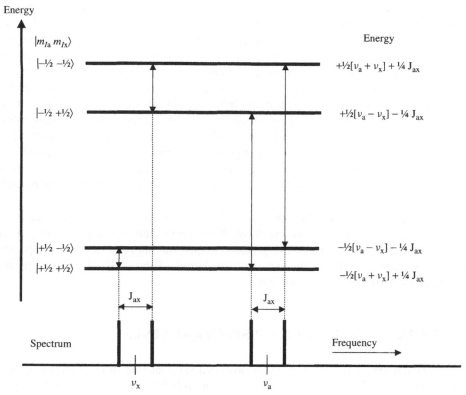

Figure 9.9   The energy levels and spectrum of an AX system ($\nu_a > \nu_x$ and $[\nu_a - \nu_x] \gg |J_{ax}|$)

A changes:

$$E(-\tfrac{1}{2} + \tfrac{1}{2}) - E(+\tfrac{1}{2} + \tfrac{1}{2}) = \nu_a - \tfrac{1}{2}J_{ax}$$

$$E(-\tfrac{1}{2} - \tfrac{1}{2}) - E(+\tfrac{1}{2} - \tfrac{1}{2}) = \nu_a + \tfrac{1}{2}J_{ax}$$

We have four lines of equal intensity equally spaced on either side of ν_a and ν_x with a separation of J_{ax} (Figure 9.9).

9.6.2   Second order spectra

If the chemical shifts of the two protons are the same, i.e. they are in identical environments as in the hydrogen molecule, we have what NMR spectroscopists call an A_2 spin system. The functions $|+\tfrac{1}{2} - \tfrac{1}{2}\rangle$ and $|-\tfrac{1}{2} + \tfrac{1}{2}\rangle$ are mixed in equal proportions and the following four spin energy levels result:

	Spin function	Energy
1	$\lvert+\tfrac{1}{2} + \tfrac{1}{2}\rangle$	$-\tfrac{1}{2}[\nu_a + \nu_b] + \tfrac{1}{4}J_{ab}$
2	$[\lvert+\tfrac{1}{2} - \tfrac{1}{2}\rangle + \lvert-\tfrac{1}{2} + \tfrac{1}{2}\rangle]/\sqrt{2}$	$+\tfrac{1}{4}J_{ab}$
3	$\lvert-\tfrac{1}{2} - \tfrac{1}{2}\rangle$	$\tfrac{1}{2}[\nu_a + \nu_b] + \tfrac{1}{4}J_{ab}$
4	$[\lvert+\tfrac{1}{2} - \tfrac{1}{2}\rangle - \lvert-\tfrac{1}{2} + \tfrac{1}{2}\rangle]/\sqrt{2}$	$-\tfrac{3}{4}J_{ab}$

Functions 1, 2 and 3 are the three components of a triplet state having $I = 1$ and $M_I = +1$, 0 and -1 respectively. Function 4 is a singlet with $I = M_I = 0$. NMR spectra in which the eigenstates of $\hat{\mathcal{H}}$ can only be described in terms of combinations of the basis functions are known as *second order* spectra. Since NMR transitions are allowed only between states in which one proton spin is changed from $+\frac{1}{2}$ to $-\frac{1}{2}$, or vice versa, we might expect to observe the four transitions $1 \leftrightarrow 2$, $1 \leftrightarrow 4$, $2 \leftrightarrow 3$ and $3 \leftrightarrow 4$. However, transitions between states of different multiplicity are not allowed (Problem 1), so transitions $1 \leftrightarrow 2$ and $2 \leftrightarrow 3$ only should occur and since they correspond to the same energy difference, ν_a, the corresponding spectral lines will appear as one. This is a general result. The NMR spectrum of a group of identical protons appears as a single line and although spin–spin coupling is present its magnitude cannot be determined from the spectrum. We expand on this point in Box 9.2 and there are more details on NMR transition probabilities in Box 9.1 and Section 9.7.

The description 'second order spectrum' also applies to the intermediate case of a smaller difference between the two chemical shifts where there is unequal mixing of two or more basis states. In the case of two protons unequal mixing produces the AB spectrum which varies smoothly between the two extremes described above. Harris[1] gives a detailed discussion of all aspects of this problem.

9.7   THE INTENSITIES OF NMR SPECTRAL LINES

Two factors determine the strength of an observed spectral line in all branches of spectroscopy. These are the magnitude of the transition probability between the initial and the final state and the relative populations of the two states. In normal electronic and infrared spectroscopy the population aspect of the intensity plays no role since the population of the state of lower energy always massively exceeds that of the state of higher energy. In NMR spectroscopy this is not the case since, as we have seen, the energy difference between the resonating states is very small and, at the start of the experiment, the number of nuclei, N_L, in the lower-energy state is only slightly larger than that in the upper state, N_U. For example, for hydrogen nuclei at room temperature in a field of 4 T we have, using data from Table 9.2 and the Boltzmann distribution:

$$N_U/N_L = \exp[-(112.9 \times 10^{-27})/(1.38062 \times 10^{-23} \times 300)]$$
$$= \exp[-0.2726 \times 10^{-4}] = 0.999972$$

So that, if we have one million nuclei in the lower state there will be almost as many in the upper state, 28 less to be exact. Therefore, when we irradiate our sample we rapidly equalise the number of nuclei in the two states and then, since the selection rule for a transition absorbing a photon and one giving out a photon are exactly the same, the number of photons absorbed will be equal to the number emitted and no net signal will be observed. This phenomenon is known as signal saturation and it presents many problems for the NMR spectroscopist. The opposing *relaxation* processes, whereby the system returns to the lower-energy state also play an important a role in determining the intensity of an NMR signal, (see Section 9.8). The second factor which determines the strength of an NMR signal, the transition dipole moment, has been outlined in Sections 8.7.2 and 9.3 and is discussed further in Box 9.1.

In the case of a molecule containing a number of NMR-active nuclei the probability of the transition $|\Psi_i\rangle \rightarrow |\Psi_f\rangle$ is proportional to the matrix element:

$$|\langle\Psi_f|\sum_r \hat{I}_{rx}|\Psi_i\rangle|^2 = |\langle\Psi_f|\tfrac{1}{2}\sum_r(\hat{I}_{r+} + \hat{I}_{r-})|\Psi_i\rangle|^2 \qquad (9.7.1)$$

where the sum extends over all the nuclei, r, involved in the initial and final wave functions. Since the operator $\hat{I}_{rx}$ operates only on the spin of nucleus r, leaving the spins other nuclei unchanged, it is clear that these must all be the same if the matrix element is to be non-zero. Thus an NMR transition between $|+\tfrac{1}{2} + \tfrac{1}{2} - \tfrac{1}{2}\rangle$ and $|\tfrac{1}{2} + \tfrac{1}{2} + \tfrac{1}{2}\rangle$ is allowed through $\hat{I}_{3x}$, but the transition $|+\tfrac{1}{2} + \tfrac{1}{2} - \tfrac{1}{2}\rangle \leftrightarrow |-\tfrac{1}{2} + \tfrac{1}{2} + \tfrac{1}{2}\rangle$ is forbidden. For the purposes of calculation it is very convenient to replace $\hat{I}_{rx}$ by $\tfrac{1}{2}(\hat{I}_{r+} + \hat{I}_{r-})$. Problem 3 provides an example of the use of the above result to determine the relative intensities of the proton NMR lines of a two-spin system.

Finally, it is noteworthy that the relative intensities of a line in a normal NMR spectrum is proportional to the number of resonating nuclei responsible for that particular line. Allowance must be made for the splitting of lines by spin–spin coupling. This simple proportionality is a great help in the interpretation of the spectra and in the quantitative analytical applications of NMR spectroscopy.

9.8 QUANTUM MECHANICS AND NMR SPECTROSCOPY

The applications of quantum mechanics in chemical NMR spectroscopy fall into three rather distinct categories:

(a) The calculation of the energy levels of a nuclear spin system in order to interpret a spectrum.
(b) The calculation of chemical shifts and coupling constants from molecular wave functions.
(c) The description of the dynamic processes associated with the Fourier transform NMR technique.

In this Chapter we have been concerned exclusively with topic (a), the analysis of experimental spectra in terms of the chemical shifts, σ, and coupling constants, J, which are regarded as parameters. But quantum mechanics also offers the possibility of calculating chemical shifts and coupling constants from the wave functions of the molecules concerned and the theoretical expressions required have been available since the 1950s. Though the methods are well known their application has had only limited quantitative success and the basic reason for this is not far to seek. The energy of an NMR transition corresponds to a frequency in the radiowave region of the spectrum and the small changes in the resonance frequency, which are brought about by chemical shifts and coupling constants, correspond to even smaller energies. The energies of the electrons in a molecule, however, correspond to frequencies in the visible and ultra-violet regions of the e-m spectrum some 8 to 10 orders of magnitude greater. Therefore, the calculation of σ and J values which are of sufficient accuracy to be useful requires extremely good molecular wave functions, much better than those required for a quantitative interpretation of electronic spectra. In general, apart from in the case of a few small molecules, we do not have such wave functions. Nevertheless, the efforts which have been made to calculate chemical shifts and coupling constants have been valuable in that they have revealed the major determining factors and made possible a deeper understanding of the phenomena. Space does not permit a further examination of the calculation of chemical shifts and coupling constants here but an introduction to the field can be found in Harris[1] and Memory.[3]

The subjects subsumed under category (c) are many and complex. The modern NMR experiment utilises the Fourier-transform (FT) technique and as such it is a dynamic experiment in which the resonating nuclei are subjected to an intricate sequence of pulses of radio-frequency radiation which vary in frequency, band width, intensity, duration etc. The result of this treatment is to produce in the spin system a non-equilibrium orientation of the nuclear spins so that the sample becomes magnetised in some specially chosen way. The final step is normally the recording of the decay of this induced magnetisation; the *free induction decay*. The response of the nuclei to the radiation pulses and the subsequent relaxation processes are all time-dependent phenomena and very advanced quantum-mechanical calculations are required to analyse them. An introduction to this type of NMR spectroscopy may be found in Harris[1] and in the first edition of Slichter's[4] text. Freeman[5] has given a very readable descriptive account with almost no mathematics. The most penetrating description of the resonance phenomenon remains the classic treatise of Abragam[6] but, naturally, it does not deal with many aspects of the modern NMR experiment for which the latest edition of Slichter[7] and Goldman[8] can be recommended.

9.9 BIBLIOGRAPHY AND FURTHER READING

1. R.K. Harris, *Nuclear Magnetic Resonance Spectroscopy*, Pitman, London, 1983.
2. J.A. Pople, W.G. Schneider and H.J. Bernstein, *High-Resolution Nuclear Magnetic Resonance*, McGraw-Hill Book Company, New York/London, 1959.
3. J.D. Memory, *Quantum Theory of Magnetic Resonance Parameters*, McGraw-Hill Book Company, New York/London, 1968.
4. C.P. Slichter, *Principles of Magnetic Resonance*, Harper & Row, New York/London, 1963.
5. R. Freeman, *Magnetic Resonance in Chemistry and Medicine*, Oxford University Press, Oxford, 2003.
6. A. Abragam, *The Principles of Nuclear Magnetism*, Oxford University Press, Oxford, 1961.
7. C.P. Slichter, *Principles of Magnetic Resonance*, 3rd edn, Springer-Verlag, New York, 1989.
8. M. Goldman, *Quantum Description of High-Resolution NMR in Liquids*, Oxford University Press, Oxford, 1988.

BOX 9.1 The NMR selection rule

To calculate the NMR transition probability, $W(x, y, z, t)$, we return to Equation (8.6.21), repeated here for convenience, in which the electric field, E_0, has been replaced by the magnetic field, B_0, because it is the magnetic field of the e-m radiation which interacts with the nuclear spins:

$$W(x, y, z, t) = B_0^2 \cdot \frac{4\pi^2}{h^2} |\langle \phi_f | \hat{V}(x, y, z) | \phi_i \rangle|^2 \cdot \frac{\sin^2\{\frac{1}{2}(\Delta\omega - \omega)T\}}{4\{\frac{1}{2}(\Delta\omega - \omega)\}^2} \qquad (8.6.21)$$

B_0 is given by the experimental conditions and the term in $\sin^2$ is the time-dependent factor, which is the same in all branches of spectroscopy. We have to evaluate the time-independent factor:

$$\frac{1}{\hbar^2}|\langle \phi_f | \hat{V}(x, y, z) | \phi_i \rangle|^2$$

In the case of NMR spectroscopy the operator $\hat{V}(x, y, z)$ takes the form:

$$-(\gamma \hbar B_{1x})\langle \Psi_f | \hat{I}_x | \Psi_i \rangle \tag{B9.1.1}$$

so that we have to calculate:

$$\frac{1}{\hbar^2}(\gamma \hbar B_{1x})^2 |\langle \Psi_f | \hat{I}_x | \Psi_i \rangle|^2 = (\gamma B_{1x})^2 |\langle \Psi_f | \hat{I}_x | \Psi_i \rangle|^2 \tag{B9.1.2}$$

where B_{1x} is the magnetic field of the applied radiation which oscillates in the x-direction and $\hat{I}_x$ is the operator for the x-component of the nuclear spin. Equation (B9.1.1) as the form of the link between the initial and final states is understandable if we first recall (Equation (9.1.7)) that the energy of interaction between a nuclear magnetic moment and an applied field is given by the product of the nuclear moment, the applied field and the cosine of the angle between them. Therefore, if a field is applied along the x-direction the interaction will be determined by the x-component of the nuclear magnetic moment. We can express this interaction with equations like (9.1.6) and (9.1.8) but with z replaced by x. That $\hat{I}_x$ links nuclear spin states having $m_I + \frac{1}{2}$ and $-\frac{1}{2}$ can be readily demonstrated by expressing $\hat{I}_x$ in terms the raising and lowering operators of Chapter 4. We have:

$$\hat{I}_+ = \hat{I}_x + i\hat{I}_y \quad \text{and} \quad \hat{I}_- = \hat{I}_x - i\hat{I}_y \tag{B9.1.3}$$

which, on addition, give:

$$\hat{I}_x = \frac{1}{2}(\hat{I}_+ + \hat{I}_-) \tag{B9.1.4}$$

The effect of operating on an angular momentum eigenfuction $|I, m_I\rangle$ with $\hat{I}_+$ and $\hat{I}_-$ is given by Equations (4.7.5) and (4.7.6):

$$\hat{I}_+|I, m_I\rangle = \sqrt{[I(I+1) - m_I(m_I+1)]} \cdot \hbar \cdot |I, m_I + 1\rangle \tag{B9.1.5}$$

$$\hat{I}_-|I, m_I\rangle = \sqrt{[I(I+1) - m_I(m_I-1)]} \cdot \hbar \cdot |I, m_I - 1\rangle \tag{B9.1.6}$$

Inserting the appropriate values of I $(\frac{1}{2})$ and m_I $(\pm\frac{1}{2})$, we easily find that:

$$\hat{I}_+|\tfrac{1}{2}, +\tfrac{1}{2}\rangle = 0 \quad \text{and} \quad \hat{I}_+|\tfrac{1}{2}, -\tfrac{1}{2}\rangle = |\tfrac{1}{2}, +\tfrac{1}{2}\rangle$$

while

$$\hat{I}_-|\tfrac{1}{2}, +\tfrac{1}{2}\rangle = |\tfrac{1}{2}, -\tfrac{1}{2}\rangle \quad \text{and} \quad \hat{I}_-|\tfrac{1}{2}, -\tfrac{1}{2}\rangle = 0$$

so that

$$\hat{I}_x|\tfrac{1}{2}, +\tfrac{1}{2}\rangle = \tfrac{1}{2}|\tfrac{1}{2}, -\tfrac{1}{2}\rangle \quad \text{and} \quad \hat{I}_x|\tfrac{1}{2}, -\tfrac{1}{2}\rangle = \tfrac{1}{2}|\tfrac{1}{2}, +\tfrac{1}{2}\rangle \tag{B9.1.7}$$

Note that $\hbar$ has not been included in these expressions because we have γ, which is the factor which converts angular momentum in units of $\hbar$ into magnetic moment (see Section 9.1).

We can now evaluate any matrix element in Equation (B9.1.2):

$$(\gamma B_{1x})^2 |\langle \tfrac{1}{2}, -\tfrac{1}{2} | \hat{I}_x | \tfrac{1}{2}, +\tfrac{1}{2} \rangle|^2 = (\gamma B_{1x})^2 \cdot \tfrac{1}{4} |\langle \tfrac{1}{2}, -\tfrac{1}{2} | \tfrac{1}{2}, -\tfrac{1}{2} \rangle|^2 = (\tfrac{1}{2} \gamma B_{1x})^2$$

[More sophisticated analyses of this problem include a line-shape function in the expression for the transition probability. This introduces a multiplicative factor of 2π and a δ function into the above equation, but that need not concern us here.]

We see that the interaction of an e-m wave with its magnetic field vibrating in the x-direction converts states from $m_I = +\tfrac{1}{2}$ to $m_I = -\tfrac{1}{2}$, or vice versa, with equal facility; i.e. both processes are equally probable. And, because of the form of Equations (B9.1.5) and (B9.1.6), $\hat{I}_x$ links any two spin states whose m_I values differ by ± 1, e.g. $+\tfrac{1}{2}$ is linked to $-\tfrac{1}{2}$ and $+\tfrac{3}{2}$ and 0 to $+1$ and -1. But note that the matrix elements for the two possible links are not, in general, equal.

Thus far we have focused our attention upon a single spin but the spin functions of interest in NMR spectroscopy are eigenstates of sets of coupled nuclei and we require the transition probabilities between such states. First we note that, since the operator $\hat{I}_{rx}$ operates solely on the spin of nucleus r, the operator required for a molecule containing several NMR-active nuclei is $\sum_r \hat{I}_{rx}$. Each term of the sum can link only initial and final eigenfunctions, which differ in the m_I value of nucleus r, while the m_I values of all other nuclei are the same. Thus $\langle +\tfrac{1}{2} + \tfrac{1}{2} + \tfrac{1}{2} | \sum_r \hat{I}_{rx} | -\tfrac{1}{2} + \tfrac{1}{2} + \tfrac{1}{2} \rangle \neq 0$ by virtue of $\hat{I}_{1x}$ but $\langle +\tfrac{1}{2} + \tfrac{1}{2} - \tfrac{1}{2} | \sum_r \hat{I}_{rx} | -\tfrac{1}{2} + \tfrac{1}{2} + \tfrac{1}{2} \rangle = 0$ because, for example, $\langle +\tfrac{1}{2} | \hat{I}_{1x} | -\tfrac{1}{2} \rangle \langle +\tfrac{1}{2} | +\tfrac{1}{2} \rangle \langle -\tfrac{1}{2} | +\tfrac{1}{2} \rangle = 0$ on account of the orthogonality of the spin functions of nucleus 3.

Because of the many and complex factors that determine the intensity of an NMR signal, e.g. relaxation rates, intensity of the exciting radiation, strength of the static magnetic field, it is very rare that anything more than relative intensity values are required and for this it is sufficient to calculate only the matrix elements $|\langle \Psi_f | \hat{I}_x | \Psi_i \rangle|^2$. But the relative intensities are vitally important to spectral interpretation, especially in the case of second order spectra, and recourse to them is frequently made.

As a final comment on the subject of the NMR selection rule we note that the transition from $m_I = +\tfrac{1}{2}$ to $m_I = -\tfrac{1}{2}$ involves a decrease of one $(h/2\pi)$ unit of the z-component of the angular momentum. Therefore, since the z-component of angular momentum must be conserved, we require a right circularly polarised photon to effect it. In an NMR spectrometer we irradiate the sample with linearly rather than circularly polarised radiation. This is not essentially different because linearly polarised radiation can always be regarded as a sum of left and right circularly polarised beams (see Section 8.2). We have taken account of this in the development above by expressing $\hat{I}_x$ in terms of $\hat{I}_+$ and $\hat{I}_-$.

BOX 9.2   Equivalent nuclei

The resonating nuclei in a molecule can frequently be divided into groups of equivalent nuclei. This simplifies the analysis of the NMR spectrum and makes possible certain generalisations about the spectrum. The nuclei must be of the same isotopic species and, when that condition is satisfied, the sub-division of equivalence into two classes is possible. In what follows we shall consider only protons, though the concept can be extended to other nuclei.

If two or more protons have identical electronic environments, as they will have if they are interchanged by one or more of the symmetry operations of the molecule, then all their properties will be the same and, in particular, they are *isochronous*, i.e. they all resonate at the same frequency. Such protons are said to be *chemically equivalent*. The six protons of benzene and the four of methane are examples. But when we consider their magnetic properties we note that each benzene proton has different coupling constants to the protons which are *ortho, meta* and *para* to it in the molecule while all six H–C–H couplings in methane are the same. The methane protons are said to be *magnetically* equivalent, but the benzene protons are not. To make the distinction clear, some authors describe the methane protons as being *completely* equivalent.

From the theoretical standpoint, the significance of the equivalence classification is that, in spectral analysis, a group of magnetically equivalent protons can be considered as a single *composite particle*. This also reveals why the coupling within a group of equivalent nuclei does not result in a splitting of spectral lines. The argument goes as follows.

If the spins of a group of magnetically equivalent nuclei are coupled, the resulting total spin angular momenta and z-components can be found by exactly the same methods as we have used elsewhere in this book for electrons. If we take the three protons, H_a, H_b and H_c of a methyl group as our example then, just as with three electrons and in accord with the Clebsch–Gordan rule, the coupled system gives rise to a quartet state and two doublets. The eigenfunctions of these states can be obtained by applying the lowering operator to the quartet function having $M_I = \frac{3}{2}$ just as in Appendix 10, i.e. we evaluate both sides of the equation:

$$\hat{I}_-|\tfrac{3}{2}, +\tfrac{3}{2}\rangle = \sum_r \hat{I}_{r-}|\tfrac{1}{2}, +\tfrac{1}{2}; \tfrac{1}{2}, +\tfrac{1}{2}; \tfrac{1}{2}, +\tfrac{1}{2}\rangle$$

(The spins of the three protons on the right-hand side of the above equation are written in the order a, b, c). We continue down the ladder of M_I values until we reach

$$|\tfrac{3}{2}, -\tfrac{3}{2}\rangle = |\tfrac{1}{2}, -\tfrac{1}{2}; \tfrac{1}{2} - \tfrac{1}{2}; \tfrac{1}{2}, -\tfrac{1}{2}\rangle.$$

We find, quoting only the m_I values of the individual spins:

Spin eigenfunction	M_I	Energy
A quartet, $I = \frac{3}{2}$		
$\|+\tfrac{1}{2} + \tfrac{1}{2} + \tfrac{1}{2}\rangle$	$+\tfrac{3}{2}$	$-3v/2 + 3J/4$
$(1/\sqrt{3})\{\|+\tfrac{1}{2} + \tfrac{1}{2} - \tfrac{1}{2}\rangle + \|+\tfrac{1}{2} - \tfrac{1}{2} + \tfrac{1}{2}\rangle + \|-\tfrac{1}{2} + \tfrac{1}{2} + \tfrac{1}{2}\rangle\}$	$+\tfrac{1}{2}$	$-v/2 + 3J/4$
$(1/\sqrt{3})\{\|-\tfrac{1}{2} - \tfrac{1}{2} + \tfrac{1}{2}\rangle + \|-\tfrac{1}{2} + \tfrac{1}{2} - \tfrac{1}{2}\rangle + \|+\tfrac{1}{2} - \tfrac{1}{2} - \tfrac{1}{2}\rangle\}$	$-\tfrac{1}{2}$	$+v/2 + 3J/4$
$\|-\tfrac{1}{2} - \tfrac{1}{2} - \tfrac{1}{2}\rangle$	$-\tfrac{3}{2}$	$+3v/2 + 3J/4$
A doublet, $I = \frac{1}{2}$		
$(1/\sqrt{2})\{\|+\tfrac{1}{2} - \tfrac{1}{2} + \tfrac{1}{2}\rangle - \|-\tfrac{1}{2} + \tfrac{1}{2} + \tfrac{1}{2}\rangle\}$	$+\tfrac{1}{2}$	$-v/2 - 3J/4$

Spin eigenfunction	M_I	Energy
$(1/\sqrt{2})\{\lvert-\tfrac{1}{2}+\tfrac{1}{2}-\tfrac{1}{2}\rangle - \lvert+\tfrac{1}{2}-\tfrac{1}{2}-\tfrac{1}{2}\rangle\}$	$-\tfrac{1}{2}$	$+\nu/2 - 3J/4$

A second doublet, $I = \tfrac{1}{2}$

$(1/\sqrt{6})\{2\lvert+\tfrac{1}{2}+\tfrac{1}{2}-\tfrac{1}{2}\rangle - \lvert+\tfrac{1}{2}-\tfrac{1}{2}+\tfrac{1}{2}\rangle - \lvert-\tfrac{1}{2}+\tfrac{1}{2}+\tfrac{1}{2}\rangle\}$	$+\tfrac{1}{2}$	$-\nu/2 - 3J/4$
$(1/\sqrt{6})\{2\lvert-\tfrac{1}{2}-\tfrac{1}{2}+\tfrac{1}{2}\rangle - \lvert-\tfrac{1}{2}+\tfrac{1}{2}-\tfrac{1}{2}\rangle - \lvert+\tfrac{1}{2}-\tfrac{1}{2}-\tfrac{1}{2}\rangle\}$	$-\tfrac{1}{2}$	$+\nu/2 - 3J/4$

It is convenient to include the energies of the spin eigenfunctions here. Examples of how they may be calculated are described later in the Box.

If we define an operator, $\hat{F}_A$, for the total spin of the group of the three magnetically equivalent protons:

$$\hat{F}_A = \sum_r \hat{\mathbf{I}}_r, \qquad (B9.2.1)$$

we can write the Hamiltonian operator (Equation (9.6.1)):

$$h^{-1}\hat{\mathcal{H}} = -\sum_r \nu_r \hat{I}_{rz} + \sum_{r<s} J_{rs}\hat{\mathbf{I}}_r \cdot \hat{\mathbf{I}}_s \qquad (9.6.1)$$

in the form:

$$h^{-1}\hat{\mathcal{H}} = -\nu \hat{F}_{Az} + J\{\hat{\mathbf{I}}_a \cdot \hat{\mathbf{I}}_b + \hat{\mathbf{I}}_b \cdot \hat{\mathbf{I}}_c + \hat{\mathbf{I}}_c \cdot \hat{\mathbf{I}}_a\}$$

where ν is the resonance frequency of the three protons and J is the H–C–H coupling constant.

$$= -\nu \hat{F}_{Az} + J\{\tfrac{1}{2}[\hat{\mathbf{I}}_a + \hat{\mathbf{I}}_b + \hat{\mathbf{I}}_c]^2 - \tfrac{1}{2}[\hat{\mathbf{I}}_a^2 + \hat{\mathbf{I}}_b^2 + \hat{\mathbf{I}}_c^2]\}$$

$$= -\nu \hat{F}_{Az} + J\{\tfrac{1}{2}\hat{F}_A^2 - \tfrac{1}{2}[\tfrac{3}{4} + \tfrac{3}{4} + \tfrac{3}{4}]\}$$

$$= -\nu \hat{F}_{Az} + \tfrac{1}{2}J\{\hat{F}_A^2 - \tfrac{9}{4}\}$$

This development can easily be generalised. For n nuclei with $I = \tfrac{1}{2}$ for example, we have:

$$h^{-1}\hat{\mathcal{H}} = -\nu \hat{F}_{Az} + \tfrac{1}{2}J\{\hat{F}_A^2 - 3n/4\} \qquad (B9.2.2)$$

Thus, the energy depends only on the total spin of the methyl group, F_A, and its z-component, F_{Az}. And the spin states are eigenfunctions of these two commuting, angular momentum operators with eigenvalues $[F_A(F_A + 1)]^{\frac{1}{2}}$ and M_A. But NMR transitions are allowed only between states with the same value of F_A^2. To see why this is so we use Equation (B9.1.4) of Box 9.1 where the NMR transition moment operator, $\hat{I}_x$, is expressed as a sum of the corresponding raising and lowering operators:

$$\hat{I}_x = \tfrac{1}{2}(\hat{I}_+ + \hat{I}_-)$$

When applied to the eigenfunctions of the operator $\hat{F}_A$, these two operators simply move us up or down the ladder of the M_I values of a single value of F_A, so an NMR transition can never change the value of F_A. But, though they differ in energy in an applied magnetic field, all the $2F_A + 1$ components of a state of total spin F_A are equally spaced and all have the same spin–spin coupling energy which therefore does not change in such a transition. Thus, the coupling has no visible effect on the NMR spectrum.

We shall now look a little more closely at the energies involved. It is a useful exercise in basic quantum-mechanical calculation to determine the energy matrix of the three methyl protons. The on-diagonal matrix elements can be calculated (in frequency units) using Equation (9.6.1), the first term of which gives the contribution of the chemical shift and the second that of the coupling constants.

As an example we first evaluate a diagonal element:

$$\langle +\tfrac{1}{2} + \tfrac{1}{2} - \tfrac{1}{2}| - \sum_r \nu_r \hat{I}_{rz} + \sum_{r<s} J_{rs} \hat{\mathbf{I}}_r \cdot \hat{\mathbf{I}}_s |+\tfrac{1}{2} + \tfrac{1}{2} - \tfrac{1}{2}\rangle$$

The chemical-shift term gives:

$$- \langle +\tfrac{1}{2} + \tfrac{1}{2} - \tfrac{1}{2}|\nu_a \hat{I}_{az} + \nu_b \hat{I}_{bz} + \nu_c \hat{I}_{cz}|+\tfrac{1}{2} + \tfrac{1}{2} - \tfrac{1}{2}\rangle$$

$$= -\nu_a \langle +\tfrac{1}{2}|\hat{I}_{az}|+\tfrac{1}{2}\rangle\langle +\tfrac{1}{2}|+\tfrac{1}{2}\rangle\langle -\tfrac{1}{2}|-\tfrac{1}{2}\rangle - \nu_b \langle +\tfrac{1}{2}|+\tfrac{1}{2}\rangle\langle +\tfrac{1}{2}|\hat{I}_{bz}|+\tfrac{1}{2}\rangle\langle -\tfrac{1}{2}|-\tfrac{1}{2}\rangle$$

$$\quad - \nu_c \langle +\tfrac{1}{2}|+\tfrac{1}{2}\rangle\langle +\tfrac{1}{2}|+\tfrac{1}{2}\rangle\langle -\tfrac{1}{2}|\hat{I}_{cz}|-\tfrac{1}{2}\rangle$$

$$= -\nu_a \langle +\tfrac{1}{2}|+\tfrac{1}{2}|+\tfrac{1}{2}\rangle\langle +\tfrac{1}{2}|+\tfrac{1}{2}\rangle\langle -\tfrac{1}{2}|-\tfrac{1}{2}\rangle - \nu_b \langle +\tfrac{1}{2}|+\tfrac{1}{2}\rangle\langle +\tfrac{1}{2}|+\tfrac{1}{2}|+\tfrac{1}{2}\rangle\langle -\tfrac{1}{2}|-\tfrac{1}{2}\rangle$$

$$\quad - \nu_c \langle +\tfrac{1}{2}|+\tfrac{1}{2}\rangle\langle +\tfrac{1}{2}|+\tfrac{1}{2}\rangle\langle -\tfrac{1}{2}|-\tfrac{1}{2}|-\tfrac{1}{2}\rangle$$

$$= -\tfrac{1}{2}(\nu_a + \nu_b - \nu_c) = -\tfrac{1}{2}\nu$$

if the protons are equivalent.

The coupling term requires a little more thought. We first replace $\hat{\mathbf{I}}_r \cdot \hat{\mathbf{I}}_s$ with the equivalent expression using the raising and lowering operators Equation (9.5.1):

$$J_{rs}\hat{\mathbf{I}}_r \cdot \hat{\mathbf{I}}_s = J_{rs}[\hat{I}_{rx}\hat{I}_{sx} + \hat{I}_{ry}\hat{I}_{sy} + \hat{I}_{rz}\hat{I}_{sz}] = J_{rs}[\hat{I}_{rz}\hat{I}_{sz} + \tfrac{1}{2}(\hat{I}_{r+}\hat{I}_{s-} + \hat{I}_{r-}\hat{I}_{s+})] \quad (9.5.1)$$

The term in Equation (9.5.1) which contains raising and lowering operators cannot give an on-diagonal matrix element because any operation with it changes the spins of two of the three nuclei and the spin functions are orthogonal. But the term $J_{rs}\hat{I}_{rz}\hat{I}_{sz}$ can give rise to on-diagonal elements because the spin functions are unchanged by it and for our particular example we find:

$$\langle +\tfrac{1}{2} + \tfrac{1}{2} - \tfrac{1}{2}| \sum_{r<s} J_{rs}\hat{I}_{rz}\hat{I}_{sz}|+\tfrac{1}{2} + \tfrac{1}{2} - \tfrac{1}{2}\rangle$$

$$= \langle +\tfrac{1}{2} + \tfrac{1}{2} - \tfrac{1}{2}|J_{ab}\hat{I}_{az}\hat{I}_{bz} + J_{bc}\hat{I}_{bz}\hat{I}_{cz} + J_{ca}\hat{I}_{cz}\hat{I}_{az}|+\tfrac{1}{2} + \tfrac{1}{2} - \tfrac{1}{2}\rangle$$

$$= \langle +\tfrac{1}{2} + \tfrac{1}{2}|J_{ab}\hat{I}_{az}\hat{I}_{bz}|+\tfrac{1}{2} + \tfrac{1}{2}\rangle\langle -\tfrac{1}{2}|-\tfrac{1}{2}\rangle$$

$$\quad + \langle +\tfrac{1}{2} - \tfrac{1}{2}|J_{bc}\hat{I}_{bz}\hat{I}_{cz}|+\tfrac{1}{2} - \tfrac{1}{2}\rangle\langle +\tfrac{1}{2}|+\tfrac{1}{2}\rangle$$

$$\quad + \langle +\tfrac{1}{2} - \tfrac{1}{2}|J_{ca}\hat{I}_{cz}\hat{I}_{az}|+\tfrac{1}{2} - \tfrac{1}{2}\rangle\langle +\tfrac{1}{2}|+\tfrac{1}{2}\rangle$$

$$= \tfrac{1}{4}(J_{ab} - J_{bc} - J_{ca}) = -\tfrac{1}{4}J$$

if the protons are equivalent.

We can find all the other on-diagonal matrix elements in the same way. They are:

	Basis state	On-diagonal matrix element of $\hat{\mathcal{H}}$
1.	$\lvert +\frac{1}{2} + \frac{1}{2} + \frac{1}{2} \rangle$	$\frac{1}{2}(-\nu_a - \nu_b - \nu_c) + \frac{1}{4}(J_{ab} + J_{bc} + J_{ca})$
2.	$\lvert +\frac{1}{2} + \frac{1}{2} - \frac{1}{2} \rangle$	$\frac{1}{2}(-\nu_a - \nu_b + \nu_c) + \frac{1}{4}(J_{ab} - J_{bc} - J_{ca})$
3.	$\lvert +\frac{1}{2} - \frac{1}{2} + \frac{1}{2} \rangle$	$\frac{1}{2}(-\nu_a + \nu_b - \nu_c) + \frac{1}{4}(-J_{ab} - J_{bc} + J_{ca})$
4.	$\lvert -\frac{1}{2} + \frac{1}{2} + \frac{1}{2} \rangle$	$\frac{1}{2}(\nu_a - \nu_b - \nu_c) + \frac{1}{4}(-J_{ab} + J_{bc} - J_{ca})$
5.	$\lvert +\frac{1}{2} - \frac{1}{2} - \frac{1}{2} \rangle$	$\frac{1}{2}(-\nu_a + \nu_b + \nu_c) + \frac{1}{4}(-J_{ab} + J_{bc} - J_{ca})$
6.	$\lvert -\frac{1}{2} + \frac{1}{2} - \frac{1}{2} \rangle$	$\frac{1}{2}(\nu_a - \nu_b + \nu_c) + \frac{1}{4}(-J_{ab} - J_{bc} + J_{ca})$
7.	$\lvert -\frac{1}{2} - \frac{1}{2} + \frac{1}{2} \rangle$	$\frac{1}{2}(\nu_a + \nu_b - \nu_c) + \frac{1}{4}(J_{ab} - J_{bc} - J_{ca})$
8.	$\lvert -\frac{1}{2} - \frac{1}{2} - \frac{1}{2} \rangle$	$\frac{1}{2}(\nu_a + \nu_b + \nu_c) + \frac{1}{4}(J_{ab} + J_{bc} + J_{ca})$

The off-diagonal matrix elements can be found in a similar manner. The first term on the right of Equation (9.6.1) operates on only one nucleus and cannot therefore link two nuclei and contribute an off-diagonal matrix element. However, the raising and lowering operators in the coupling term (Equation (9.5.1)) can. Since $\hat{I}_+\lvert +\frac{1}{2}\rangle = 0$, $\hat{I}_+\lvert -\frac{1}{2}\rangle = \lvert +\frac{1}{2}\rangle$, $\hat{I}_-\lvert +\frac{1}{2}\rangle = \lvert -\frac{1}{2}\rangle$ and $\hat{I}_-\lvert -\frac{1}{2}\rangle = 0$ we have:

$$\frac{1}{2}\sum_{r<s} J_{rs}(\hat{I}_{r+}\hat{I}_{s-} + \hat{I}_{r-}\hat{I}_{s+})\lvert +\tfrac{1}{2} + \tfrac{1}{2} - \tfrac{1}{2} \rangle$$

$$= \tfrac{1}{2}\{J_{ab}(\hat{I}_{a+}\hat{I}_{b-} + \hat{I}_{a-}\hat{I}_{b+}) + J_{bc}(\hat{I}_{b+}\hat{I}_{c-} + \hat{I}_{b-}\hat{I}_{c+})$$

$$+ J_{ca}(\hat{I}_{c+}\hat{I}_{a-} + \hat{I}_{c-}\hat{I}_{a+})\}\lvert +\tfrac{1}{2} + \tfrac{1}{2} - \tfrac{1}{2} \rangle$$

$$= \tfrac{1}{2}J_{bc}\lvert +\tfrac{1}{2} - \tfrac{1}{2} + \tfrac{1}{2} \rangle + \tfrac{1}{2}J_{ca}\lvert -\tfrac{1}{2} + \tfrac{1}{2} + \tfrac{1}{2} \rangle$$

The basis functions are orthogonal so this result means that we have off-diagonal matrix elements of the form:

$$\mathcal{H}_{32} = \tfrac{1}{2}\langle +\tfrac{1}{2} - \tfrac{1}{2} + \tfrac{1}{2}\lvert J_{bc}\rvert +\tfrac{1}{2} + \tfrac{1}{2} - \tfrac{1}{2} \rangle = \tfrac{1}{2}J_{bc}$$

and

$$\mathcal{H}_{42} = \tfrac{1}{2}\langle -\tfrac{1}{2} + \tfrac{1}{2} + \tfrac{1}{2}\lvert J_{ca}\rvert +\tfrac{1}{2} + \tfrac{1}{2} - \tfrac{1}{2} \rangle = \tfrac{1}{2}J_{ca}$$

The other off-diagonal elements are:

$$\mathcal{H}_{67} = \tfrac{1}{2}\langle -\tfrac{1}{2} + \tfrac{1}{2} - \tfrac{1}{2}\lvert J_{bc}\rvert -\tfrac{1}{2} - \tfrac{1}{2} + \tfrac{1}{2} \rangle = \tfrac{1}{2}J_{bc},$$

$$\mathcal{H}_{57} = \tfrac{1}{2}\langle +\tfrac{1}{2} - \tfrac{1}{2} - \tfrac{1}{2}\lvert J_{ca}\rvert -\tfrac{1}{2} - \tfrac{1}{2} + \tfrac{1}{2} \rangle = \tfrac{1}{2}J_{ca},$$

$$\mathcal{H}_{34} = \tfrac{1}{2}\langle +\tfrac{1}{2} - \tfrac{1}{2} + \tfrac{1}{2}\lvert J_{ab}\rvert -\tfrac{1}{2} + \tfrac{1}{2} + \tfrac{1}{2} \rangle = \tfrac{1}{2}J_{ab}$$

$$\mathcal{H}_{56} = \tfrac{1}{2}\langle +\tfrac{1}{2} - \tfrac{1}{2} - \tfrac{1}{2}\lvert J_{ab}\rvert -\tfrac{1}{2} + \tfrac{1}{2} - \tfrac{1}{2} \rangle = \tfrac{1}{2}J_{ab}$$

The eigenfunctions and energies

Two eigenfunctions and energies can be written down immediately:

$$\Psi_1 = |+\tfrac{1}{2} + \tfrac{1}{2} + \tfrac{1}{2}\rangle \quad E_1 = \tfrac{1}{2}(-\nu_a - \nu_b - \nu_c) + \tfrac{1}{4}(J_{ab} + J_{bc} + J_{ca})$$

$$\Psi_8 = |-\tfrac{1}{2} - \tfrac{1}{2} - \tfrac{1}{2}\rangle \quad E_8 = \tfrac{1}{2}(\nu_a + \nu_b + \nu_c) + \tfrac{1}{4}(J_{ab} + J_{bc} + J_{ca})$$

To obtain the remaining solutions to our problem we must diagonalise two 3×3 matrices, one of which is:

	$\lvert+\tfrac{1}{2}+\tfrac{1}{2}-\tfrac{1}{2}\rangle$	$\lvert+\tfrac{1}{2}-\tfrac{1}{2}+\tfrac{1}{2}\rangle$	$\lvert-\tfrac{1}{2}+\tfrac{1}{2}+\tfrac{1}{2}\rangle$
$\langle+\tfrac{1}{2}+\tfrac{1}{2}-\tfrac{1}{2}\rvert$	$\tfrac{1}{2}(-\nu_a - \nu_b + \nu_c)$ $+\tfrac{1}{4}(J_{ab} - J_{bc} - J_{ca})$	$\tfrac{1}{2}J_{bc}$	$\tfrac{1}{2}J_{ca}$
$\langle+\tfrac{1}{2}-\tfrac{1}{2}+\tfrac{1}{2}\rvert$	$\tfrac{1}{2}J_{bc}$	$\tfrac{1}{2}(-\nu_a + \nu_b - \nu_c)$ $+\tfrac{1}{4}(-J_{ab} - J_{bc} + J_{ca})$	$\tfrac{1}{2}J_{ab}$
$\langle-\tfrac{1}{2}+\tfrac{1}{2}+\tfrac{1}{2}\rvert$	$\tfrac{1}{2}J_{ca}$	$\tfrac{1}{2}J_{ab}$	$\tfrac{1}{2}(\nu_a - \nu_b - \nu_c)$ $+\tfrac{1}{4}(-J_{ab} + J_{bc} - J_{ca})$

A general solution of this problem is clearly rather difficult and it illustrates well how rapidly the complexity of the analysis of an NMR spectrum increases with the number of spins in the system. A general solution is possible if certain approximations are made and analyses along these lines are described in many texts. We can briefly mention two.

The ABX spectrum

If the chemical shift, i.e. ν value, of one proton, 'c' say, is very different from those of protons 'a' and 'b' and the difference is also much larger than the coupling constants J_{ca} and J_{bc} we have an ABX spin system and the spectrum may be regarded as second order in AB and first order in X (Section 9.6). In this approximation J_{ca} and J_{bc} are regarded as zero where they occur in off-diagonal matrix elements, but not in the on-diagonal, and the matrix above reduces to a 2×2, which is easy to solve, and a single eigenvalue belonging to the spin function $|+\tfrac{1}{2} + \tfrac{1}{2} - \tfrac{1}{2}\rangle$.

The A₃ spectrum

If the three protons are magnetically equivalent the above matrix reduces to:

	$\lvert+\tfrac{1}{2}+\tfrac{1}{2}-\tfrac{1}{2}\rangle$	$\lvert+\tfrac{1}{2}-\tfrac{1}{2}+\tfrac{1}{2}\rangle$	$\lvert-\tfrac{1}{2}+\tfrac{1}{2}+\tfrac{1}{2}\rangle$
$\langle+\tfrac{1}{2}+\tfrac{1}{2}-\tfrac{1}{2}\rvert$	$-\tfrac{1}{2}\nu - \tfrac{1}{4}J$	$\tfrac{1}{2}J$	$\tfrac{1}{2}J$
$\langle+\tfrac{1}{2}-\tfrac{1}{2}+\tfrac{1}{2}\rvert$	$\tfrac{1}{2}J$	$-\tfrac{1}{2}\nu - \tfrac{1}{4}J$	$\tfrac{1}{2}J$
$\langle-\tfrac{1}{2}+\tfrac{1}{2}+\tfrac{1}{2}\rvert$	$\tfrac{1}{2}J$	$\tfrac{1}{2}J$	$-\tfrac{1}{2}\nu - \tfrac{1}{4}J$

This matrix is quite simple to diagonalise, but it is much easier to determine the energies of the spin eigenfunctions, which have been obtained using the raising and lowering operators, by the procedure described below. For example, the energy of the component $(1/\sqrt{2})\{|+\frac{1}{2} - \frac{1}{2} + \frac{1}{2}\rangle - |-\frac{1}{2} + \frac{1}{2} + \frac{1}{2}\rangle\}$ of the first spin doublet is:

$$(1/2)\langle\{\{\langle+\frac{1}{2} - \frac{1}{2} + \frac{1}{2}| - \langle-\frac{1}{2} + \frac{1}{2} + \frac{1}{2}|\}|\hat{\mathcal{H}}|\{|+\frac{1}{2} - \frac{1}{2} + \frac{1}{2}\rangle - |-\frac{1}{2} + \frac{1}{2} + \frac{1}{2}\rangle\}\rangle$$

$$= (1/2)\{\langle+\frac{1}{2} - \frac{1}{2} + \frac{1}{2}|\hat{\mathcal{H}}|+\frac{1}{2} - \frac{1}{2} + \frac{1}{2}\rangle + \langle-\frac{1}{2} + \frac{1}{2} + \frac{1}{2}|\hat{\mathcal{H}}|-\frac{1}{2} + \frac{1}{2} + \frac{1}{2}\rangle$$

$$- \langle+\frac{1}{2} - \frac{1}{2} + \frac{1}{2}|\hat{\mathcal{H}}|-\frac{1}{2} + \frac{1}{2} + \frac{1}{2}\rangle - \langle-\frac{1}{2} + \frac{1}{2} + \frac{1}{2}|\hat{\mathcal{H}}|+\frac{1}{2} - \frac{1}{2} + \frac{1}{2}\rangle\}$$

$$= (1/2)\{-\frac{1}{2}\nu - \frac{1}{4}J - \frac{1}{2}\nu - \frac{1}{4}J - \frac{1}{2}J - \frac{1}{2}J\} = -\frac{1}{2}\nu - \frac{3}{4}J$$

To calculate the NMR transition probability, P, between the two states of the first doublet above we must evaluate (Equation (9.7.1)):

$$\langle\sqrt{\tfrac{1}{2}}\{\langle+\frac{1}{2} - \frac{1}{2} + \frac{1}{2}| - \langle-\frac{1}{2} + \frac{1}{2} + \frac{1}{2}|\}|\sum_r \hat{I}_{rx}|\sqrt{\tfrac{1}{2}}\{|-\frac{1}{2} + \frac{1}{2} - \frac{1}{2}\rangle - |+\frac{1}{2} - \frac{1}{2} - \frac{1}{2}\rangle\}\rangle$$

and square the result:

$$= \tfrac{1}{2}\langle+\frac{1}{2} - \frac{1}{2} + \frac{1}{2}|\sum_r \hat{I}_{rx}|-\frac{1}{2} + \frac{1}{2} - \frac{1}{2}\rangle + \tfrac{1}{2}\langle-\frac{1}{2} + \frac{1}{2} + \frac{1}{2}|\sum_r \hat{I}_{rx}|+\frac{1}{2} - \frac{1}{2} - \frac{1}{2}\rangle$$

$$- \tfrac{1}{2}\langle+\frac{1}{2} - \frac{1}{2} + \frac{1}{2}|\sum_r \hat{I}_{rx}|+\frac{1}{2} - \frac{1}{2} - \frac{1}{2}\rangle - \tfrac{1}{2}\langle-\frac{1}{2} + \frac{1}{2} + \frac{1}{2}|\sum_r \hat{I}_{rx}|-\frac{1}{2} + \frac{1}{2} - \frac{1}{2}\rangle$$

$$= \tfrac{1}{4}\{\langle+\frac{1}{2} - \frac{1}{2} + \frac{1}{2}|\sum_r (\hat{I}_{r+} + \hat{I}_{r-})|-\frac{1}{2} + \frac{1}{2} - \frac{1}{2}\rangle$$

$$+ \langle-\frac{1}{2} + \frac{1}{2} + \frac{1}{2}|\sum_r (\hat{I}_{r+} + \hat{I}_{r-})|+\frac{1}{2} - \frac{1}{2} - \frac{1}{2}\rangle$$

$$- \langle+\frac{1}{2} - \frac{1}{2} + \frac{1}{2}|\sum_r (\hat{I}_{r+} + \hat{I}_{r-})|+\frac{1}{2} - \frac{1}{2} - \frac{1}{2}\rangle$$

$$- \langle-\frac{1}{2} + \frac{1}{2} + \frac{1}{2}|\sum_r (\hat{I}_{r+} + \hat{I}_{r-})|-\frac{1}{2} + \frac{1}{2} - \frac{1}{2}\rangle\}$$

Only the third and fourth of the above four integrals are non-zero. To illustrate their evaluation consider the third:

$$\langle+\frac{1}{2} - \frac{1}{2} + \frac{1}{2}|\sum_r (\hat{I}_{r+} + \hat{I}_{r-})|+\frac{1}{2} - \frac{1}{2} - \frac{1}{2}\rangle$$

$$= \langle+\frac{1}{2} - \frac{1}{2} + \frac{1}{2}|(\hat{I}_{a+} + \hat{I}_{a-})|+\frac{1}{2} - \frac{1}{2} - \frac{1}{2}\rangle$$

$$+ \langle+\frac{1}{2} - \frac{1}{2} + \frac{1}{2}|(\hat{I}_{b+} + \hat{I}_{b-})|+\frac{1}{2} - \frac{1}{2} - \frac{1}{2}\rangle$$

$$+ \langle+\frac{1}{2} - \frac{1}{2} + \frac{1}{2}|(\hat{I}_{c+} + \hat{I}_{c-})|+\frac{1}{2} - \frac{1}{2} - \frac{1}{2}\rangle$$

$$= \langle+\frac{1}{2}|(\hat{I}_{a+} + \hat{I}_{a-})|+\frac{1}{2}\rangle\langle-\frac{1}{2}|-\frac{1}{2}\rangle\langle+\frac{1}{2}|-\frac{1}{2}\rangle$$

$$+ \langle+\frac{1}{2}|+\frac{1}{2}\rangle\langle-\frac{1}{2}|(\hat{I}_{b+} + \hat{I}_{b-})|-\frac{1}{2}\rangle\langle+\frac{1}{2}|-\frac{1}{2}\rangle$$

$$+ \langle+\frac{1}{2}|+\frac{1}{2}\rangle\langle-\frac{1}{2}|-\frac{1}{2}\rangle\langle+\frac{1}{2}|(\hat{I}_{c+} + \hat{I}_{c-})|-\frac{1}{2}\rangle$$

Of these three expressions, six terms in total, only the last is non-zero. The others are zero because of the orthogonality of the spin functions. The term containing $\hat{I}_{c+}$ in the last evaluates as:

$$\langle +\tfrac{1}{2}|+\tfrac{1}{2}\rangle \langle -\tfrac{1}{2}|-\tfrac{1}{2}\rangle \langle +\tfrac{1}{2}|+\tfrac{1}{2}\rangle = 1$$

The fourth term also contributes 1 so that the total transition probability is proportional to $\{\tfrac{1}{4}(-1-1)\}^2 = \tfrac{1}{4}$.

PROBLEMS FOR CHAPTER 9

1. Use the information in Box 9.1 to show that transitions between the singlet state of two identical protons, $(1/\sqrt{2})\{|+\tfrac{1}{2} - \tfrac{1}{2}\rangle - |-\tfrac{1}{2} + \tfrac{1}{2}\rangle\}$, and the triplet states, $|+\tfrac{1}{2} + \tfrac{1}{2}\rangle$, $|-\tfrac{1}{2} - \tfrac{1}{2}\rangle$ and $(1/\sqrt{2})\{|+\tfrac{1}{2} - \tfrac{1}{2}\rangle + |-\tfrac{1}{2} + \tfrac{1}{2}\rangle\}$, are forbidden. That is, show that $\langle \Psi_S|\hat{I}_x|\Psi_T\rangle = 0$.

2. Show that for a system of n nuclei the matrix of the Hamiltonian operator given in Equation (9.6.1) can have off-diagonal matrix elements only between basis states which have the same total z-component of the nuclear spin, i.e. have the same value of $m_{I1} + m_{I2} + m_{I3} + \cdots + m_{In}$.

3. Starting from the matrix in Section 9.6 and the general method for diagonalising a 2×2 matrix from Appendix 3, show that the energies and wave functions of a system of two nuclei, 'a' and 'b', each with $I = \tfrac{1}{2}$ are ($J_{ab} \to J$ to simplify the notation):

Wave function	Energy (as frequency)
$\Psi_1 = |+\tfrac{1}{2} + \tfrac{1}{2}\rangle$	$-\tfrac{1}{2}[\nu_a + \nu_b] + \tfrac{1}{4}J$
$\Psi_2 = -\sin\omega\,|+\tfrac{1}{2} - \tfrac{1}{2}\rangle + \cos\omega\,|-\tfrac{1}{2} + \tfrac{1}{2}\rangle$	$-\tfrac{1}{2}\{[\nu_a - \nu_b]^2 + J^2\}^{\tfrac{1}{2}} - \tfrac{1}{4}J$
$\Psi_3 = +\cos\omega\,|+\tfrac{1}{2} - \tfrac{1}{2}\rangle + \sin\omega\,|-\tfrac{1}{2} + \tfrac{1}{2}\rangle$	$+\tfrac{1}{2}\{[\nu_a - \nu_b]^2 + J^2\}^{\tfrac{1}{2}} - \tfrac{1}{4}J$
$\Psi_4 = |-\tfrac{1}{2} - \tfrac{1}{2}\rangle$	$+\tfrac{1}{2}[\nu_a + \nu_b] + \tfrac{1}{4}J$

Where

$$\sin\omega = \left[\frac{\tfrac{1}{2}\{[\nu_a - \nu_b]^2 + J^2\}^{\tfrac{1}{2}} + \tfrac{1}{2}[\nu_a - \nu_b]}{\{[\nu_a - \nu_b]^2 + J^2\}^{\tfrac{1}{2}}}\right]^{\tfrac{1}{2}}$$

and

$$\cos\omega = \left[\frac{\tfrac{1}{2}\{[\nu_a - \nu_b]^2 + J^2\}^{\tfrac{1}{2}} - \tfrac{1}{2}[\nu_a - \nu_b]}{\{[\nu_a - \nu_b]^2 + J^2\}^{\tfrac{1}{2}}}\right]^{\tfrac{1}{2}}$$

Show that the four lines of the AB spectrum have frequencies of $\pm\tfrac{1}{2}(D \pm J)$ relative to the mean frequency $\tfrac{1}{2}[\nu_a + \nu_b]$, where $D = \{[\nu_a - \nu_b]^2 + J^2\}^{\tfrac{1}{2}}$.

Show that the intensities of the four possible transitions are proportional to:

Transition	Intensity	Transition	Intensity
$\Psi_1 \leftrightarrow \Psi_2$	$\frac{1}{4}(1 - \sin 2\omega)$	$\Psi_4 \leftrightarrow \Psi_2$	$\frac{1}{4}(1 - \sin 2\omega)$
$\Psi_1 \leftrightarrow \Psi_3$	$\frac{1}{4}(1 + \sin 2\omega)$	$\Psi_4 \leftrightarrow \Psi_3$	$\frac{1}{4}(1 + \sin 2\omega)$

4. The nucleus ^{14}N has $I = 1$ and $\gamma = 1.934 \times 10^7$ $T^{-1} s^{-1}$. The corresponding data for ^1H are $I = \frac{1}{2}$ and $\gamma = 26.75 \times 10^7$ $T^{-1} s^{-1}$. Use Equation (9.6.1) to construct the 6×6 matrix analogous to that in Section 9.6 for the nuclear spin energy levels of the molecule N–H in a magnetic field. [You should find that the matrix is blocked out into two 1×1 and two 2×2 matrices with off diagonal elements of $\sqrt{2}J$.] J, the $^{14}N-^1H$ coupling constant, has not been reported for N–H, but from other measurements of the coupling between directly bonded N and H, e.g. $[NH_4]^+$, it is expected to be about 50 Hz.

Assuming that the spectrum is first order, calculate the transition probabilities and predict the ^1H NMR spectrum of the molecule.

Using the values of the magnetogyric ratios and J given above and a magnetic field for which protons resonate in the region of 100 MHz, show that the assumption that the spectrum is first order is fully justified.

CHAPTER 10

Infrared Spectroscopy

10.0	Introduction	289
10.1	The origin of the infrared spectra of molecules	290
10.2	Simple harmonic motion	290
10.3	The quantum-mechanical harmonic oscillator	293
	10.3.1 Quantisation of the energy	293
	10.3.2 Zero-point energy	294
	10.3.3 Vibrational eigenfunctions	294
10.4	Rotation of a diatomic molecule	294
	10.4.1 Eigenfunctions of the rigid rotator	297
10.5	Selection rules for vibrational and rotational transitions	297
	10.5.1 A semi-classical view of the selection rules	301
	10.5.2 Infrared intensities	301
10.6	Real diatomic molecules	302
10.7	Polyatomic molecules	303
	10.7.1 Normal co-ordinates, normal vibrations, vibrational eigenfunctions and eigenvalues	303
	10.7.2 Vibrations of real polyatomic molecules	305
	10.7.3 Characteristic group frequencies	308
	10.7.4 Large molecules	308
10.8	Anharmonicity	309
	10.8.1 Fermi resonance	309
	10.8.2 Vibrational angular momentum and the Coriolis interaction	311
10.9	The *ab-initio* calculation of IR spectra	316
10.10	The special case of near infrared spectroscopy	317
10.11	Bibliography and further reading	317
	Problems for Chapter 10	324

10.0 INTRODUCTION

There are numerous books dealing, in whole or in part, with the theory and practice of infrared (IR) spectroscopy[1-3] and an authoritative account of the historical development of the subject has recently been given by Sheppard.[4] Brand's[5] historical review of dispersive spectroscopy also contains much of theoretical as well as practical interest. In keeping with

The Quantum in Chemistry   R. Grinter
© 2005 John Wiley & Sons, Ltd

the title of the book, the purpose of the present chapter is to highlight those aspects of IR spectroscopy in which quantisation, and hence quantum mechanics, plays an important role. We shall find that the theory of IR spectroscopy leans quite heavily on classical concepts and classical mechanics.

It is difficult to pin-point the first application of wave mechanics to the harmonic oscillator problem, since the theory of oscillation has been inextricably entwined with the classical and quantum-mechanical description of radiation since before the turn of the 20th century until the present day. It is clear, however, that the first application of matrix mechanics to the one-dimensional oscillator was published by Max Born, Werner Heisenberg and Ernst P. Jordan (1902–1980) in 1925. The matrix mechanics of the symmetric rotor was first described by David M. Dennison (1900–1976) in 1926; the wave mechanics of the same system by Fritz Reiche (1883–1962) and, independently, by Ralph de L. Kronig (1904–) and Isador I. Rabi (1898–1988) in 1927.

10.1   THE ORIGIN OF THE INFRARED SPECTRA OF MOLECULES

If we have a group of N atoms which are not bonded together then each can move, independently of the others, in one of three directions, i.e. in the x-, y- or z-direction, and the N atoms are said to possess 3N *degrees of freedom*. If these N atoms are bonded together, then the molecule as a whole can move in three directions (translation) and it can also rotate about axes directed along x, y and z. (If the molecule is linear it can only rotate about the two directions perpendicular to the line of atoms since a rotation of the line itself implies no movement of any of the atomic masses.) Thus, a non-linear molecule has three degrees of translational freedom and three degrees of rotational freedom and, since degrees of freedom cannot be destroyed, the remaining $3N-6$ ($3N-5$ for a linear molecule) degrees of freedom must be vibrational, i.e. they must be movements of the atoms with respect to each other in which the position of the centre of mass of the molecule remains stationary. The most important vibrations from the point of view of chemical spectroscopy are those which involve the changing of bond lengths and angles since these properties are directly related to molecular structure and the strengths of chemical bonds. But a molecule obeys the laws of quantum mechanics. Therefore the energy associated with these vibrations is quantised and when a molecule changes its vibrational energy levels radiation in the infrared region may be emitted or absorbed and the observation of this process, almost invariably in absorption, is IR spectroscopy. In discussing this problem it is helpful first to consider vibration from the point of view of classical mechanics.

10.2   SIMPLE HARMONIC MOTION

A sphere of mass, *m*, attached by two equal, mass-less springs to two immovable walls is shown in Figure 10.1. If the mass is displaced slightly from its position of rest (the equilibrium position) on release it will vibrate about the equilibrium position and, if no energy was dissipated in the springs and the surrounding air etc., this vibrational motion would continue indefinitely. In order to be quantitative, we assume that the *restoring force, F*, which returns the mass to the equilibrium position, x_o, is proportional to its displacement, x, from that position, according to the law discovered by Robert Hooke

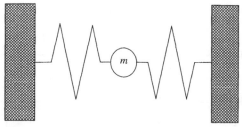

Figure 10.1 A vibrating mass, m

(1635–1703). The constant of proportionality, the *force constant*, is k. Then, since force = mass × acceleration:

$$F = -kx = m\partial^2 x/\partial t^2 \tag{10.2.1}$$

The negative sign indicates that a displacement in the positive x-direction generates a force in the negative x-direction, and vice versa. The solution of Equation (10.2.1) is well known, it is:

$$x = \mathcal{X}\sin([k/m]^{\frac{1}{2}}t + \phi) \equiv \mathcal{X}\sin(\omega_c t + \phi) \equiv \mathcal{X}\sin(2\pi v_c t + \phi) \tag{10.2.2}$$

In Equation (10.2.2) v_c is the classical vibrational frequency, ω_c the corresponding angular frequency, $\mathcal{X}$ is the amplitude of the vibration and ϕ is an angle, known as the *phase* angle, which depends upon what point in the vibrational cycle we choose to call the starting point. The subscript "c" serves to remind us that v_c or ω_c is the *classical* vibrational frequency, which we shall take over into the quantum-mechanical analysis shortly. In terms of the fundamental properties of the oscillator, k and m, the frequency is given by Equation (10.2.3):

$$v_c = (\tfrac{1}{2}\pi)[k/m]^{\frac{1}{2}} \tag{10.2.3}$$

The restoring force is the negative derivative of the potential energy, V, and therefore:

$$-kx = F = -\partial V/\partial x \Rightarrow \partial V = kx\partial x$$

Integration gives:

$$V = \tfrac{1}{2}kx^2 + C$$

so that setting the constant of integration to zero, because we are interested only in relative energies, and using Equation (10.2.3) we have:

$$V = 2\pi^2 m v_c^2 x^2 \tag{10.2.4}$$

Thus, the graph of the potential energy against displacement is a parabola (Figure 10.2). Motion described by Equations (10.2.1) to (10.2.4) is known as simple harmonic motion and our moving mass is a *harmonic oscillator*.

We can extend the above model to something very much like a diatomic molecule if we envisage two masses, m_1 and m_2 connected by a spring of force constant k (Figure 10.3). In this dumb-bell model of a diatomic molecule the two masses move, in phase, away from and towards each other while the centre of mass of the system remains fixed. In order to treat the two masses as one we introduce the concept of reduced mass, μ, which is defined by Equation (10.2.5):

$$\mu = m_1 m_2/(m_1 + m_2) \tag{10.2.5}$$

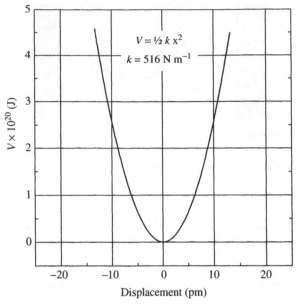

Figure 10.2 Energy versus internuclear distance for a simple harmonic oscillator

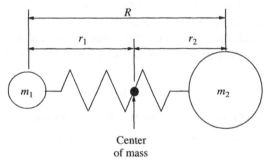

Figure 10.3 The dumb-bell model of a diatomic molecule

Using the reduced mass we obtain (see Box 10.1) equations exactly analogous to Equations (10.2.1) to (10.2.4) above where, in each case, m is replaced by μ. $\mathcal{X}$ and x are replaced by $\mathcal{R}$ and $R - r_e$ respectively, where r_e is the distance between the centres of the two masses at equilibrium and R is the distance at any other time. $\mathcal{R}$ is the amplitude of the oscillation in the sense that the maximum value of R is $r_e + \mathcal{R}$ and the minimum value is $r_e - \mathcal{R}$. Thus, Equation (10.2.2) is replaced by:

$$R - r_e = \mathcal{R}\sin(\omega_c t + \phi) \equiv \mathcal{R}\sin(2\pi v_c t + \phi) \qquad (10.2.6)$$

The change of the bond length over the vibrational cycle is illustrated in Figure 10.4 in a grossly exaggerated form. The rate of change of R is v_R, where:

$$v_R = \partial R/\partial t = \mathcal{R}\omega_c \cos(\omega_c t + \phi) \qquad (10.2.7)$$

and oscillates between the values of zero and the maximum of $\mathcal{R}\omega_c$ as the total energy changes from purely potential to purely kinetic twice during each cycle. Only one

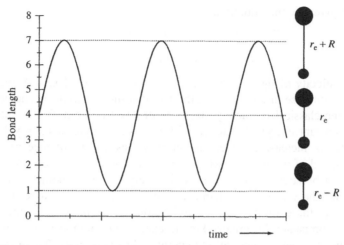

Figure 10.4 The change of bond length with time for a vibrating diatomic molecule

vibrational frequency:

$$\omega_c = 2\pi \nu_c = \sqrt{(k/\mu)} \tag{10.2.8}$$

is possible but the energy associated with it can be changed by changing the amplitude and any energy is possible, including zero: no displacement, masses at rest.

10.3 THE QUANTUM-MECHANICAL HARMONIC OSCILLATOR

The classical energy of the dumb-bell model of a diatomic molecule is (Box 10.1):

$$E = V + T = \tfrac{1}{2}k(R - r_e)^2 + (\mu v_R)^2/2\mu \tag{10.3.1}$$

Transforming to quantum-mechanical operators in the usual way (Chapter 3), the Hamiltonian operator, $\hat{\mathcal{H}}$, is given by Equation (10.3.2):

$$\hat{\mathcal{H}} = \tfrac{1}{2}k(R - r_e)^2 - (h^2/8\pi^2\mu)\partial^2/\partial R^2 \tag{10.3.2}$$

and the Schrödinger equation, $\hat{\mathcal{H}}\Psi = E\Psi$, is:

$$\hat{\mathcal{H}}\Psi = \tfrac{1}{2}k(R - r_e)^2\Psi - (h^2/8\pi^2\mu)\partial^2\Psi/\partial R^2 = E\Psi \tag{10.3.3}$$

As it turns out, this is a tedious equation to solve, but the method of solution is described in detail by Pauling and Wilson[6] and in many other texts. We shall content ourselves with examining the solutions of the equation, i.e. the eigenvalues and eigenfunctions of the harmonic oscillator which is our model for the bond-stretching vibration of a diatomic molecule. The most important differences between classical and quantum mechanics in this problem are described in the remainder of Section 10.3.

10.3.1 Quantisation of the energy

In classical mechanics the energy of the harmonic oscillator can take any value, depending upon the displacement from the equilibrium position which initiates the motion. In quantum mechanics the energy levels are quantised. They are characterised by the quantum number, υ and given by the equation:

$$E_{\upsilon} = (\upsilon + \tfrac{1}{2})h\nu_{c} \quad \text{in energy units} \qquad (10.3.4)$$

$$\approx (\upsilon + \tfrac{1}{2})\nu/c \quad \text{in wave numbers} \qquad (10.3.4a)$$

where c is the velocity of light, $\upsilon = 0, 1, 2, \ldots$. In Equation (10.3.4a), the subscript c, having served its purpose in emphasising the fact that the quantum-mechanical frequency is the same as the classical frequency, has been dropped to simplify the notation.

There is a constant energy difference of $h\nu$ between adjacent energy levels. As the quantum number υ increases, the energy increases as a consequence of the increased amplitude and speed of the motion, while the frequency of oscillation remains the same. This point should be stressed. The increase in energy of a harmonic oscillator, in both the classical- and the quantum-mechanical descriptions, arises as a result of increased potential energy at the extremes of the motion, due to an increase in the amplitude of the vibration, accompanied by greater kinetic energy, i.e. atomic masses moving faster, at the equilibrium geometry where all the energy is kinetic energy. There is no change of frequency. These points are illustrated quantitatively in Box 10.2.

10.3.2 Zero-point energy

The fact that, although υ can be zero, the energy can never be lower than $\tfrac{1}{2}h\nu$ is a remarkable departure from the classical situation where an energy of zero is possible. This quantity of energy, which the quantum-mechanical oscillator cannot lose, is known as the *zero-point energy*. It is responsible for many of the cases where the properties of solids differ from those predicted by classical mechanics, especially at very low temperatures. Note also that it is a necessary consequence of the Heisenberg uncertainty principle (Section 3.10). If the two masses forming the oscillator were stationary with a separation of r_{e} we would know the positions and the momenta (zero) of each of them simultaneously, and that is not allowed.

10.3.3 Vibrational eigenfunctions

The final important difference between the classical and the quantum-mechanical harmonic oscillator concerns the eigenfunctions of the latter (Figure 10.5). The probability of finding the oscillator with a particular distance, R, between the masses is a rather complicated function (Box 10.3). This is compared with the classical function for $\upsilon = 12$ in Figure 10.6. Whereas the ends of the classical motion are precisely defined, the quantum-mechanical probability function is not and allows the masses to go outside the rigid boundaries which confine the classical oscillator. But the two probability functions become increasingly similar as energy increases Figures 10.5 and 10.6.

10.4 ROTATION OF A DIATOMIC MOLECULE

In Section 10.1 we noted that linear molecules have two degrees of rotational freedom and non-linear molecules three. At this point we must expand a little on these facts

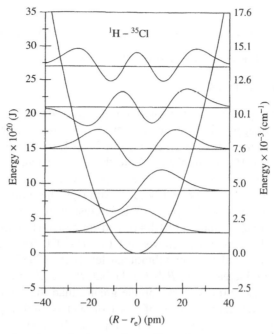

Figure 10.5 The eigenfunctions of a harmonic oscillator for $\nu = 0$ to 4

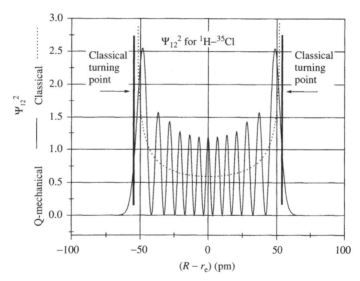

Figure 10.6 A comparison of the classical and quantum-mechanical wave functions of the harmonic oscillator for $\nu = 12$

since they have important consequences for IR spectroscopy. In solution, and especially in the gaseous phase, all molecules rotate about the three (two in the case of linear molecules) mutually perpendicular axes through their centre of mass. The energy, E_J, and angular momentum, A_J, associated with these motions are quantised according to the equations (Box 10.1):

$$E_J = h^2 J(J+1)/8\pi^2 \mu R^2 = h^2 J(J+1)/8\pi^2 I \text{ in energy units} \qquad (10.4.1)$$

$$\approx h J(J+1)/8\pi^2 c I \equiv B J(J+1) \text{ in wave numbers} \qquad (10.4.1a)$$

$$A_J = [J(J+1)]^{\frac{1}{2}} h/2\pi \qquad (10.4.2)$$

where I is the *moment of inertia*, B the *rotational constant* and J a quantum number that can take all positive integer values, including zero. And, since we are dealing here with angular momentum, each level has a degeneracy of $2J+1$, the levels being characterised by the quantum number M_J or simply M. Note that since J can equal zero, the lowest energy state is also zero, in contrast to the zero-point energy of the oscillator. But the uncertainty principle is not contravened because although the atoms are stationary they could be anywhere in their circular paths. It is interesting to compare the energy-level spacing for the vibrational (Equation (10.3.4a)) and rotational (Equation (10.4.1a)) levels of a diatomic molecule. Experiment shows that for ^{1}H–^{35}Cl, for example, the approximately constant spacing of the lower vibrational levels is $\sim$2989.5 cm^{-1}. The spacing of the rotational levels increases as $B J(J+1)$ but $B \approx 10.6$ cm^{-1} so that the vibrational spacing is some two orders of magnitude greater than the rotational (Figure 10.7). Thus the energy-level diagram consists of a set of widely and approximately equally spaced vibrational levels, each of which is accompanied by a set of much more closely, but unequally, spaced rotational levels. Therefore, if the selection rules allow it, the IR absorption band associated with a change of vibrational quantum number will have, closely

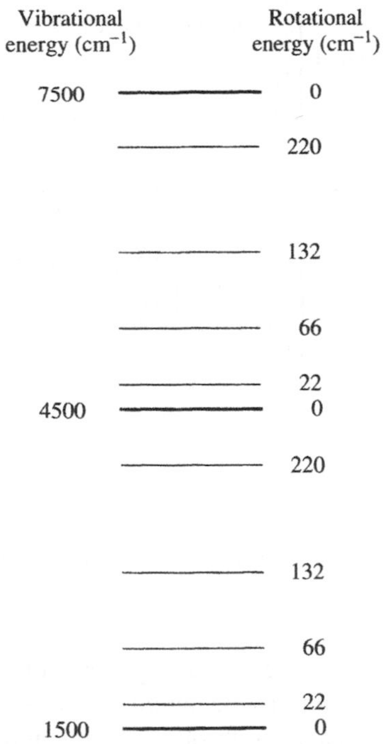

Vibrational energy (cm^{-1})		Rotational energy (cm^{-1})
7500	————	0
	————	220
	————	132
	————	66
	————	22
4500	————	0
	————	220
	————	132
	————	66
	————	22
1500	————	0

Figure 10.7 Lower vibrational and rotational energy levels of ^{1}H–^{35}Cl

spaced on either side of it, a *fine structure* of bands due to changes in rotational quantum number. This fine structure may or may not be resolved. Note that the energy-level scale is purely schematic. The energy of a particular level is the sum of the vibrational and rotational energies and only very high rotational levels of vibrational level $\upsilon = n$ overlap the stack of rotational levels based on the next vibrational level, $\upsilon = n + 1$. But the overlap provides an important pathway for the relaxation of molecules in excited vibrational levels. Note also that the spacing of the levels is inversely proportional to the molecular mass.

10.4.1 Eigenfunctions of the rigid rotator

When the atomic masses are expressed in terms of the reduced mass, the Hamiltonian for the rigid rotator is identical to the kinetic energy operator of the one-electron atom (Box 10.1). The eigenfunctions are the spherical harmonics, $Y_{l,m}(\theta, \phi)$, which are described in detail in Appendix 5. When describing the angular momentum of the rigid rotator it is conventional to replace l and m by J and M (see Section 4.8).

10.5 SELECTION RULES FOR VIBRATIONAL AND ROTATIONAL TRANSITIONS

We assume the idealised model of a *rigid rotator*, where there is no increase in bond length as the speed of rotation increases and hence no interaction between rotation and vibration. In such a model the vibrational and rotational motions are quite independent and we can consider the vibrational and rotational energy levels and their wave functions separately. This assumption, together with the Born-Oppenheimer approximation (Section 6.3), enables us to write the molecular wave functions as products of electronic, vibrational and rotational wave functions:

$$\Psi_{\text{total}} = \psi_{\text{elec}} \cdot \psi_{\text{vib}} \cdot \psi_{\text{rot}} \tag{10.5.1}$$

and the selection rules for the molecule as products of the individual electronic, vibrational and rotational selection rules.

According to Section 8.7, we require the time-independent interaction between the electric field of the radiation, E, and the transition dipole moment between the initial and final rotation-vibration states of the molecule. We start with the first line of Equation (8.7.7), reproduced here as Equation (10.5.2) and use the first line, rather than the second line, of that equation because we wish to introduce molecular rotation which will affect our ability to take the terms in M_0 and M_k' outside the integrations:

$$M_{\text{i,f}} = \langle \psi^{\text{f}}_{\text{vib}} | M_0(e^f e^i) | \psi^{\text{i}}_{\text{vib}} \rangle + \langle \psi^{\text{f}}_{\text{vib}} | \sum_{k=1}^{3N-6} Q_k M_k'(e^f e^i) | \psi^{\text{i}}_{\text{vib}} \rangle \tag{10.5.2}$$

It is useful to make a number of modification to Equation (10.5.2). The superscripts f (= final) and i (= initial) will be replaced by the single (′) and double (″) prime marks respectively; the notation normally used in this subject and the appendage $(e^f e^i)$ will be dropped, since it is assumed throughout this chapter that we are dealing only with the ground electronic state. Also, since we are focussing attention on a diatomic molecule, there is only one normal co-ordinate, the bond stretching represented by r. With these modifications Equation (10.5.2) becomes:

$$M = \langle \psi_{\text{vib}'} \psi_{\text{rot}'} |M_0| \psi_{\text{vib}''} \psi_{\text{rot}''} \rangle + \langle \psi_{\text{vib}'} \psi_{\text{rot}'} |rM'| \psi_{\text{vib}''} \psi_{\text{rot}''} \rangle \qquad (10.5.3)$$

We shall find it important to identify the quantum numbers associated with each of the vibrational (ν-) and rotational (J, M) eigenfunctions, so we now write Equation (10.5.3) in the form:

$$M = \langle \psi_{\nu'} \psi_{J',M'} |M_0| \psi_{\nu''} \psi_{J'',M''} \rangle + \langle \psi_{\nu'} \psi_{J',M'} |rM'| \psi_{\nu''} \psi_{J'',M''} \rangle \qquad (10.5.4)$$

The dipole moment, M_0, and its derivative with respect to r, M', are constants with respect to r, since they are evaluated at a fixed value of r, the equilibrium internuclear distance. However, they are vectors in the r-direction in space which changes as the molecule tumbles. Therefore, they are functions of the co-ordinates which describe their spatial orientation. We adopt a polar co-ordinate system and call the internuclear direction r and use θ and ϕ to describe the orientation of M_0 and M', which we shall write as $\text{f}(\theta, \phi) = \sin\theta \cdot \cos\phi$, $\sin\theta \cdot \sin\phi$ and $\cos\theta$ for the x, y and z directions respectively (Appendix 7). The consideration of molecular rotation also requires the co-ordinates, θ and ϕ. Thus, the vibrational wave functions, $\psi_{\nu'}(r)$ and $\psi_{\nu''}(r)$, are functions of r while the rotational functions, $\psi_{J',M'}(\theta, \phi)$ and $\psi_{J'',M''}(\theta, \phi)$ depend only upon θ and ϕ (Box 10.1). We now have:

$$M = \langle \psi_{\nu'}(r)\psi_{J',M'}(\theta, \phi) |M_0\text{f}(\theta, \phi)| \psi_{\nu''}(r)\psi_{J'',M''}(\theta, \phi) \rangle$$

$$+ \langle \psi_{\nu'}(r)\psi_{J',M'}(\theta, \phi) |rM'\text{f}(\theta, \phi)| \psi_{\nu''}(r)\psi_{J'',M''}(\theta, \phi) \rangle$$

$$= \langle \psi_{\nu'}(r)|\psi_{\nu''}(r) \rangle M_0 \langle \psi_{J',M'}(\theta, \phi)|\text{f}(\theta, \phi)| \psi_{J'',M''}(\theta, \phi) \rangle$$

$$+ \langle \psi_{\nu'}(r)|r| \psi_{\nu''}(r) \rangle M' \langle \psi_{J',M'}(\theta, \phi)|\text{f}(\theta, \phi)|\psi_{J'',M''}(\theta, \phi) \rangle \qquad (10.5.5)$$

where $\langle \rangle$ implies integration only over the co-ordinate(s) specified in the wave functions. We distinguish two cases.

Case 1: $\psi_{\nu'}(r) = \psi_{\nu''}(r)$

This is the case of a change of rotational energy level without a change of vibrational level; a pure rotational transition. Since the vibrational wave functions are normalised $\langle \psi_\nu(r)|\psi_\nu(r) \rangle = 1$. But $\langle \psi_\nu(r)|r|\psi_\nu(r) \rangle = 0^{\ddagger}$ so that only the first term in Equation (10.5.5) can contribute to the transition dipole moment, provided that neither M_0 nor $\langle \psi_{J',M'}(\theta, \phi)|\text{f}(\theta, \phi)|\psi_{J'',M''}(\theta, \phi) \rangle$ are equal to zero. The first condition requires that the molecule in its equilibrium nuclear configuration has a permanent dipole moment so that only heteronuclear diatomic molecules can show a pure rotation spectrum. The second condition for a pure rotation spectrum is that the third component of the product, the integration over θ and ϕ:

$$\langle \psi_{J',M'}(\theta, \phi)|\text{f}(\theta, \phi)|\psi_{J'',M''}(\theta, \phi) \rangle \qquad (10.5.6)$$

does not equal zero. This will provide the rotational selection rule. We first note (Appendix 5) that, apart from some multiplying constants, the three functions $\text{f}(\theta, \phi)$ are three spherical harmonics just like the three p-orbitals, $\Psi_{2\text{px}}$, $\Psi_{2\text{py}}$ and $\Psi_{2\text{pz}}$, which may also be

---

$\ddagger$ r is an odd function (it changes sign on passing through r_e) but the square of any wave function must be even. Therefore, the integrand $\psi_\nu(r)|r|\psi_\nu(r)$ is an odd function of r and must be zero.

written in the forms $Y_{1,+1}$, $Y_{1,-1}$ and $Y_{1,0}$. The second formulation is the more convenient for our present purposes and we use:

$$f_+(\theta, \phi) = \sin\theta \cdot \exp(+i\phi), \quad f_-(\theta, \phi) = \sin\theta \cdot \exp(-i\phi) \quad \text{and} \quad f_0(\theta, \phi) = \cos\theta$$

Since the rotational eigenfunctions are also spherical harmonics (Box 10.1) the required integral turns out to be the integral over θ and ϕ of a product of three spherical harmonics. The integration over ϕ is simple; we require that $M'' - M' + M_f = 0$, where M_f is the M-value of the particular choice of $f(\theta, \phi)$, i.e. $+1$, -1 or 0 for f_+, f_- and f_0 respectively. When this condition is satisfied the integral over ϕ has a value of 2π, otherwise it is zero. Since $M_f = 0$ or ± 1 this condition is usually written in the form:

$$M' - M'' \equiv \Delta M = 0, \pm 1 \tag{10.5.7}$$

In molecules with no unpaired electrons the effects of the selection rule on M are only seen in the presence of an electric field, which partially lifts the degeneracy of the M levels; a phenomenon known as the Stark effect (Johannes Stark (1874–1957)). The integration over θ is more complicated, but use of two formulae given in Appendix 5 simplifies the problem. The required formulae are:

$$\cos\theta \cdot \Theta_{J,|M|}(\theta) = A(J, M)\Theta_{J+1,|M|}(\theta) + B(J, M)\Theta_{J-1,|M|}(\theta) \tag{10.5.8}$$

and

$$\sin\theta \cdot \Theta_{J,|M|-1}(\theta) = C(J, M)\Theta_{J+1,|M|}(\theta) - D(J, M)\Theta_{J-1,|M|}(\theta) \tag{10.5.9}$$

where $A(J, M)$, $B(J, M)$, $C(J, M)$ and $D(J, M)$ are functions of J and M only. When the operator $f_0(\theta, \phi)$ is used we have, from Equation (10.5.7), the condition $\Delta M = 0$ and the matrix element requires integration over two Θ functions and $\cos\theta$; explicitly:

$$\text{Int.} = 2\pi \langle \Theta_{J',M'}(\theta)| \cos\theta |\Theta_{J'',M'}(\theta)\rangle$$

Replacing $\cos\theta|\Theta_{J'',M'}(\theta)\rangle$ with the right-hand side of Equation (10.5.8) gives:

$$\text{Int.} = 2\pi \{A(J, M)\langle\Theta_{J',M'}(\theta)|\Theta_{J''+1,M'}(\theta)\rangle + B(J, M)\langle\Theta_{J',M'}(\theta)|\Theta_{J''-1,M'}(\theta)\rangle\}$$

Therefore, since the Θ functions are orthogonal and normalised, the integral is zero unless $J' = J'' \pm 1$. When the operators $f_+(\theta, \phi)$ or $f_-(\theta, \phi)$ are used we have the condition $\Delta M = \pm 1$ and the matrix element requires integration over two Θ functions and $\sin\theta$. For that integral we apply Equation (10.5.9), which conveniently combines the sine function and the change of M value. Again, the orthonormality of the Θ functions leads to the selection rule:

$$\Delta J \equiv J' - J'' = \pm 1 \tag{10.5.10}$$

Case 2: $\quad \psi_{o'}(r) \neq \psi_{o''}(r)$

This is the case in which there is a change of vibrational state which may, or may not, be accompanied by a change of rotational state. Because of the orthogonality of the vibrational wave functions, the first term in Equation (10.5.5) is zero and the transition moment is given by the second, which implies that the transition moment will be zero unless $M' \neq 0$, i.e. the vibration must change the dipole moment of the molecule. The first term in the product is the integration over r:

$$\langle \psi_{o'}(r)|r|\psi_{o''}(r)\rangle$$

which will provide the vibrational selection rule. Using a result from Box 10.3 to express $r\psi_{\upsilon''}(r)$ in terms of the functions $\psi_{\upsilon''-1}(r)$ and $\psi_{\upsilon''+1}(r)$ we obtain:

$$\langle\psi_{\upsilon'}(r)|r|\psi_{\upsilon''}(r)\rangle = P\langle\psi_{\upsilon'}(r)|\psi_{\upsilon''+1}(r)\rangle + Q\langle\psi_{\upsilon'}(r)|\psi_{\upsilon''-1}(r)\rangle \quad (10.5.11)$$

where P and Q are constants. Since the vibrational eigenfunctions are orthogonal, the first integral on the right-hand side will be zero unless $\upsilon' = \upsilon'' + 1$, and the second integral will be zero unless $\upsilon' = \upsilon'' - 1$, which immediately gives us the selection rule:

$$\Delta\upsilon = \pm 1 \qquad (10.5.12)$$

The integrals over θ and ϕ are identical in both terms of Equation (10.5.5). Therefore, the rotational selection rules in the case of vibration-rotation spectra are the same as they are for pure rotation spectra; in gas-phase spectra rotational transitions are frequently seen as fine structure surrounding a vibrational absorption band. Spectroscopists call the transitions with $\Delta J = -1$, 0 and $+1$ the *P-*, *Q-* and *R-branches* of the spectrum respectively. With regard to the intensity of rotational transitions we must also take note of the degeneracy associated with the angular momentum. A rotational state characterised by the quantum number J is $(2J + 1)$-fold degenerate (Chapter 4) so that the population of such a state, and hence the intensity of transitions for which it is the initial state, depends upon J as well as the Boltzmann factor.

Equations (10.5.7), (10.5.10) and (10.5.12) are the selection rules for the rotational and vibrational spectra of the idealised diatomic molecule. It is worth reiterating that for an allowed rotational transition the molecule must have a ***permanent*** dipole moment, whereas for an allowed vibrational transition the vibration must be accompanied by a ***change*** in dipole moment.

In conclusion, three remarks should be made concerning the above selection rules and the derivation of them:

- We have developed this quantum-mechanical analysis of the vibrational transition probability in the context of a diatomic molecule executing its bond-stretching vibration. But the treatment is equally valid for any molecular vibration that is an eigenfunction of the vibrational Hamiltonian operator, i.e. a normal mode (see Section 10.7.1), and for which M' is not zero. However, it should be noted that real molecules are never harmonic oscillators (Sections 10.6 and 10.8) and therefore they deviate from that ideal behaviour.10.8

- Many other subtle aspects of these selection rules emerge when a group-theoretical analysis is made. For further details specialist texts, e.g. Herzberg[7] and Hollas,[8] should be consulted.

- Similarly, the selection rules for the changes in vibrational and rotational levels that accompany a change of electronic state are an important and complex area of spectroscopy from which most of our knowledge of the shapes of molecules in their electronically excited states is derived. The books by Herzberg[7] and Hollas[8] give comprehensive treatments of this subject.

Finally, we note that the selection rules for the rotational transitions of the rigid rotator we have found are those which we should expect in view of the established angular momentum of the electric dipole photon (see Section 8.4.5) Since, following convention,

J' characterises the upper state and J'' the lower, a transition with $\Delta J = +1$ corresponds to the addition to the rotator of one unit of angular momentum while $\Delta J = -1$ corresponds to a loss of angular momentum by the molecule. The selection rule on M, $\Delta M = \pm 1$, reflects the two possible z-components of the photon's angular momentum.

10.5.1   A semi-classical view of the selection rules

We have two conditions to fulfil if there is to be a transition between two vibrational energy levels as a result of the absorption of e-m radiation by a molecule:

- The frequency of the radiation, v, must be equal to the vibrational frequency of the molecule, v.
- The Bohr–Einstein condition must be satisfied, i.e. the product of h and v must be equal to the energy gap between two vibrational energy levels.

These two requirements can be simultaneously satisfied. If a molecule in any vibrational state has a vibrational frequency of v then the gap, ΔE, between two adjacent energy levels, quantum numbers v and $v + 1$, is (Equation (10.3.4)):

$$\Delta E = hv\{(v + 1 + \tfrac{1}{2}) - (v + \tfrac{1}{2})\} = hv \qquad (10.5.13)$$

Thus the two conditions are satisfied for an absorption process in which the quantum number v is changed by $+1$, which means that transitions between any two adjacent levels of the harmonic oscillator model of a diatomic molecule can be stimulated by e-m radiation of frequency v; but only, of course, if the oscillation of the molecule produces an oscillating electric dipole. A homonuclear diatomic molecule such as hydrogen (H_2) or nitrogen (N_2) has no electric dipole moment and cannot therefore interact with e-m radiation in the IR or microwave regions as a consequence of its vibrational, or rotational, motion.

A semi-classical interpretation of the selection rules for the rotational spectrum of a rotating dipole similar to that given above for the vibrational spectrum of the harmonic oscillator is not possible. This is because, in contrast to vibrational frequency where both classical and quantum-mechanical definitions exist, there is no exact quantum-mechanical definition of rotational frequency. Therefore, we do not expect to be able to express the true (quantum-mechanical) selection rule in classical terms. There is a precise quantum-mechanical description of angular momentum, but it does not imply the motion of a particle in a closed path and at a particular frequency about some central axis, as we can see when we think of the orbital angular momentum of a p electron or electron spin.

10.5.2   Infrared intensities

As in all branches of spectroscopy, the intensity of a transition depends not only upon the transition probability but also upon the population of molecules in the initial, ψ'', and final, ψ', states. This can be readily determined from Boltzmann's equation. For $^1H^{35}Cl$ at room temperature (300 K) for example, the $v = 0$ to $v = 1$ absorption band is observed at 2885.9 cm^{-1}, which is equal to 5733×10^{-23} J. Thus, according to Boltzmann:

$$N(v = 1)/N(v = 0) = \exp(-5733 \times 10^{-23}/300 \times 1.381 \times 10^{-23}) = 0.9785 \times 10^{-6}$$

Therefore, in a sample of one million molecules, there is only one molecule in the $\upsilon = 1$ vibrational state and 999 999 in the $\upsilon = 0$ state. This is a very favourable situation for absorption spectroscopy. With such a large excess of molecules in the lowest state problems due to saturation do not occur in the infrared: contrast NMR spectroscopy. However, for vibrations of low frequency and energy, there can be a significant population of molecules in the $\upsilon = 1$ or $\upsilon = 2$ states and transitions from these states to higher values of υ are sometimes observed; especially with samples at high temperatures. Such bands are known as *hot bands*.

10.6 REAL DIATOMIC MOLECULES

Though the model described in Sections 10.3 to 10.5 is a very useful starting point, a moment's reflection reveals that there are important differences between the harmonic oscillator and a real diatomic molecule. We know, for example, that if we supply sufficient energy to the molecule it will vibrate with such a large amplitude that the atoms separate at the extreme of the vibration and the molecule decomposes. Similarly, though it is not quite so obvious, at very small bond lengths the potential energy of the molecule increases more rapidly than the parabolic curve indicates. A rather general expression for the potential $V(R)$ as a function of the internuclear distance, R, can be obtained by expanding $(R - r_e)$ as a Taylor series about r_e:

$$V(R) = V(r_e) + (R - r_e)\partial V(r_e)/\partial R + (\tfrac{1}{2!})(R - r_e)^2 \partial^2 V(r_e)/\partial R^2$$
$$+ (\tfrac{1}{3!})(R - r_e)^3 \partial^3 V(r_e)/\partial R^3 + \cdots \qquad (10.6.1)$$

If we *define* the potential energy to be zero at the equilibrium bond length, the first term in the expansion is zero. The second term is zero because there is no change of $V(r_e)$ with R at r_e (Figure 10.5), so the first non-zero term is the quadratic term:

$$V(R) = \tfrac{1}{2}(R - r_e)^2 \partial^2 V(r_e)/\partial R^2 \qquad (10.6.2)$$

from which we see, by comparison with Equation (10.3.1), that the force constant k = $\partial^2 V(r_e)/\partial R^2$ and we know that this potential function gives rise to harmonic oscillation. Inclusion of the higher terms provides a representation of the *anharmonic oscillator*; this is sometimes necessary to explain more subtle effects in vibrational spectroscopy. Some examples are discussed in Section 10.8.

In practice, the $V(R)$: R relationship proposed by Morse (Equation (10.6.3) and Figure 10.8) is frequently used. It provides a simple but quite accurate approximation to the potential curve of a real diatomic molecule for small values of $r_e - R$:

$$V(R) = D_e[1 - \exp\{a(r_e - R)\}]^2 \qquad (10.6.3)$$

D_e is the energy difference between the dissociation energy and the lowest point of the V versus R curve (not the lowest vibrational level which gives the experimental dissociation energy, D_0) and a is a constant. It is possible, though difficult, to solve the Schrödinger equation for a Morse potential energy curve. Two aspects of the solution are of particular interest in the present context:

1. The new energy levels are given by the formula:

$$\text{E} = (\upsilon + \tfrac{1}{2})h\nu - (\upsilon + \tfrac{1}{2})^2 x_e h\nu$$

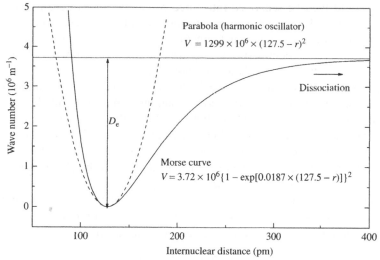

Figure 10.8 The Morse curve for a diatomic molecule

from which it is clear that they are now no longer equally spaced. x_e is the *anharmonicity constant*; it is a measure of the departure of the real molecule from the ideal harmonic oscillator.

2. The strict selection rule of the harmonic oscillator no longer applies and transitions having $\Delta v = \pm 2, \pm 3 \ldots$ are now possible, in addition to the original $\Delta v = \pm 1$.

   Though the transitions with $\Delta v = \pm 1$ remain by far the strongest, transitions having $\Delta v = +2$ are frequently seen in infrared absorption spectra where they are known as *overtones*. Their frequencies are normally just less than twice the frequency of the $\Delta v = +1$ transitions. Transitions having $\Delta v > 2$ are much rarer and the intensity of overtones decreases rapidly with increasing Δv. For example, the first overtone ($v'' = 0 \rightarrow v' = 2$) of hydrogen chloride (HCl) has only 1.6 % of the intensity of the fundamental ($v'' = 0 \rightarrow v' = 1$). *Combination bands*, transitions in which two fundamental vibrations are simultaneously excited, are also seen with a much-reduced intensity.

10.7   POLYATOMIC MOLECULES

The number of vibrations, 3N−5 or 3N−6, increases rapidly with the number of atoms (N), in a molecule and the possibility of observing vibrations of different types arises. In water, for example, there are vibrations associated with the stretching of the two O–H bonds and with the changing of the H–Ô–H bond angle. In benzene there is a vibration in which all six C–C bonds grow longer or shorter simultaneously and the C_6 hexagon is said to execute a *breathing* vibration. We need to consider the actual forms of the possible vibrations of a polyatomic molecule in more detail.

10.7.1   Normal co-ordinates, normal vibrations, vibrational eigenfunctions and eigenvalues

In Section 5.2 we noted that the expression of Schrödinger's equation for the hydrogen atom in polar rather than Cartesian co-ordinates was a great aid to its solution because the equation could then be separated into three equations, one in each of the three co-ordinates r, θ and ϕ. The eigenfunctions are then found as products of the solutions of the three separate equations, the radial function, $R_{n,l}(r)$ and the two angular functions, $\Theta_{l,m}(\theta)$ and $\Phi_m(\phi)$. We adopt a similar approach to find the solutions of Schrödinger's equation for the vibrations of a polyatomic molecule.

There is a set of co-ordinates in which the wave equation is separable into 3N−6 (3N−5) separate eigenvalue–eigenfunction equations. They are called the *normal co-ordinates* and are usually given the symbol Q_n. When the atoms move along a normal co-ordinate they are said to be executing a *normal mode* of vibration and there are no cross-terms in the Hamiltonian operator for the potential or kinetic energies between any two different normal modes. Since there are no cross-terms in the Hamiltonian the normal modes do not interact with each other, which is why the wave equation can be separated. Most importantly, the normal modes and their associated energies are the eigenfunctions, $\psi(Q_n)$, and eigenvalues, E_n, of the wave equation for the vibrational problem, just like those of the harmonic oscillator which we have studied above:

$$\hat{\mathcal{H}}_n \psi(Q_n) = E_n \psi(Q_n) \qquad (10.7.1)$$

And, in just the same way, we usually consider them to be harmonic, although we know that this is only an approximation. Each normal mode must belong to one of the symmetry species (irreducible representations) of the molecular point group[‡] and they are not too difficult to envisage. A normal mode describes the movement of *all* the atoms in the molecule, including the possibility that one or more atoms do not move. When the atoms are vibrating in a normal mode they all move in phase, though not necessarily with the same amplitudes, and pass through the equilibrium nuclear configuration and the turning points of their motion simultaneously. Therefore, although the motion may be complex, it can be thought of in terms of a single displacement in which all atoms are simultaneously involved and the basic theory of the motion can be expressed in the same form as that of the simple bond stretching of a diatomic molecule. Thus, the eigenfunctions $\psi(Q_n)$ have exactly the same algebraic form as those for the diatomic molecule, which are described in Box 10.3, with R replaced by Q_n. And just as for the hydrogen-atom problem, the solution for the whole molecule is given by the product of the 3N−6 (3N−5) solutions:

$$\Psi_{\text{total}} = \psi(Q_1) \cdot \psi(Q_2) \cdots \psi(Q_{3N-6}) \equiv \prod_n \psi(Q_n) \qquad (10.7.2)$$

However, unlike the hydrogen-atom case, each of the 3N−6 (3N−5) separated eigen-value – eigenfunction equations is of exactly the same form as Equation (10.7.1) and the total vibrational energy of the molecule is the sum of the 3N−6 (3N−5) E_n values, i.e. the sum of the energies of the individual modes, taking account of the number of quanta (υ_n) of energy in that mode:

$$E_{\text{total}} = \sum_n E_n = \sum_n \left(\upsilon_n + \frac{1}{2}\right) h\nu_n \qquad (10.7.3)$$

[‡] Group theory has many important and valuable applications to IR spectroscopy and any text on the theory of molecular vibrations will discuss these in detail.

10.7.2 Vibrations of real polyatomic molecules

In general, we do not observe the vibrations of individual bonds. Thus, using water as our example, we do not see a vibration which corresponds to an oscillation in the length of one O–H bond. The simple mathematical reason for this is that such a vibration is not a normal mode. Furthermore, a change in the length of one O–H bond affects the other O–H bond because they share the oxygen atom and also because any change of bond length introduces some change in the electron distribution in a molecule. The two O–H stretching vibrations are therefore *coupled* and we do not observe their independent frequencies in the infrared spectrum of water. When two systems, ϕ_a and ϕ_b say, which obey the laws of quantum mechanics are coupled then a mixing takes place and two new systems with the wave functions, ψ_+ and ψ_- are formed, where:

$$\psi_+ = \cos\theta.\phi_a + \sin\theta.\phi_b$$

$$\psi_- = \sin\theta.\phi_a - \cos\theta.\phi_b$$

We note that since $\cos^2\theta + \sin^2\theta = 1$, the new mixed functions are always normalised (Section 4.5). The mixing of two vibrations depends upon two factors, the magnitude of the interaction causing the mixing and the separation in energy (frequency) between the mixing vibrations. The mixing is at a maximum when the two vibrations to be mixed have exactly the same energy, in which case the two new vibrations contain equal quantities of the old combined in-phase and out-of-phase. The O–H stretching vibrations of water are illustrated in Figure 10.9, where the angle bending vibration has been included for completeness. The movements of the atoms are inversely proportional to their masses and must be such as to keep the centre of mass of the molecule stationary. The observed wave numbers of the bands show some important general characteristics. The symmetric stretching (3832 cm^{-1}) is lower than the asymmetric (3942 cm^{-1}) and the bending (648 cm^{-1}) is much lower than the two stretchings. This reflects the different force constants which are much larger for bond stretching than for angle bending (Table 10.1).

What cannot be detected without more detailed analysis of the data is the fact that there is a mixing of the symmetric stretching and the bending vibration. A change in the O–H bond length changes the force constant for the H–Ô–H angle bending, and vice versa. Therefore, the two types of vibration are coupled, which is the same as saying that the bond-stretching and angle-bending vibrations are not themselves normal modes because they are linked by energy terms in the Hamiltonian operator. We can view this as the process of diagonalising a 2×2 matrix (Appendix 3), in which the basis states, i.e. the two modes, are combined in such a way as to reduce the interaction to zero and

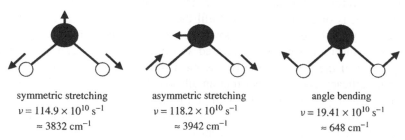

symmetric stretching	asymmetric stretching	angle bending
$\nu = 114.9 \times 10^{10} \text{ s}^{-1}$	$\nu = 118.2 \times 10^{10} \text{ s}^{-1}$	$\nu = 19.41 \times 10^{10} \text{ s}^{-1}$
$\approx 3832 \text{ cm}^{-1}$	$\approx 3942 \text{ cm}^{-1}$	$\approx 648 \text{ cm}^{-1}$

Figure 10.9 The vibrations of the water molecule

Table 10.1   The bond-stretching and angle-bending force constants[a] for some small molecules

Molecule	Bond[b] stretching	Angle[c] bending	Molecule	Bond[b] stretching	Angle[c] bending
H_2O	H–O = 780	HOH = 70	H_2S	H–S = 430	HSH = 45
NH_3	H–N = 650	HNH = 50	CH_4	C–H = 540	HCH = 45
CO_2	C=O = 1550	OCO = 60	SO_2	S=O = 995	OSO = 80

[a] Many more force constants are tabulated in sources such as E.B. Wilson Jr, J.C. Decius and P.C. Cross, *Molecular Vibrations*, McGraw-Hill, New York, 1955, and G. Herzberg, *Infrared and Raman Spectra of Polyatomic Molecules*, Van Nostrand, New York, 1945.
[b] The units of the bond-stretching force constants are Nm^{-1}.
[c] The units of the angle-bending force constants are Nm.
The reason for the different units of the two types of force constant is the following. The change of a bond length or bond angle are measured by δl and $\delta\theta$ respectively. $\delta\theta$ is measured in radians and therefore has no units. The force constant relates the potential energy due to a change in bond length or bond angle to the square of the change (Equation (10.2.4)), i.e. $V = \frac{1}{2}k_l(\delta l)^2$ or $\frac{1}{2}k_\theta(\delta\theta)^2$, so that in order for the dimensions of the potential energy to be the same (mass length^2 time^{-2}) in both cases, the force constant for angle bending is usually multiplied by the lengths of the two bonds which define the angle.

generate the two genuine normal vibrational modes. But there will only be mixing if the symmetry of the two vibrations is the same and we can see that this is the case in this particular example if we visualise the changing dipole moment associated with each vibration. In a static water molecule the dipole moment lies along a line in the plane of the molecule bisecting the H–Ô–H bond angle. When the molecule performs either symmetric stretching or angle-bending vibration the change in the dipole moment lies along the same direction. With the asymmetric stretching vibration, on the other hand, the change in the dipole moment is directed at right-angles to the static dipole moment. This indicates that the asymmetric stretching vibration is of a different kind from the other two vibrations; spectroscopists say it belongs to a different *symmetry species*. Vibrations of different symmetry species cannot interact and mix, but those of the same species can and usually do. As before, the degree of their mixing depends upon the interaction between them and the difference in their energies. In the case of the stretching/bending vibrations of water the mixing is small, only 1–2 % of the one is found in the other. But it is important to recognise that this phenomenon exists and that a band observed in an infrared spectrum cannot always be assigned to the stretching of just one bond or a single angle-bending vibration.

The fact that the direction of the changing dipole moment of the water molecule is different for different vibrations has important applications. A molecule can only interact with e-m radiation that is polarised in the same direction as a changing dipole moment in the molecule. Therefore, if a molecule is irradiated with plane polarised light it will absorb the light only if it is orientated so as to fulfil the above condition. If the structure of a solid sample (a crystal, for example) is known then the orientation of the sample necessary for a particular band to appear in the infrared absorption spectrum can be used to assist in the assignment of the band. Similarly, the metal-surface selection rule (Section 8.7.3) depends for its efficacy upon the fact that an allowed IR transition can be related to the direction of a changing dipole moment within a molecule.

A typical spectrum of a small molecule which illustrates several of the most important points about infrared spectra is that of bromochloromethane (CH_2BrCl) (Figure 10.10). The assignment of the bands is given in Table 10.2. Two carbon–halogen stretching

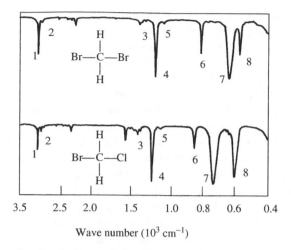

Wave number (10^3 cm^{-1})

Figure 10.10 The infrared absorption spectra of dibromomethane and bromochloromethane

Table 10.2 The assignment of the infrared spectra of dibromomethane and bromochloromethane (shown in Figure 10.10)

	CH_2Br_2[†]		CH_2BrCl[†]
1. CH_2 asymmetric stretch	3065		3060
2. CH_2 symmetric stretch	3003		2986
3. CH_2 scissors	1390		1407
4. CH_2 wagging	1192		1227
5. CH_2 twisting	1095		1130
6. CH_2 rocking	813		850
7. CBr_2 asymmetric stretch	639	C–Cl stretch	730
8. CBr_2 symmetric stretch	578	C–Br stretch	605
9. CBr_2 scissors	169	CBrCl scissors	229

[†]Band positions in wave numbers (cm^{-1})

vibrations are visible. The two C–H stretching vibrations interact, as the O–H stretches in water do, to form asymmetric and symmetric combinations which occur at 3060 and 2986 cm^{-1} respectively; the asymmetric stretch at the higher wave number has the greater intensity because it entails a greater change of dipole moment than the symmetric stretch. These are the highest wave number bands in the spectrum. The C–Cl stretch is found at 730 cm^{-1} and the C–Br even lower at 605 cm^{-1}. The decrease in wave number in the sequence C–H > C–Cl > C–Br reflects the fact that the frequency, and therefore the wave number, is inversely proportional to the reduced mass of the two atoms forming the bond (Equation (10.3.3)), though changes in the force constant also play a part. The effect of the atomic mass can also be seen in the angle bending where the H–C–H scissoring vibration is found at 1407 cm^{-1} while the Cl–C–Br scissoring is off the scale at 229 cm^{-1}. Two other significant bands can be identified in this spectrum. The intense CH_2 wagging at 1227 cm^{-1} and the less intense CH_2 rocking at 850 cm^{-1}.

The names of these three types of vibration arise in the following way. Imagine that the plane containing the carbon, chlorine and bromine atoms is at right-angles to the plane containing the carbon and two hydrogen atoms. If the atoms in the Cl–C–Br plane remain stationary then the vibration in which the H–C–H bond angle opens and closes is the H–C–H scissoring. If the H–C–H plane moves to and fro in such a way that the two hydrogens move towards and then away from a halogen atom, this is the wagging vibration. If, on the other hand, the H–C–H plane moves from side-to-side at right-angles to the Cl–C–Br plane, then this is the rocking vibration.

Note that all $3N - 6 = 9$ bands are allowed by the selection rules and all are seen but with very different intensities, which reflect the magnitude of the change in dipole moment associated with the different modes of vibration. The strong absorption of the C–Cl (7) and C–Br (8) stretching vibrations is a consequence of the large dipole moments of these bonds, which therefore undergo large changes during the vibrational cycle. But the change is very small for the CH_2 scissoring (3) and twisting (5) vibrations. Indeed, unassigned bands stronger than the last two can be seen in the spectrum, one at approximately 1560 cm^{-1}, for example. Such a band may be due to an impurity in the sample or to an overtone or a combination band (Section 10.6). The sum of the wave numbers of the strong bands 6 and 7 is 1580 cm^{-1}, which suggests that the 1560 cm^{-1} band may be due to the combined excitation of these two vibrations. Compare the spectrum just described with that of dibromomethane and note the effect of increasing the mass of one halogen atom on the frequencies.

10.7.3   Characteristic group frequencies

In the last section the spectra of CH_2BrCl and CH_2Br_2 were compared. In Table 10.2 there is a close similarity in the wave numbers of each of the six vibrations attributed to the CH_2 group. The frequencies of vibrations such as these, usually involving atoms on the periphery of a molecule, do not change much when the remainder of the molecule is changed and they therefore constitute a valuable method in molecular structure determination. They are known as characteristic group frequencies and lists of them can be found in many books devoted to IR spectroscopy.[1,9-11] In this context vibrations involving hydrogen, e.g. the O–H stretch in alcohols ($\sim 3600 \text{ cm}^{-1}$), the N–H stretching of the NH_2 group ($\sim 3400 \text{ cm}^{-1}$) are useful and easy to identify because, like the C–H vibrations, they lie near the upper extremity of the spectrum. Similarly, the bands of C–Cl ($\sim 725 \text{ cm}^{-1}$), C–Br ($\sim 650 \text{ cm}^{-1}$) and C–I ($\sim 550 \text{ cm}^{-1}$), which are found near the other spectral extreme, provide evidence of the presence of halogen atoms. A band that occurs in the centre of the spectral range ($1750 - 1600 \text{ cm}^{-1}$) is the C=O stretch of aldehydes and ketones; it is particularly well known.

10.7.4   Large molecules

Everything that has been said above concerning the water and bromochloromethane molecules is equally important in the infrared spectra of the larger molecules which are of interest to chemists. But as the number of atoms in a molecule rises the spectrum becomes increasingly difficult to assign and interpret. Moreover, when many atoms are linked together in a complex three-dimensional structure changes of individual bond angles and bond lengths are not possible. In benzene, for example, a single C–C–C bond angle cannot change without some corresponding change in other C–C–C angles or C–C bond lengths. Complicated vibrations involving many simultaneous changes of bond angles and bond lengths are known as *skeletal* vibrations and, even when their exact nature is obscure, they give rise to a series of infrared absorption bands which are highly characteristic of the molecule. This *fingerprint* is very important in qualitative analysis and structure determination.

10.8 ANHARMONICITY

We have frequently noted above that although the concept of the harmonic oscillator is a good model for molecular vibrations, all real molecules depart from this ideal and the effects of their anharmonicity must be taken into account when we attempt to interpret some of the more subtle, but none the less important, effects seen in IR spectroscopy. We can distinguish two different types of anharmonicity.

Electrical anharmonicity has to do with the problem of correctly representing the molecule as an array of electrical charges which can interact with e-m radiation. In Equation (10.5.2) the transition dipole moment is written in terms of the dipole moment plus its first derivative with respect to the normal co-ordinates; in mathematical parlance, the Taylor series expansion of the dipole moment. But these are simply the first two terms in the series and if higher derivatives were included they would allow transitions having $\Delta \upsilon > 1$. Thus, the effect of electrical anharmonicity is primarily to allow bands to appear which we would not have expected.

Mechanical anharmonicity, from which the adjective mechanical is often omitted, describes the fact that the potential energy curve associated with a molecular vibration is not parabolic in the co-ordinate, i.e. is not of the form of Equation (10.2.4). The effects of mechanical anharmonicity are seen in the vibrational energy levels and hence band frequencies. Also, in some rather special cases where the anharmonicity results in a mixing of vibrational states, there can be quite dramatic effects upon band intensities. Some classic examples of mechanical anharmonicity are examined in the following two sections. Both involve carbon dioxide.

10.8.1 Fermi resonance

The assumption of harmonic vibrations provides such a simplification that it is invariably the starting point for the analysis of IR spectra. However, the effects of anharmonicity can be large and we are sometimes forced to recognise them. The vibrational spectrum of carbon dioxide provides a good example. Unfortunately, this most famous of examples involves Raman spectroscopy,[1,2] which is not discussed in this book, rather than IR. However, our primary objective is to illustrate how anharmonicity can bring about the mixing of vibrational energy states and the fact that the effects of this mixing are seen here in the Raman rather than IR spectrum is of secondary importance.

The bands seen in the IR and Raman spectra of carbon dioxide (CO_2) under conditions of low resolution are listed in Table 10.3. For this linear molecule we expect to see

Table 10.3 The vibrational spectrum of carbon dioxide under conditions of low resolution

Mode	$\nu(cm^{-1})$	Description	R/IR active
ν_1	1340	symmetric bond stretching	R
ν_2	667	angle bending	IR
ν_3	2349	asymmetric bond stretching	IR

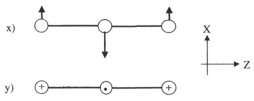

Figure 10.11 The two components, ν_{2x} and ν_{2y}, of the doubly degenerate angle-bending vibration of carbon dioxide

four ($3 \times 3 - 5$) fundamental vibrations and we do because the angle bending is doubly degenerate. (The O=C=O angle can be deformed in either of two planes at right-angles to each other (Figure 10.11).) However, when the spectra are measured under higher resolution the 1340 cm^{-1} band in the Raman spectrum is found to consist of two bands of approximately equal intensity centred at 1285 and 1388 cm^{-1}. This observation appears to be at odds with theory.

The following is a simplified version of the explanation suggested by Enrico Fermi in 1931, since when very many similar examples have been reported. Although he knew ν_2 and ν_3 from infrared spectroscopy, Fermi had no value for ν_1 so he determined the force constant, k, for the C=O bond stretching from ν_3 and using this and the atomic masses he estimated that the symmetric stretch should lie somewhere near 1230 cm^{-1}; close to the predicted first overtone of ν_2 at approximately $2 \times \nu_2 = 1334$ cm^{-1}. Furthermore, Fermi knew from the symmetry of the overtone that it could interact with ν_1. The mechanism of the interaction is the anharmonicity, in particular cubic terms (Equation (10.6.1)) which are neglected in the simple, harmonic oscillator theory. The formulation of the energy matrix (Appendix 3) for this interaction is slightly complicated by the fact that ν_2 is a doubly degenerate vibration and thus its first overtone may correspond to three different excitations. Since ν_3 had no part to play in the calculation, Fermi omitted it and wrote his wave functions in the form ($\nu_1, \nu_{2x}, \nu_{2y}$), where ν_{2x} and ν_{2y} are the two degenerate components of ν_2. In this notation the singly excited ν_1 state is (100) and it may, in principle, interact with the three doubly excited ν_2 states (020), (002) and (011). However, symmetry forbids an interaction between (100) and (011) and the energy matrix reduces to:

$\hat{\mathcal{H}}$	(100)	(020)	(002)
(100)	ν_1	α	α
(020)	α	$2\nu_2$	0
(002)	α	0	$2\nu_2$

(α is the interaction for which Fermi estimated a value of 40 cm^{-1}.)

Rather than evaluating the matrix immediately, Fermi simplified it by combining the two overtone functions into their sum and difference:

$$\Psi_+ = \sqrt{\tfrac{1}{2}}\{(020) + (002)\} \quad \text{and} \quad \Psi_- = \sqrt{\tfrac{1}{2}}\{(020) - (002)\}$$

which reduces the matrix to the form:

$\hat{\mathcal{H}}$	(100)	Ψ_+	Ψ_-
(100)	ν_1	$\sqrt{2}\alpha$	0
Ψ_+	$\sqrt{2}\alpha$	$2\nu_2$	0
Ψ_-	0	0	$2\nu_2$

We require the eigenvalues and eigenfunctions of the 2×2 matrix. Each will be a mixture of the Raman-allowed excitation (100) and Ψ_+ giving two new excited states to which Raman transitions are allowed by virtue of their component of (100). They can easily be found using the method described in Appendix 3 if the values of α, ν_1 and ν_2 are known. Since he had no reliable value of ν_1, Fermi simply noted that if $\nu_1 \approx 2\nu_2$ we expect the two new vibrational states to be separated by $2 \times \sqrt{2}\alpha = 113$ cm^{-1} and the corresponding spectral bands to be approximately equal in intensity. These results are both in good agreement with experiment where the two Raman bands are almost equal in intensity and the observed $\Delta\nu$ is 103 cm^{-1}. In the light of the value of $\Delta\nu$ Fermi suggested that his estimate of $\nu_1 \approx 1230$ cm^{-1} was probably about 100 cm^{-1} too low.

In the last part of his paper Fermi explains the presence of two further, very weak Raman bands in terms of exactly the same type of mixing among groups of three states: (110), (030) and (012) in the case of one band and (101), (003) and (021) for the other. In each of these states the molecule is already excited with at least one quantum of ν_2 and there are only 7 % of such molecules in the gas at room temperature, hence the very low intensity.

Finally, we might note that, although a quantitative solution of the carbon dioxide problem requires a quantum-mechanical analysis, we can see in a simple classical view why there is an interaction between the vibrations ν_1 and ν_2 and why resonance is possible when $\nu_1 \approx 2\nu_2$. As the molecule executes the angle-bending vibration, ν_2, the oxygen atoms swing from one side of the linear O=C=O configuration to the other and back again, passing through the linear configuration twice in each vibrational cycle. The centrifugal force arising from this motion over an arc of a circle is at a maximum when the molecule is linear and acts in such a way as to extend both C=O bonds simultaneously. If the frequency of this effect, $2\nu_2$, coincides with that of the symmetric bond stretching, ν_1, then we have the ideal conditions for resonance.

10.8.2 Vibrational angular momentum and the Coriolis interaction

The mechanics of bodies undergoing simultaneous linear and rotational motion, was first investigated by Gaspard de Coriolis (1792–1843), a French physicist who studied the movements of air and water on the surface of the earth. The eponymous force which his work revealed is responsible, among other things, for the whirling motions of tropical

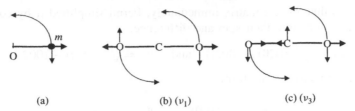

(a)   (b) (v_1)   (c) (v_3)

Figure 10.12 Coriolis forces on a single mass, (a) the symmetric stretching vibration of carbon dioxide (b) and the asymmetric stretching vibration (c)

storms. It also has some very interesting effects upon the spectra of molecules; but first we should clarify its origin.

Consider an atom of mass, m, which is part of a molecule rotating about an axis through the centre O and perpendicular to the plane of the paper (Figure 10.12(a)). The rotating molecule has angular momentum so that if in the course of a vibration the mass moves to the right, increasing its distance from O, the speed of movement of the mass must be reduced to conserve that momentum. The slowing of the mass will appear to an observer riding on the molecule, who will not be aware of the rotational motion as we are unaware that the earth is rotating, as the response to a force in the direction indicated by the downwards arrow. This is the Coriolis force and it has the potential to cause an interaction between the vibration and rotation of a molecule, as the following examples will show. Several types of Coriolis interaction can be distinguished.

Type 1. Taking carbon dioxide as our example with z as the internuclear axis we first examine the Coriolis forces arising from the symmetric stretching vibration, v_1, (Figure 10.12(b)). The heavy, horizontal arrows represent the *velocities* of the oxygen atoms; the carbon atom is stationary. The molecule is rotating, end-over-end, and the lighter, vertical arrows show the resulting Coriolis forces which, in this case, produce a coupling to the rotation rather than the vibration of the molecule. This has a very small effect upon the rotational energy levels but is not normally considered to be a Coriolis effect.

Type 2. We continue with our carbon dioxide example and examine the effect of Coriolis forces on the asymmetric stretching v_3. In this case (Figure 10.12(c)) the directions of the Coriolis forces on each atom are such as to promote the bending vibration, v_2, but with the frequency of v_3. v_2 is a doubly degenerate mode and in order to appreciate fully the effect of the Coriolis coupling it is necessary to examine it in more detail. We again define the two components of the vibration, v_{2x} and v_{2y}, to occur in the xz and yz planes (Figure 10.11). Inspection of the figure shows that a rotation of $\pi/2$ about the z axis, in either sense, converts one of the angle-bending vibrations into the other. This suggests that there is angular momentum about the z-axis associated with the vibration and we can show that this is indeed the case in the following manner. We first use Box 10.3 to write down the harmonic oscillator eigenfunctions, Ω_ν, for the normal co-ordinate q with $\nu = 0 - 2$:

$$\Omega_0 = [\gamma/\pi]^{\frac{1}{4}} \exp(-\tfrac{1}{2}\gamma q^2)$$

$$\Omega_1 = [\gamma/\pi]^{\frac{1}{4}} \sqrt{(2\gamma)} q \exp(-\tfrac{1}{2}\gamma q^2)$$

$$\Omega_2 = [\gamma/\pi]^{\frac{1}{4}} (\sqrt{2}\gamma q^2 - 1/\sqrt{2}) \exp(-\tfrac{1}{2}\gamma q^2)$$

In accordance with the concept of the normal mode or normal co-ordinate (Section 10.7.1), we can regard the vibrational modes in the figure as displacements along a single co-ordinate x or y, with the proviso that, to keep the centre of gravity stationary, a displacement of the carbon atom of $^{12}C^{16}O_2$ by $-x$ is accompanied by displacement of the two oxygen atoms by $+\frac{1}{2}(12/16)x = +3x/8$, and similarly for y. There are simultaneous, equal and opposite displacements of the oxygen atoms in the z direction.

We can now write down the appropriate functions to describe the three degenerate vibrationally-excited states, Γ, which can be formed by adding two vibrational quanta to the ν_2 vibration. Each is a product of two harmonic oscillator eigenfunctions characterised by their quantum numbers (ν_{2x}, ν_{2y}) as subscripts, the co-ordinates, q, being replaced by x or y, as required:

$$\Gamma_{11} = [\gamma/\pi]^{\frac{1}{2}} \, 2\gamma xy \exp(-\tfrac{1}{2}\gamma[x^2 + y^2])$$

$$\Gamma_{20} = [\gamma/\pi]^{\frac{1}{2}} \left(\sqrt{2}\gamma x^2 - 1/\sqrt{2}\right) \exp(-\tfrac{1}{2}\gamma[x^2 + y^2])$$

$$\Gamma_{02} = [\gamma/\pi]^{\frac{1}{2}} \left(\sqrt{2}\gamma y^2 - 1/\sqrt{2}\right) \exp(-\tfrac{1}{2}\gamma[x^2 + y^2])$$

We know from Chapters 3 and 4 that if we are looking for angular momentum then a polar co-ordinate system is more useful than a Cartesian and we use the plane-polar system in which:

$$r^2 = x^2 + y^2, \quad x = r\cos\phi \quad \text{and} \quad y = r\sin\phi$$

Substituting r and ϕ for x and y we have:

$$\Gamma_{11} = [\gamma/\pi]^{\frac{1}{2}} \, 2\gamma r^2 \sin 2\phi \exp(-\tfrac{1}{2}\gamma r^2)$$

$$\Gamma_{20} = [\gamma/\pi]^{\frac{1}{2}} \left(\sqrt{2}\gamma r^2 \cos^2\phi - 1/\sqrt{2}\right) \exp(-\tfrac{1}{2}\gamma r^2)$$

$$\Gamma_{02} = [\gamma/\pi]^{\frac{1}{2}} \left(\sqrt{2}\gamma r^2 \sin^2\phi - 1/\sqrt{2}\right) \exp(-\tfrac{1}{2}\gamma r^2)$$

We shall look for functions which are eigenfunctions of the operator for the z-component of angular momentum, $\hat{l}_z = -i\hbar\partial/\partial\phi$ (see Chapters 3 and 4). None of the above functions can satisfy this requirement since they each give a different function when differentiated once with respect to ϕ. But we can attempt to find the functions we seek by making combinations. Provided that we ensure that they are normalised, any new functions we form will remain eigenfunctions of the Hamiltonian operator, $\hat{\mathcal{H}}$, because all three functions, Γ_{11}, Γ_{20} and Γ_{02}, are degenerate. Γ_{20} and Γ_{02} are functions of $\cos^2\phi$ and $\sin^2\phi$ respectively and we know that $\cos^2\phi + \sin^2\phi = 1$. Therefore, the normalised sum:

$$\Psi_0 = \sqrt{\tfrac{1}{2}}\{\Gamma_{20} + \Gamma_{02}\} = [\gamma/\pi]^{\frac{1}{2}} (\gamma r^2 - 1) \exp(-\tfrac{1}{2}\gamma r^2)$$

does not contain ϕ and is an angular momentum eigenfunction with an eigenvalue of zero, hence the subscript of 0 qualifying the new function. This is the state which we called Ψ_+ in the previous section. Having taken the sum of Γ_{20} and Γ_{02} we must also take their difference:

$$\Psi_- = \sqrt{\tfrac{1}{2}}\{\Gamma_{20} - \Gamma_{02}\} = [\gamma/\pi]^{\frac{1}{2}} \gamma r^2 \cos 2\phi \exp(-\tfrac{1}{2}\gamma r^2)$$

We note that Γ_{11} and Ψ_- are functions of $\sin 2\phi$ and $\cos 2\phi$ respectively which is reminiscent of Chapter 3 (Table 3.1), where we found that the functions $(1/\sqrt{\pi})\sin(n\phi)$ and $(1/\sqrt{\pi})\cos(n\phi)$ were not eigenfunctions of the angular momentum operator $(\hat{l}_z = -i\hbar\partial/\partial\phi)$ but that eigenfunctions such as:

$$\frac{1}{\sqrt{2\pi}} \cdot \exp(\pm in\phi) = \frac{1}{\sqrt{2}} \cdot \left\{ \frac{1}{\sqrt{\pi}}\cos(n\phi) \pm i\frac{1}{\sqrt{\pi}}\sin(n\phi) \right\}$$

could be formed from normalised combinations of them. The new functions remained eigenfunctions of the energy operator, $\hat{\mathcal{H}}$. This suggests a similar approach here[‡] so we form:

$$\Psi_{+2} = \sqrt{\tfrac{1}{2}}\{\Psi_- + i\Gamma_{11}\} = [\gamma/2\pi]^{\frac{1}{2}}\gamma r^2 \exp(-\tfrac{1}{2}\gamma r^2)\exp(+2i\phi)$$

and:

$$\Psi_{-2} = \sqrt{\tfrac{1}{2}}\{\Psi_- - i\Gamma_{11}\} = [\gamma/2\pi]^{\frac{1}{2}}\gamma r^2 \exp(-\tfrac{1}{2}\gamma r^2)\exp(-2i\phi)$$

We now have:

$$\hat{l}_z\Psi_{+2} = -i\hbar\frac{\partial}{\partial\phi}\Psi_{+2} = +2\hbar\Psi_{+2}$$

$$\hat{l}_z\Psi_0 = -i\hbar\frac{\partial}{\partial\phi}\Psi_0 = 0$$

and:

$$\hat{l}_z\Psi_{-2} = -i\hbar\frac{\partial}{\partial\phi}\Psi_{-2} = -2\hbar\Psi_{-2}$$

These are three eigenfunctions of $\hat{l}_z$ with eigenvalues, characterised by the quantum number, symbol l, of $+2$, 0 and -2 respectively. All excitations of the bending vibration give rise to angular momentum about the z axis so it is important to include l in the notation for the vibrational state of the carbon dioxide molecule given in the form $(\upsilon_1, \upsilon_2{}^l, \upsilon_3)$. The functions Ψ_{+2} and Ψ_{-2} correspond to a slightly bent carbon dioxide molecule, which is rotating around the z axis thereby giving rise to a moment of inertia where previously there was none. The state Ψ_0 has no angular momentum about z and the differing interactions of the three functions lifts the triple degeneracy of the (0, 2, 0) state. To a first approximation (see below) the sense of rotation makes no difference to the rotational energy levels due to the end-over-end rotation of the molecule, but there is a difference between $l = 0$ and $l = \pm 2$ since the correction, ΔE, to the total energy, rotational plus vibrational, is given by:

$$\Delta E = (g_{22} - B)l^2 \qquad (10.8.1)$$

The first term in Equation (10.8.1) is the correction to the vibrational energy in which g_{22} is the anharmonicity constant (Section 10.6) for the degenerate vibration ν_2. The second term is the correction to the rotational energy in which B is the rotational constant defined by Equation (10.4.1a). Since the values of g_{22} and B for carbon dioxide are

---

[‡] When forming combinations by adding and subtracting functions, addition corresponds to the in-phase combination (phase angle = 0) while subtraction corresponds to the out-of-phase combination (phase angle = $\pm\pi$). When the combination also involves i we are combining functions which differ in phase by $+\pi/2$ $(+i)$ or $-\pi/2$ $(-i)$. Compare the Argand diagram in Appendix 8.

-0.62 and 0.39 cm^{-1} respectively, the separation of the $l = 0$ and $l = 2$ levels would be expected to be of the order of -4.0 cm^{-1} but this cannot be determined directly from the spectrum because, as we have seen above, the (02^00) level is in Fermi resonance with the (10^00), which accounts for the anomalously large observed splitting of ~ 51 cm^{-1}. Furthermore, the resonance interaction causes the two states to mix in almost a $1 : 1$ ratio (see below) so that it is no longer correct to speak of either of them as being (02^00) or (10^00).

Type 3. In the last example of Coriolis coupling we examined the angular momentum associated with the doubly excited ν_2 vibration. In fact, every overtone of ν_2 has angular momentum and the two components of the singly excited ν_2 state also undergo a Coriolis interaction as a consequence of their angular momentum about the z axis for which $l = \pm 1$ (see problem 5). As shown in Figure 10.13, the Coriolis forces on the atoms convert one component of (01^10) into the other. This is an example where, even if the energy of interaction (off-diagonal matrix element) is small, the resulting splitting of the energy levels can be large because, prior to the interaction, they were degenerate. It is frequently called a first order Coriolis perturbation and it removes the degeneracy of the pairs of $\pm l$ levels. This effect is known as l-type doubling. Thus the two (01^10) energy levels, which are degenerate in the absence of vibrational angular momentum, are split into two. And in the case of the doubly excited ν_2, the degeneracy of $(02^{+2}0)$ and $(02^{-2}0)$ is removed so that all three states have different energies.

To see in a simple way how this may come about, we note that since carbon dioxide is a linear molecule there is no moment of inertia or angular momentum associated with rotation about the internuclear axis. But the angle-bending vibration removes this linearity and the molecule becomes, momentarily, capable of angular momentum with respect to rotation about that axis. This perturbs the energy levels of the molecule, especially, but not exclusively, the rotational levels. Unfortunately, there appear to be no experimental observations of l-type doubling in carbon dioxide in the literature though there are for other molecules, e.g. hydrogen cyanide (HCN). This may well be because the splittings are very small and normally measurable only in microwave spectroscopy but the carbon dioxide molecule gives no microwave spectrum since it has no permanent dipole moment.

The energy levels associated with some of the interactions which we have discussed above are represented in Figure 10.14. The figure is illustrative rather than exact because the energy levels are determined by many subtle effects of anharmonicity which have not been discussed above and these factors result in a slightly larger value of the Fermi-resonance interaction (α in Section 10.8.1) than we find in the following. On the sides of the diagram the estimated positions of the levels in the absence of the Fermi resonance and Coriolis coupling are shown; the three degenerate double excitations of ν_2 on the left at

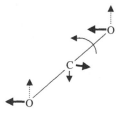

Figure 10.13 Coriolis forces on the nuclei of a carbon dioxide molecule rotating about the internuclear axis

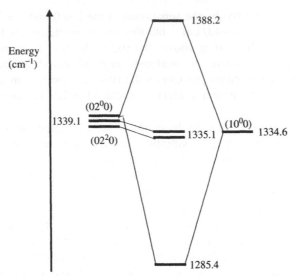

Figure 10.14 The effect of Fermi resonance and Coriolis coupling on some vibrational energy levels of carbon dioxide (Not to scale)

1339.1 cm^{-1} and the singly excited ν_1 on the right at 1334.6 cm^{-1}. The Fermi resonance interaction of (02^00) and (10^00) mixes these levels and displaces them to 1285.4 cm^{-1} and 1388.2 cm^{-1}, where they are observed experimentally. $\alpha = 51$ cm^{-1} and the vibrational wave functions after the mixing are found to be:

$$\Psi(1388.2) = 0.72(02^00) + 0.69(10^00)$$

and

$$\Psi(1285.4) = 0.69(02^00) - 0.72(10^00)$$

The Coriolis interaction of rotation and vibration lowers the two (02^20) levels about 4 cm^{-1} to the experimental value of 1335.1 cm^{-1}. The loss of the $(02^{+2}0)/(02^{-2}0)$ degeneracy is not shown.

10.9   THE *AB-INITIO* CALCULATION OF IR SPECTRA

In the second half of the 20th century the quantity of IR-spectral data increased rapidly and the quantitative analysis of that information, i.e. the determination of the force constant for each normal mode, grew in importance. The determination of force constants was of interest not only for a complete assignment of the spectrum but also for the light which they could throw upon the strength of chemical bonds and on the fundamental thermodynamic properties of the molecules. Most calculations employed the FG-matrix method proposed by Edgar Bright Wilson (1908–1992) and when, in the 1960s, they were programmed for computers many important advances were rapidly achieved. Bands in the spectrum could then be quantitatively ascribed to mixtures of the vibrations of chemically significant groupings of atoms in a molecule or to the whole molecule in the case of skeletal vibrations. The force constants found could be related to the strength of bonds, and the rigidity, i.e. resistance to deformation, of the molecule. The mechanics used in the calculations, which were based upon an input of atomic masses and trial normal modes and force constants, was purely classical. The trial force constants were empirical values deduced from those in the literature for similar molecules and groups; they were not calculated.

More recently, as the power of density-functional methods and *ab-initio* molecular orbital calculations has increased, molecular vibrational energy levels have been derived using force constants calculated quantum mechanically from molecular wave functions and the use of normal modes as a basis has been superseded by expressing the motions of the atoms in Cartesian co-ordinates. Very good theoretical IR spectra are now being obtained by this method; sufficiently good, in fact, to assign an absolute configuration to each member of a pair of optical isomers by comparison of their theoretical and experimental vibrational circular dichroism (VCD) spectra.[12] (A VCD IR spectrum is an IR spectrum in which the differential absorption of left and right circularly polarised radiation (Section 8.2.2), rather than the total absorption, is recorded against wave number.) However, the assignment of the spectral bands to particular bond-angle or bond-length changes does not emerge as naturally as with the earlier approach.

10.10 THE SPECIAL CASE OF NEAR INFRARED SPECTROSCOPY[13]

Measurements in the region of the infrared nearest to the visible, i.e. the wavelength range 8×10^{-7} to 3×10^{-6} m justify a brief mention in this chapter for two reasons. From the theoretical standpoint we should note that the absorption of e-m radiation in this region is entirely due to the overtones and combination bands of vibrations involving hydrogen, rather than to fundamental vibrations. There are a very large number of such combinations and overtones and though they are weak bands they provide strong spectra when neat samples are measured in transmission or reflectance. Generally, the spectral bands are broad and lack the distinctive features of mid infrared spectra. The second reason concerns the applications of NIR spectroscopy which are of a very special nature. When analysed by means of rather sophisticated statistical methods, NIR spectra have proved uniquely valuable in quantitative analysis, especially the analysis of foods and agricultural products; e.g. the amount of water and protein in grains, flour, malt and beans.

10.11 BIBLIOGRAPHY AND FURTHER READING

1. N.B. Colthup, L.H. Daley and S.E. Wiberley, *Introduction to Infrared and Raman Spectroscopy*, 3rd edn, Academic Press, Boston, 1990.
2. G. Herzberg, *Infrared and Raman Spectra of Polyatomic Molecules*, D. Van Nostrand Inc., Princeton, 1945.
3. P.R. Griffiths and J.A. de Haseth, *Fourier Transform Infrared Spectroscopy*, Wiley, New York, 1986.
4. N. Sheppard, The Historical Development of Experimental Techniques in Vibrational Spectroscopy, in *Handbook of Vibrational Spectroscopy* Vol. 1, (eds J.M. Chalmers and P.R. Griffiths), Wiley, Chichester, 2002.
5. J.C.D. Brand, *Lines of Light*, Gordon and Breach, United Kingdom, 1995.
6. L. Pauling and E.B. Wilson, *Introduction to Quantum Mechanics*, McGraw-Hill Inc., New York, 1935.
7. G. Herzberg, *Spectra of Diatomic Molecules*, D. Van Nostrand Inc., New York, 1950.

8. J.M. Hollas, *High Resolution Spectroscopy*, Butterworths, London, 1982.
9. L.J. Bellamy, *Infrared Spectra of Complex Molecules*, 3rd edn, Chapman and Hall, London, 1975.
10. K. Nakamoto, *Infrared and Raman Spectra of Inorganic and Coordination Compounds*, 4th edn, Wiley, New York, 1986.
11. G. Socrates, *Infrared Charateristic Group Frequencies*, 2nd edn, Wiley, Chichester, 1994.
12. A. Aamouche, F.J. Devlin and P.J. Stephens, *J. Amer. Chem. Soc.*, **122**, 2346 (2000).
13. B.G. Osborne, T. Fearn and P.H. Hindle, *Practical NIR Spectroscopy with Applications in Food and Beverage Analysis*, 2nd edn, Wiley, New York, 1963.

BOX 10.1   The reduced mass and the Hamiltonian operator for the vibration and rotation of a diatomic molecule

Vibration

Consider first the case of the vibrating dumb-bell model of a diatomic molecule (Figure B10.1.1). The kinetic energy, T, depends upon the masses of the two nuclei and the speed with which they are moving, v_{a1} and v_{a2}. The subscript 'a' indicates that the velocity with which we are concerned is that directed along the internuclear axis:

$$T = \tfrac{1}{2}m_1 v_{a1}^2 + \tfrac{1}{2}m_2 v_{a2}^2 \qquad (B10.1.1)$$

The potential energy (V) depends upon the difference between the distance between the two nuclei (R) and the equilibrium distance (r_e). Both atoms are subject to a restoring force equal to $k(R - r_e)$, which gives rise to a parabolic potential energy of the form:

$$V = \tfrac{1}{2}k(R - r_e)^2 \qquad (B10.1.2)$$

The occurrence of two masses in the equation for T presents problems and an analysis of the motion of the molecule would be easier if the two masses could be expressed as one quantity. That objective can be realised in the following way.

The masses m_1 and m_2 oscillate in phase on either side of the centre of mass in such a way that at all times:

$$m_1 r_1 = m_2 r_2 \qquad (B10.1.3)$$

The amplitude of the motion of the lighter mass is greater than that of the heavier by the ratio of their masses.

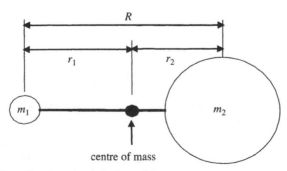

Figure B10.1.1   The vibrating dumb-bell model

At any moment in time the bond length is R and:

$$R = r_1 + r_2 \tag{B10.1.4}$$

Using the above equations we can express r_1 and r_2 in terms of R and m_1 and m_2:

$$r_1 = m_2 R/(m_1 + m_2) \quad \text{and} \quad r_2 = m_1 R/(m_1 + m_2) \tag{B10.1.5}$$

Since the centre of mass is stationary, the velocity of each mass is:

$$v_{a1} = \frac{\partial r_1}{\partial t} = \left(\frac{m_2}{m_1 + m_2}\right) \cdot \frac{\partial R}{\partial t} \equiv \left(\frac{m_2}{m_1 + m_2}\right) \cdot v_R \tag{B10.1.6}$$

and

$$v_{a2} = \frac{\partial r_2}{\partial t} = \left(\frac{m_1}{m_1 + m_2}\right) \cdot \frac{\partial R}{\partial t} \equiv \left(\frac{m_1}{m_1 + m_2}\right) \cdot v_R$$

where v_R is the rate of change of the bond length. Inserting the above results into Equation (B10.1.1) we have:

$$
\begin{aligned}
T &= \frac{m_1}{2} \cdot \left(\frac{m_2}{m_1 + m_2}\right)^2 \cdot v_R^2 + \frac{m_2}{2} \cdot \left(\frac{m_1}{m_1 + m_2}\right)^2 \cdot v_R^2 \\
&= \frac{1}{2}\left(\frac{m_1 m_2}{m_1 + m_2}\right)^2 \cdot \frac{v_R^2}{m_1} + \frac{1}{2}\left(\frac{m_1 m_2}{m_1 + m_2}\right)^2 \cdot \frac{v_R^2}{m_2} \\
&= \frac{1}{2}\left(\frac{m_1 m_2}{m_1 + m_2}\right)^2 \cdot \left(\frac{m_1 + m_2}{m_1 m_2}\right) \cdot v_R^2 \\
&\equiv \frac{1}{2\mu} \cdot (\mu v_R)^2 \quad \text{or} \quad \frac{1}{2} \cdot \mu v_R^2 \tag{B10.1.7}
\end{aligned}
$$

where

$$\mu = m_1 m_2/(m_1 + m_2) \tag{B10.1.8}$$

is the *reduced mass* which replaces m while v_R replaces v in the standard expression for the kinetic energy of a moving mass.

Thus, the total vibrational energy (E_{vib}) is given by:

$$E_{\text{vib}} = V + T = \tfrac{1}{2}k(R - r_e)^2 + \tfrac{1}{2}(\mu v_R)^2/\mu \tag{B10.1.9}$$

so that, replacing μv_R by $-(ih/2\pi)\partial/\partial R$, the energy operator $\hat{\mathcal{H}}$ is:

$$\hat{\mathcal{H}} = \frac{k}{2} \cdot (R - r_e)^2 - \frac{h^2}{8\pi^2 \mu} \cdot \frac{\partial^2}{\partial R^2} \tag{B10.1.10}$$

Rotation

There are three properties of interest which are associated with the end-over-end rotation of the dumb-bell model (Figure B10.1.2).

The moment of inertia (I):

$$I = m_1 r_1^2 + m_2 r_2^2 = \mu R^2 \tag{B10.1.11}$$

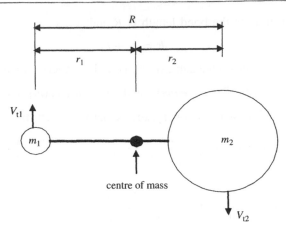

Figure B10.1.2   The rotating dumb-bell model

where we have used Equation (B10.1.5), and the angular momentum, A, (see Chapter 4):

$$A = m_1 r_1 v_{t1} + m_2 r_2 v_{t2}.$$ (B10.1.12)

The subscript 't' indicates that we are concerned with the *tangential* velocity along the circular path taken by each atom around the molecular centre of mass. The velocity is the length of the path divided by the time taken to complete one revolution which is the inverse of the frequency of rotation, v_{rot}, so that we have:

$$v_{t1} = 2\pi r_1 v_{rot} \quad \text{and} \quad v_{t2} = 2\pi r_2 v_{rot}$$ (B10.1.13)

Using this result we readily find that:

$$A = 2\pi v_{rot}(m_1 r_1{}^2 + m_2 r_2{}^2) = 2\pi v_{rot}(I_1 + I_2) = 2\pi v_{rot} I$$ (B10.1.14)

and using Equation (B10.1.11):

$$A = 2\pi v_{rot}\mu R^2 \quad \text{or} \quad \omega I$$ (B10.1.15)

where $\omega = 2\pi v_{rot}$ is the angular velocity in radians s^{-1}.

Finally, a very similar development shows that the kinetic energy (there is no potential energy) associated with the rotational motion, E_{rot}, is given by:

$$E_{rot} = 2\pi^2 v_{rot}^2 \mu R^2 = 2\pi^2 v_{rot}^2 I = \tfrac{1}{2}\omega^2 I$$ (B10.1.16)

Equations (B10.1.11), (B10.1.15) and (B10.1.16) show the close and constant relationship between I, A and E_{rot} in classical mechanics. We expect that relationship to persist in quantum mechanics so that the quantisation of the angular momentum and energy will go hand-in-hand, as they do for the one-electron atom. Indeed, we have here two particles rotating about their centre of mass, exactly as in the one-electron atom, apart from the fact that the distance between the two particles is constant and their relative potential energy may be regarded as zero in the rigid rotator approximation. (Actually, the bond becomes longer as the frequency of rotation increases, but we shall not consider such refinements to the theory here.) Therefore, the Hamiltonian operator is that for a system with kinetic energy only and in Section 3.13.4 it was shown that for motion in the x-direction it is:

$$\hat{\mathcal{H}} = -(h^2/8\pi^2 m)\partial^2\Psi/\partial x^2$$

Since we are dealing here with three-dimensional motion, $\hat{\mathcal{H}}$ becomes:

$$\hat{\mathcal{H}} = -(h^2/8\pi^2 m)\{\partial^2\Psi/\partial x^2 + \partial^2\Psi/\partial y^2 + \partial^2\Psi/\partial z^2\}$$

and the Schrödinger equation is:

$$- (h^2/8\pi^2\mu)\nabla^2\Psi = E\Psi \qquad (B10.1.17)$$

in which the three derivatives of the wave function have been written in the standard abbreviation as $\nabla^2\Psi$ and m has been replaced by the reduced mass, μ. The eigenvalues of this equation are deduced in many texts. They are:

$$E_J = \frac{h^2 J(J+1)}{8\pi^2\mu R^2} = \frac{h^2 J(J+1)}{8\pi^2 I} \approx \frac{h J(J+1)}{8\pi^2 cI} \text{ in wave numbers}$$

$$(B10.1.18)$$

where J is a quantum number which can take all positive integer values, including zero.

Using Equation (B10.1.15) we obtain the equation for the quantised values of the angular momentum as:

$$A_J = [J(J+1)]^{\frac{1}{2}} h/2\pi \qquad (B10.1.19)$$

The eigenfunctions of the energy or angular momentum operators are the same and since this is a problem with spherical symmetry they are almost always expressed in polar coordinates (Appendix 7). They are, in fact, the spherical harmonics which are described in detail in Appendix 5. There, l is used for the quantum number which we have called J here. The quantum number m_l which characterises the z-component of the angular momentum in spherical harmonic functions is called M_J or simply M in rotational spectroscopy and we should note that every level with total angular momentum characterised by J is $(2J + 1)$-fold degenerate which has important implications for the intensities of the rotational fine structure of the IR absorption bands.

BOX 10.2   Some numerical data concerning the vibration of the ^{1}H–^{35}Cl molecule (assuming that it behaves as an harmonic oscillator)

The relative molecular masses are:

$$^1\text{H} = 1.0078250 \Rightarrow 1.6735 \times 10^{-27} \text{ kg}$$

$$^{35}\text{Cl} = 34.968853 \Rightarrow 58.067 \times 10^{-27} \text{ kg}$$

which give (Equation (10.3.5)) a reduced mass, μ, of 1.6267×10^{-27} kg.

The equilibrium internuclear distance, r_e, is 127.5 pm and the $\upsilon = 0$ to $\upsilon = 1$ transition is found at 298 974 m^{-1} which corresponds to a frequency, ν, of 8.9630×10^{13} Hz.

Entering these data in Equation (10.3.3) with m replaced by μ, we find a force constant, k, of 515.9 Nm^{-1}.

The energy of the lowest vibration energy level of the molecule, E_0, is $\frac{1}{2}h\nu = 2.9695 \times 10^{-20}$ J.

When the H–Cl bond is at its maximum (or minimum) length all this energy is potential energy so that at this point Equation (10.3.1) can be used to calculate $|R - r_e|_{\max}$, the maximum extension (compression) of the bond. We find:

$$|R - r_e|_{\max} = \sqrt{\frac{2E_0}{k}} = 1.0729 \times 10^{-11} \text{ m} \qquad \text{(B10.2.1)}$$

This value of $|R - r_e|_{\max}$ for maximum extension (compression) is $\mathcal{R}$, the amplitude of the vibration, and the percentage change in bond length for this vibration, the zero-point vibration, is $\mathcal{R}/r_e = 0.084$ or 8.4 %.

The rate of change of the internuclear distance, υ_R, is given by Equation (10.2.7) and its maximum value will be that where the cosine function takes its maximum value of 1.0.

This will occur when the bond length is at its equilibrium value ($R = r_e$), which is the instant at which all the energy is kinetic energy, the potential energy being zero. Thus:

$$\upsilon_R(\max) = 2\pi\nu\mathcal{R} = 6.0424 \times 10^3 \text{ m s}^{-1} \qquad \text{(B10.2.2)}$$

The rate of change of the bond length is composed of two terms, the velocity of the hydrogen and chlorine atoms:

$$\upsilon_R = \upsilon_H + \upsilon_{Cl} \qquad \text{(B10.2.3)}$$

Furthermore, in order that the centre of mass does not move, the motions of the two atoms must satisfy the equation:

$$\upsilon_H/\upsilon_{Cl} = m_{Cl}/m_H \qquad \text{(B10.2.4)}$$

With the aid of Equation (B10.2.3) we can eliminate υ_{Cl} and obtain, $\upsilon_H = 5.8731 \times 10^3$ m s^{-1}. The value of υ_{Cl} is found to be 1.6927×10^2 m s^{-1}.

These are the maximum velocities achieved by the two atoms in the lowest vibrational state. They correspond to the situation in which all the vibrational energy is kinetic energy and the calculation can be checked by evaluating the equation:

$$E_0 = \frac{1}{2}\left(m_H\upsilon_H^2 + m_{Cl}\upsilon_{Cl}^2\right) \qquad \text{(B10.2.5)}$$

The results of the above calculations for energy levels with quantum numbers $\upsilon = 0$ to 5 are given in the table below. To emphasise the fact that the H–Cl molecule is *not*, in fact, an harmonic oscillator, the actual values of the energy levels are given in the last column of the table.

ν	E_{ν} (ho) $(10^{-20}$ J)	$\mathcal{R}$ $(10^{-11}$ m)	$\mathcal{R}/r_e$ %	$\nu_H(\max)^{\dagger}$ $(10^4$ m s$^{-1})$	$\nu_{Cl}(\max)$ $(10^2$ m s$^{-1})$	E_{ν} (obs) $(10^{-20}$ J)
0	2.969	1.0729	8.4	0.5873	1.6972	2.866
1	8.908	1.8584	14.6	1.0173	2.9318	8.393
2	14.847	2.3992	18.8	1.3133	3.7849	13.715
3	20.786	2.8387	22.3	1.5539	4.4784	18.382
4	26.725	3.2188	25.2	1.7619	5.0780	23.745
5	32.644	3.5585	27.9	1.9479	5.6139	28.455

$^{\dagger}10^4$ m s^{-1} = 22 374 mph.

Note how $\mathcal{R}$, the amplitude of the vibration, increases and recall that the forces which oppose the compression of the bond are not the same as those which oppose its extension. This removes the exact proportionality between potential energy and bond length causing the potential energy curve to deviate from the ideal parabolic form of the harmonic oscillator. This is manifested in the increasing difference between E(obs) and E(ho).

BOX 10.3 Harmonic oscillator eigenfunctions and the Hermite polynomials

The solutions of Schrödinger's equation for the harmonic oscillator are of the form:

$$\psi_{\nu}(r) = N_{\nu} H_{\nu}(\sqrt{\gamma} r) \cdot \exp(-\tfrac{1}{2}\gamma r^2) \qquad (B10.3.1)$$

where $r = R - r_e$, $\gamma = 4\pi^2 \mu \nu_c / h$ and N_{ν} is a normalising constant given by:

$$N_{\nu} = (\gamma/\pi)^{\frac{1}{4}} (2^{\nu} \cdot \nu!)^{-\frac{1}{2}}$$

The functions $H_{\nu}(\sqrt{\gamma} r)$ are the Hermite polynomials. The first few members of the series have the following forms:

$$H_0(\sqrt{\gamma} r) = 1 \qquad H_1(\sqrt{\gamma} r) = 2(\sqrt{\gamma} r) \qquad H_2(\sqrt{\gamma} r) = 4(\sqrt{\gamma} r)^2 - 2$$

$$H_3(\sqrt{\gamma} r) = 8(\sqrt{\gamma} r)^3 - 12(\sqrt{\gamma} r) \qquad H_4(\sqrt{\gamma} r) = 16(\sqrt{\gamma} r)^4 - 48(\sqrt{\gamma} r)^2 + 12$$

$$H_5(\sqrt{\gamma} r) = 32(\sqrt{\gamma} r)^5 - 160(\sqrt{\gamma} r)^3 + 120(\sqrt{\gamma} r)^2$$

and $\psi_{12}(r)$, which is plotted in Figure 10.6, contains $H_{12}(\sqrt{\gamma} r)$ which is:

$$H_{12}(\sqrt{\gamma} r) = 4096(\sqrt{\gamma} r)^{12} - 135168(\sqrt{\gamma} r)^{10} + 1520640(\sqrt{\gamma} r)^8$$
$$- 7096320(\sqrt{\gamma} r)^6 + 13305600(\sqrt{\gamma} r)^4 - 7983360(\sqrt{\gamma} r)^2 + 665280$$

There is a particularly useful recursion formula which enables a member of the series to be calculated from the two preceding members. It is:

$$H_{\nu+1}(\sqrt{\gamma} r) = 2\sqrt{\gamma} r H_{\nu}(\sqrt{\gamma} r) - 2\nu H_{\nu-1}(\sqrt{\gamma} r) \qquad (B10.3.2)$$

If we multiply throughout by $\exp(-\frac{1}{2}\gamma r^2)$ and use Equation (B10.3.1) we can find a recursion formula for the harmonic oscillator eigenfunctions:

$$\frac{\psi_{\upsilon+1}(r)}{N_{\upsilon+1}} = \frac{2\sqrt{\gamma}r\psi\upsilon(r)}{N_\upsilon} - \frac{2\upsilon H_{\upsilon-1}(r)}{N_{\upsilon-1}}$$

which can be rearranged to give a result, which we find very useful in Section 10.5:

$$r\psi\upsilon(r) = \{N_\upsilon/2\sqrt{(\gamma)}\}\{[\psi_{\upsilon+1}(r)/N_{\upsilon+1}] + [2\upsilon\psi_{\upsilon-1}(r)/N_{\upsilon-1}]\}$$

$$(B10.3.3)$$

PROBLEMS FOR CHAPTER 10

1. Confirm Equation (10.2.3) by differentiating Equation (10.2.2) with respect to t and substituting the result into Equation (10.2.1).

2. Use Equation (10.2.8) to complete the following table:

Molecule	X–H stretch[a] (m^{-1})	Reduced mass[b] (atomic mass units)	Force constant (N/m^{-1})
XH_n			
OH_2	365 200		
NH_3	333 400		
CH_4	291 400		

[a]The wave number given is that of the symmetric stretching, i.e. the vibration in which all n X–H bonds extend or contract simultaneously.
[b]1 a.m.u. $= 1.6605 \times 10^{-27}$ kg.

3. By expanding the exponential (Box 3.1) of the Morse function show that the force constant for harmonic motion (k in Equation (10.3.1)) can be expressed as:

$$k = 2D_e a^2$$

4. Under conditions of low resolution the IR spectrum of gaseous H–Cl at 25 °C shows a very strong, broad absorption band centered at 2990 cm^{-1}. Under high resolution the band can be resolved into a large number of bands; those near the centre of the spectrum have the following wave numbers (cm^{-1}):

2906.8, 2927.6, 2948.4, 2969.2, 3010.8, 3031.6, 3052.4 and 3073.2.

Draw an energy-level diagram to interpret these data and calculate the H–Cl bond length. (Answer 128.6 pm).

Note: *The experimental data above have been slightly modified in order to remove the small effects of complicating phenomena which have not been described in this chapter.*

5. Following the method described in Section 10.8.2, write down the functions Γ_{10} and Γ_{01} for the vibrational state of carbon dioxide, which has one quantum of energy in the bending vibration ν_2. By converting to plane-polar co-ordinates and forming normalised, complex combinations of Γ_{10} and Γ_{01}, show that the vibrational angular momentum of the state may be characterised by the quantum numbers $l = \pm 1$.

6. In Figure 10.10, the C–Cl (7) and C–Br (8) bands of CH_2BrCl are approximately equal in intensity but the two C–Br bands of CH_2Br_2 are very different. Suggest why.

CHAPTER 11
Electronic Spectroscopy

11.0	Introduction	327
11.1	Atomic and molecular orbitals	328
11.2	The spectra of covalent molecules	329
	11.2.1 $\pi \rightarrow \pi^*$ transitions	329
	11.2.2 $n \rightarrow \pi^*$ transitions	330
	11.2.3 Transition-metal complexes	330
11.3	Charge transfer (CT) spectra	330
11.4	Many-electron wave functions	332
11.5	The $1s^1 2s^1$ configuration of the helium atom; singlet and triplet states	333
	11.5.1 The energies of the $1s \rightarrow 2s$ excited states of the helium atom	335
	11.5.2 The one-electron energies; operator $-\frac{1}{2}\nabla_1^2 - \frac{1}{2}\nabla_2^2 - 2/r_1 - 2/r_2$	335
	11.5.3 The two-electron, i.e. electron-repulsion, energy; operator $1/r_{12}$	336
	11.5.4 The total energies of singlet and triplet state	338
	11.5.5 Electron repulsion in the triplet and singlet states of the excited helium atom: a diagrammatic illustration	339
	11.5.6 Summary of Section 11.5	341
11.6	The π-electron spectrum of benzene	341
11.7	Selection rules	344
	11.7.1 Electron spin (multiplicity) and transition probability	344
	11.7.2 Spatial aspects of transition probability for an allowed electronic transition	346
	11.7.3 The vibrational factor in the transition probability	347
11.8	Slater determinants (Appendix 6)	348
11.9	Bibliography and further reading	348
	Problems for Chapter 11	348

11.0 INTRODUCTION

Electronic spectroscopy is the oldest of the spectroscopies and has played a central role in the development of our understanding of the structure of atoms and molecules. The work of Bunsen and Kirchhoff in the 1860s demonstrated that the absorption and emission of light provided a characteristic bar code for each element which could be used for

The Quantum in Chemistry   R. Grinter
© 2005 John Wiley & Sons, Ltd

qualitative analysis. Subsequent developments have made atomic spectroscopy one of the most important quantitative analytical methods available to us. At the turn of the century, however, the sharp line spectra of atoms were of central theoretical interest because of the challenge presented by their interpretation. As we have seen in Chapter 2, this puzzle was finally solved by quantum mechanics and since that time it has been clear that no real understanding of the electronic spectra of atoms or molecules is possible without it. In the first half of the 20th century technological advances made the systematic study of the electronic spectra of molecules possible and the interpretation of the spectra became intimately linked with the quantum-mechanical theory of the chemical bond.

The significance of electronic spectroscopy in modern science often lies more in its wide range of applications than in the fundamentals of the subject itself. Thus, the time scale of electronic spectroscopy is very short (Section 8.8), so it finds application in measuring the rates of very fast processes such as photosynthesis. The transition probability of electronic transitions can be very high (Section 8.7) and electronic spectroscopy is therefore a very sensitive, quantitative tool in analytical chemistry, both in direct application and as a detection method in chromatography. The discrimination and precision of electronic spectroscopy has provided much of our data on the composition of extra-terrestrial bodies and the velocities at which they are receding from us. These are highly specialised subjects which we can take no further here. The purpose of the remainder of this chapter is to set out some of the theoretical concepts which underlie our current understanding of atomic and molecular electronic spectroscopy. Since the basic features of the spectroscopy of atoms and of the transition metal ions have been described in Chapters 5 and 7 respectively, in this chapter attention will be focused on the electronic spectra of molecules, apart from the use of the helium atom to illustrate some important principles. In Sections 11.2 and 11.3 we review the major classes of such spectra in a purely qualitative manner. Following that we outline the quantum-mechanical approach to the quantitative interpretation of molecular electronic spectra.

11.1 ATOMIC AND MOLECULAR ORBITALS

The atomic orbitals of Schrödinger's hydrogen atom (Chapter 5) and the molecular orbitals developed soon afterwards by Hund and Mulliken (Chapter 6) have proved to be indispensable starting points in the interpretation of molecular electronic spectra. The simplest approach is to identify the energy of an observed absorption band with the energy difference between the orbital which the electron initially occupied (ϕ_i) and the orbital which it finally occupies (ϕ_f):

$$\Delta E(i \to f) = E_f - E_i \qquad (11.1.1)$$

Although, as we shall see later, there are fundamental problems with this over-simplified view of electronic spectroscopy, it has nevertheless served as a useful concept on which to construct a broad view of the types of electronic spectra which we might expect atoms and molecules to have. Furthermore, more sophisticated theories almost always begin from an orbital view point and must return to orbitals when an explicit calculation of a quantity, e.g. a transition probability (Sections 8.7 and 11.7.2) is required to interpret a spectrum.

11.2 THE SPECTRA OF COVALENT MOLECULES

As we have seen in Section 6.11, the molecular orbitals of diatomic molecules may be described as σ and π bonding, σ^* and π^* antibonding and nonbonding, n, depending upon the type and phase of their atomic orbital overlap. The same essential features may be recognised in larger molecules and a generalised energy-level diagram based on them can be drawn (Figure 11.1). On the basis of this diagram, in a simple orbital view we expect that electronic transitions will increase in energy in the order $n \rightarrow \pi^*, \pi \rightarrow \pi^*, \pi \rightarrow \sigma^* \approx \sigma \rightarrow \pi^*, \sigma \rightarrow \sigma^*$. This corresponds well with the experimental observations, though transitions involving σ and σ^* orbitals are rarely seen since they lie in the far ultraviolet, well outside the range of conventional laboratory spectrometers which are normally limited to wavelengths greater than 200 nm, i.e. energies less than $\sim$600 k J mol^{-1} or $\sim$6.3 eV. The vast majority of spectra observed and used in chemical applications where transition metal ions are not involved are therefore either n $\rightarrow \pi^*, \pi \rightarrow \pi^*$ or charge-transfer transitions (see Section 11.3).

11.2.1 $\pi \rightarrow \pi^*$ transitions

As far as spectroscopy is concerned, the archetypal π-electron systems are the linear conjugated polyenes and the aromatic hydrocarbons. The Hückel theory (Section 12.1) is the simplest way of obtaining the relative energies of the π-electron molecular orbitals of such molecules. It is also the starting point for methods which introduce electron repulsion into the scheme. As one would expect from the behaviour of an electron on a ring or in a box (Sections 3.3 and 3.6), the spacing of the energy levels diminishes as the length of the electron's path, linear or circular, increases. In particular, the gap between the highest occupied MO (HOMO) and the lowest unoccupied MO (LUMO) decreases, which is reflected in the electronic spectrum by a shift of the lowest energy absorption band to longer wavelengths, i.e. lower energies;[1,2] a discussion of the spectra of linear conjugated molecules appears in Box 3.6.

The difference between the energies of the LUMO and the HOMO, as calculated by the Hückel theory, correlates well with the observed first absorption band in the case of linear conjugated systems and quite well for the aromatic hydrocarbons (Section 12.1.5, Figure 12.4). However, the apparent success of these correlations owes much to the similarity of the series of molecules compared; attempts to improve the correlation or extend the spectral interpretation to other bands or other molecules was not successful. It was

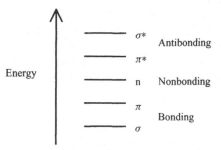

Figure 11.1 The order of energy levels in small covalent molecules

soon realised that the problem lay in the lack of explicit consideration of electron repulsion and, with the advent of computers, a systematic analysis of the spectra of the aromatic hydrocarbons provided a satisfying interpretation of their spectra; this is described in detail in Section 11.6.

11.2.2   n → π* transitions

By definition, n or lone-pair electrons play little or no role in chemical bonding and therefore their energies remain approximately constant from molecule to molecule. They are close to those in the isolated atom and lie in the energetic centre of the energy level diagram (Figure 11.1). Thus, in a simple orbital energy view an n-orbital constitutes a HOMO of constant energy and the variation in the position of an n → π* transition is a function of the changing energy of the π* LUMO, which decreases as the length of the conjugated chain increases. This concept has provided a clear interpretation of the n → π* transitions observed in the spectra of conjugated carbonyl compounds.[1,2] More sophisticated analysis of such spectra was hampered by the fact that the n-orbitals could not be readily incorporated into a π-electron theory that included electron repulsion. However, later methods, based upon the same principles but including the σ as well as the π electrons, provided a more versatile and comprehensive interpretation of the spectra of organic molecules.

11.2.3   Transition-metal complexes

The interpretation of the spectra of the transition metal complexes is very specialised and differs significantly from that of covalent molecules in several important respects. Furthermore, the theoretical analysis of the spectra is intimately linked with the explanation of other properties of these compounds, notably their magnetism. Therefore, the spectra of transition metal complexes are discussed in Chapter 7, which is devoted to all aspects of the electronic structure of these compounds.

11.3   CHARGE TRANSFER (CT) SPECTRA

Many, though not all, electronic transitions involve a redistribution of charge. (But note that a redistribution of charge is not *essential* for a transition dipole moment, Appendix 9.) Normally, electrons move from one part of a molecule to another and the process is described as *intramolecular* charge transfer. An example of this is the complex ion $[Fe^{II}(2,2'\text{-bipyridene})_3]^{2+}$ which absorbs strongly at 522 nm. ($\varepsilon = 8700 \, l \, mol^{-1} \, cm^{-1}$). In the transition responsible for the absorption the electron is transferred from the ferrous iron atom to the bipyridene ligands, leaving the iron in the ferric state. This is a metal-to-ligand charge transfer (MLCT) and such transitions find important applications in analytical chemistry on account of their specificity and sensitivity. LMCT transitions are also important, e.g. the intense purple of the $[MnO_4]^-$ ion.

In extreme cases the charge is essentially transferred from one molecule to another; *intermolecular* charge transfer. A very dramatic example of this is the blood-red solution produced when a colourless solution of hexamethyl benzene is added to a faintly yellow solution of chloranil. The electron is transferred from the chloranil to the hexamethyl benzene.

Outside the range of the human eye are the charge transfer to solvent (CTTS) transitions of some inorganic ions, especially the halide ions. In these transitions an electron is transferred from the halide ion to the water molecules in its immediate vicinity. This much is agreed, but there remains some uncertainty concerning the exact nature of the orbital in which the electron ends its journey.

If the intensity of the charge-transfer band of a solution of two substances, an electron donor D and an electron acceptor A, is studied as a function of the concentrations of A and D, it can be readily established that there is an equilibrium between separate A and D and a weak complex of the two, $A^- - D^+$, in which there is an appreciable transfer of charge from D to A:

$$A + D \leftrightarrow A^- - D^+ \tag{11.3.1}$$

The complex $A^- - D^+$ is usually very weakly bonded and can only be isolated in exceptional circumstances. It is the charge transfer process itself which is responsible for the stability of the complex as R.S. Mulliken pointed out in 1952. If we describe the states on either side of the above equilibrium with wave functions $\phi(A + D)$ and $\phi(A^- - D^+)$, then Mulliken proposed that the ground state of the complex (Ψ_g) would be a state which was largely $\phi(A + D)$ with a small admixture of $\phi(A^- - D^+)$, while for the excited state (Ψ_e) the reverse would be the case. He wrote the ground and excited state wave functions as:

$$\Psi_g = \cos\theta \cdot \phi(A + D) + \sin\theta \cdot \phi(A^- - D^+) \tag{11.3.2a}$$

and

$$\Psi_e = \sin\theta \cdot \phi(A + D) - \cos\theta \cdot \phi(A^- - D^+) \tag{11.3.2b}$$

Use of the sine and cosine functions as multiplying coefficients ensures that the two wave functions are orthogonal and normalised, provided that it is assumed that the two basis functions, $\phi(A + D)$ and $\phi(A^- - D^+)$, are also orthonormal; even though this is not usually the case. Mulliken's wave functions show how the ground state of the complex is stabilised by an admixture of the charge transfer state in a way which is essentially quantum mechanical.

Mulliken also estimated the energy of the charge transfer absorption band (ΔE) as the ionisation energy of the donor (I_D) minus the electron affinity of the acceptor (E_A) and other contributing terms (Ξ) with an equation of the form:

$$\Delta E = I_D - E_A - \Xi \tag{11.3.3}$$

The equation predicts that a graph of ΔE against I_D for a series of donors and a single acceptor, or the corresponding plot for a series of acceptors and a single donor will be a straight line and this is found to be approximately true in practice. There are however, considerable deviations from the expected linearity and slope which are the result of the fact that the last term, Ξ, also varies with A and D. This suggests that we examine it in a little more detail.

The contributions made by the ionisation energy and the electron affinity (recall that it is the unconventional way in which electron affinity is defined which results in it appearing in this equation with the negative sign) are clear. There are two major contributions to Ξ. The first is the Coulombic attraction (E_C) between the negative charge created on the acceptor and the positive charge on the donor. The second part is the quantum mechanical or *resonance* stabilisation (E_R), which results from the mixing of the two basis functions, $\phi(A + D)$ and $\phi(A^- - D^+)$ and makes Ψ_g lower in energy than $\phi(A + D)$ and Ψ_e higher

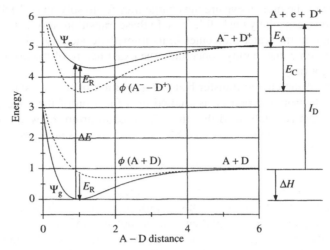

Figure 11.2   Energy curves for a charge-transfer complex

than $\phi(A^- - D^+)$. How the interaction of the basis states $\phi(A + D)$ and $\phi(A^- - D^+)$ splits their energy curves apart when the composite states (Equations (11.3.2a) and (11.3.2b)) are formed is illustrated in Figure 11.2. In its essence this is just the same as the stabilisation which results when two hydrogen atoms combine their 1s wave functions to form bonding, σ_{1s}, and antibonding, σ^*_{1s}, molecular orbitals.

11.4   MANY-ELECTRON WAVE FUNCTIONS

The electrons themselves have played a rather insignificant role in the brief survey of electronic spectroscopy above. But, with the notable exception of the hydrogen atom and some ions of the early members of the periodic table, atoms and molecules contain many electrons each of which feels the repulsion of all the other electrons in the system (Section 5.7). These electron-repulsion energies are large and require explicit consideration. If the two electrons of the helium atom are placed opposite each other in the first Bohr orbit of the atom, the energy of the repulsion between them is 4.35×10^{-18} J or 2.62×10^3 kJ mol^{-1}; the energy of the attraction of each to the doubly charged nucleus is, of course, four times as large. Although this is a very artificial model, the energies determined are of the same order of magnitude as those obtained by more sophisticated calculations (Table 11.2) and, clearly, the repulsion energy cannot be ignored.

Since the energy due to interelectronic repulsion is so large, the description of any one electron depends upon the positions of all the others and the wave function describing the electronic state of an atom or molecule must therefore be a wave function which includes all the electrons. Thus, the wave function of the helium atom in its ground state, configuration 1s^2, must be something of the form $\Psi(\text{He g.s.}) = \phi_{1s}\alpha(1) \cdot \phi_{1s}\beta(2)$. We take the product, and not the sum, of the two atomic wave functions because wave functions represent probabilities. But, as it stands, this is not sufficient. In Section 2.8 we saw that we cannot ascribe a trajectory or path to an electron and one consequence of this is that we cannot 'keep track' of an electron. Since we cannot follow an electron along a path we cannot ever be sure which of many electrons we are observing; we cannot distinguish them with a label as we can identical classical objects. It is important to understand that this is a fundamental problem, not a practical one, and our helium atom wave function must therefore contain the second possible way of assigning electrons 1 and 2 to orbitals $\phi_{1s}\alpha$ and $\phi_{1s}\beta$, i.e. $\phi_{1s}\beta(1) \cdot \phi_{1s}\alpha(2)$.

We now have two ways of combining these two parts of the wave function:

or:

$$\Psi_+(\text{He g.s.}) = \sqrt{\tfrac{1}{2}}\{\phi_{1s}\alpha(1) \cdot \phi_{1s}\beta(2) + \phi_{1s}\beta(1) \cdot \phi_{1s}\alpha(2)\} \tag{11.4.1a}$$

$$\Psi_-(\text{He g.s.}) = \sqrt{\tfrac{1}{2}}\{\phi_{1s}\alpha(1) \cdot \phi_{1s}\beta(2) - \phi_{1s}\beta(1) \cdot \phi_{1s}\alpha(2)\} \tag{11.4.1b}$$

The two results are very different. (The factor $\sqrt{\tfrac{1}{2}}$ ensures that the wave function remains normalised.) If we take the positive sign the exchange of electrons does not change the sign of Ψ(He g.s.), but if we take the negative sign the sign changes:

$$\hat{\mathcal{P}}_{1,2}\Psi_+(\text{He g.s.}) = \sqrt{\tfrac{1}{2}}\{\phi_{1s}\alpha(2) \cdot \phi_{1s}\beta(1) + \phi_{1s}\beta(2) \cdot \phi_{1s}\alpha(1)\} = \Psi_+(\text{He g.s.})$$

but:

$$\hat{\mathcal{P}}_{1,2}\Psi_-(\text{He g.s.}) = \sqrt{\tfrac{1}{2}}\{\phi_{1s}\alpha(2) \cdot \phi_{1s}\beta(1) - \phi_{1s}\beta(2) \cdot \phi_{1s}\alpha(1)\} = -\Psi_-(\text{He g.s.})$$

where $\hat{\mathcal{P}}_{1,2}$ is the operator for the exchange of electrons 1 and 2. The wave function Ψ_+ (He g.s.) does not change sign and is said to be symmetric with respect to electron exchange while the wave function Ψ_- (He g.s.) changes sign and is antisymmetric with respect to that operation. Quantum mechanics tells us that both possibilities cannot co-exist in nature, but does not say which one is, in fact, found. However, experiment shows that it is the function which is antisymmetric with respect to electron exchange, Ψ_-(He g.s.), which is found in our universe and this gives rise to a law which may be stated:

Many-electron wave functions are antisymmetric with respect to electron exchange.

This is, in fact, just another way of stating the Pauli principle because if two electrons could occupy the same atomic orbital with the same spin then:

$$\Psi_-(\text{He g.s.}) = \sqrt{\tfrac{1}{2}}\{\phi_{1s}\alpha(2) \cdot \phi_{1s}\alpha(1) - \phi_{1s}\alpha(2) \cdot \phi_{1s}\alpha(1)\} = 0$$

and the wave function vanishes.

The fact that atomic and molecular orbitals can be occupied by not more than two electrons with their spins paired has a profound influence upon the structure of atoms and molecules and thus upon the whole of chemistry. It also plays an important role in electronic spectroscopy and it is this aspect of the Pauli principle which we must examine in this chapter. We again use the example of the helium atom.

11.5   THE $1s^12s^1$ CONFIGURATION OF THE HELIUM ATOM; SINGLET AND TRIPLET STATES

Though it is not always possible, we find it very convenient here to separate the electronic wave functions of the $1s^12s^1$ configuration of the helium atom into spin and space parts. For the space part, there are two ways of describing the assignment of the two electrons, 1 and 2, to the two atomic orbitals (ϕ_{1s} and ϕ_{2s}):

$$\phi_{1s}(1) \cdot \phi_{2s}(2) \text{ and } \phi_{2s}(1) \cdot \phi_{1s}(2)$$

Four combinations of the spin functions of the two electrons are possible:

$$\alpha(1) \cdot \alpha(2), \quad \beta(1) \cdot \beta(2), \quad \alpha(1) \cdot \beta(2) \text{ and } \beta(1) \cdot \alpha(2)$$

As we have seen above, quantum mechanics is quite adamant about the effect upon a wave function of the exchange of identical particles. A wave function must be clearly symmetric or antisymmetric with respect to such an operation. The two space wave functions above change one into the other and hence have no identifiable symmetry in this respect. They also imply that we can assign a known electron to a particular orbital. Both these facts are quite unacceptable and must be removed by taking the two combinations:

$$\Psi_+ = \sqrt{\tfrac{1}{2}}\{\phi_{1s}(1) \cdot \phi_{2s}(2) + \phi_{2s}(1) \cdot \phi_{1s}(2)\} \qquad (11.5.1a)$$

and

$$\Psi_- = \sqrt{\tfrac{1}{2}}\{\phi_{1s}(1) \cdot \phi_{2s}(2) - \phi_{2s}(1) \cdot \phi_{1s}(2)\} \qquad (11.5.1b)$$

Ψ_+ is clearly symmetric with respect to electron exchange while Ψ_- is antisymmetric and these functions are potentially acceptable provided that, when they are combined with acceptable spin functions, the total wave function is antisymmetric. When we check the spin functions for symmetry with respect to electron exchange we see that the first two are symmetric while the last two change one into the other. We resolve the problem, as before, by taking normalised combinations of the problem functions and obtain a symmetric (Φ_+) and an antisymmetric (Φ_-) spin function:

$$\Phi_+ = \sqrt{\tfrac{1}{2}}\{\alpha(1) \cdot \beta(2) + \beta(1) \cdot \alpha(2)\} \qquad (11.5.2a)$$

$$\Phi_- = \sqrt{\tfrac{1}{2}}\{\alpha(1) \cdot \beta(2) - \beta(1) \cdot \alpha(2)\} \qquad (11.5.2b)$$

We should also note that Φ_+ and Φ_- are eigenfunctions of $\hat{S}$, the operator for the total spin angular momentum (Chapter 4), whereas $\alpha(1) \cdot \beta(2)$ and $\beta(1) \cdot \alpha(2)$ are not. A complete wave function for the $1s^1 2s^1$ configuration of the helium atom must be a product of a spin function and a space function, so we have eight possible functions and the requirement for antisymmetry applies to the *total wave function*, space × spin, not to the individual space or spin parts (if they can be separated as they are in this example). It is easy to check for antisymmetry because the product of two symmetric or two antisymmetric functions gives a symmetric function, whereas an antisymmetric function results from combining a symmetric with an antisymmetric function. The possibilities are set out in Table 11.1. We find that there is one acceptable wave function ($^1\Psi$) formed from the symmetric space and the antisymmetric spin part and three acceptable functions ($^3\Psi$) formed from the antisymmetric space and the symmetric spin parts. The three functions are said to form the three components of a *triplet* whilst the other function is a *singlet*:

$$^1\Psi = \tfrac{1}{2}\{\phi_{1s}(1) \cdot \phi_{2s}(2) + \phi_{2s}(1) \cdot \phi_{1s}(2)\} \{\alpha(1) \cdot \beta(2) - \beta(1) \cdot \alpha(2)\} \quad (11.5.3a)$$

$$\alpha(1) \cdot \alpha(2)$$

$$^3\Psi = \tfrac{1}{2}\{\phi_{1s}(1) \cdot \phi_{2s}(2) - \phi_{2s}(1) \cdot \phi_{1s}(2)\}\{\alpha(1) \cdot \beta(2) + \beta(1) \cdot \alpha(2)\} \quad (11.5.3b)$$

$$\beta(1) \cdot \beta(2)$$

Table 11.1 The possible combinations of space and spin functions for the 1s^{1}2s^1 configuration of the helium atom

Spin wave functions	Space wave functions	
	Symmetric Ψ_+	Antisymmetric Ψ_-
Symmetric $\alpha(1) \cdot \alpha(2)$ $\Phi_- = \sqrt{\frac{1}{2}}\{\alpha(1) \cdot \beta(2) + \beta(1) \cdot \alpha(2)\}$ $\beta(1) \cdot \beta(2)$ Antisymmetric	3 unacceptable symmetric functions	3 acceptable antisymmetric functions
$\Phi_- = \sqrt{\frac{1}{2}}\{\alpha(1) \cdot \beta(2) - \beta(1) \cdot \alpha(2)\}$	1 acceptable antisymmetric function	1 unacceptable symmetric function

Furthermore, recalling our discussion of angular momentum in Chapter 4, it is clear that the spin functions $\alpha(1) \cdot \alpha(2)$, $\sqrt{\frac{1}{2}}\{\alpha(1) \cdot \beta(2) + \beta(1) \cdot \alpha(2)\}$ and $\beta(1) \cdot \beta(2)$ have z-components of spin angular momentum of $+1$, 0 and -1 respectively. The three symmetric spin functions are therefore the three components of a spin system having a total spin quantum number, S, equal to 1. The singlet is a spin state with $S = 0$. These two states are the lowest two excited states of the helium atom and from the point of view of electronic spectroscopy we need to know their energies.

11.5.1 The energies of the 1s → 2s excited states of the helium atom

Having deduced the wave functions which describe two of the excited states of the helium atom we should now be able to calculate the energies of the states. The most significant contributions to the energies are the kinetic energy, the attraction between the nucleus and the electrons and the interelectronic repulsion. In atomic units (Appendix 1) the Hamiltonian operator is:

$$\hat{\mathcal{H}} = -\frac{\nabla_1^2}{2} - \frac{\nabla_2^2}{2} - \frac{2}{r_1} - \frac{2}{r_2} + \frac{1}{r_{12}} \qquad (11.5.4)$$

In this Hamiltonian the first two terms are the operators for the kinetic energies of electrons one and two respectively. The third and fourth terms represent the attraction of each electron for the nucleus which has a charge of $+2$. The first four terms each involve only one electron and are therefore often known as the *one-electron* terms or operators. Since r_{12} is the distance between the electrons, the last term is the interelectronic repulsion; it is clearly a *two-electron* term.

11.5.2 The one-electron energies; operator $-\frac{1}{2}\nabla_1{}^2 - \frac{1}{2}\nabla_2{}^2 - 2/r_1 - 2/r_2$

If electron repulsion is neglected, the electronic energy is that of two electrons occupying the 1s and 2s atomic orbitals of a hydrogen-like atom which has a nuclear charge of $Z = +2$. This energy, potential plus kinetic, is given by the formula below, in which Z is the nuclear charge and n the principal quantum number (Section 5.5):

$$E(Z, n) = -\frac{1}{2} \sum_{n=1,2} (Z/n)^2 = -\frac{1}{2}\{(4/1) + (4/4)\} = -2.5E_H \text{ (Hartrees)}$$

Note that the energy is negative. The system is more stable than the isolated particles at rest would be.

11.5.3 The two-electron, i.e. electron-repulsion, energy; operator $1/r_{12}$

The triplet wave function where the two electrons have α spin components is:

$$\Psi_{\alpha\alpha} = \sqrt{\tfrac{1}{2}}\{\phi_{1s}(1) \cdot \phi_{2s}(2) - \phi_{2s}(1) \cdot \phi_{1s}(2)\} \times \alpha(1)\alpha(2) \qquad (11.5.5)$$

and the electron repulsion energy integral to be evaluated (ER_{trip}) is:

$$ER_{trip} = \tfrac{1}{2}\langle\phi_{1s}(1) \cdot \phi_{2s}(2) - \phi_{2s}(1) \cdot \phi_{1s}(2)|1/r_{12}|\phi_{1s}(1) \cdot \phi_{2s}(2) - \phi_{2s}(1) \cdot \phi_{1s}(2)\rangle$$
$$\times \langle\alpha(1)\alpha(2)|\alpha(1)\alpha(2)\rangle$$

Note that the integration implied by $\langle\rangle$ must be carried out for both electrons and over both their space and their spin functions. These two types of integration can be separated because $1/r_{12}$ only operates on the spatial part of the wave function. We consider the integration term-by-term. Taking the first term on the left of the operator with the first on the right in the above integral we have:

$$\langle\phi_{1s}(1) \cdot \phi_{2s}(2)|1/r_{12}|\phi_{1s}(1) \cdot \phi_{2s}(2)\rangle \times \langle\alpha(1)\alpha(2)|\alpha(1)\alpha(2)\rangle$$
$$= \langle\phi_{1s}(1) \cdot \phi_{2s}(2)|1/r_{12}|\phi_{1s}(1) \cdot \phi_{2s}(2)\rangle \times \langle\alpha(1)\alpha(1)\rangle \times \langle\alpha(2)\alpha(2)\rangle$$
$$= \langle\phi_{1s}(1) \cdot \phi_{2s}(2)|1/r_{12}|\phi_{1s}(1) \cdot \phi_{2s}(2)\rangle \equiv J_{1s,2s}$$

where the third line follows from the second because the spin functions are ortho-normal and $J_{1s,2s}$ is simply a symbol for the electron repulsion integral which precedes it. Later, we shall give it a numerical value. If we evaluate the integral formed from the second term on the left and the second on the right we get an extremely similar result, with electron 1 now in the 2s orbital and electron 2 in the 1s. Since all electrons are identical, this term must have exactly the same value as the one to which we have assigned the symbol $J_{1s,2s}$. We shall return to give a physical meaning to the integral when we have evaluated another of a different form. The first term on the left and the second on the right give the following contribution to the integral:

$$- \langle\phi_{1s}(1) \cdot \phi_{2s}(2)|1/r_{12}|\phi_{2s}(1) \cdot \phi_{1s}(2)\rangle \times \langle\alpha(1)\alpha(2)|\alpha(1)\alpha(2)\rangle$$
$$= -\langle\phi_{1s}(1) \cdot \phi_{2s}(2)|1/r_{12}|\phi_{2s}(1) \cdot \phi_{1s}(2)\rangle \times \langle\alpha(1)\alpha(1)\rangle \times \langle\alpha(2)\alpha(2)\rangle$$
$$= -\langle\phi_{1s}(1) \cdot \phi_{2s}(2)|1/r_{12}|\phi_{2s}(1) \cdot \phi_{1s}(2)\rangle \equiv -K_{1s,2s}$$

The second term on the left and the first on the right give a similar result so that we finally obtain:

$$ER_{trip} = \tfrac{1}{2}\{J_{1s,2s} + J_{1s,2s} - K_{1s,2s} - K_{1s,2s}\} = J_{1s,2s} - K_{1s,2s} \qquad (11.5.6)$$

Exactly the same result is obtained for the other two triplet wave functions. It is appropriate at this point to attempt to obtain a physical picture of the meaning of these two-electron repulsion integrals. $J_{1s,2s}$ is quite straightforward but it may be helpful if we express the integral to be evaluated in a more familiar form:

$$J_{1s,2s} = \int_2 \int_1 \{\phi^*_{1s}(1) \cdot \phi_{1s}(1)|1/r_{12}|\phi^*_{2s}(2) \cdot \phi_{2s}(2)\}\, dv_1 dv_2$$

Note that the integration has to be carried out over all three spatial co-ordinates of both electrons, i.e. six integrations in total, and the two wave functions (ϕ_{1s} and ϕ_{2s}) and the distance between the electrons (r_{12}) must all be expressed in the same co-ordinate system. Since the probability of finding electron 1 at a particular point in space is given by $\phi_{1s}^*(1)\phi_{1s}(1)$, we have in the expression for $J_{1s,2s}$ the probability distributions for electrons 1 and 2, the former in ϕ_{1s} and the latter in ϕ_{2s}. The operator is the inverse distance between the two electrons. Thus, the physical interpretation of the integral is that we take the product of an element of the distribution of electron 1 and an element of the distribution of electron 2, divide by the distance between them and sum up (integrate) all such terms. This is exactly the procedure we would adopt if we wished to calculate the repulsion of two charge distributions in classical electrostatics (Figure 11.3). Thus $J_{1s,2s}$ represents a classical Coulombic repulsion energy and is therefore known as a *Coulomb* integral.

But what of $K_{1s,2s}$? In this integral the distributions of the two electrons are not given by the product of an atomic orbital function and its complex conjugate but by the product of one atomic orbital function and the complex conjugate of another! Once we have accepted, as we must do, that $\phi^*_{1s}(1)\phi_{2s}(1)$ also describes the spatial distribution of electron 1 and $\phi^*_{2s}(2)\phi_{1s}(2)$ that of electron 2, then we can view the integration to obtain the contribution to the electron repulsion in just the same manner as we viewed the calculation of $J_{1s,2s}$. The fundamental difference lies in the fact that the description of an electron distribution as a product of two different functions is an entirely quantum-mechanical concept; it has no classical counterpart. The existence of integrals such as $K_{1s,2s}$ is one of the most important reasons for the difference between the results of classical and quantum mechanical calculations. Because it arises as a result of the fact that, in quantum mechanics, we have to allow for the exchange of electrons between orbitals, $K_{1s,2s}$ is usually called an *exchange* integral.

We can now calculate the contribution of electron repulsion to the energy of the singlet state. We have to evaluate:

$$\text{ER}_{\text{sing}} = \tfrac{1}{4}\langle\phi_{1s}\alpha(1)\cdot\phi_{2s}\beta(2) - \phi_{2s}\beta(1)\cdot\phi_{1s}\alpha(2) - \phi_{1s}\beta(1)\cdot\phi_{2s}\alpha(2) + \phi_{2s}\alpha(1)\cdot\phi_{1s}\beta(2)$$

$$|1/r_{12}|\phi_{1s}\alpha(1)\cdot\phi_{2s}\beta(2) - \phi_{2s}\beta(1)\cdot\phi_{1s}\alpha(2) - \phi_{1s}\beta(1)\cdot\phi_{2s}\alpha(2) + \phi_{2s}\alpha(1)\cdot\phi_{1s}\beta(2)\rangle$$

In this integration terms of exactly the same form as the $J_{1s,2s}$ and $K_{1s,2s}$ found above arise. But among the 16 terms there are others of a type we have not yet encountered.

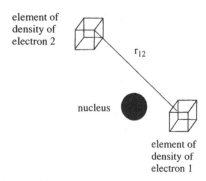

Figure 11.3 The Coulomb integral J$_{1s,2s}$

For example, the first term on the left of $|1/r_{12}|$ and the second on the right give:

$$- \langle \phi_{1s}\alpha(1) \cdot \phi_{2s}\beta(2)|1/r_{12}|\phi_{2s}\beta(1) \cdot \phi_{1s}\alpha(2)\rangle$$

$$= -\langle \phi_{1s}(1)\phi_{2s}(2)|1/r_{12}|\phi_{2s}(1)\phi_{1s}(2)\rangle \times \langle \alpha(1)\beta(2)|\beta(1)\alpha(2)\rangle$$

$$= -\langle \phi_{1s}(1)\phi_{2s}(2)|1/r_{12}|\phi_{2s}(1)\phi_{1s}(2)\rangle \times \langle \alpha(1)|\beta(1)\rangle \times \langle \beta(2)\alpha(2)\rangle = 0$$

where the integral vanishes because the spin functions are orthogonal.

The final result for the electron repulsion energy of the singlet state is:

$$\text{ER}_{\text{sing}} = \tfrac{1}{4}\{4J_{1s,2s} + 4K_{1s,2s}\} = J_{1s,2s} + K_{1s,2s} \qquad (11.5.7)$$

11.5.4 The total energies of singlet and triplet state

Fortunately, the integrals $J_{1s,2s}$ and $K_{1s,2s}$ are not particularly difficult to evaluate. They depend upon the nuclear charge (Z), which determines the extension in space of the two spherical atomic orbitals, and are found to be:

$$J_{1s,2s} = Z(17/81) = 0.420\ E_H \quad \text{and} \quad K_{1s,2s} = Z(16/729) = 0.044\ E_H$$

Armed with these figures we can complete the first three rows of Table 11.2. It is immediately clear that, because $K_{1s,2s}$ enters the singlet energy expression with a positive sign and the triplet energy expression with a negative sign, the calculated energies of the two states should differ by $2K_{1s,2s}(\Delta E = 0.088\ E_H = 0.384 \times 10^{-18}\ \text{J})$; the triplet being the lower. Qualitatively, this fact is completely substantiated by experiment, though we shall see below that a quantitative comparison exposes a weakness in our theory. The difference in energy of the singlet and triplet states arising from the same electronic configuration is quite general and has very widespread implications. Since the vast majority of covalent molecules have all their electrons paired in their ground state they have singlet ground states. However, the excitation of any one of those electrons can give rise to either a singlet (spins still paired) or a triplet (spins parallel) state, with the triplet state always $2K$ lower in energy than the singlet. Since the selection rule strongly forbids transitions between the singlet ground state and a triplet excited state (Section 11.7), transitions to triplet states are much less frequent than those to singlet states. But if a molecule does enter a triplet excited state then its transition to the ground state is forbidden and it may remain in the excited state for a time that can be as long as seconds. The singlet excited state, on the other hand, returns to the ground state in a time of the order of 10^{-9} s. The exceptional life of the excited triplet state offers the energy-rich molecule a long time in which to react and much photochemistry proceeds from this fact.

Table 11.2 The energies of the $1s^1 2s^1$ states of the helium atom

	Triplet		Singlet	
	(E_H)	10^{18}(J)	(E_H)	10^{18}(J)
One-electron energy	−2.500	−10.899	−2.500	−10.899
Coulomb energy, $J_{1s,2s}$	+0.420	+1.831	+0.420	+1.831
Exchange energy, $K_{1s,2s}$	−0.044	−0.192	+0.044	+0.192
Total (sum first three rows)	−2.124	−9.260	−2.036	−8.876
Experimental energy	−2.175	−9.487	−2.146	−9.360
∴ Charge correlation	−0.051	−0.227	−0.110	−0.484

A very important fact must be stressed. The differing energies of the two states with different electron pairings, singlet (spins antiparallel) and triplet (spins parallel) is only an *indirect* result of the different spin pairings. Because of the Pauli principle the spatial distributions of the electrons in the two states differ; because of that their electron repulsion energies differ and hence their total energies differ. The singlet/triplet energy difference is **not** a consequence of a magnetic interaction between the two electrons. These thoughts lead us on to consider the last two rows of Table 11.2.

The true energies of the helium singlet and triplet states can be obtained from the electronic spectrum of the atom. The last row of the table gives the difference between theory and experiment, which arises as follows. Since the electrons are charged they move such that they keep out of each other's way, i.e. their motions are correlated. Expressed in another way, the 1s and 2s orbitals of a two-electron atom differ from those of a one-electron atom because the motion of one electron is modified by the repulsion of the other. We do not fully allow for this fact when we calculate $K_{1s,2s}$ and $J_{1s,2s}$ using ϕ_{1s} and ϕ_{2s} since these are one-electron orbitals with a *fixed*, spherical distribution. A comparison of the singlet-triplet energy difference ($=0.029\ E_H$) and $2\ K_{1s,2s}$ ($=0.088\ E_H$) already reveals that there is a problem. This modification of the electronic motion gives rise to the *charge correlation energy*, which is the major reason for the discrepancy between theory and experiment. Note that the correlation energy of the singlet is approximately twice that of the triplet. This is due to the fact that in the triplet state the electrons are already kept more apart than in the singlet, so the effect of correlation is smaller.

11.5.5 Electron repulsion in the triplet and singlet states of the excited helium atom[3]: a diagrammatic illustration

The differing electron distributions which result in the differing energies of the singlet and triplet excited states of helium are illustrated in Figures 11.4(a) and 11.4(b). The following normalised, singlet and triplet radial functions form the basis of our plot:

$$\Psi_{\text{sing}} = 8\{\exp(-2r_1)\cdot[1-r_2]\exp(-r_2) + \exp(-2r_2)\cdot[1-r_1]\exp(-r_1)\} \tag{11.5.8}$$

$$\Psi_{\text{trip}} = 8\{\exp(-2r_1)\cdot[1-r_2]\exp(-r_2) - \exp(-2r_2)\cdot[1-r_1]\exp(-r_1)\} \tag{11.5.9}$$

Since the atomic orbitals involved are spherically symmetrical, the radial co-ordinates of the two electrons (r_1 and r_2) give all the information we require. To represent the probability of finding an electron at a particular radius (r), i.e. in the shell between r and $r + \delta r$ which increases as r^2, we multiply each of the above functions by r_1 and r_2 and square the result. Thus, the functions plotted are $\{r_1 \times r_2 \times \Psi_{\text{sing}}\}^2$ and $\{r_1 \times r_2 \times \Psi_{\text{trip}}\}^2$. r_1 is plotted on the vertical and r_2 on the horizontal axis; both in atomic units. Any point on either diagram therefore represents the momentary situation in which the distance of each electron from the nucleus is given by the projection of the point upon the relevant axis; the probability that such a situation will pertain can be determined from Equations (11.5.8) or (11.5.9) and ascribed to each point. If sufficient point probabilities are calculated for any particular atomic state a set of contours may be drawn, as shown in the figures. The most notable difference between the two figures is the zero probability along the $r_1 = r_2$ line in the triplet in contrast to the marked peak at $r_1 = r_2 \approx 0.5\ a_0$ in the singlet. Clearly, the probability that $r_1 = r_2$ is much higher in the case of the singlet, which accounts for the higher electron repulsion in that state. Of course, two electrons

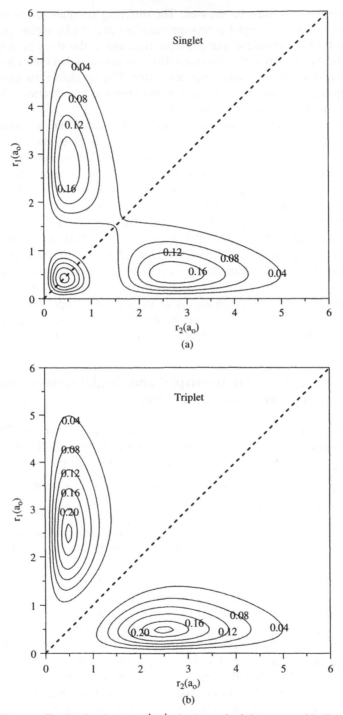

Figure 11.4 Electron distribution in the $1s^1 2s^1$ singlet and triplet states of helium

having the same value of their radial co-ordinate can reduce their mutual repulsion by moving in such a way that they are positioned on opposite sides of the nucleus. And in the real atom that is what they do. But this *angular correlation* cannot take place in our simple mathematical model because our orbitals are defined to be spherically symmetrical and do not permit a non-spherical electron distribution. To introduce an element of angular correlation we would have to use p-orbitals in our wave functions. Calculations using p-orbitals do indeed result in a reduction in energy.

It is also possible to include the correlation energy in a more explicit manner, just as James and Coolidge did for the hydrogen molecule (Section 6.8).

11.5.6 Summary of Section 11.5

The reader may feel that the arcane problem of the excited states of the helium atom has received far too much attention in the preceding paragraphs. It is true that the problem is not one with which many chemists are concerned, but it does raise a number of very important points which may be summarised as follows:

1. In quantitative work it is not sufficient to consider that the energy of an excited state is given simply by the difference in energy between two orbitals, molecular or atomic. Electron repulsion terms are very large and must be taken into account.

2. Because of the influence of electron spin on electron spatial functions, i.e. the Pauli principle, electron repulsion in a triplet state is always somewhat less than in the corresponding singlet state. Therefore, the energy of the triplet state is always lower, as Hund first noted.

3. The energy difference between singlet and triplet is due to electron repulsion. It is therefore very much larger than any energy which might arise because of a direct interaction of the electron spins through their magnetic properties.

4. The fact that singlet-triplet transitions are strictly forbidden (in molecules where spin-orbit coupling is small) means that, once created, triplet states can have a very long lifetime which is important in photochemistry.

11.6 THE π-ELECTRON SPECTRUM OF BENZENE

The $\pi \rightarrow \pi^*$ electronic spectrum of benzene provides a good illustration of the way in which interelectronic repulsion affects the spectra of systems of more general chemical interest. The Hückel molecular orbital theory of the π-electron structure of benzene (Section 12.1) gives an energy-level scheme for the molecule in which both the highest occupied MOs and the lowest unoccupied MOs are a degenerate pair (Figure 11.5). More sophisticated theories give the same result since the π-electron energy level pattern is determined by symmetry. The forms of the six molecular orbitals, in order of increasing energy (β is a negative quantity!), are:

$$\psi_1 = \{1/\sqrt{6}\}\{\phi_1 + \phi_2 + \phi_3 + \phi_4 + \phi_5 + \phi_6\} \qquad E = \alpha + 2\beta$$
$$\psi_2 = \{\tfrac{1}{2}\}\{\phi_2 + \phi_3 - \phi_5 - \phi_6\} \qquad\qquad E = \alpha + \beta \qquad \text{HOMO}$$

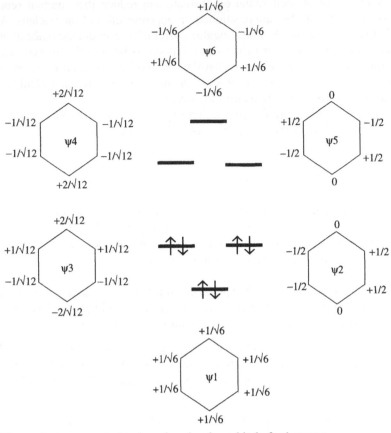

Figure 11.5 π-electron energy levels and molecular orbitals for benzene

$$\psi_3 = \{1/\sqrt{12}\}\{2\phi_1 + \phi_2 - \phi_3 - 2\phi_4 - \phi_5 + \phi_6\} \qquad E = \alpha + \beta \qquad \text{HOMO}$$

$$\psi_4 = \{1/\sqrt{12}\}\{2\phi_1 - \phi_2 - \phi_3 + 2\phi_4 - \phi_5 - \phi_6\} \qquad E = \alpha - \beta \qquad \text{LUMO}$$

$$\psi_5 = \{\tfrac{1}{2}\}\{-\phi_2 + \phi_3 - \phi_5 + \phi_6\} \qquad\qquad\quad E = \alpha - \beta \qquad \text{LUMO}$$

$$\psi_6 = \{1/\sqrt{6}\}\{\phi_1 - \phi_2 + \phi_3 - \phi_4 + \phi_5 - \phi_6\} \qquad\quad E = \alpha - 2\beta$$

In a simplistic view therefore, the spectrum of benzene at lower energies would consist of four transitions from the two HOMOs to the two LUMOs, all having the same energy. The experimental finding (Figure 11.6), however, bears no relation to this prediction. Three singlet bands are seen with maxima at approximately 256 nm (38 000 cm^{-1}), 200 nm (48 000 cm^{-1}) and 180 nm (54 500 cm^{-1}); the last is doubly degenerate. How does this arise? The discrepancy is directly attributable to the effect of electron repulsion on the energy levels of the molecule, which is best considered in two stages. First, as we have seen above (Section 11.5), the energy of an excited electronic state depends not only upon the energies of the orbital the electron leaves and that which it enters but also on the Coulomb and exchange integrals between those orbitals. In a general formulation[2] we write the energy difference between an excited singlet state ($^1\Omega_{i \rightarrow k}$), in which an electron

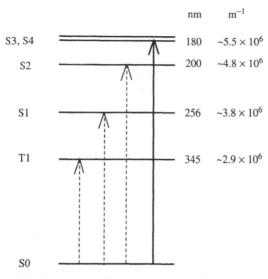

The dashed and solid vertical lines indicate
forbidden and allowed transitions respectively.

Figure 11.6   The π-electron spectrum of benzene

has been promoted from molecular orbital Ψ_i to Ψ_k, and the ground state ($^1\Omega_0$) as:

$$E(^1\Omega_{i\to k}) = \langle ^1\Omega_{i\to k}|\hat{\mathcal{H}}|^1\Omega_{i\to k}\rangle - \langle ^1\Omega_0|\hat{\mathcal{H}}|^1\Omega_0\rangle = E_k - E_i - J_{i,k} + 2K_{i,k} \quad (11.6.1)$$

where E_i and E_k are the molecular orbital energies, which also contain electron-repulsion terms, and $J_{i,k}$ and $K_{i,k}$ are the corresponding Coulomb and exchange integrals. The evaluation of the energy difference (Equation (11.6.1)) is difficult, but by making a series of justifiable approximations in the calculation of all the electron-repulsion terms the following energies may be determined:

$$E(^1\Omega_{2\to 4}) = E(^1\Omega_{3\to 5}) = 5.923 \text{ eV}$$

and

$$E(^1\Omega_{2\to 5}) = E(^1\Omega_{3\to 4}) = 6.127 \text{ eV}$$

These results show immediately that the four transitions do not all have the same energy. And there is a second effect of electron repulsion which we must take into account. The states $|^1\Omega_{i\to k}\rangle$ are not eigenfunctions of the Hamiltonian operator ($\hat{\mathcal{H}}$), which includes electron-repulsion terms, and this causes a mixing of the states through off-diagonal matrix elements (Appendix 3), which may be calculated using the formula[2]:

$$\langle ^1\Omega_{i\to k}|\hat{\mathcal{H}}|^1\Omega_{j\to l}\rangle$$
$$= 2\langle \Psi_j(1)\Psi_k(2)|1/r_{12}|\Psi_l(1)\Psi_i(2)\rangle - \langle \Psi_j(1)\Psi_k(2)|1/r_{12}|\Psi_i(1)\Psi_l(2)\rangle \quad (11.6.2)$$

Using the same approximate electron-repulsion terms as before we calculate that:

$$\langle ^1\Omega_{2\to 4}|\hat{\mathcal{H}}|^1\Omega_{3\to 5}\rangle = 1.023 \text{ eV and } \langle ^1\Omega_{3\to 4}|\hat{\mathcal{H}}|^1\Omega_{2\to 5}\rangle = -0.818 \text{ eV}$$

All other off-diagonal terms are found to be zero. We can now set up an energy matrix (Appendix 3) for the interaction of the four lowest π-electron excited states of benzene.

It is:

$\hat{\mathcal{H}}/eV$	$|{}^1\Omega_{2\to4}\rangle$	$|{}^1\Omega_{3\to5}\rangle$	$|{}^1\Omega_{3\to4}\rangle$	$|{}^1\Omega_{2\to5}\rangle$
$\langle{}^1\Omega_{2\to4}|$	5.923	1.023	0.0	0.0
$\langle{}^1\Omega_{3\to5}|$	1.023	5.923	0.0	0.0
$\langle{}^1\Omega_{3\to4}|$	0.0	0.0	6.127	-0.818
$\langle{}^1\Omega_{2\to5}|$	0.0	0.0	-0.818	6.127

We can diagonalise the two 2×2 sub-matrices separately to obtain the following four energies. Since the interactions occur only between states with equal on-diagonal matrix elements we readily find:

$${}^1B_{2u}(\alpha,{}^1L_b); \sqrt{\tfrac{1}{2}}\{{}^1\Omega_{2\to4} - {}^1\Omega_{3\to5}\}\ E = 5.923 - 1.023 = 4.90\ \text{eV} \approx 39\,500\ \text{cm}^{-1}$$

$${}^1B_{1u}(p,{}^1L_a); \sqrt{\tfrac{1}{2}}\{{}^1\Omega_{3\to4} + {}^1\Omega_{2\to5}\}\ E = 6.127 - 0.818 = 5.31\ \text{eV} \approx 43\,000\ \text{cm}^{-1}$$

$${}^1E_{1u}(\beta,{}^1B); \sqrt{\tfrac{1}{2}}\{{}^1\Omega_{2\to4} + {}^1\Omega_{3\to5}\}\ E = 5.923 + 1.023 = 6.95\ \text{eV} \approx 56\,000\ \text{cm}^{-1}$$

$${}^1E_{1u}(\beta',{}^1B); \sqrt{\tfrac{1}{2}}\{{}^1\Omega_{3\to4} - {}^1\Omega_{2\to5}\}\ E = 6.127 + 0.818 = 6.95\ \text{eV} \approx 56\,000\ \text{cm}^{-1}$$

The symbols ${}^1B_{2u}$ etc. are the symmetry labels of the states; α, p, β and β' are labels for the bands proposed by E. Clar; and ${}^1L_a$, ${}^1L_b$ and ${}^1B$ are those proposed by J.R. Platt. All are widely used by practical spectroscopists. These results are in reasonable, quantitative agreement with experiment. They illustrate the magnitude of the electron repulsion energy and bear witness to the fact that no satisfactory interpretation of electronic spectra is possible without taking it explicitly into account. Murrell[2] has given a detailed description of the methods outlined above for the interpretation of the spectra of benzene, the aromatic hydrocarbons and other π-electron systems.

11.7 SELECTION RULES

Under conditions of high resolution, absorption of light, which results in a change of electronic state accompanied by a fine structure due to changes in vibrational and rotational quantum number, may be observed in gas-phase molecules. The detailed analysis of this structure has provided much valuable information about the geometry of small molecules in their excited electronic states. However, in general chemical spectroscopy, which is usually performed on samples in solution, only vibrational fine structure is resolved; and that only in a small percentage of samples. Therefore, we shall confine our attention here to the selection rules for electronic transitions and the associated vibrational fine structure. Consider first the effect of electron spin on transition probability, using the helium atom as an example.

11.7.1 Electron spin (multiplicity) and transition probability

In Section 11.4 (Equation (11.4.1b)) we found that the correct wave function for the helium atom in its ground electronic state, which is a singlet state, is:

$$\begin{aligned}
{}^1\Psi_i &= \sqrt{\tfrac{1}{2}}\{\phi_{1s}\alpha(1)\cdot\phi_{1s}\beta(2) - \phi_{1s}\beta(1)\cdot\phi_{1s}\alpha(2)\} \\
&= \{\phi_{1s}(1)\phi_{1s}(2)\}\ \sqrt{\tfrac{1}{2}}\{\alpha(1)\beta(2) - \beta(1)\alpha(2)\}
\end{aligned} \tag{11.7.1}$$

The superscript '1' has been introduced to indicate the singlet state and the subscript 'i' because it forms the initial state in a spectroscopic transition. The corresponding wave functions for excited states in which an electron has been promoted from a 1 s to a 2p orbital (compare Equations (11.5.3)) are:

$$^1\Psi_f = \sqrt{\tfrac{1}{2}}\ \{\phi_{1s}(1) \cdot \phi_{2p}(2) + \phi_{2p}(1) \cdot \phi_{1s}(2)\}\sqrt{\tfrac{1}{2}}\{\alpha(1) \cdot \beta(2) - \beta(1) \cdot \alpha(2)\} \quad (11.7.2a)$$

$$\alpha(1) \cdot \alpha(2)$$

$$^3\Psi_f = \sqrt{\tfrac{1}{2}}\{\phi_{1s}(1) \cdot \phi_{2p}(2) - \phi_{2p}(1) \cdot \phi_{1s}(2)\}\sqrt{\tfrac{1}{2}}\{\alpha(1) \cdot \beta(2) + \beta(1) \cdot \alpha(2)\} \quad (11.7.2b)$$

$$\beta(1) \cdot \beta(2)$$

where Equation (11.7.2a) is the singlet and Equation (11.7.2b) the triplet.

As an example we shall evaluate the transition moments $\langle ^1\Psi_f|\hat{M}|^1\Psi_i\rangle$ and $\langle ^3\Psi_f|\hat{M}|^1\Psi_i\rangle$ where $\hat{M} = e(r_1 + r_2)$ (Appendix 9) paying particular attention to the influence of the electron spins. For the singlet excited state:

$$M_S = e\langle ^1\Psi_f|r_1 + r_2|^1\Psi_i\rangle$$

$$= e\sqrt{\tfrac{1}{2}}\langle\phi_{1s}(1) \cdot \phi_{2p}(2) + \phi_{2p}(1) \cdot \phi_{1s}(2)|r_1 + r_2|\phi_{1s}(1)\phi_{1s}(2)\rangle$$

$$\times \tfrac{1}{2}\langle\alpha(1)\beta(2) - \beta(1)\alpha(2)|\alpha(1)\beta(2) - \beta(1)\alpha(2)\rangle$$

We can evaluate the space and spin parts of M separately and for the space part we find:

$$= \langle\phi_{1s}(1)|r_1|\phi_{1s}(1)\rangle\langle\phi_{2p}(2)|\phi_{1s}(2)\rangle + \langle\phi_{2p}(1)|r_1|\phi_{1s}(1)\rangle\langle\phi_{1s}(2)|\phi_{1s}(2)\rangle$$

$$+ \langle\phi_{2p}(2)|r_2|\phi_{1s}(2)\rangle\langle\phi_{1s}(1)|\phi_{1s}(1)\rangle + \langle\phi_{1s}(2)|r_2|\phi_{1s}(2)\rangle\langle\phi_{2p}(1)|\phi_{1s}(1)\rangle$$

$$= 0 + \langle\phi_{2p}(1)|r_1|\phi_{1s}(1)\rangle + \langle\phi_{2p}(2)|r_2|\phi_{1s}(2)\rangle + 0 \text{ because } \langle\phi_m(1)|\phi_n(1)\rangle = \delta_{mn}$$

$$= 2\langle\phi_{2p}(1)|r_1|\phi_{1s}(1)\rangle \text{ because all electrons are identical.}$$

The integration over spin gives:

$$\langle\alpha(1)|\alpha(1)\rangle\langle\beta(2)|\beta(2)\rangle - \langle\alpha(1)|\beta(1)\rangle\langle\beta(2)|\alpha(2)\rangle$$

$$- \langle\beta(1)|\alpha(1)\rangle\langle\alpha(2)|\beta(2)\rangle + \langle\beta(1)|\beta(1)\rangle\langle\alpha(2)|\alpha(2)\rangle$$

$$= 1 - 0 - 0 + 1 = 2$$

We finally have:

$$M_S = e\sqrt{\tfrac{1}{2}} \times 2\langle\phi_{2p}(1)|r_1|\phi_{1s}(1)\rangle \times \tfrac{1}{2} \times 2 = e\sqrt{2}\langle\phi_{2p}(1)|r_1|\phi_{1s}(1)\rangle$$

For the triplet final state:

$$M_T = e\langle ^3\Psi_f|r_1 + r_2|^1\Psi_i\rangle$$

$$= e\sqrt{\tfrac{1}{2}}\langle\phi_{1s}(1) \cdot \phi_{2p}(2) - \phi_{2p}(1) \cdot \phi_{1s}(2)|r_1 + r_2|\phi_{1s}(1)\phi_{1s}(2)\rangle$$

$$\times \tfrac{1}{2}\langle\alpha(1)\beta(2) + \beta(1)\alpha(2)|\alpha(1)\beta(2) - \beta(1)\alpha(2)\rangle$$

The space part of the integral gives:

$$= \langle\phi_{1s}(1)|r_1|\phi_{1s}(1)\rangle\langle\phi_{2p}(2)|\phi_{1s}(2)\rangle - \langle\phi_{2p}(1)|r_1|\phi_{1s}(1)\rangle\langle\phi_{1s}(2)|\phi_{1s}(2)\rangle$$
$$+ \langle\phi_{2p}(2)|r_2|\phi_{1s}(2)\rangle\langle\phi_{1s}(1)|\phi_{1s}(1)\rangle - \langle\phi_{1s}(2)|r_2|\phi_{1s}(2)\rangle\langle\phi_{2p}(1)|\phi_{1s}(1)\rangle$$
$$= 0 - \langle\phi_{2p}(1)|r_1|\phi_{1s}(1)\rangle + \langle\phi_{2p}(2)|r_2|\phi_{1s}(2)\rangle - 0 = 0$$

Furthermore, the spin part of the integral also gives zero since:

$$\langle\alpha(1)|\alpha(1)\rangle\langle\beta(2)|\beta(2)\rangle - \langle\alpha(1)|\beta(1)\rangle\langle\beta(2)|\alpha(2)\rangle$$
$$+ \langle\beta(1)|\alpha(1)\rangle\langle\alpha(2)|\beta(2)\rangle - \langle\beta(1)|\beta(1)\rangle\langle\alpha(2)|\alpha(2)\rangle$$
$$= 1 - 0 + 0 - 1 = 0$$

Therefore, on both counts we find that the singlet $\leftrightarrow$ triplet transition is forbidden and we can say that, provided that spin-orbit coupling can be neglected, any transition in which there is a change of multiplicity is forbidden. It is not always possible to separate the space and spin functions as we have done here, but the result is quite general and can be demonstrated using appropriate Slater determinants (Section 11.8 and Appendix 6) as wave functions.

11.7.2   Spatial aspects of transition probability for an allowed electronic transition

To proceed further with the spatial aspects of molecular transition moments we build upon the discussion of transition probability in Appendix 9 and Sections 8.6 and 8.7. The case of an allowed electronic transition, which may or may not be accompanied by changes in vibrational levels, is Case 2 of the discussion in Section 8.7.1 where we have deduced that the transition dipole moment ($M_{i,f}$) is given by Equation 8.7.10 which we repeat here for convenience:

$$M_{i,f} = \langle\psi^f_{vib}|\psi^i_{vib}\rangle \cdot M_0(e^f e^i) \qquad (11.7.3)$$

$M_0(e^f e^i)$ is the value of the integral $\langle\psi^f_{elec}|\hat{M}|\psi^i_{elec}\rangle$ evaluated at the equilibrium nuclear configuration. $\hat{M}$ is the dipole moment operator and it is important to note that the integral extends over the electronic co-ordinates only. The integral $\langle\psi^f_{vib}|\psi^i_{vib}\rangle$ is the overlap of the vibrational wave functions of the initial and final states, which is important in the analysis of the vibrational structure of small molecules but rarely in the spectra of large molecules.

To illustrate the calculation of the electronic matrix element we take the case of the transition from a highest occupied (HOMO) to a lowest unoccupied (LUMO) π-electron molecular orbital of benzene.

The π-electron coefficients of the MOs ψ_2 and ψ_5 are shown in Figure 11.7. The y-component of the transition moment integral, T_y, between these two orbitals is:

$$\left\langle\psi_2\left|\sum_k q_k y_k\right|\psi_5\right\rangle = e\langle\tfrac{1}{2}(\phi_2 + \phi_3 - \phi_5 - \phi_6)|y_2 + y_3 + y_5 + y_6|\tfrac{1}{2}(-\phi_2 + \phi_3 - \phi_5 + \phi_6)\rangle$$

where e is the electronic charge. Since the atomic orbitals (ϕ_i) are normalised, if we neglect overlap, i.e. assume that $\langle\phi_i|\phi_j\rangle = \delta_{i,j}$, we can separate the integral into four parts and write:

$$T_y = \tfrac{1}{4}e\{\langle\phi_2|y_2|-\phi_2\rangle + \langle\phi_3|y_3|\phi_3\rangle + \langle-\phi_5|y_5|-\phi_5\rangle + \langle-\phi_6|y_6|\phi_6\rangle\}$$

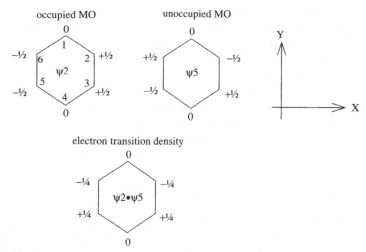

Figure 11.7 The electron transition density for a $\pi \rightarrow \pi^*$ transition in benzene

If the co-ordinate origin is the centre of the ring then $y_2 = y_6 = -y_3 = -y_5$ and if we set $y = \frac{1}{2}R_y$, where R is the C–C bond length and the subscript indicates that the moment lies in the y-direction, then:

$$T_y = \tfrac{1}{4}e\ \{\langle\phi_2|\tfrac{1}{2}R_y| - \phi_2\rangle + \langle\phi_3| - \tfrac{1}{2}R_y|\phi_3\rangle + \langle-\phi_5| - \tfrac{1}{2}R_y| - \phi_5\rangle + \langle-\phi_6|\tfrac{1}{2}R_y|\phi_6\rangle\}$$

$$= \tfrac{1}{4}e\ \{-4 \times \tfrac{1}{2}R_y\} = -\tfrac{1}{2}eR_y$$

In developing the above expression for T_y we have assumed that the π-electron transition density in the region of each carbon atom k, $\langle\phi_k|\phi_k\rangle$, is located at the position of the atomic nucleus. A similar evaluation of T_x for the $2 \rightarrow 5$ transition results in a transition moment of zero in that direction. For the $3 \rightarrow 4$ transition we find $T_y = +\tfrac{1}{2}eR_y$ and $T_x = 0$. In the previous section we found that two of the final benzene excited states ($^1B_{1u}$ and $^1E_{1u}$) were formed from the sum and difference of the states $^1\Omega_{3\rightarrow4}$ and $^1\Omega_{2\rightarrow5}$ respectively, so that for the transition to $^1B_{1u}$ the transition moment is proportional to $+\tfrac{1}{2}eR_y - \tfrac{1}{2}eR_y = 0$, i.e. the transition is forbidden. But for the transition to $^1E_{1u}$ the transition moment is proportional to $+\tfrac{1}{2}eR_y - (-\tfrac{1}{2}eR_y) = eR_y$, i.e. the transition is allowed. To obtain an absolute value for the transition moment in the y-direction we would need to evaluate eR_y at the equilibrium nuclear configuration and the vibrational overlap integral, which is discussed in the following Section. However, such calculations are rarely performed because the quality of the electronic wave functions seldom justifies more than a semi-quantitative estimate of the transition moment, the most valuable information being the direction of the moment, which is usually very easy to find and in the case of symmetrical molecules is possible using group theory alone.

11.7.3 The vibrational factor in the transition probability

The vibrational factor ($\langle\psi^f{}_{vib}|\psi^i{}_{vib}\rangle$) in $M_{i,f}$ has been discussed in Section 8.7 and Box 8.2 and will not be taken further here. We simply note that the Franck–Condon principle applies to the vibrational structure of polyatomic molecules, just as it does to diatomics, though the analysis is naturally more complicated. The calculation of vibrational overlap integrals presents formidable problems, not the least of which is the necessity of knowing the ground and excited state potential surfaces. Such calculations have rarely been carried out and then only on molecules much smaller than benzene. In the particular case of benzene, the y-polarised transition for which a non-zero transition moment was found above is one of the pair which take the molecule to the $^1E_{1u}$ state. The other member of the pair is x-polarised. The high symmetry of the molecule is responsible for the fact that the transitions to the two states of lowest energy ($^1B_{1u}$ and $^1B_{2u}$) are forbidden. However, vibrations which remove this symmetry by distorting the ring can make the transitions allowed as described under Case 3 of Section 8.7.1. Both of these forbidden bands show vibrational fine structure which has been studied in detail. Murrell gives an account of this work.[2]

11.8   SLATER DETERMINANTS (APPENDIX 6)

It is clear that there are n! (factorial n) ways of placing n electrons in n orbitals so that the proper description of a many-electron atom or molecule leads to a very complicated wave function. Slater showed that a many-electron wave function with all the required antisymmetry could be written most concisely in the form of a determinant. We shall not work with Slater's methods here; Appendix 6 contains a brief introduction. The principal reason for mentioning Slater determinants here is to emphasise that only a many-electron wave function, which may be written as a single Slater determinant or as a sum of such determinants, is an adequate description of an atom or molecule as far as its electronic spectrum is concerned.

11.9   BIBLIOGRAPHY AND FURTHER READING

The material in Section 11.5 owes a great deal to J.W. Linnett. For an excellent discussion and illustration of electron repulsion and correlation in the helium atom see his book.[3]

It is unfortunate that there appear to be very few, if any, modern treatments of the theory of the electronic spectra of molecules. The sources 1–2 and 4–6, however, present very thorough discussions of the topic and contain all the essential details.

1. S.F. Mason, *Quarterly Reviews of the Chemical Society*, **15**, 287 (1961).
2. J.N. Murrell, *The Theory of the Electronic Spectra of Organic Molecules*, Methuen, London, 1963.
3. J.W. Linnett, *Wave Mechanics and Valency*, Methuen, London, 1960.
4. H.H. Jaffé and M. Orchin, *Theory and Applications of Ultraviolet Spectroscopy*, Wiley, New York, 1962.
5. C. Sandorfy, *Electronic Spectra and Quantum Chemistry*, Prentice Hall, Englewood Cliffs, 1964.
6. G. Herzberg, *Electronic Spectra of Polyatomic Molecules*, Van Nostrand, New York, 1966.

Details concerning the evaluation of one- and two-electron matrix elements between wave functions expressed as Slater determinants are given by

7. R. Daudel, R. Lefebvre and C. Moser, *Quantum Chemistry*, Interscience, New York, 1959.

PROBLEMS FOR CHAPTER 11

1. Confirm Equation (11.5.7) for the electron-repulsion energy of the singlet state of the excited helium atom.

2. In Section 11.7.1 the transition probability for the transition from $^1\Psi_i$ to $^3\Psi_f$ in the helium atom is calculated using the function $\sqrt{\frac{1}{2}}\{\alpha(1)\beta(2) + \beta(1)\alpha(2)\}$ as the spin function for the excited state. Repeat the calculation using the spin function $\alpha(1)\alpha(2)$.

3. Following the calculation of transition moments for benzene in Section 11.7.2, calculate the transition dipole moments for the $3 \to 5$ and $2 \to 4$ transitions.

4. Confirm, following the calculation in Section 11.7.2, that the inclusion of overlap terms in the calculation of the transition dipole moment of benzene results in a small contribution to T_y which opposes that due to the other terms.

5. The electronic spectrum of the bromide ion in acetonitrile shows two absorption bands at energies corresponding to $45\,900$ cm^{-1} and $49\,300$ cm^{-1}. The iodide ion gives bands at $40\,800$ cm^{-1} and $48\,500$ cm^{-1}. These bands have been assigned to charge transfer from the halide ion to the solvent. Explain why two CT bands are seen and show that this assignment is in agreement with the fact that the bands of lowest energy in the atomic spectra of bromine and iodine are found at 3685 cm^{-1} and 7603 cm^{-1} respectively.

   [Hint: What is the lowest energy state (i.e. ground electronic state) of a halogen atom and what separates it from the state next highest in energy? If necessary, see Section 5.9 for further information.]

6. The $\pi \to \pi^*$ absorption band of ethene ($H_2C{=}CH_2$) is found at $\lambda = 162$ nm but it moves a few nanometres to longer wavelengths when the planarity of the molecule is removed by twisting about the double bond, usually by steric effects. It is difficult to separate the effect of the twisting from that of the substituent groups causing it, but we can estimate the former in the following way.

   Given that the π-electron molecular orbitals of ethene as calculated by the Hückel method (Section 12.1) are $\pi = \sqrt{\frac{1}{2}}\{\phi_1 + \phi_2\}$ and $\pi^* = \sqrt{\frac{1}{2}}\{\phi_1 - \phi_2\}$, determine their energies by calculating $\langle\pi|\hat{\mathcal{H}}|\pi\rangle$ and $\langle\pi^*|\hat{\mathcal{H}}|\pi^*\rangle$ using the Hückel approximations; $\langle\phi_1|\hat{\mathcal{H}}|\phi_1\rangle = \langle\phi_2|\hat{\mathcal{H}}|\phi_2\rangle = \alpha$ and $\langle\phi_1|\hat{\mathcal{H}}|\phi_2\rangle = \langle\phi_2|\hat{\mathcal{H}}|\phi_1\rangle = \beta$. $[E(\pi) = \alpha + \beta$ and $E(\pi^*) = \alpha - \beta]$. Convert the value of λ to energy and from that figure determine β. (We show in Section 12.1 that this primitive approach to the calculation of absorption energies, neglecting electron repulsion, can be quite successful e.g. Figure 12.4). There is a well-established relationship from which the value of β at a particular angle (β_θ) can be calculated from its value at $\theta = 0$. It is:

$$\beta_\theta = \beta_o \cos(\theta)$$

   Now, assuming that nothing apart from β changes, calculate the shift in the absorption maximum which results from a twist of $10°$. You should find a value of approximately 2 nm.

CHAPTER 12
Some Special Topics

12.0	Introduction	352
12.1	The Hückel molecular orbital (HMO) theory	352
	12.1.1 The basis of Hückel's approach	352
	12.1.2 The method	353
	12.1.3 Hückel's assumptions	354
	12.1.4 Determination of HMO energies and AO coefficients	354
	12.1.5 Applications of HMO energies	357
	12.1.6 Applications of HMO coefficients	361
	12.1.7 Some final comments on the Hückel theory	362
12.2	Magnetism in chemistry	363
	12.2.1 Magnetic susceptibility: diamagnetism and paramagnetism	364
	12.2.2 Magnetic susceptibility: ferromagnetism and antiferromagnetism	365
	12.2.3 Magnetic fields and dipoles: some definitions	365
	12.2.4 The magnetic effect of electronic orbital motion	366
	12.2.5 The consequences of chemical bonding	368
	12.2.6 The magnetic effect of electron spin	369
	12.2.7 Magnetism in practice	370
	12.2.8 Systems of interacting molecular magnets	374
	12.2.9 A note of warning	376
	12.2.10 An application	377
12.3	The band theory of solids	378
	12.3.1 The tight binding approximation	378
	12.3.2 The electron–gas (free-electron) approximation	381
	12.3.3 Molecular and ionic solids	386
	12.3.4 Applications	387
	12.3.5 Metals, insulators and semiconductors	387
	12.3.6 Optical properties of solids	390
	12.3.7 Mechanical properties of solids	390
12.4	Bibliography and further reading	390
	Problems for Chapter 12	397

The Quantum in Chemistry   R. Grinter
© 2005 John Wiley & Sons, Ltd

12.0 INTRODUCTION

This chapter is concerned with three important topics which, although they might have been accommodated in other chapters, are probably best discussed separately. Each of them has had a very significant impact upon our understanding of chemical structure and the concepts upon which they are based are now so deeply imbedded in our descriptions of molecular and material properties that they form a part of the everyday language of chemistry. They also provide some of the best examples of the way in which quite simple quantum mechanical concepts can enhance our understanding of the properties of large molecules, materials and metals that are of paramount interest to biologists, materials scientists and engineers in the wider world of applied science and technology.

12.1 THE HÜCKEL MOLECULAR ORBITAL (HMO) THEORY

The Hückel theory finds a place here for several reasons, which are worth noting at the outset. Its place would be justified because it was the first molecular orbital theory that could be applied to large molecules. It is a theory that can be implemented without the aid of a computer and is therefore an ideal test bed on which to study some of the basic principles and procedures of the MO theory. In addition, some 30 years after its birth it formed the starting point for a very successful theoretical interpretation of the electronic spectra of the aromatic hydrocarbons, which included electron repulsion terms, and for the extended Hückel theory, which has been widely used by organo-metallic chemists as an aid to the interpretation of the chemical reactions and structures of their compounds.

12.1.1 The basis of Hückel's approach

We saw in Section 6.15.2 how the formation of sp^2 hybrids between the 2s and 2p AOs of a carbon atom gives rise to three hybridised orbitals lying in a plane and ideally suited to forming 3 σ bonds to neighbouring atoms with bond angles in the plane of approximately $120°$. The remaining, unhybridised 2p AO is available for forming a π-electron structure. The geometries and special properties of the aromatic hydrocarbons and conjugated polyenes may be interpreted in terms of this type of hybridisation of their carbon atoms. In 1931 Erich Hückel (1896–1980) proposed that the energies and wave functions of the π-electron system could be calculated using an approach in which the σ-electron system was disregarded, apart from the role which it plays in establishing the molecular geometry. Though obviously not a complete theory, Hückel's approach has been very successful because so many of the distinguishing chemical and physical properties of the conjugated hydrocarbons are π-electron properties. Hückel's justification of this approach was based on the following reasoning:

- The σ bonds between sp^2-hybidised carbon atoms and their neighbours all lie in the molecular plane formed by the carbon atoms and they are symmetrical with respect to reflection in that plane; i.e. the phase of the wave function which describes a particular σ bond is the same above the plane as below it. In contrast, the phase of the unhybridised 2p AO which lies at $90°$ to the plane changes sign upon reflection in the plane so that any molecular orbitals formed with such 2p AOs will be antisymmetric. Therefore, there can be no bonding between σ and π orbitals in planar conjugated molecules.

- The σ electrons occupy the space in the plane of the molecule whilst the π electrons are concentrated above and below that plane. The two types of electrons move in different regions of space.

- The energies of the σ electrons involved in strong σ bonds are, on the whole, lower than those of the π electrons which are involved in the weaker π bonds.

The first of the above reasons, that involving symmetry, is the most powerful argument for treating the σ and π electrons separately.

12.1.2 The method

Thus, the problem which we have is to find the set of π-electron MOs (Ψ_a) formed as linear combinations of carbon $2p_z$ AOs, one from each carbon atom; i.e. what are the values of the coefficients, C_{ai}, in the expression:

$$\Psi_a = C_{a1}\phi_1 + C_{a2}\phi_2 + C_{a3}\phi_3 + \cdots + C_{an}\phi_n \equiv \sum_i C_{ai}\phi_i$$

In order to find the best values of these coefficients we use the criterion, provided by the variation theorem (Appendix 2), that the best wave function we can obtain is the one which gives the lowest energy. Therefore, we first write down the expectation value of the energy (E_a) calculated with the wave function (Ψ_a) as:

$$E_a = \langle \Psi_a | \hat{\mathcal{H}} | \Psi_a \rangle / \langle \Psi_a | \Psi_a \rangle = \left\langle \sum_i C_{ai}\phi_i \middle| \hat{\mathcal{H}} \middle| \sum_j C_{aj}\phi_j \right\rangle \middle/ \left\langle \sum_i C_{ai}\phi_i \middle| \sum_j C_{aj}\phi_j \right\rangle$$

$$= \sum_i C_{ai} \sum_j C_{aj} \langle \phi_i | \hat{\mathcal{H}} | \phi_j \rangle \middle/ \sum_i C_{ai} \sum_j C_{aj} \langle \phi_i | \phi_j \rangle$$

$$\equiv \sum_i C_{ai} \sum_j C_{aj} H_{ij} \middle/ \sum_i C_{ai} \sum_j C_{aj} S_{ij}$$

Note that there are n^2 terms in the numerator and the denominator of the above expression for E_a. The expressions H_{ij} and S_{ij} are simply symbols for the integrals $\langle \phi_i | \hat{\mathcal{H}} | \phi_j \rangle$ and $\langle \phi_i | \phi_j \rangle$ respectively and at some stage in the calculation they have to be evaluated. However, before we do that we apply the criterion that the coefficients we require are those which give the lowest value of E_a. We do this by differentiating the above expression for E_a and setting the result to zero. This procedure is described in detail in Appendix 2. Here we need only note that the process leads to a set of n simultaneous linear equations all equal to zero:

$$C_{a1}(H_{11} - E_a S_{11}) + C_{a2}(H_{12} - E_a S_{12}) + C_{a3}(H_{13} - E_a S_{13}) + \cdots + C_{an}(H_{1n} - E_a S_{1n}) = 0$$

$$C_{a1}(H_{21} - E_a S_{21}) + C_{a2}(H_{22} - E_a S_{22}) + C_{a3}(H_{23} - E_a S_{23}) + \cdots + C_{an}(H_{2n} - E_a S_{2n}) = 0$$

$$\vdots \qquad\qquad \vdots \qquad\qquad \vdots \qquad\qquad \vdots$$

$$C_{a1}(H_{n1} - E_a S_{n1}) + C_{a2}(H_{n2} - E_a S_{n2}) + C_{a3}(H_{n3} - E_a S_{n3}) + \cdots + C_{an}(H_{nn} - E_a S_{nn}) = 0$$

In order to solve these equations Hückel made some drastic assumptions.[1] When we apply the theory we shall see how far these were justified in terms of the insight which the Hückel theory gives us. First, we describe each one in turn.

12.1.3   Hückel's assumptions

$S_{ij} = \langle \phi_i | \phi_j \rangle$ is an overlap integral, it is a measure of the extent to which the $2p_z$ AOs on carbon atoms i and j occupy the same region of space and is therefore closely related to the π-electron bonding between atoms i and j. When $i = j$ we have the overlap of an AO with itself which is unity for a normalised wave function. Therefore, $S_{ii} = 1.0$. The overlap integrals where $i \neq j$ are not difficult to calculate and it is found that for adjacent carbon atoms $S_{ij} \approx 0.25$. For carbon atoms separated by two bonds the value of S_{ij} is much smaller and it drops off rapidly with distance. Hückel set all overlap, even that between neighbouring carbon atoms, to zero. This is quite a drastic assumption.

The integrals $H_{ij} = \langle \phi_i | \hat{\mathcal{H}} | \phi_j \rangle$ are very complex since the Hamiltonian operator consists of many terms representing the energy of a single π-electron moving in the averaged field of all the σ-electrons and all the other π-electrons. But its essential physical content is the energy of interaction between the two AOs, ϕ_i and ϕ_j which, where ϕ_i is adjacent to ϕ_j, represents the π-electron bonding energy between the two carbon atoms. The integral is often called the *resonance* integral and because it represents the bonding energy between two carbon atoms it is a negative quantity. When $i = j$ the integral represents the energy of an electron in a carbon $2p_z$ AO, but in the molecule rather than in a free carbon atom. Hückel assumed that the value of H_{ij} for carbon atoms separated by two or more bonds was sufficiently small to be neglected. He did not attempt to calculate H_{ii} and H_{ij} (i adjacent to j) but expressed them in parametric form as indicated in Table 12.1. With these assumptions the secular equations become much simpler but, since the simplifications depend upon the particular geometry of the molecule (which atoms are adjacent to which other atoms), the simplification is not readily appreciated when it is expressed in a general form. Therefore, let us assume that we are dealing with the special case of the butadiene molecule Figure 12.1.

12.1.4   Determination of HMO energies and AO coefficients

For butadiene Hückel's approximations lead to the following secular equations. Note that, in Hückel theory, no allowance is made for the fact that the two end bonds of the molecule

Table 12.1   Hückel's approximations

	$i = j$	i adjacent to j	i not adjacent to j
S_{ij} (overlap integral)	1.0	0.0	0.0
H_{ij} (interaction integral)	α	β	0.0

Figure 12.1   *trans* Butadiene

are not the same as the central bond. Note also, that the equations for the *cis* isomer are *exactly the same* as those for the *trans*.

$$
\begin{aligned}
C_{a1}(\alpha - E_a) &+ C_{a2}\beta & & & &= 0 \\
C_{a1}\beta &+ C_{a2}(\alpha - E_a) &+ C_{a3}\beta & & &= 0 \\
&C_{a2}\beta &+ C_{a3}(\alpha - E_a) &+ C_{a4}\beta & &= 0 \\
& &C_{a3}\beta &+ C_{a4}(\alpha - E_a) & &= 0
\end{aligned}
$$

To proceed further it is useful to divide all four equations by β. Since β is an energy this step is tantamount to adopting β as our energy unit and expressing all other energies in terms of it. Setting $(\alpha - E_a)/\beta = x$ we then have:

$$
\begin{aligned}
C_{a1}x &+ C_{a2} & & &= 0 \\
C_{a1} &+ C_{a2}x &+ C_{a3} & &= 0 \\
&C_{a2} &+ C_{a3}x &+ C_{a4} &= 0 \\
& &C_{a3} &+ C_{a4}x &= 0
\end{aligned}
$$

These are four simultaneous equations from which we hope to determine the values of the coefficients C_{a1}, C_{a2}, C_{a3} and C_{a4}. We should also note that, since we do not know E_a, x is also unknown. There is a *trivial* solution to our problem, $C_{a1} = C_{a2} = C_{a3} = C_{a4} = 0.0$; but this is clearly not what we require. From the theory of simultaneous equations we know that in order for there to be a *non-trivial* solution the determinant formed by the factors which multiply the coefficients C_{ai} in the above equations must be equal to zero:

$$
\begin{vmatrix} x & 1 & 0 & 0 \\ 1 & x & 1 & 0 \\ 0 & 1 & x & 1 \\ 0 & 0 & 1 & x \end{vmatrix} = x \times \begin{vmatrix} x & 1 & 0 \\ 1 & x & 1 \\ 0 & 1 & x \end{vmatrix} - 1 \times \begin{vmatrix} 1 & 1 & 0 \\ 0 & x & 1 \\ 0 & 1 & x \end{vmatrix}
$$

$$
= x \times x \times \begin{vmatrix} x & 1 \\ 1 & x \end{vmatrix} - x \times 1 \times \begin{vmatrix} 1 & 1 \\ 0 & x \end{vmatrix} - 1 \times 1 \times \begin{vmatrix} x & 1 \\ 1 & x \end{vmatrix} + 1 \times 1 \times \begin{vmatrix} 0 & 1 \\ 0 & x \end{vmatrix}
$$

$$
= x^2(x^2 - 1) - x^2 - 1(x^2 - 1) = x^4 - 3x^2 + 1 = 0.0
$$

At this stage in a HMO calculation we are always confronted with the problem of finding the roots of a polynomial, the order of which is the number of carbon atoms. In the present case, the easiest way to a solution is to first set $y = x^2$ and solve for y, which gives $y = 0.38197$ and 2.61803. The positive and negative square roots of these values of y give the roots $x = (\alpha - E_a)/\beta = \pm 0.61803$ and ± 1.61803. Or in terms of α and β; $E_a = \alpha \pm 0.61803\beta$ and $\alpha \pm 1.61803\beta$. Note that although we began by looking for the lowest value of E_a we have found not one but four values for that quantity. Naturally, we could simply take the lowest one, but it is interesting to consider the significance of the others. We shall return to that subject later. Our immediate task is to determine the coefficients which go with each of the energy values. To do this we return to the secular equations expressed in terms of x and substitute *one* of the values of x which we have found into *all four* equations. Let us choose $x = -0.61803$ giving Equation (12.1.1):

$$
\begin{aligned}
-0.61803C_{a1} &+ C_{a2} & & &= 0 \\
C_{a1} &- 0.61803C_{a2} &+ C_{a3} & &= 0 \\
&C_{a2} &- 0.61803C_{a3} &+ C_{a4} &= 0 \\
& &C_{a3} &- 0.61803C_{a4} &= 0
\end{aligned} \tag{12.1.1}
$$

The first equation provides a simple relationship between C_{a1} and C_{a2} and the last does the same for C_{a3} and C_{a4}. By substituting these results into equations 2 and 3 any three coefficients can be expressed in terms of the other one. For example, $C_{a2} = 0.61803C_{a1}$; $C_{a3} = -0.61803C_{a1}$; $C_{a4} = -C_{a1}$. But note that the absolute values of the C_{ai} are not found, only their relative values. This is because the equations are all equal to zero, i.e. they are *homogeneous*. Fortunately, we have one further criterion which must be satisfied and by means of which we can fix the absolute values of the coefficients. The MO Ψ_a must be normalised if it to be satisfactory from the physical viewpoint:

$$\langle \Psi_a | \Psi_a \rangle = \sum_i C_{ai} \sum_j C_{aj} \langle \phi_i | \phi_j \rangle \equiv \sum_i C_{ai} \sum_j C_{aj} \delta_{ij} = \sum_i (C_{ai})^2$$

where we have used the fact that, in Hückel theory, the overlap integrals are equal to unity when $i = j$ and zero otherwise. Using the normalisation condition and the relative values of the coefficients given above their absolute values can be determined. When this procedure is repeated for each of the values of x the results given in Table 12.2 are found.

As a check of the calculation of the MO coefficients, the energy of the corresponding orbital can be calculated:

$$\langle \Psi_2 | \hat{\mathcal{H}} | \Psi_2 \rangle = \langle 0.6015\phi_1 + 0.3718\phi_2 - 0.3718\phi_3 - 0.6015\phi_4 | \hat{\mathcal{H}} | 0.6015\phi_1$$
$$+ 0.3718\phi_2 - 0.3718\phi_3 - 0.6015\phi_4 \rangle$$
$$= \alpha\{2.0 \times (0.6015 \times 0.6015 + 0.3718 \times 0.3718)\}$$
$$+ \beta\{2.0 \times (0.6015 \times 0.3718 - 0.3718 \times 0.3718 + 0.3718 \times 0.6015)\}$$
$$= \alpha + 0.6180\beta$$

Although we set out to determine one energy value, the lowest, we regard the four energies above and the associated MOs (wave functions) as four possible π-electron energy levels for butadiene.

The above equations for the MOs and the energy-level diagram (Figure 12.2) illustrate a number of important points:

- Since β is a negative quantity, because it represents the π-electron bonding energy between two adjacent carbon atoms, Ψ_1 is the MO of lowest energy and Ψ_4 the highest.

- The order of the energy levels follows the number of nodes, i.e. changes in sign between one coefficient and the next. Thus Ψ_1 has no nodes but Ψ_4 has three.

- The term α is common to all the energies which split out from that value in a symmetrical pattern as illustrated in Figure 12.2. α is the energy of a non-bonded $2p_z$ electron of carbon and MOs of energy greater than α are π-antibonding while those below α are π-bonding.

Table 12.2 The Hückel orbitals and energies for butadiene

$\Psi_4 =$	$0.3718\phi_1$	$-0.6015\phi_2$	$+0.6015\phi_3$	$-0.3718\phi_4$ $\alpha - 1.6180\beta$
$\Psi_3 =$	$0.6015\phi_1$	$-0.3718\phi_2$	$-0.3718\phi_3$	$+0.6015\phi_4$ $\alpha - 0.6180\beta$
$\Psi_2 =$	$0.6015\phi_1$	$+0.3718\phi_2$	$-0.3718\phi_3$	$-0.6015\phi_4$ $\alpha + 0.6180\beta$
$\Psi_1 =$	$0.3718\phi_1$	$+0.6015\phi_2$	$+0.6015\phi_3$	$+0.3718\phi_4$ $\alpha + 1.6180\beta$

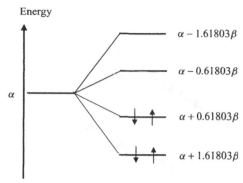

Figure 12.2 π-Electron energy levels for butadiene

- Each carbon atom contributes one π electron and before bonding each has an energy of α. All four of them can enter bonding orbitals and stabilise the molecule. The stabilisation due to the π-electrons (ΔE_π) is:

$$\Delta E_\pi = 4\alpha - 2 \cdot (\alpha + 1.61803\beta) - 2 \cdot (\alpha + 0.61803\beta) = -4.47212\beta$$

The energies and coefficients, illustrated above for the case of butadiene, are the primary results of a Hückel theory calculation. In the next section we examine some of the ways in which these data can be used to interpret experimental results.

12.1.5 Applications of HMO energies

The energies of individual MOs

As a general rule, π-bonding is weaker than σ-bonding and π-antibonding less antibonding than σ. Therefore the π-electron energy levels of a molecule lie near the energetic centre of gravity and, in particular, the highest occupied MO (HOMO) is usually a π-electron orbital. The major exception to this is the presence of non-bonding MOs in carbonyl compounds and heterocycles. Similarly, the lowest unoccupied MO (LUMO) is almost invariably a π orbital and the energies of these *frontier orbitals* may be expected to correlate with the experimental ionisation energy and electron affinity of the molecule. If these properties cannot be measured directly, the reduction and oxidation energies of the molecule can sometimes be measured by electrochemical techniques; in the case of the aromatic hydrocarbons this has been done very successfully. All that need be said here about these methods is that the molecule (M) is dissolved in an electrically conducting solution (e.g. acetonitrile containing tetraethylammonium perchlorate) which is placed in an electrochemical cell where the negative potential required to form M^-, or the positive potential required to form M^+, is measured. The potentials so measured correlate well with the Hückel HOMO and LUMO energies as Figure 12.3 shows.[2] The slopes of the two correlations also enable us to determine values of β of 287 kJ mol^{-1} and 197 kJ mol^{-1} for the reduction and oxidation potentials respectively. We shall return to comment upon the differing values of β given by the various correlations with experimental data in Section 12.1.7.

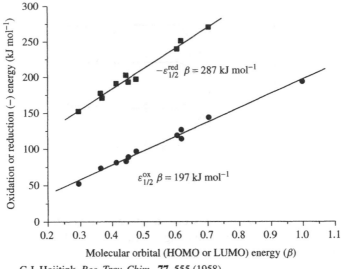

G.J. Hoijtink, *Rec. Trav. Chim.*, **77**, 555 (1958).
Reproduced by permission of Wiley-VCH.

Figure 12.3 Correlation of Hückel HOMO and LUMO energies with polarographic oxidation and reduction potentials[2]

MO energy differences

It is tempting to think that the energy difference between the HOMO and LUMO will be a measure of the energy of the lowest energy band in the electronic spectrum of the molecule. However, as we have seen in Chapter 11, this is an oversimplified view since it does not allow for the very significant energies which arise from interelectronic repulsion. Nevertheless, the LUMO–HOMO energy difference (ΔE) does make a very significant contribution to the corresponding electronic spectral band and we do find a good correlation between calculated and observed ΔE in many cases. The example of the *para*-band (^1L_a) of the aromatic hydrocarbons is shown in Figure 12.4.[3] A value of $\beta = 257$ kJ mol^{-1} can be deduced from the slope of the graph.

Total π-electron energy

The hydrocarbons where electron delocalisation is possible are distinguished from their brethren that do not have delocalised π-electron systems by the important property of additivity of bond energies. This can be seen when we compare heats of combustion. Where there is no delocalisation, each bond type, C–H, C–C, C=C etc., can be assigned a heat of combustion and the total heat of combustion of the molecule can be written as a sum of the individual contributions from each bond. For a molecule with electron delocalisation, such as benzene, this is not possible, the molecule being found to be more stable, i.e. to have a smaller heat of combustion, than the sum of the individual bond contributions, $3xE$(C–C) $+ 3xE$(C=C) $+ 6xE$(C–H):[‡] Table 12.3 lists some typical data illustrating this discrepancy.[4]

[‡] The thermochemical argument is explained in more detail in Box 12.1.

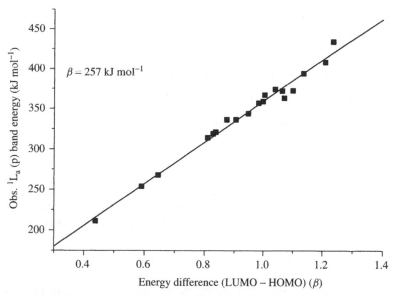

$\beta = 257$ kJ mol^{-1}

Energy difference (LUMO − HOMO) (β)

Obs. ^1L_a (p) band energy (kJ mol^{-1})

E. Heilbronner and J.N. Murrell, *J. Chem. Soc.*, 2611 (1962).
Reproduced by permission of The Royal Society of Chemistry.

Figure 12.4 Correlation of the LUMO−HOMO energy difference with the energy of the *para* band of some aromatic hydrocarbons[3]

Table 12.3 Delocalisation energies (DLE) of some aromatic hydrocarbons

Compound	Heat of Combustion (kJ mol^{-1})		Difference (kJ mol^{-1})	Hückel DLE (β)
	Calculated	Observed		
Benzene	3 451	3 302	149	2.00
Naphthalene	5 484	5 229	255	3.68
Diphenyl	6 631	6 333	298	4.38
Anthracene	7 513	7 154	359	5.32
Phenanthrene	7 516	7 101	415	5.45
Pyrene	8 401	7 946	455	6.51
Chrysene	9 554	9 048	506	7.19
Perylene	10 430	9 902	528	8.25

Naturally, the discrepancy has been associated with the fact that we do not believe there are three double and three single carbon-carbon bonds in benzene. Rather, we know from experiment that all the carbon-carbon bonds are the same and that the molecular structure should be represented by something intermediate between, or some combination of, the two Kekulé structures which we can draw. This is the cause of the additional stability of the molecule over and above that of one of the Kekulé structures. This extra stability, the difference between the experimental combustion energy and the sum of bond contributions, is called the *resonance* or *delocalisation* energy. Some comparisons of theory and experiment are listed in Table 12.3.

The heat of combustion for a particular Kekulé structure of localised single and double bonds was calculated by Klages from the following values for individual bonds, which he deduced from experimental calorimetric data: E(C−H) = 226 kJ mol^{-1}, E(C−C) =

206 kJ mol^{-1} and $E(\text{C=C}) = 491 \text{ kJ mol}^{-1}$ with some additional corrections for special structural features, e.g. 4 kJ mol^{-1} for a benzene ring.[‡] Thus the calculated heat of combustion for benzene is:

$$6E(\text{C–H}) + 3E(\text{C=C}) + 3E(\text{C–C}) + 4 \text{ (for ring)}$$

$$6 \times 226 + 3 \times 491 + 3 \times 206 + 4 = 3451 \text{ kJ mol}^{-1}$$

But   experimental heat of combustion $= 3302 \text{ kJ mol}^{-1}$

Therefore   delocalisation energy $= 149 \text{ kJ mol}^{-1}$

The energies of the three HMO levels occupied by the six π electrons of benzene are $\alpha + 2\beta$, $\alpha + \beta$ and $\alpha + \beta$ and the occupied π orbital of ethene has an energy of $\alpha + \beta$. Therefore, the theoretical value with which this experimental delocalisation energy is to be compared, the Hückel π-electron energy of benzene minus that of three ethene molecules, is:

$$6.0\alpha + 8.0\beta - 3(2.0\alpha - 2.0\beta) = 2.0\beta$$

A graph of calculated versus experimental delocalisation energies from which a value of $\beta = 64 \text{ kJ mol}^{-1}$ can be derived is shown in Figure 12.5.

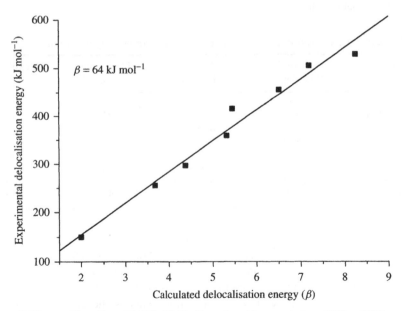

F. Klages, *Chem. Ber.*, **82**, 358 (1949). Reproduced by permission of Wiley-VCH.

Figure 12.5 Correlation of calculated and experimental delocalisation energies for some aromatic hydrocarbons[4]

[‡] There are many subtle corrections of this form to allow for special aspects of the structures. Therefore, the data in Table 12.3 cannot be reproduced using only the three bond energies given. For more information see F. Klages, *Chem. Ber.*, **82**, 358 (1949).

12.1.6 Applications of HMO coefficients

Charge densities and bond orders

The square of the AO coefficient C_{ai} measures the participation of the particular carbon $2p_z$ AO, ϕ_i on carbon atom i, in the MO Ψ_a and, in particular, the proportion of the electron delocalised in Ψ_a which may be said to be present in ϕ_i. Since each MO normally accommodates two electrons, the charge density in ϕ_i (q_{ii}), which is due to the occupation of Ψ_a, is $2C_{ai}^2$. For the total charge density on carbon atom i (Q_{ii}) this must be summed over all occupied orbitals:

$$Q_{ii} = 2 \sum_{a \text{ occ.}} (C_{ai})^2$$

The bond order, P_{ij}, is obtained in an analogous manner:

$$P_{ij} = 2 \sum_{a \text{ occ.}} C_{ai} C_{aj}$$

The meaning of the bond order may be appreciated by noting, firstly, that for a significant value of P_{ij} both C_{ai} and C_{aj} must also have significant values. Secondly, if the two AO coefficients have the same sign they make a positive contribution to the bond order whereas if they are of opposite signs the contribution is negative. Since a change of sign on going from atom i to atom j implies a node between the two atoms for that MO, negative contributions to P_{ij} are antibonding contributions while positive contributions are bonding. Therefore, a large positive bond order indicates a strong π-bond between the two atoms concerned while a negative value indicates an antibond. The charge densities and bond orders for butadiene calculated from the MOs in Table 12.2 are given in Table 12.4.

Two points about Table 12.4 are worthy of comment:

- The charge density in the ground state is 1.0 on each carbon atom. This is quite generally true for a large class of compounds, which are known as *alternant*. All the aromatic hydrocarbons that contain no five- or seven-membered rings and all the linear polyenes are alternant hydrocarbons.

- The bond order of the terminal bonds is twice that of the central bond, in stark contradiction to the Hückel approximation of the same resonance integral (β) for all C–C bonds.

As far as applications are concerned, it seems natural to assume that positively charged reactants (electrophiles) will preferentially attack those atoms of a molecule where the

Table 12.4 π-electron charge densities and bond orders for butadiene

MO	q_{11}	p_{12}	q_{22}	p_{23}
2 electrons in Ψ_4	+0.2764	−0.4472	+0.7236	−0.7236
2 electrons in Ψ_3	+0.7236	−0.4472	+0.2764	+0.2764
2 electrons in Ψ_2	+0.7236	+0.4472	+0.2764	−0.2764
2 electrons in Ψ_1	+0.2764	+0.4472	+0.7236	+0.7236
	Q_{11}	P_{12}	Q_{22}	P_{23}
Ground state[a]	+1.0000	+0.8944	+1.0000	+0.4472

[a]MOs Ψ_1 and Ψ_2 only are occupied in the ground electronic state.

electron density (Q) is high while negatively charged reactants (nucleophiles) will seek out atoms where Q is low. This assumption is based upon the simple concept that the Coulombic energy of the two reacting molecules will be lowered by the approach of oppositely charged atoms and raised by the converse. This simple analysis contains much more than a grain of truth, but the progress of a chemical reaction from reactants to products is a very complicated process in which the major role is played by the energy differences between the reactants, the transition state(s) and the products. For molecules to which the Hückel theory is applicable, and for many others, Klopman[5] has shown that the interaction of atom r of a nucleophile (R) with atom s of an electrophile (S) may be expressed in the form:

$$\Delta E = -Q_r Q_s / \varepsilon D + 2(C_r^{HOMO} C_s^{LUMO} \beta)^2 / (E_r^{HOMO} - E_s^{LUMO})$$

ΔE is the energy change due to the interaction of r and s. The first term on the right is the Coulombic term representing the interaction between the total charges (assumed to be opposite) on the atoms r and s which are separated by a distance D. ε is the dielectric constant of the solvent. The second term contains the product of the coefficient of the HOMO of the nucleophile at the interacting atom (r) and the coefficient of the LUMO of the electrophile at atom s multiplied by β. The denominator is the energy difference between the HOMO and LUMO. The many and varied applications of this equation, and others like, it are largely responsible for the explosive growth in our understanding of organic chemical reactions which took place in the second half of the 20th century. The results leave no room for doubt as to the value of the Hückel MO coefficients in the interpretation of organic chemical reactions. This is far too extensive a subject for us to enter into here but, fortunately, Fleming[6] has provided us with a splendid and readable account to which interested readers should turn for further information and confirmation of the bald statement in the last sentence.

Our last example of a correlation of Hückel theory with experiment will also serve to illustrate the way in which the theory can be extended to take account of π-electron systems involving atoms other than carbon, such as nitrogen and oxygen. If we call the hetero-atom X, then we have to ascribe to it appropriate values of α_X and β_X. Since β_C is our unit of energy, we define these two quantities in terms of β_C and two new parameters, h_X and k_X, as follows; allowing only for C–X bonds:

$$\alpha_X = \alpha_C + h_X \beta_C \qquad \beta_X = k_X \beta_C$$

A good discussion of h_X and k_X and a table of values may be found in Streitwieser.[7]

The correlation in question is that of π-electron bond order with the carbonyl stretching frequency in aldehydes, ketones and quinones. Since the mass of the atoms and the σ-bonding are a constant feature, we expect the frequency of the C=O stretching vibration to increase linearly with the bond order. Figure 12.6,[8] in which the bond orders were calculated with $h_O = 1.2$ and $k_O = 2.0$, confirms this expectation.

12.1.7   Some final comments on the Hückel theory

The simplicity of the Hückel theory and the wide range of important chemical compounds to which it is relevant ensured its extensive application from the outset. For such a simple theory its success has been remarkable. However, it is only fair to say that the success

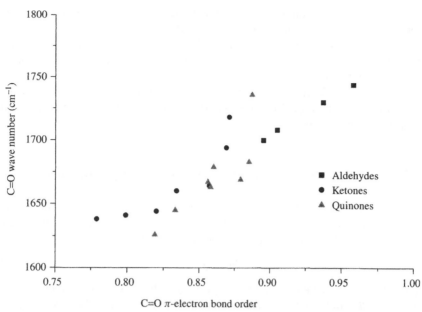

G. Berthier, B. Pullman and J. Pontis, J. *Chim. Phys.*, **49**, 367 (1952). Reproduced by permission of Wiley-VCH.

Figure 12.6   Correlation of C=O π bond order with carbonyl stretching frequencies[5]

is in an important sense a result of the parameterisation of the theory and of the fact that our expectations of the results are not overly high. This is illustrated most clearly in the different values of β which result from the applications described above. We do not find, and for such a simple theory we should not expect to find, the same value of β from each application. This is because β is a composite parameter representing the electron-nucleus attraction, the interelectronic repulsion and the kinetic energy. These three forms of energy make different contributions to the experimental energies with which the results of the Hückel calculations are compared. This subject has been discussed in detail by Murrell, Kettle and Tedder[9] who, taking the subject further, show how the inclusion of the all-important electron repulsion terms makes the theory far more quantitative; but also far more complex!

12.2   MAGNETISM IN CHEMISTRY

The fact that many atomic nuclei possess magnetic properties made possible the chemical applications of nuclear magnetic resonance spectroscopy (Chapter 9), which has had such a profound effect upon the determination of molecular structure. But the magnetic moments of nuclei are very small and the major, macroscopic magnetic phenomena, some of which have been known since antiquity, are due to the spin and orbital motion of electrons. In this section we shall be concerned only with 'electronic' magnetism. Our task is to study the way in which the magnetic properties of atoms manifest themselves in molecules and extended solids, and how the magnetic behaviour of such materials can be used to extract information about their electronic structures. The study of magnetism dates back to the 19th century when the classical phenomena of *paramagnetism, diamagnetism* and *ferromagnetism* were identified and defined. More recently, especially in the second half the 20th century, the study of the magnetic properties of chemical compounds has been stimulated by two factors. The growing understanding of the connection between magnetism and chemical structure has shown that magnetism can be a valuable tool in the elucidation of the finer details of molecular electronic structure, and the use of magnetic materials in many technological devices has provided a strong applied-science stimulus to research into chemical magnetism.

12.2.1   Magnetic susceptibility: diamagnetism and paramagnetism[10]

If a substance is placed in a *magnetic field* of strength H (units $A\,m^{-1}$) then a *magnetic moment* of magnitude M ($A\,m^{-1}$) will be induced in that substance. The relationship between H and M, which is known as the *magnetisation*, is:

$$M = \chi H \qquad (12.2.1)$$

where χ (chi) is the *volume magnetic susceptibility*, which is dimensionless.

The *magnetic flux density* within the substance, B (Tesla $\sim T = J\,C^{-1}\,s\,m^{-2}$), is related to the strength of the applied field and the magnetisation by:

$$B = \mu_o(H + M) = \mu_o(1 + \chi)H \qquad (12.2.2)$$

where μ_o is the *vacuum permeability* which is *defined* to have the value $4\pi \times 10^{-7}\,J\,C^{-2}\,m^{-1}\,s^2$. If $\chi > 0$ H and M add to increase B within the magnetised substance which is then said to be *paramagnetic*. If $\chi < 0$ the substance is said to be *diamagnetic*.

Diamagnetism is found in all materials. The electrons are always in motion and their motion constitutes an electric current. When the material is subjected to a magnetic field, the motion of the electrons is perturbed and, according to Lenz's law, in such a way that the very small magnetic field which results from this perturbation of the electric current opposes the field which caused it. Thus, the induced magnetisation opposes and therefore reduces the magnetic field within the material which is the characteristic behaviour of a diamagnetic substance. The effect is largely independent of temperature.

Though ever-present, the diamagnetism can be overwhelmed by the magnetic effects of unpaired electrons, if there are any, giving rise to paramagnetism. The magnetic properties of a purely diamagnetic substance are *induced* by an applied magnetic field, but in the case of a paramagnetic substance each molecule or atom already has a *permanent magnetic moment* and the observed bulk magnetic moment arises from the partial alignment of these molecular (atomic) moments by the applied field. Two characteristics of paramagnetic materials stand out. Firstly, the induced moment (M) adds to the magnetic field (H) and, secondly, paramagnetism, in contrast to diamagnetism, is very temperature-dependent. The reason for the marked temperature dependence is not far to seek. The individual atomic/molecular magnets of which the material is composed tend to align themselves with the applied field, thereby increasing B (Box 12.2). But this alignment is opposed by the thermal motion of the atoms/molecules which increases with increasing temperature. Thus, paramagnetism decreases as temperature increases according to a law discovered by Pierre Curie (1859–1906):

$$\chi = C/T \qquad (12.2.3)$$

C is the Curie constant of the material and T is the absolute temperature. Though many paramagnetic substances follow the Curie law many others follow the Curie-Weiss (Pierre Weiss, 1865–1940) law:

$$\chi = C/(T - \theta) \qquad (12.2.4)$$

where θ is the Weiss constant. The susceptibility of paramagnets shows little or no dependence upon the applied field.

Temperature-independent paramagnetism (t.i.p.) is a form of paramagnetism which, as its name implies, does not depend on temperature. Unlike normal paramagnetism, this form of paramagnetism is not the result of the alignment of magnetic moments already present in the substance. Like diamagnetism, it is the result of orbital electronic motion *induced* by the applied field. T.i.p. differs from diamagnetism in that while the latter involves the use of ground-state orbitals only, the former requires the use of excited-state orbitals and is therefore of significance only in those molecules or atoms where there are excited states of low energy, such as are sometimes found in complexed transition-metal ions (see Chapter 7).

12.2.2 Magnetic susceptibility: ferromagnetism and antiferromagnetism

The Curie or Curie-Weiss behaviour is typical of simple paramagnetic substances but there are other forms of paramagnetism which show a more complicated dependence upon temperature and applied field. These more complex behaviours arise where there is some degree of interaction between the permanent, individual molecular or atomic magnetic moments, so that if thermal motion is sufficiently reduced this interaction causes the individual moments to orient themselves, one with respect to the other. If the mutual orientation of the moments is parallel (Figure 12.7a), the bulk magnetic moment becomes very large and the material is said to be ferromagnetic. If the mutual orientation is antiparallel (Figure 12.7b), the material is antiferromagnetic. In each case there is a transition temperature below which the mutual interaction of the magnetic moments is dominant and Curie-Weiss behaviour is replaced by ferro- or antiferro-magnetism. This characteristic temperature is known as the Curie temperature in the former case and the Néel (Louis Néel, 1904–2000) temperature in the latter.

To obtain quantitative relationships between observed magnetic properties and molecular parameters we must now define some of the terms which we shall use when we derive expressions for the magnetic moments due to electronic orbital motion and spin; first some definitions.

12.2.3 Magnetic fields and dipoles: some definitions

A magnetic field (H) is a vector quantity and the direction of the vector, by definition, is the direction in which a hypothetical, isolated north pole would move under the influence

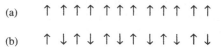

Figure 12.7 The relative orientations of magnetic moments in ferromagnetic (a) and antiferromagnetic (b) materials

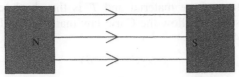

Figure 12.8 The direction of a magnetic field is the direction in which a hypothetical, isolated north pole would move

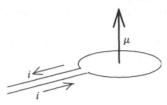

Figure 12.9 A magnetic dipole, μ, is a vector representing the magnetic field due to a current flowing in a small loop. (The current is a flow of positive charge)

of the field. Thus, the field can be represented by lines of force which carry arrows indicating its direction (Figure 12.8).

The definition of a magnetic dipole (μ) is discussed in Box 12.2. It is a vector formed by a current (i) circulating in a small loop (area A) (Figure 12.9) and its magnitude is given by:

$$\mu = iA \tag{12.2.5}$$

The vector μ is perpendicular to the plane of the current loop and points in the direction in which a right-handed screw advances when turned in the direction of the current flow. The units of μ are JT^{-1} or Am^2.

In Box 12.3 the Langevin relationship between the measurable, macroscopic property (χ) of a substance and the magnetic moment (μ) of the molecules composing that substance is derived. We must now obtain expressions for μ in terms of the orbital motion and spin of the electrons so as to establish the vital, quantitative link between the macroscopic observable, susceptibility, and atomic and molecular electronic structure.

12.2.4 The magnetic effect of electronic orbital motion

We are interested in the magnetic fields, which are created by the orbital motion and spin of the electron, and the way in which these fields interact with applied magnetic fields. As we know from Section 2.7, the Bohr theory of the hydrogen atom is not a fully fledged quantum-mechanical theory. Nevertheless, we can gain much insight into our present problem by means of that theory and we now use it to calculate the magnetic dipole generated by an electron moving in a circular Bohr orbit. We make use of the result from classical electromagnetic theory that the magnetic dipole, often called magnetic dipole moment (μ) which arises from a current (i) flowing in a closed loop is given by Equation (12.2.5). The vector (μ) is perpendicular to the plane of the current loop and its direction is related to the direction of the current as shown in Figure 12.10[‡].

[‡] By convention, the current in a circuit is taken to flow from regions of positive potential to regions of negative potential. When the current is being carried by electrons they flow in the opposite direction.

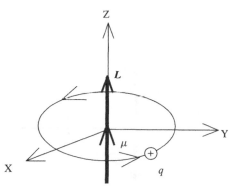

Figure 12.10   The angular momentum (L) and magnetic dipole moment (μ) due to a current (i) produced by an orbiting positive charge (q)

If the current flows from +x to +y the magnetic dipole vector points in the +z direction of a right-handed coordinate system. Thus, if the current is the result of an orbiting positive charge, the dipole moment vector is parallel to the angular momentum vector, L, (Chapter 4) of the orbiting charge. A positive charge (q) moving with velocity (v) in a circular orbit (radius r), produces a current (i) (charge per second) of:

$$i = qv/2\pi r \qquad (12.2.6)$$

so that, since $A = \pi r^2$:

$$\mu = qvr/2 \qquad (12.2.7)$$

The classical angular momentum, L, of an orbiting particle of mass m_p is $m_p vr$ so that we can also write the magnetic moment in terms of the orbital angular momentum, L, and we find that:

$$\mu = qL/2m_p \qquad (12.2.8)$$

in which the radius of the orbit no longer features so that the expression applies to an orbit of any radius.

Here we are interested in the case of an electron in a Bohr orbit to which the above expression applies if we substitute $-e$ (the electronic charge) for q and m_e (the electronic mass) for m_p. The change of the sign of the charge means that the angular momentum and magnetic moment are now anti-parallel (Figure 12.11) and:

$$\mu = -eL/2m_e \qquad (12.2.9)$$

We now recall that Bohr quantised the angular momentum of his allowed orbits, a result which arises naturally in the full quantum mechanical analysis of Schrödinger who found (Chapter 4) that the allowed values of angular momentum (L') are given, in units of $h/2\pi$, by the quantum number (l) through the formula:

$$L' = \{l(l + 1)\}^{\frac{1}{2}} h/2\pi \qquad (12.2.10)$$

(L' is used to avoid confusion with the quantum number L in further development). Thus, the possible values of μ are:

$$\mu = -\{l(l + 1)\}^{\frac{1}{2}} eh/4\pi m_e \equiv -\{l(l + 1)\}^{\frac{1}{2}} \mu_B, \qquad (12.2.11)$$

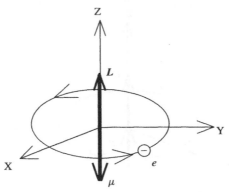

Figure 12.11   The angular momentum (L) and the magnetic dipole moment (μ) due to an orbiting electron

where $eh/4\pi m_e (\equiv \mu_B)$ is the *Bohr magneton* which, in SI units, has the value 9.27408×10^{-24} JT^{-1} (A m^2). It is the fundamental constant which relates the magnetic dipole generated by an orbiting electron to its quantised angular momentum. Thus, the magnetic moment is also quantised. In a many-electron atom the total magnetic moment may be calculated using the above formula but replacing l by L, the quantum number for the total orbital angular momentum:

$$\mu = -\{L(L+1)\}^{\frac{1}{2}} eh/4\pi m_e \equiv -\{L(L+1)\}^{\frac{1}{2}} \mu_B \qquad (12.2.12)$$

Though the theory just described is a simple one, the above result agrees exactly with that obtained by a full quantum-mechanical treatment.

12.2.5   The consequences of chemical bonding

When atoms form covalent chemical bonds the freedom of movement of the electrons, and hence their orbital angular momenta, can be much reduced and this also reduces the corresponding magnetic moments. This *quenching* of the orbital angular momentum which results from bonding is very difficult to calculate; it has proved convenient to allow for this phenomenon by introducing a parameter, g_l, known as the orbital *g-factor* which can be determined by experiment and, in favourable cases, calculated. With this modification Equation (12.2.11) reads:

$$\mu = -g_l \{l(l+1)\}^{\frac{1}{2}} \mu_B \qquad (12.2.13)$$

and in a free atom, $g_l = 1.0$.

We know from Chapter 4 that if we select a particular direction in space, almost invariably the z-direction, then the orbital angular momentum vector characterised by the quantum number l can have $2l + 1$ orientations with respect to the chosen axis, each characterised by a value of the quantum number m_l, which takes all the integer values from $+l$ to $-l$ and gives the value of the z-component of the angular momentum (l_z) through the equation:

$$l_z = m_l h/2\pi \qquad (12.2.14)$$

The corresponding equation for the z-component of the magnetic moment is:

$$\mu_z = -g_l m_l \mu_B \tag{12.2.15}$$

The positive (negative) values of m_l which characterise electrons with positive (negative) z-components of angular momentum correspond, in the simple view described above, to electrons orbiting the nucleus in opposite directions. The magnetic moments due to pairs of electrons occupying orbitals with equal and opposite values of m_l, e.g. a full shell of 10 d-electrons, cancel and such atoms can have no orbital magnetic moment. The presence of magnetism due to orbital electronic motion is therefore a characteristic of atoms with incompletely filled electron sub-shells.

12.2.6   The magnetic effect of electron spin

It seems quite obvious that an orbiting electron can produce a magnetic moment and, as we have seen above, the magnitude of the moment can be readily calculated using classical physics. Less obviously, the spin of the electron also gives rise to a magnetic moment, though this cannot be calculated using a classical approach. Indeed, it was only when Dirac introduced relativity into quantum mechanics that the exact quantitative description of a phenomenon which had been well known experimentally for about 30 years was found. The application of the relativistic theory is very difficult, especially for systems containing more than one electron. Fortunately however, both theory and experiment show that an excellent quantitative description of the magnetic moment due to electron spin may be obtained by a simple extension of the theory derived for the orbital motion. Thus, the magnetic moment due to electron spin angular momentum and its z-component may be written:

$$\mu = -g_s \{s(s+1)\}^{\frac{1}{2}} \mu_B \tag{12.2.16}$$

and

$$\mu_z = -g_s m_s \mu_B \tag{12.2.17}$$

The value of the spin angular momentum is not changed by chemical bonding. However, we retain the g-factor because we find that the correct relativistic treatment of the magnetic moment due to electron spin shows that spin angular momentum gives rise to a magnetic moment which, for a free electron, is 2.002322 times larger than that produced by an orbital angular momentum of the same magnitude. Thus, for magnetic moments due to spin the g-factor is 2.002322. For most purposes, and certainly for ours, it is quite sufficient to set $g_s = 2.0$. As in the case of orbital magnetic moments, the moment arising from a number of electrons can be obtained using the above formulae and replacing s and m_s with S and M_s, the quantum numbers for the total spin and its z-component, respectively. If the electron spins are antiparallel, i.e. if they are paired and have equal and opposite m_s values, then their spin magnetic moments cancel. Herein lies the essence of one of the simplest but most important uses of magnetic measurements in the elucidation of molecular electronic structure—magnetism is a sensitive indicator of the number of unpaired electrons.

Where there are both orbital and spin contributions, the total magnetic moment is the vector sum of the two.

12.2.7   Magnetism in practice

So far, so good; we have obtained the basic formulae with which, in principle, the magnetic moments of atoms and molecules may be calculated. Unfortunately, their application is made difficult by the fact that magnetically interesting species have ground electronic states with unfilled electron shells and are consequently multiply degenerate. For example, the ground state of an atom having a single d electron outside closed shells is ^2D, which is ten-fold degenerate; this degeneracy may be partially lifted by spin-orbit coupling to give a four-fold degenerate $^2D_{\frac{3}{2}}$ state and a six-fold degenerate $^2D_{\frac{5}{2}}$. In a transition-metal ion these degeneracies may be lifted further by the crystal field of the surrounding ligands (see Chapter 7). Therefore, the Boltzmann population of the components of the ground state must be taken into account when the magnetic properties of the species are calculated and formulae applicable to general cases can only be given when accompanied by caveats regarding the relative magnitudes of any ground state splittings and kT. The problem is discussed in detail in Van Vleck's classic treatise.[11] We illustrate some aspects of the problem here by examining the magnetic properties of an atom with one d electron under various conditions relevant to the discussion of Section 7.6.

First, we consider a gas-phase atom or ion with one d electron outside closed shells. All 10 orbitals available to the electron have well defined spin and orbital angular momenta and their magnetic properties are summarised in Table 12.5.

The magnetisation, M, of an assembly of N d^1 atoms can now be calculated, for a flux density B, using the equations derived in Box 12.3. We multiply the magnetic moment of each state by its population, sum the result and divide by the total population. As an example of the contributions to numerator and denominator, consider $|+2, +\frac{1}{2}\rangle$. In the numerator, this state is entered as a magnetic moment of -3 Bohr magnetons multiplied by a population factor of $\exp(-3B/kT)$. The population factor also appears in the denominator. States having magnetic moments of zero are also populated so they appear in the denominator, though not in the numerator. The complete expression is:

$$M = N\mu = \frac{N\{3(e^{3x} - e^{-3x}) + 2(e^{2x} - e^{-2x}) + 2(e^{x} - e^{-x})\}}{\{e^{3x} + e^{-3x} + e^{2x} + e^{-2x} + 2(e^{x} + e^{-x}) + 2e^{0}\}} \tag{12.2.18}$$

Table 12.5   The magnetic moments (μ in Bohr magnetons) of the states of a single d electron and their energies (E) in a flux density (B)

$|\psi\rangle$	$\hat{l}_z|\psi\rangle(h/2\pi)$	$2\hat{s}_z|\psi\rangle(h/2\pi)$	μ/μ_B	E/J
$\lvert -2, -\frac{1}{2}\rangle$	-2	-1	$+3$	$-3B$
$\lvert -1, -\frac{1}{2}\rangle$	-1	-1	$+2$	$-2B$
$\lvert -2, +\frac{1}{2}\rangle$	-2	$+1$	$+1$	$-1B$
$\lvert 0, -\frac{1}{2}\rangle$	0	-1	$+1$	$-1B$
$\lvert -1, +\frac{1}{2}\rangle$	-1	$+1$	0	0
$\lvert +1, -\frac{1}{2}\rangle$	$+1$	-1	0	0
$\lvert 0, +\frac{1}{2}\rangle$	0	$+1$	-1	$+1B$
$\lvert +2, -\frac{1}{2}\rangle$	$+2$	-1	-1	$+1B$
$\lvert +1, +\frac{1}{2}\rangle$	$+1$	$+1$	-2	$+2B$
$\lvert +2, +\frac{1}{2}\rangle$	$+2$	$+1$	-3	$+3B$

where x $\equiv$ B/kT. The pairs of exponential terms with equal and opposite exponents can be taken together as sinh or cosh so the final equation for M is:

$$M = \frac{N\{3\sinh(3x) + 2\sinh(2x) + 2\sinh(x)\}}{\{\cosh(3x) + \cosh(2x) + 2\cosh(x) + 1\}} \qquad (12.2.19)$$

This function, known as a Brillouin function, is plotted in the form M/N versus B/T in Figure 12.12 (graph 1). Two aspects of the graph deserve comment.

Where $B/T < 1.0$, i.e. where B is small and T is large, the graph is linear. This is the Curie region where χ is proportional to $1/T$. We shall comment on it further below. As B/T increases, however, curvature sets in and the graph becomes horizontal at $B/T \approx 8$. This is known as *saturation*; it arises when the states are split so far apart by the field and/or the temperature is so low that only the lowest state is occupied. For d^1 this is $|-2, -\frac{1}{2}\rangle$ with a magnetic moment of $+3\mu_B$, as the asymptotic value of the graph confirms. Experimental measurements of the magnetisation of transition-metal and rare-earth compounds show excellent quantitative agreement with this theory.[12]

Much of the experimental data on magnetic properties in the literature has been measured at modest fields ($<0.5T$) and temperatures not far from 25 °C, with the result that the determination of magnetic susceptibilities and their use in inferring electronic configurations etc. has usually been based upon measurements made in the Curie region of the magnetisation curve. In view of the dependence of susceptibility on temperature these results are now of limited value. In Box 12.3 the slope, C, of the magnetisation curve in

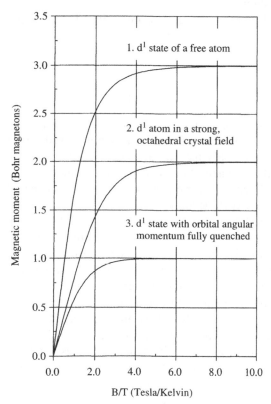

Figure 12.12 Brillouin functions for a single d electron

this region is shown to be $N\mu_0\mu^2/3k$ and this relationship might be used to determine the required magnetic moment. Alternatively, since x $\ll$ 1.0 in this region, we may expand the exponential functions in Equation (12.2.18) to just two terms and evaluate M, which we then find to be:

$$M = N\mu = N(30x/10) = 3NB/kT \qquad (12.2.20)$$

Therefore:

$$\chi = M\mu_0/B = 3N\mu_0/kT \qquad (12.2.21)$$

which shows the expected linear dependence of χ upon $1/T$. From Box 12.3 we also have:

$$\chi = N\mu_0\mu^2/3kT \qquad (12.2.22)$$

and equating the two results we find $\mu = 3\mu_B$. The moment is in Bohr magnetons because these are the units of μ used in Table 12.5. Normally, we would calculate μ for the free d^1 atom directly using the quantum numbers for the angular momentum of the ^2D state and a formula (Equation (12.2.23)) derived by Van Vleck[11] for the case where multiplet splittings are small compared with kT. This certainly applies here since we are assuming that spin-orbit coupling is zero. We also assume that $g_l = 1$ and $g_s = 2$:

$$\mu = [\{L(L+1)\} + 4\{S(S+1)\}]^{\frac{1}{2}}\mu_B \qquad (12.2.23)$$

With this formula, for a ^2D state where $L = 2$ and $S = \frac{1}{2}$, we find $\mu = 3\mu_B$, in agreement with the result above.

When multiplet splittings are large compared with kT, Van Vleck[11] has shown that:

$$\mu = -g_J\{J(J+1)\}^{\frac{1}{2}}\mu_B \qquad (12.2.24)$$

where:

$$g_J = 1 + \{J(J+1) + S(S+1) - L(L+1)\}/2J(J+1) \qquad (12.2.25)$$

Two other magnetisation curves for a d^1 atom are shown in Figure 12.12. Graph 2 is obtained if the atom is subject to a strong, octahedral crystal field when, because of the large crystal-field splitting, only the d_{xy}, d_{xz} and d_{yz} orbitals are occupied (see problem 4). Graph 3 shows the magnetisation of a free d^1 atom which has only spin angular momentum, all orbital angular momentum having been quenched. This is a case which we are unlikely to meet experimentally.

To see how a Curie-Weiss dependence of susceptibility on temperature might arise we can consider the case of a d^1 ion in the field of eight equal charges arranged at the corners of a tetragonal antiprism. To picture such an arrangement, imagine first that you have eight charges arranged at the corners of a cube. Now grasp two opposite square faces and pull them apart so that the remaining four faces are now rectangular. The resulting body is a tetragonal prism. Finally, to obtain the tetragonal antiprism rotate one square face through 90° relative to the other. This arrangement of charges can be adjusted to give a d-orbital energy-level scheme such as in Figure 12.13, in which D is an arbitrary energy. The magnetic moments (μ) of the d-electron states and their energies (E) in such a crystal field are listed in Table 12.6. Setting x $= B/kT$ and y $= D/kT$ and expanding exponential functions of x and y to two terms only we find:

$$\mu = \{20x(1-y) + 8x(1-2y) + 2x\}/\{4(1-y) + 4(1-2y) + 2\} \qquad (12.2.26)$$

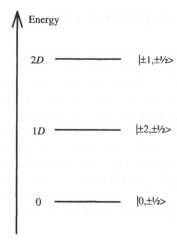

Figure 12.13 The energy levels of the d states of a single d electron in a tetragonal antiprismatic crystal field

Table 12.6 The magnetic moments (μ in Bohr magnetons) of the states of a single d electron and their energies (E) in a tetragonal antiprismatic crystal field and a flux density (B)

$|\psi\rangle$	$(\hat{l}_z + 2\hat{s}_z)|\psi\rangle(h/2\pi)$	μ/μ_B	E/J
$\lvert +2, +\frac{1}{2}\rangle$	$+3$	-3	$D + 3B$
$\lvert +2, -\frac{1}{2}\rangle$	$+1$	-1	$D + 1B$
$\lvert -2, +\frac{1}{2}\rangle$	-1	$+1$	$D - 1B$
$\lvert -2, -\frac{1}{2}\rangle$	-3	$+3$	$D - 3B$
$\lvert +1, +\frac{1}{2}\rangle$	$+2$	-2	$2D + 2B$
$\lvert +1, -\frac{1}{2}\rangle$	0	0	$2D$
$\lvert -1, +\frac{1}{2}\rangle$	0	0	$2D$
$\lvert -1, -\frac{1}{2}\rangle$	-2	$+2$	$2D - 2B$
$\lvert 0, +\frac{1}{2}\rangle$	$+1$	-1	$+1B$
$\lvert 0, -\frac{1}{2}\rangle$	-1	$+1$	$-1B$

Multiplying out the numerator and denominator, neglecting terms in x × y which are second order, we obtain:

$$M = N\mu = N\{15x/(5 - 6y)\} = \{15NB/kT\}/\{5 - 6D/kT\} \qquad (12.2.27)$$

Therefore:

$$\chi = M\mu_0/B = \{15N\mu_0/kT\}/\{5 - 6D/kT\}$$

and multiplying top and bottom by $T/5$ we find:

$$\chi = \{3N\mu_0/k\} \cdot 1/\{T - 6D/5k\} \qquad (12.2.28)$$

which shows that χ has a Curie-Weiss temperature dependence with $\theta = 6D/5k$.

12.2.8   Systems of interacting molecular magnets

In Section 12.2.2 the phenomena of ferromagnetism and antiferromagnetism were briefly mentioned. These properties are the result of interactions between molecular or atomic magnets and interest in materials which show them has grown rapidly because of their technological applications. Here we can address only the most basic aspect of this subject; the mechanism of the interaction. It should be clearly understood at the outset that the interaction is *not* the magnetic dipole-dipole interaction of the constituent molecular magnets. A much-simplified but instructive calculation will make this clear. Iron, a well-known ferromagnetic material, has a body-centred cubic (bcc) lattice in which every atom is 2.48×10^{-10} m from its eight nearest neighbours. The individual magnetic moment of the iron atoms is 2.22 μ_B and the Curie temperature (T_C) below which the ferromagnetic behaviour is observed is 1043 K. The energy of interaction (E_{int}) of two magnetic dipoles (μ_1 and μ_2) distance R apart is:

$$E_{int} = \{\mu_0/4\pi\}\{(\mu_1 \cdot \mu_2)/R^3 - 3(\mu_1 \cdot R)(\mu_2 \cdot R)/R^5\} \qquad (12.2.29)$$

If the dipoles are orientated at 90° to R:

$$E_{int} = \{\mu_0/4\pi\}\{(\mu_1 \cdot \mu_2)/R^3\} \qquad (12.2.30)$$

This interaction will be a maximum if μ_1 is parallel to μ_2 so that, for the particular case of iron:

$$E_{max} = \{(2.22 \times 9.27 \times 10^{-24})^2/(2.48 \times 10^{-10})^3\} \times 10^{-7} = 27.77 \times 10^{-25} \text{ J}$$

Such an interaction in a bcc lattice will give rise to a band of energy levels, each level corresponding to a different relative orientation of the magnetic moments, with a spread of 16 times the above figure, i.e. 44.43×10^{-24} J (see Section 12.3.1).[‡] Now, consider the population of this band of energy levels at 100 K, far below T_C, where the iron should be clearly ferromagnetic. According to the Boltzmann law, at 100 K for every 100 iron atoms in the lowest state in the band there will be approximately 97 in the highest state. And, of course, all the intermediate levels will be correspondingly highly populated. Therefore, even at 100 K all relative orientations of the magnetic moments of the iron atoms will be approximately equally populated, there will be no overall ordering of the moments and no ferromagnetism. We should also note that if there was significant excess population in the lowest state the iron would be antiferromagnetic! Clearly, the energy associated with magnetic dipole-dipole interaction is several orders of magnitude too small to be responsible for maintaining magnetic order up to 1043 K and, furthermore, the order which it would produce if it were stronger would not be a ferromagnetic one. What, then, is the mechanism of this strong interaction?

Section 11.5.3 contains an extensive discussion of the reasons for the difference in the energies of the two excited states of helium which have the same configuration, $1s^1 2s^1$, but differ in the relative orientations of the electron spins. The triplet with parallel spins is significantly lower in energy than the singlet in which the spins are paired. In that case too, the difference in energy is not a consequence of the direct magnetic interaction

---

[‡] This is an over-estimate of the band width since it is not possible to have an arrangement in which the magnetic moments of the central iron atom and each of its eight nearest neighbours are all parallel to each other and perpendicular to the line joining each pair. However, this is not important since we shall show that the interaction is much too small anyway.

of the electron spin magnetic moments but arises in an indirect way from the effects of inter-electronic repulsion, a very much more potent force.

In order to understand more clearly why the interaction between the magnetic centres has the characteristics of a dipole-dipole interaction, even though it is not, we first return to the helium atom problem, following van Vleck.[11] The symmetric (Φ_+) and antisymmetric (Φ_-) spin functions for the two electron are:

$$\Phi_+ = \sqrt{\tfrac{1}{2}}\{\alpha(1)\cdot\beta(2) + \beta(1)\cdot\alpha(2)\} \tag{12.2.31a}$$

and

$$\Phi_- = \sqrt{\tfrac{1}{2}}\{\alpha(1)\cdot\beta(2) - \beta(1)\cdot\alpha(2)\} \tag{12.2.31b}$$

It is sufficient for our present purpose to consider only the two spin functions above and neglect the other two symmetric functions, $\alpha(1)\cdot\alpha(2)$ and $\beta(1)\cdot\beta(2)$.

The corresponding space functions are:

$$\Psi_+ = \sqrt{\tfrac{1}{2}}\{\phi_{1s}(1)\cdot\phi_{2s}(2) + \phi_{2s}(1)\cdot\phi_{1s}(2)\} \tag{12.2.32a}$$

and

$$\Psi_- = \sqrt{\tfrac{1}{2}}\{\phi_{1s}(1)\cdot\phi_{2s}(2) - \phi_{2s}(1)\cdot\phi_{1s}(2)\} \tag{12.2.32b}$$

Because, according to Pauli, a total wave function must be antisymmetric with respect to the exchange of electrons 1 and 2, the spin and space functions must be combined as follows to form the triplet ($|T\rangle$) and the singlet ($|S\rangle$) state wave functions:

$$|S\rangle = \Psi_+ \cdot \Phi_- \tag{12.2.33a}$$

and

$$|T\rangle = \Psi_- \cdot \Phi_+ \tag{12.2.33b}$$

The energies of the triplet and singlet states differ because the inter-electronic repulsion in the two states is different. We can express this result in terms of the matrix of the operator for inter-electronic repulsion, e^2/r_{12}:

e^2/r_{12}	$|T\rangle$	$|S\rangle$	
$\langle T|$	$J_{1s,2s} - K_{1s,2s}$	0	(12.2.34)
$\langle S|$	0	$J_{1s,2s} + K_{1s,2s}$	

$J_{1s,2s}$ and $K_{1s,2s}$ are the Coulomb and exchange electron repulsion integrals respectively (Section 11.5.3) and we recall that the operator e^2/r_{12} operates solely on the space functions, Ψ_+ and Ψ_-.

We now examine the operation of the spin operator $\hat{\mathbf{S}}_1 \cdot \hat{\mathbf{S}}_2$ on the spin functions, Φ_+ and Φ_-. First we express the operator in terms of the components of the vectors $\hat{\mathbf{S}}_1$ and $\hat{\mathbf{S}}_2$:

$$\hat{\mathbf{S}}_1 \cdot \hat{\mathbf{S}}_2 = \hat{S}_{1z}\hat{S}_{2z} + \hat{S}_{1x}\hat{S}_{2x} + \hat{S}_{1y}\hat{S}_{2y} = \hat{S}_{1z}\hat{S}_{2z} + \tfrac{1}{2}\{\hat{S}_{1+}\hat{S}_{2-} + \hat{S}_{1-}\hat{S}_{2+}\} \tag{12.2.35}$$

Bearing in mind that an operator with subscript 1 (2) operates only on electron 1 (2), we can tabulate the results of the above spin-component operators on the two spin functions:

$$
\begin{array}{ccc}
 & |\alpha(1) \cdot \beta(2)\rangle & |\beta(1) \cdot \alpha(2)\rangle \\
\hat{S}_{1z}\hat{S}_{2z} & -\frac{1}{4}|\alpha(1) \cdot \beta(2)\rangle & -\frac{1}{4}|\beta(1) \cdot \alpha(2)\rangle \\
\frac{1}{2}\hat{S}_{1+}\hat{S}_{2-} & 0 & \frac{1}{2}|\alpha(1) \cdot \beta(2)\rangle \\
\frac{1}{2}\hat{S}_{1-}\hat{S}_{2+} & \frac{1}{2}|\beta(1) \cdot \alpha(2)\rangle & 0
\end{array}
$$

Using the results in the above table and Equation (12.2.35) we find that $\hat{S}_1 \cdot \hat{S}_2|T\rangle = +\frac{1}{4}|T\rangle$ and $\hat{S}_1 \cdot \hat{S}_2|S\rangle = -\frac{3}{4}|S\rangle$ and we can also express this result as a matrix:

$$
\begin{array}{ccc}
\hat{S}_1 \cdot \hat{S}_2 & |T\rangle & |S\rangle \\
\langle T| & +\frac{1}{4} & 0 \\
\langle S| & 0 & -\frac{3}{4}
\end{array}
\tag{12.2.36}
$$

Again we recall that, although we have placed the symbol for the complete wave function $|T\rangle$ and $|S\rangle$ at the head of the columns and rows of the matrix, the matrix elements in this case are derived solely from the spin functions. The matrices 12.2.34 and 12.2.36 can be linked with a matrix (Equation (12.2.37a)):

$$
-2K_{1s,2s}\hat{S}_1 \cdot \hat{S}_2 = e^2/r_{12} - J_{1s,2s} + \tfrac{1}{2}K_{1s,2s}
\tag{12.2.37a}
$$

or, in terms of the complete matrices (Equation (12.2.37b)):

$$
\begin{bmatrix} -K/2 & 0 \\ 0 & 3K/2 \end{bmatrix} = \begin{bmatrix} J - K & 0 \\ 0 & J + K \end{bmatrix}
$$
$$
+ \begin{bmatrix} -J + K/2 & 0 \\ 0 & -J + K/2 \end{bmatrix}
\tag{12.2.37b}
$$

where the subscripts 1s,2s on J and K have been omitted to simplify the notation. The rules of matrix algebra require that the equality holds for every set of corresponding matrix elements, e.g. for $(1, 1)$, $-K/2 = J - K - J + K/2$. This result shows that the eigenvalues of the operator e^2/r_{12}, which are the different energies of the parallel (symmetric spin function) and anti-parallel (antisymmetric spin function) electron pairings, plus a constant term $(-J + K/2)$ are equal to the eigenvalues of the operator $-2 K\hat{S}_1 \cdot \hat{S}_2$. Remarkably, the spin operator $\hat{S}_1 \cdot \hat{S}_2$ mimics the operator e^2/r_{12}, which is in fact responsible for the energy difference between the parallel and antiparallel coupling of the magnetic moments, and provides a much simpler way of calculating the *relative* energies of the coupled levels than would otherwise be the case. An important question remains. The above analysis is based on the very simple system of just two electrons in the helium atom; does it apply to other systems with many more electrons? Dirac has shown that it does. In a many-electron molecule or crystal the spins of any pair of electrons, i and j, may be regarded as being coupled together by an energy of the form $-2 K_{i,j}\hat{S}_i \cdot \hat{S}_j$. Note that the subscripts on K relate to the spatial wave functions of the electrons i and j so that the value of the coupling energy will always depend upon the distribution of the electrons involved.

12.2.9   A note of warning

The reader's attention is drawn to some potential causes of confusion. The symbol $K_{i,j}$ has been used here in the expression for the coupling for the sake of consistency with the discussion of the helium atom in Chapter 11. But in the magnetochemistry literature J is invariably used in this context. It is most important to note that the magnetochemist's J is our K and not the Coulomb integral which we have called J. Unfortunately, there is also a lack of consistency in the form of the coupling expression used in applications; some authors neglect the factor of 2 and/or the negative sign.

12.2.10 An application

As an example of the above theory we consider the case of a triangular arrangement of three interacting centres (Figure 12.14). Systems of this form, in which the magnetism derives from transition-metal ions linked by their ligands, have been extensively studied. In order to limit the size of the problem we consider the case where the magnetism at each centre derives from one unpaired spin only, i.e. $S_z = \pm\frac{1}{2}$. The Hamiltonian operator is:

$$\hat{\mathcal{H}} = -2\{K_{12}\hat{\mathbf{S}}_1 \cdot \hat{\mathbf{S}}_2 + K_{23}\hat{\mathbf{S}}_2 \cdot \hat{\mathbf{S}}_3 + K_{31}\hat{\mathbf{S}}_3 \cdot \hat{\mathbf{S}}_1\} \qquad (12.2.38)$$

and by expressing $\hat{\mathbf{S}}_1 \cdot \hat{\mathbf{S}}_2$ in terms of $\hat{S}_z$, $\hat{S}_+$ and $\hat{S}_-$, as we have done above, we obtain the 8×8 matrix of $\hat{\mathcal{H}}$ as two 1×1 and two 3×3 matrices:

	$|\alpha\alpha\alpha\rangle$		$|\beta\beta\beta\rangle$
$\langle\alpha\alpha\alpha|$	$-\frac{1}{2}(K_{12} + K_{23} + K_{31})$	$\langle\beta\beta\beta|$	$-\frac{1}{2}(K_{12} + K_{23} + K_{31})$

	$|\alpha\alpha\beta\rangle$	$|\alpha\beta\alpha\rangle$	$|\beta\alpha\alpha\rangle$
$\langle\alpha\alpha\beta|$	$\frac{1}{2}(-K_{12} + K_{23} + K_{31})$	$-K_{23}$	$-K_{31}$
$\langle\alpha\beta\alpha|$	$-K_{23}$	$\frac{1}{2}(+K_{12} + K_{23} - K_{31})$	$-K_{12}$
$\langle\beta\alpha\alpha|$	$-K_{31}$	$-K_{12}$	$\frac{1}{2}(+K_{12} - K_{23} + K_{31})$

	$|\beta\beta\alpha\rangle$	$|\beta\alpha\beta\rangle$	$|\alpha\beta\beta\rangle$
$\langle\beta\beta\alpha|$	$\frac{1}{2}(-K_{12} + K_{23} + K_{31})$	$-K_{23}$	$-K_{31}$
$\langle\beta\alpha\beta|$	$-K_{23}$	$\frac{1}{2}(+K_{23} + K_{12} - K_{31})$	$-K_{12}$
$\langle\alpha\beta\beta|$	$-K_{31}$	$-K_{12}$	$\frac{1}{2}(+K_{12} - K_{23} + K_{31})$

When $K_{12} = K_{23} = K_{31} \equiv K$ these four matrices give four-fold degenerate eigenvalues of $+3K/2$ and $-3K/2$. When $K_{12} = K_{23} \equiv K$ and $K_{31} = fK$, where f is an arbitrary parameter, we find two-fold degenerate eigenvalues of $(4 - fK)/2$ and $3fK/2$

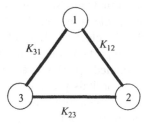

Figure 12.14 Three interacting magnetic centres

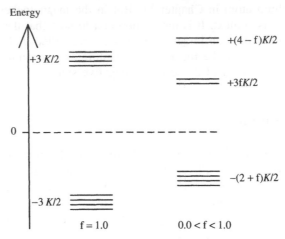

Figure 12.15   The energy levels of three interacting magnetic centres

and a four-fold degenerate eigenvalue of $-(2+f)K/2$. Energy-level schemes for $f = 1.0$ and $0.0 < f < 1.0$ are illustrated in Figure 12.15.

12.3   THE BAND THEORY OF SOLIDS

The electrical properties of conductors, semiconductors and insulators are arguably the material properties most important to modern technology. The basic quantum-mechanical description of the electronic structures of solids and the interpretation of their properties by that means has played a vital role in the development of the technology. This is a branch of solid state physics which we can only touch upon here. We shall emphasise the elementary principles of the quantum-mechanical description of the bonding in solids and find it to be a logical extension to infinite arrays of interacting atoms of the concepts and methods already described in our treatment of discrete molecules. We approach the problem from two different viewpoints.

12.3.1   The tight binding approximation

Consider the sodium atom; the ionisation energy of the 3s electron is only 5.14 eV. To remove the next electron 52.43 eV are required which, even allowing for the fact that this is the removal of an electron from Na^+, shows how high the energy of the 3s electron is compared with the 2p electron, i.e. the 3s electron is much less tightly bound to its atomic core than the 2p electron. It is clear from these figures that the unpaired 3s electron must play an important role in the bonding in sodium metal and, although we shall find that the unoccupied 3p orbitals are also involved, the concept of the interaction of the singly occupied 3s orbitals provides our starting point. We now recall the molecular orbital theory of the hydrogen molecule (Section 6.5), where two hydrogen 1s AOs of energy E_{1s} combine to form a bonding and an antibonding MO of energy $E_{1s} + \beta$ and $E_{1s} - \beta$ respectively. β is the interaction between pairs of adjacent orbitals and is a negative quantity. If we make the reasonable assumption that the interaction of two sodium atoms is confined to the interaction of their two 3s AOs we can draw an energy-level diagram of exactly the same type for Na_2. We now ask what would be the result of adding further sodium atoms to form linear Na_3, Na_4, ... Na_N molecules, with $\beta \neq 0$ only between adjacent atoms. This is a problem for which there is an algebraic solution; the energies are given by the formula:

$$E_m = E_{3s} + 2\beta \cos[m\pi/(N+1)] \quad m = 1, 2, 3, \ldots N \qquad (12.3.1)$$

where m is a quantum number characterising the energy level and E_m increases with increasing m. The equation above applies to a linear chain of carbon $2p_z$ AOs such as we have in the Hückel theory and the interaction (β) has exactly the same meaning here as it did in Section 12.1. In the limit of a very large value of N we find the lowest and highest energy levels, E_1 and E_N, at:

$$E_1 = E_{3s} + 2\beta \cos[\pi/(N+1)] \approx E_{3s} + 2\beta \cos[0] = E_{3s} + 2\beta$$

$$E_N = E_{3s} + 2\beta \cos[N\pi/(N+1)] \approx E_{3s} + 2\beta \cos[\pi] = E_{3s} - 2\beta \qquad (12.3.2)$$

All the other N-2 energy levels must fit in between these two outer limits. They form a *band* of closely spaced levels. The extension of this concept to a three-dimensional array of sodium atoms is quite straightforward. In a row of sodium atoms each atom has two nearest neighbours and the width of the band is $2 \times 2\beta = 4\beta$; as above. For a simple cubic lattice each atom has six nearest neighbours, two along each of the three Cartesian axes, and the total spread of the band is $6 \times 2\beta = 12\beta$. In fact, at normal temperatures and pressures sodium metal has a body-centred cubic lattice with eight nearest-neighbour atoms and the spread of the band is 16β. The subject is clearly explained by Kittel.[13] A 1-mm cube of sodium contains about 10^{20} atoms, each of which provides one 3s AO, so that the number of energy levels in the band, which experiment shows to be approximately 4.8×10^{-19} J from top to bottom, is extremely large and they form, to all intents and purposes, an energy continuum. If this was the only contribution to the bonding in sodium metal we would now place the N electrons from N sodium atoms in the lower N/2 orbitals of the band and our description of the bonding would be complete. But the 3s orbitals are not the only ones which overlap as the sodium atoms approach each other as a more detailed study reveals.

An approximate calculated band structure for sodium metal as a function of interatomic distance (r) is shown in Figure 12.16. As r is reduced, the unoccupied 3p orbitals interact first and a band begins to form; the highest and lowest levels of the band are indicated by dotted and solid lines respectively. But since sodium has no 3p electrons the formation of this band generates no bonding. However, as r is reduced further the 3p band broadens and the 3s orbitals, which have one electron per atom, begin to interact and bonding commences. At $r \approx 45$ nm the highest level of the 3s band crosses the lowest level of the 3p band and the equilibrium value of r (r_o) follows at 37 nm. At r_o the lower part of the 3p band is low enough to be populated by some of the electrons from the 3s band and the Na–Na bonds formed have both 3p and 3s character, i.e. they are hybrid bonds (Section 6.15). There is no significant interaction of the 2p orbitals at r_o, and, *a fortiori*, of the 2s and 1s, which is why they have not been included in the diagram.

As with sodium, in general a number of different atomic orbitals may be expected to be involved in metal bonding. In a typical first-row transition metal, for example, 4s, 4p and 3d AOs contribute to band formation and bonding. Each type of orbital will form a band

The band structure of sodium metal

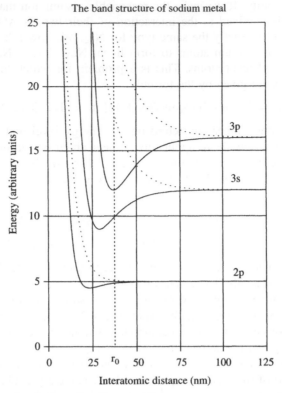

Figure 12.16   The band structure of sodium metal

with its energetic centre at roughly the energy of the corresponding AO in the isolated atom, and an idealised band structure might appear as represented in Figure 12.17(a).

However, we need to add more information to this bare picture. In the first instance we must indicate on our diagram the occupation of the energy levels by electrons. We show this (Figure 12.17(b)) by shading the bands, or parts of bands, which are filled with electrons. The energy at the border between occupied and unoccupied levels is known as the Fermi level, after Enrico Fermi whose name we have already encountered.

We must also recognise that the width of a band, i.e. the energy separation between the highest and lowest energy level, depends upon the interaction (β), which will not be the same for the three types of AOs. The most important factors determining the magnitude of β are:

- The interatomic distance; the greater the distance the smaller the interaction and the narrower the band.

- The closeness with which the electron is bound to the atom; the more tightly the electron is confined to the atom the smaller the interaction and the narrower the band.

Finally, no indication of the number of levels at any particular energy is given in Figures 12.17(a) and 12.17(b). Generally, the number of levels at the extremes of the band energy will be low. As we move towards the centre of the band the number of

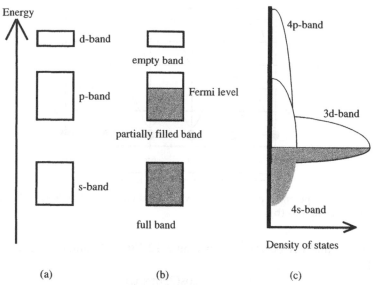

Figure 12.17 Band structure for a typical first-row transition metal

levels per energy unit, the *density of states*, will increase. Thus, the band structure of a typical transition metal might appear as in Figure 12.17(c).

12.3.2 The electron–gas (free-electron) approximation

An alternative approach to a theoretical description of the structure of metals was suggested by Arnold Sommerfeld (1868–1951) in 1928. It is based on a perturbed, free-electron model. If we again think of sodium as our example we can imagine that the 3s electrons, one from each atom, are effectively detached from their atomic cores and form a *sea* or *gas* of mobile electrons free to roam, almost at will, through the metal. They are not completely free because they move in the periodic potential field of the positively charged cores, which makes important modifications to their energies and wave functions. For the sake of simplicity we shall consider a one-dimensional model developing further the description of an electron in a linear box we initiated in Sections 3.6 and 3.7.

We first make some modifications to the wave functions described there. We noted that it was only possible to have electrons moving freely in one dimension if there were no boundaries to restrict the motion and that, under those conditions, the wave functions were eigenfunctions of both the operator for linear momentum and of the Hamiltonian. Pairs of wave functions that have equal energies and correspond to electrons moving in the positive and negative x-direction with equal and opposite linear momenta can be identified. However, we did not normalise these wave functions, and we must address that problem here. Also, because we are interested in the effect upon the electrons of the periodic field of the atomic cores, we need to choose a form of our x co-ordinate that reflects the arrangement of the cores. Our one-dimensional model consists of a series of positively charged atomic cores positioned at equal intervals (a) along the x co-ordinate. The potential energy of a free electron (V) will be lowest when it is at a core and

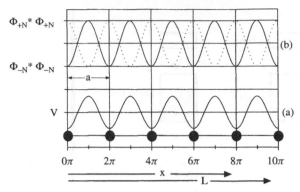

Figure 12.18 The core potential of a linear array of metal atoms (a) Stable $\Phi_+{}^*\Phi_+$ and unstable $\Phi_-{}^*\Phi_-$ electron distributions (b)

highest when it is midway between them Figure 12.18(a). A suitable algebraic form for the potential is:

$$V = -\mathrm{v} \cdot \cos(2N\pi x/L) \qquad (12.3.3)$$

where L is a length which contains N cores (L = Na) and is assumed to be very much larger than a, so that the energy levels of the system when $V = 0$ are $E_n = n^2 h^2 / 8 m_e L^2$ and still very closely spaced.

Following Section 3.7.1, but expressing the co-ordinate in a manner compatible with 12.3.3, the pairs of normalised wave functions which we shall use are:

$$\Psi_{\pm n} = (1/2L)^{\frac{1}{2}} \exp\{\pm i n\pi x/L\} \qquad (12.3.4)$$

where $\pm n$ is the quantum number and $i = \sqrt{(-1)}$ (Appendix 8). The wave functions are normalised to unity and orthogonal within any section of the system between $\alpha - L$ and $\alpha + L$, where α is an arbitrary point along the chain of cores:

$$\int_{\alpha-L}^{\alpha+L} \Psi_n{}^* \Psi_n \, dx = \frac{1}{2\,L} \int_{\alpha-L}^{\alpha+L} e^{-i\pi nx/L} \cdot e^{+i\pi nx/L} \, dx$$

$$= \frac{1}{2\,L} \int_{\alpha-L}^{\alpha+L} 1 \, dx = \frac{1}{2\,L}[\alpha + L - \alpha + L] = 1$$

Since L is defined as the length containing a specific number of atomic cores, the relative position with respect to the cores of the point $\alpha + L$ is exactly the same as that of $\alpha - L$, independent of the value of α. Therefore, in choosing these limits we are simply recognising the periodic nature of the sequence of cores or lattice; i.e. we are using *periodic boundary conditions*.

We must now examine the effect of the potential (V) upon and the wave functions and energies of the mobile electrons. In order to approach the essentials of the problem as directly as possible we eschew a general analysis and concentrate upon a very special case. What is the effect of this potential upon the wave functions and energies of a pair of electrons having equal but opposite values of n, i.e. two electrons of equal energy (linear momentum) moving in opposite directions along x?

We can view the mathematics either as an application of degenerate perturbation theory (Appendix 4) or as the construction of two new functions from a basis set of the two functions (Appendix 2):

$$\Psi_{+n} = (1/2L)^{\frac{1}{2}} \exp\{+in\pi x/L\}$$

and

$$\Psi_{-n} = (1/2L)^{\frac{1}{2}} \exp\{-in\pi x/L\} \tag{12.3.5}$$

We require the matrix elements $\langle \Psi_{\pm n}|\hat{V}|\Psi_{\pm n}\rangle$ and we must remember that $\langle \Psi|$ implies the complex conjugate of $|\Psi\rangle$. As an example of a diagonal element we have:

$$\langle \Psi_{+n}|\hat{V}|\Psi_{+n}\rangle = \frac{-v}{2L} \int_{\alpha-L}^{\alpha+L} e^{-i\pi nx/L} \cdot \cos\left(\frac{2N\pi x}{L}\right) \cdot e^{+i\pi nx/L} \, dx$$

$$= \frac{-v}{2L} \int_{\alpha-L}^{\alpha+L} 1 \cdot \cos\left(\frac{2N\pi x}{L}\right) dx$$

$$= \frac{-v}{2L} \cdot \frac{L}{2N\pi} \left[\sin\left(\frac{2N\pi x}{L}\right)\right]_{\alpha-L}^{\alpha+L}$$

$$= \frac{-v}{2N\pi} \cdot \cos\left(\frac{2N\pi\alpha}{L}\right) \cdot \sin 2N\pi = 0$$

The result of zero is scarcely surprising since we know that the wave functions $\Psi_{\pm n}$ describe electrons uniformly distributed along x and therefore subject equally to the positive maxima and negative minima of the potential V. As an example of one of the off-diagonal matrix elements we evaluate:

$$\langle \Psi_{-n}|\hat{V}|\Psi_{+n}\rangle = \frac{-v}{2L} \int_{\alpha-L}^{\alpha+L} e^{+i\pi nx/L} \cdot \cos\left(\frac{2N\pi x}{L}\right) \cdot e^{+i\pi nx/L} \, dx$$

$$= \frac{-v}{2L} \int_{\alpha-L}^{\alpha+L} e^{+i2\pi nx/L} \cdot \cos\left(\frac{2N\pi x}{L}\right) dx$$

$$= \frac{-v}{2L} \int_{\alpha-L}^{\alpha+L} \left\{\cos\left(\frac{2n\pi x}{L}\right) + i\sin\left(\frac{2n\pi x}{L}\right)\right\} \cdot \cos\left(\frac{2N\pi x}{L}\right) dx$$

$$= \frac{-v}{4L} \int_{\alpha-L}^{\alpha+L} \left\{\cos\left(\frac{[n+N]2\pi x}{L}\right) + \cos\left(\frac{[n-N]2\pi x}{L}\right)\right\} dx$$

$$\frac{-iv}{4L} \int_{\alpha-L}^{\alpha+L} \left\{\sin\left(\frac{[n+N]2\pi x}{L}\right) - \sin\left(\frac{[n-N]2\pi x}{L}\right)\right\} dx$$

If $n \neq N$ each of the four integrals above gives zero! However, if $n = N$ the integration of the first and third terms is essentially unchanged and the fourth goes to zero even before the integration is performed. But the second term now becomes the integral of $\cos(0) = 1$ and we have:

$$\frac{-v}{4L} \int_{\alpha-L}^{\alpha+L} \cos(0) \, dx = \frac{-v}{4L}[x]_{\alpha-L}^{\alpha+L} = \frac{-v}{2}$$

For the pair of wave functions $(\Psi_{\pm N})$ we can now express the total energy, kinetic plus potential, in the form of the Hamiltonian matrix:

$$\begin{array}{ccc} \hat{\mathcal{H}} & |\Psi_{+N}\rangle & |\Psi_{-N}\rangle \\[4pt] \langle\Psi_{+N}| & N^2h^2/8m_eL^2 & -v/2 \\[4pt] \langle\Psi_{-N}| & -v/2 & N^2h^2/8m_eL^2 \end{array}$$

By subtracting E from each diagonal element and multiplying the resulting matrix out as a determinant (Appendix 6) we obtain the characteristic equation of the matrix:

$$(N^2h^2/8m_eL^2 - E)^2 - v^2/4 = 0$$

from which we find the eigenvalues of the total energy as $N^2h^2/8m_eL^2 \pm v/2$. Since the diagonal elements of the above matrix are equal, the two functions will be equally involved in both new eigenfunctions which, when normalised, will therefore be of the form:

$$\Phi_{+N} = \sqrt{\tfrac{1}{2}}(\Psi_{+N} + \Psi_{-N})$$

and

$$\Phi_{-N} = \sqrt{\tfrac{1}{2}}(\Psi_{+N} - \Psi_{-N}) \tag{12.3.6}$$

We can find which of our two energies belongs to which eigenfunction simply by evaluating the integrals $\langle\Phi_{+N}|\hat{\mathcal{H}}|\Phi_{+N}\rangle$ and $\langle\Phi_{-N}|\hat{\mathcal{H}}|\Phi_{-N}\rangle$ using the matrix elements from the 2×2 matrix of $\hat{\mathcal{H}}$ above. For example:

$$\langle\Phi_{+N}|\hat{\mathcal{H}}|\Phi_{+N}\rangle = \tfrac{1}{2}\langle\Psi_{+N} + \Psi_{-N}|\hat{\mathcal{H}}|\Psi_{+N} + \Psi_{-N}\rangle$$

$$= \tfrac{1}{2}\{\langle\Psi_{+N}|\hat{\mathcal{H}}|\Psi_{+N}\rangle + \langle\Psi_{-N}|\hat{\mathcal{H}}|\Psi_{-N}\rangle + \langle\Psi_{-N}|\hat{\mathcal{H}}|\Psi_{+N}\rangle + \langle\Psi_{+N}|\hat{\mathcal{H}}|\Psi_{-N}\rangle\}$$

$$= \tfrac{1}{2}\{N^2h^2/8m_eL^2 + N^2h^2/8m_eL^2 - \tfrac{1}{2}v - \tfrac{1}{2}v\} = N^2h^2/8m_eL^2 - \tfrac{1}{2}v$$

Similarly:

$$\langle\Phi_{-N}|\hat{\mathcal{H}}|\Phi_{-N}\rangle = N^2h^2/8m_eL^2 + \tfrac{1}{2}v$$

We see that the effect of V has been to combine the wave functions of electrons travelling with equal velocities in opposite directions in equal proportions, converting running to stationary waves. The result of the mixing has been to raise the energy of one combination and to lower the energy of the other. The origin of this effect of the introduction of V is clear when we examine the forms of the new wave functions:

$$\Phi_{+N} = \sqrt{\tfrac{1}{2}}(\Psi_{+N} + \Psi_{-N}) = (1/2L)^{\tfrac{1}{2}}[\exp\{+iN\pi x/L\} + \exp\{-iN\pi x/L\}]$$

$$= (1/2L)^{\tfrac{1}{2}}[\cos\{N\pi x/L\} + i\sin\{N\pi x/L\} + \cos\{N\pi x/L\} - i\sin\{N\pi x/L\}]$$

$$= (1/2L)^{\tfrac{1}{2}}[2\cos\{N\pi x/L\}]$$

The corresponding electron density is:

$$\Phi_{+N}{}^*\Phi_{+N} = (1/L)[2\cos^2\{N\pi x/L\}] = (1/L)[1 + \cos\{2N\pi x/L\}]$$

Similarly:

$$\Phi_{-N}{}^*\Phi_{-N} = (1/L)[1 - \cos\{2N\pi x/L\}]$$

The argument of the cosine in the expressions for the electron density is exactly the same as that in the potential V and the variation of the density with x is therefore very similar to that of V. Figure 12.18(b) shows that $\Phi_{+N}{}^*\Phi_{+N}$ peaks at the position of the cores where V is at its lowest whilst the reverse is true for $\Phi_{-N}{}^*\Phi_{-N}$. This is the reason for the energy difference of v between the two functions. The periodic potential of the cores has combined two travelling waves with quantum numbers $+N$ and $-N$ to give two standing waves of different energies and has created an energy gap, or band gap, in the continuum (L is assumed to be very large) of energy levels at that point. A more general analysis shows that we expect off-diagonal matrix elements of V to occur between all pairs of wave functions of the form Ψ_m and Ψ_n if $n - m = \pm 2N$, since such cases will also give rise to the integral of $\cos(0) = 1$. Thus, the matrix of $|\Psi_{+N}\rangle$ and $|\Psi_{-N}\rangle$ shown above is simply a part of a much larger matrix, involving all the unperturbed states (Ψ_n). The following is a small section of it, centred on the matrix of $|\Psi_{+N}\rangle$ and $|\Psi_{-N}\rangle$; $\Omega = h^2/8m_e L^2$:

$\hat{\mathcal{H}}$	$\lvert\Psi_{+(N-1)}\rangle$	$\lvert\Psi_{-(N-1)}\rangle$	$\lvert\Psi_{+N}\rangle$	$\lvert\Psi_{-N}\rangle$	$\lvert\Psi_{+(N+1)}\rangle$	$\lvert\Psi_{-(N+1)}\rangle$
$\langle\Psi_{+(N-1)}\rvert$	$(N-1)^2\Omega$	0	0	0	0	$-v/2$
$\langle\Psi_{-(N-1)}\rvert$	0	$(N-1)^2\Omega$	0	0	$-v/2$	0
$\langle\Psi_{+N}\rvert$	0	0	$N^2\Omega$	$-v/2$	0	0
$\langle\Psi_{-N}\rvert$	0	0	$-v/2$	$N^2\Omega$	0	0
$\langle\Psi_{+(N+1)}\rvert$	0	$-v/2$	0	0	$(N+1)^2\Omega$	0
$\langle\Psi_{-(N+1)}\rvert$	$-v/2$	0	0	0	0	$(N+1)^2\Omega$

We see that the matrix is blocked out (Appendix 6) into the single central 2×2 matrix and pairs of identical 2×2 matrices of the form:

$\hat{\mathcal{H}}$	$\lvert\Psi_{+(N-k)}\rangle$	$\lvert\Psi_{-(N+k)}\rangle$
$\langle\Psi_{+(N-k)}\rvert$	$(N-k)^2\Omega$	$-v/2$
$\langle\Psi_{-(N+k)}\rvert$	$-v/2$	$(N+k)^2\Omega$

and

$\hat{\mathcal{H}}$	$\lvert\Psi_{-(N-k)}\rangle$	$\lvert\Psi_{+(N+k)}\rangle$
$\langle\Psi_{-(N-k)}\rvert$	$(N-k)^2\Omega$	$-v/2$
$\langle\Psi_{+(N+k)}\rvert$	$-v/2$	$(N+k)^2\Omega.$

The eigenvalues of these matrices are easily found to be:

$$E_{\pm k} = [N^2 + k^2]\Omega \pm [4N^2k^2\Omega^2 + v^2/4]^{1/2}$$

In the cases where $k \neq 0$ the two free-electron functions, which are mixed, do not have exactly the same energy and so their displacement from their original positions will be smaller than the maximum of v/2. The displacement will decrease with increasing k and there will no longer be equal mixing of the two wave functions. Thus, as we move into a band and away from the band edges the standing waves with zero electron momentum gradually metamorphose into running waves with increasing electron momentum.

One can envisage other values of n and m for which $n - m = \pm 2N$, e.g. $(3N + k) - (N + k) = 2N$. But the energies of the corresponding wave functions will be very different and to a good approximation the existence of off-diagonal elements in the energy matrix

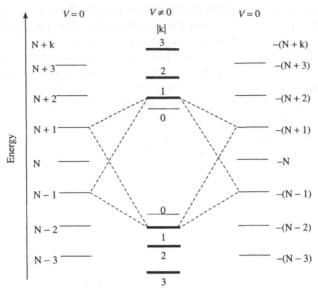

Figure 12.19   The formation of a band gap by a periodic potential, V

may be ignored. The effect of a periodic perturbation on the energy levels of a free electron is illustrated in Figure 12.19. The stacks of energy levels at the sides of the diagram are those of the unperturbed system ($V = 0$) immediately above and below E_N. The degenerate pairs of states are plotted separately according to the sign of their linear momentum. The perturbed levels ($V \neq 0$) are plotted in the centre of the diagram. The two with k = 0 are singly degenerate, but the remainder occur in degenerate pairs and are represented by heavier lines. Of the four levels characterised by a value of $|k| \neq 0$, one of each degenerate pair arises from the V-induced mixing of $\Psi_{+(N+k)}$ and $\Psi_{-(N-k)}$ and the other two from the mixing of $\Psi_{-(N+k)}$ and $\Psi_{+(N-k)}$. The dashed lines, which indicate the connection between perturbed and unperturbed states, have been limited to the case of $|k| = 1$ in the interests of the clarity of the figure.

Figure 12.19 is artificial in a number of respects, most notably in the large spacing between the levels, which in a realistic view would be much closer together. But it does show how the effect of the potential V is to create a gap of v, centred on E_N, in the energy levels with a consequent increase in the number of levels (density of states) in the regions immediately outside the range $E_N \pm v/2$. This situation is unchanged and the gap will not decrease if we now imagine the energy levels to be such that there is no discernible space between them when $V = 0$.

Finally, we must note that in a real, three-dimensional metal the cores will form periodic potentials in many directions, and not necessarily along the three Cartesian co-ordinates, and the band structure becomes very complicated. The treatment of a real metal crystal lattice requires more advanced theoretical methods.

12.3.3   Molecular and ionic solids

The tight-binding and free-electron descriptions of solids seem to be naturally applicable to single crystals composed of one type of atom only; a crystal of sodium (free-electron) or of diamond (tight-binding) for example. The differing physical properties of these two examples can be related to the delocalised electrons in the sodium and the localised electrons in diamond. In the case of molecular solids, however, a distinction must be drawn between the bonding of the atoms of one molecule to those of another, the intramolecular bonding, and the intermolecular bonding within each molecule. For the lighter elements, e.g. solid hydrogen, phosphorus or benzene, where the intermolecular bonds are far stronger than the intramolecular, the band theory has little to offer. But as we ascend the periodic table the difference between the strengths of the intra- and inter-molecular bonds decreases and band models become more applicable.

For ionic solids too, the tight-binding version of the band theory is appropriate. If we take sodium chloride as our example and assume that the transfer of one electron from sodium to chlorine is complete, then the 3s and 3p orbitals of the Cl^- form 3s and 3p bands, both of which will be full because the chloride ion has a full complement of 3s and 3p electrons. Detailed calculations show that at the equilibrium internuclear distance these two bands and the empty Na^+ 3s band are very narrow and that the gap between the highest filled band, Cl^- 3p, and lowest unfilled band, Na^+ 3s, is large, approximately 82×10^4 J mol^{-1}. The narrowness of the bands confirms the concept of an ionic bond, which is largely electrostatic in nature since it is atomic orbital overlap, and covalent bonding, which causes the energy spread of a band. The large gap between a full *valence band*, i.e. a band formed from the valence atomic orbitals of the bonded atoms, and the empty band above it explains why sodium chloride is an insulator, *vide infra*.

12.3.4 Applications

It is only fair to say that the simplicity of the basic principles of the above theoretical descriptions belies the difficulty of using them quantitatively. For quantitative purposes many more sophisticated interpretations of the basic ideas have been developed.[13-15]. But we might bear in mind here that it was in their introduction to a review of the theory of metals that Wigner and Seitz made the remarks quoted in Section 6.2. For our present purposes, these theories are more important for the light they shed on the reasons underlying the general properties of solids, and for the way in which they show how the bonding in discrete molecules and extended arrays of atoms can be explained with the same concepts. The types of solids to which the above theories are most applicable are summarised in Table 12.7.

12.3.5 Metals, insulators and semiconductors

The varying behaviour of solids with respect to the conduction of an electric current is one of their most well-known and technologically important properties. The range

Table 12.7 Areas of application of the free-electron and tight-binding theories

Free-electron approximation with V very small	Akali metals
Free-electron approximation with larger V	Solids with s and p valence electrons outside closed shells
Tight-binding approximation	Transition-metal d-electrons (heavier TMs)
	Valence electrons of non-metals (diamond)
	Ionic solids

Table 12.8 The effect of temperature on the electrical conductivity of solids

Solid	Conductivity, κ (Sm^{-1})	Effect of temperature, T
Metals and metallic alloys	$\kappa = 10^6 - 10^8$	κ rises with falling T and may become infinite at very low T
Semiconductors	$\kappa = 10^{-7} - 10^5$	κ rises with rising T
Insulators	$\kappa = 10^{-20} - 10^{-10}$	κ rises with rising T

of conductivity (κ in S m^{-1}) and the effect of temperature upon it are summarised in Table 12.8.

The band theory provides a convincing interpretation of these data in terms of the occupation of bands and the gap between full and empty bands (Figure 12.20). The running electron wave functions immediately suggest a mechanism for the transfer of electrons from one end of a solid to the other. We must remember, though, that in every direction the wave functions are paired so that, if the band is full, transport of electrons in either direction is not possible, since there is no way in which an excess of electrons travelling in a particular direction can be generated. A solid in which every band is either completely full or completely empty is therefore an insulator (Figure 12.20(a)). But the gap between the full and empty bands must be large enough to prevent electrons occupying the latter as a consequence of the Boltzmann distribution, *vide infra*.

However, if a band is only partially occupied the application of a voltage to the ends of the metal creates a situation in which the energies of the electrons at the two ends of the metal are unequal. The electrons in the higher-energy orbitals just below the Fermi level react to this situation by moving to previously unoccupied orbitals, just above the Fermi level, where they can respond to the applied voltage by moving in the appropriate direction. Thus an excess of electrons moving through the metal in the direction of the applied voltage, i.e. an electric current, is generated. Consequently, we see that an incompletely filled band, known as a *conduction band*, is required for metallic electrical conductivity (Figure 12.20(b)). Note that in this case the size of any band gap is unimportant.

When the band gap of a nominal insulator is so small that the Boltzmann distribution places some electrons in the higher, empty band (Figure 12.20(c)) there can be electrical

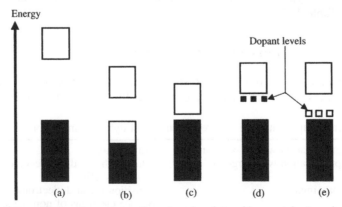

Figure 12.20 Schematic band structures for: a) an insulator, b) a metal, c) an intrinsic semiconductor, d) an n-type semiconductor and e) a p-type semiconductor

conduction and this depends critically upon the population of the higher band. Such sub-stances are known as intrinsic semiconductors; germanium and silicon are two important examples. The conductivity of an intrinsic semiconductor can be enhanced by the addition of very small quantities of impurities, which increase the number of electrons in the con-duction band. Such a substance is known as an n-type extrinsic semiconductor because the impurity adds negatively charged particles (electrons) to the conduction band.

The IV-V semiconductors in which an element from group IV of the periodic table, e.g. germanium (the host), is doped with an element from group V, e.g. antimany, form a good example. The dopant atoms substitute host atoms in the crystal lattice, which remains unchanged because the quantity of dopant is very small, of the order of 1 dopant atom to 10^9 host atoms. Since the dopant atoms are so far apart there is no interaction between them and their electronic energy levels are like those of an isolated atom perturbed by the host. Thus, the dopant levels are much narrower than the representation in Figure 12.20(d) suggests. But the important aspect of the situation is that the valence electrons of the dopant are higher in energy than those of the host and lie just below the vacant conduction band; they are thus able to contribute conducting electrons to that band. The band structure of an extrinsic p-type semiconductor is shown in Figure 12.20(e). In this case the dopant is an element from an earlier group in the periodic table; it has vacant electronic energy levels just above the full valence band of the pure semiconductor. These vacant levels are able to accept some of the electrons from the top of the full band below them leaving vacancies, usually regarded as positive holes (hence p-type), in it. Again, because the band is now no longer completely full, electrons are able to respond to an imposed voltage and the conductivity is increased.

The different effects of a change of temperature on the conductivity of a semiconductor and a metal may be understood in the following way. Generally, the energies associated with the movement of atoms or atomic cores, e.g. vibration, are small compared with those of electronic energy levels. However, if the temperature of a semiconductor is sufficiently high it may be possible for the Boltzmann distribution to take some electrons at the top of a full valence band above the band gap to a previously empty conduction band. Once there they can move in response to the applied voltage and, as the temperature is increased, the conductivity of the semiconductor increases. The effect of temperature on a metallic conductor arises in a different manner. Conducting electrons proceeding through a metal must run the gauntlet of the positive cores which impede their motion. As the temperature is raised the increasing movement of the cores makes this more difficult and the travelling electrons collide more frequently with the cores, so reducing the current. This effect outweighs the increase in conductivity that arises because the electrons populate higher levels of the conduction band as a result of the Boltzmann distribution. Conversely, as the temperature is reduced metallic conductivity increases and at very low temperatures, 20 K or lower, the resistance of most metals and metallic alloys falls to zero; they become superconductors. This is surprising since we know that vibrational motion persists, even at absolute zero, giving rise to the zero-point energy. Superconductivity therefore requires a degree of coupling between the motions of the electrons and the cores. Theoreticians have found that synchronous movements of the cores combine to assist the passage of pairs of electrons through the lattice. Where a group of positive cores have been drawn slightly closer together by the passage of an electron, the resulting region of enhanced positive charge attracts a second electron to follow the same path.

12.3.6   Optical properties of solids

We are concerned here with the optical properties of pure solids, e.g. metals or diamond; the properties of composite solids such as porcelain require a consideration of the components of the substance and the way in which those components interact. The most characteristic optical property of the metals is their high reflectivity and opacity, both of which arise directly from the presence of the electrons, which are highly mobile and not confined to specific interatomic regions as they are in molecules. When the oscillating field of a light wave strikes a metal the mobile electrons respond by oscillating at the frequency of the incident light. The moving electrons produce an oscillating electric field and themselves become a light source of the same frequency as the incident light. We perceive this light emitted by the moving electrons as a reflection from the metal surface. Being very free, the electrons can respond to light of a broad range of energies and light of all visible frequencies is reflected. Furthermore, since there are so many electrons no light beam can penetrate far into a metal without being reflected; metals are therefore opaque. However, at high frequencies in the ultraviolet range a limit is reached where the electrons can no longer respond at the frequency of the light. At this wavelength the light passes through the metal without interacting with the electrons and the metal becomes transparent.

A few metals appear coloured. The reddish hue of copper ($[Ar]3d^{10}4s^1$), for example, is a consequence of the fact that electrons can be raised from the 3d valence band to the 4s conduction band by light from the blue end of the visible spectrum. This is a normal absorption process of the kind described in Chapters 8 and 11. Thus, the light reflected from the surface of a piece of copper has less blue light than the incident light and the metal therefore appears red. The same phenomenon is found with most other metals but we do not notice it because the light absorbed lies outside the visible range.

12.3.7   Mechanical properties of solids

There are many other aspects of the solid state–thermal conductivity, heat capacity, magnetism–which cannot even be touched upon here. They lie more in the realm of solid state physics.[13,14] However, a few brief comments on the mechanical properties of solids and their interpretation in terms of the electronic structure may be in order. The bonds in metals, for which the primary bonding forces are a sea of delocalised electrons, are not directional. Thus, when metals are deformed their atoms are able to form new bonds with their new neighbours and metals can be easily deformed and withstand considerable distortion before they rupture. On the other hand, solids such as diamond with highly directional covalent bonds are the hardest substances which we know. They are very difficult to deform but if sufficient force to break the bonds is applied they shatter. Ionic solids are brittle. The force required to move an ion with respect to its neighbours is fairly large, but once the regular ionic array is disrupted the repulsion of ions of like charge destroys the crystal.

12.4   BIBLIOGRAPHY AND FURTHER READING

1. E. Hückel, *Z. Physik*, **70**, 204(1931); **72**, 310 (1931); **76**, 628 (1932); **83**, 632 (1933).
2. G.J. Hoijtink, *Rec. Trav. Chim.*, **77**, 555 (1958).

3. E. Heilbronner and J.N. Murrell, *J. Chem. Soc.*, 2611 (1962).
4. F. Klages, *Chem. Ber.*, **82**, 358 (1949).
5. G. Klopman, *J. Am. Chem. Soc.*, **90**, 223 (1968).
6. I. Fleming, *Frontier Orbitals and Organic Chemical Reactions*, Wiley, New York, 1976.
7. A. Streitwieser, *Molecular Orbital Theory for Organic Chemists*, Wiley, New York, 1961.
8. G. Berthier, B. Pullman and J. Pontis, *J. Chim. Phys.*, **49**, 367 (1952).
9. J.N. Murrell, S.F.A. Kettle and J.M. Tedder, *Valence Theory*, 2nd edn, Wiley, New York, 1970.
10. The units of electrical and magnetic properties are often confusing and difficult to handle. The article by T.I. Quickenden and R.C. Marshall, *J. Chem. Ed.*, **49**, 114 (1972), is very helpful.
11. J.H. Van Vleck, *The Theory of Electric and Magnetic Susceptibilities*, Oxford University Press, 1932.
12. W.E. Henry, *Phys. Rev.*, **88**, 559 (1952).
13. C. Kittel, *Introduction to Solid State Physics*, 7th edn, Wiley, New York, 1996.
14. N.W. Ashcroft and N.D. Mermin, *Solid State Physics*, Saunders College Publishing, Philadelphia, 2001.
15. A.P. Sutton, *Electronic Structure of Materials*, Oxford University Press, 1993.

BOX 12.1   The enthalpy of combustion and the delocalisation energy of benzene

The enthalpies of combustion of carbon and hydrogen atoms are -1110 kJ mol^{-1} and -361 kJ mol^{-1} respectively, so the combustion of six moles of carbon atoms and six moles of hydrogen atoms produces 8826 kJ of energy. The calculated energy evolved in the combustion of one mole of 'Kekulé' benzene is 3451 kJ, considerably less than that produced by the combustion of the constituent atoms because more than half of the energy released is used to break the C=C, C–C and C–H bonds. The energy released by the combustion of one mole of 'real' benzene is even less, indicating that the benzene molecule is yet more stable than the Kekulé structure implies (Figure B12.1.1). The difference of 149 kJ mol^{-1} is the resonance or delocalisation energy.

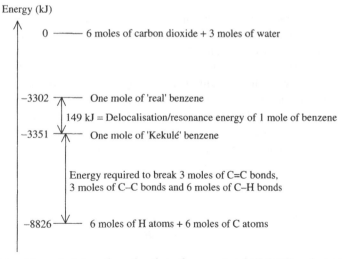

Figure B12.1.1   The enthalpies of combustion of one mole of 'Kekulé' and one mole of 'real' benzene (not to scale)

BOX 12.2   The orientation of a magnetic dipole in a magnetic field

In Section 12.2.1 we stated, without further justification, that when atomic or molecular magnets are placed in a magnetic field they align their magnetic dipoles with the applied field, thereby increasing B. This statement may appear counter-intuitive, especially so when one considers the case with which it has a great deal in common – the orientation of molecules with a permanent dipole moment in an electric field. In that case the orientation of lowest energy is where the molecules align themselves so that their positive ends point towards the negative pole of the applied field and their negative ends towards the positive pole, thereby opposing and reducing that field. The origin of this apparent dilemma lies in the fact that isolated positive and negative charges are real entities, whereas isolated north and south poles are figments of our imagination. If we take a bar magnet and saw it in half we obtain two bar magnets, each with a north and south pole, and we know that this remains true even when we continue the process until our remaining magnets are individual atoms or molecules. But bodies of atomic or macroscopic size, charged with an electrical charge of one sign only, can be readily generated in the laboratory. Therefore, when we calculate the energy of a magnetic dipole in a magnetic field we have to consider the exact form of the dipole carefully.

Since there are no such things as isolated north and south magnetic poles we cannot represent a magnetic dipole in an applied magnetic field simply by replacing the positive and negative signs of the corresponding electric dipole–electric field problem by N and S respectively. In fact, the origin of all magnetism is the circulation of electric charge and the only true representation of a magnetic dipole is the magnetic field produced by a current flowing around a small loop. The energy of the resulting magnetic dipole with respect to its orientation in an applied field must be determined by calculating the forces exerted by the field on the current-carrying components of the loop.

A plan view of a rectangular loop (area A) perpendicular to the plane (xz) of the paper and carrying a flow of positive charge (i) which enters the paper at + and emerges again at · is shown in Figure B12.2.1. The magnetic moment (μ) produced by the current is perpendicular to the plane of the loop (compare Figure 12.10) and makes an angle of θ with the direction (z) of an applied field (B). Since the currents on opposite sides of the loop flow in opposite directions there is no net force on the loop, if the field is uniform. But Fleming's left-hand rule tells us that there are forces (F), as indicated in the figure, acting on the ascending and decending conductors of the loop which are parallel to y. The loop is therefore subject to a torque (τ), which tends to align the moment (μ) with the field (B). It is not difficult to show[1,2] that:

$$\tau = iAB\sin\theta = \mu B\sin\theta$$

or in vector notation:

$$\tau = \mu \times B$$

The corresponding forces acting on the two horizontal components of the loop are vertical and opposed and produce no torque. It is clear from the figure that the torque acts in such a way as to align μ and B and we see that the presence of the

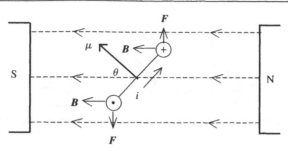

Figure B12.2.1 The forces on a current loop in a magnetic field

paramagnetic atom or molecule does indeed enhance the applied field, which was the main objective of this discussion.

We can also obtain an expression for the potential energy (E) because in order to increase the energy by dE we must perform work (dW) against the torque and:

$$dE = dW = \tau \, d\theta = \mu B \sin \theta \, d\theta$$

Integration gives:

$$E = -\mu B \cos \theta + \text{constant}$$

Since we are interested here only in relative energies we set the constant to zero so that:

$$E = -\mu B \cos \theta \quad \text{or} \quad E = -\boldsymbol{\mu} \cdot \boldsymbol{B}$$

Finally, we should note that although we have considered only the case of a rectangular loop, the result applies to a loop of any shape and the magnetic moment is always iA, where i is the current and A the area of the loop.

REFERENCES

1. R.P. Feynman, R.B. Leighton and M. Sands, *The Feynman Lectures on Physics*, Addison-Wesley Publishing Co., London, 1964.
2. I.S. Grant and W.R. Phillips, *Electromagnetism*, 2nd edn, Wiley, Chichester, 1990.

BOX 12.3   The connection between experimental magnetic susceptibility and atomic/molecular magnetic moment – a classical treatment

In order to measure the macroscopic magnetic susceptibility of a substance, and to derive information about its electronic structure from such a measurement, we must normally study the interaction of the substance with an imposed magnetic field. The potential energy (E) of a magnetic dipole (μ) subjected to a magnetic field (**H**) (Figure B12.3.1) is given (Box 12.2) by the equation:

$$E = -\mu B \cos \theta = -\boldsymbol{\mu} \cdot \boldsymbol{B}$$

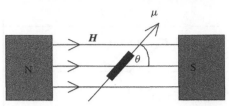

Figure B12.3.1   The energy of a magnetic dipole (μ) in a magnetic field (H)

where B is the magnetic flux density at the position of the dipole and θ is the angle between the magnetic field and the dipole. This important result assumes that the applied field *(H)* and the resultant flux density *(B)* are parallel.

When a sample substance that consists of molecules each having a magnetic dipole moment (μ) is placed in a magnetic field the molecules will be subject to two forces; the tendency for the individual molecular dipoles to orientate themselves along the direction of the magnetic field and the opposing thermal motion. The measured macroscopic magnetic properties of the material will depend upon the resulting distribution of orientations.

The fraction of dipoles (dN) lying at an angle between θ and $\theta + d\theta$ is equal to the fraction of the surface of a sphere of unit radius which lies within the same range of θ. This is a band of radius $\sin\theta$, i.e. length $2\pi \sin\theta$, and width $d\theta$ running around the surface of the sphere, (Figure B12.3.2).

Dipoles orientated at different values of θ have different potential energies and the fraction of molecules with dipoles lying at an angle between θ and $\theta + d\theta$ is determined by the Boltzmann population appropriate to that particular energy. The dipoles with this orientation have energy $(-B\mu\cos\theta)$ so that their fractional population (dN_θ) depends upon the Boltzmann factor ($\exp\{B\mu\cos\theta/kT\}$) and the orientational factor ($\{2\pi \sin\theta\}$):

$$dN_\theta \propto \exp\{B\mu\cos\theta/kT\} \cdot \{2\pi \sin\theta\} \, d\theta$$

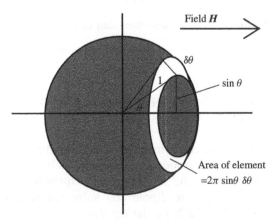

Figure B12.3.2   The area element for magnetic dipoles lying at an angle between θ and $\theta + d\theta$ to the magnetic field

where k is Boltzmann's constant and T is the absolute temperature. We can replace the proportionality sign with a constant (A), which depends upon the number of molecules in the sample, and the relative areas of the band and the sphere:

$$dN_\theta = A \cdot \exp\{B\mu \cos\theta/kT\} \cdot \{2\pi \sin\theta\}\, d\theta$$

Since the contribution to the macroscopic dipole moment of each member of this group of molecules is $\mu \cos\theta$ their contribution (dM_θ) to the total macroscopic magnetic moment (M) will be:

$$dM_\theta = A \cdot \exp\{B\mu \cos\theta/kT\} \cdot \{2\pi \sin\theta\} \cdot \{\mu \cos\theta\}\, d\theta$$

M can now be obtained by integrating this expression over all possible molecular orientations, i.e. from 0 to π. To determine the average value of μ (μ_{av}) we divide by the total number of molecules, which we can find by integrating the expression for dN_θ over the same range of θ:

$$\mu_{av} = \frac{\displaystyle\int_0^\pi A \cdot \exp\{B\mu \cos\theta/kT\} \cdot \{2\pi \sin\theta\} \cdot \{\mu \cos\theta\}\, d\theta}{\displaystyle\int_0^\pi A \cdot \exp\{B\mu \cos\theta/kT\} \cdot \{2\pi \sin\theta\}\, d\theta}$$

$$= \frac{\displaystyle\mu\int_0^\pi \exp\{B\mu \cos\theta/kT\} \cdot \sin\theta \cdot \cos\theta\, d\theta}{\displaystyle\int_0^\pi \exp\{B\mu \cos\theta/kT\} \cdot \sin\theta\, d\theta}$$

These integrals can be simplified by writing $s = B\mu/kT$, $t = \cos\theta$ (whence $\sin\theta\, d\theta = -dt$) and replacing the integration limits accordingly to give:

$$\mu_{av} = \frac{\displaystyle\mu\int_{+1}^{-1} -t \cdot \exp\{st\}\, dt}{\displaystyle\int_{+1}^{-1} -\exp\{st\}\, dt} = \frac{\mu[\exp\{st\} \cdot (st-1)/s^2]_{+1}^{-1}}{[\exp\{st\}/s]_{+1}^{-1}}$$

$$= \mu\left[\frac{\exp\{s\} + \exp\{-s\}}{\exp\{s\} - \exp\{-s\}} - \frac{1}{s}\right] = \mu\left[\coth\{s\} - \frac{1}{s}\right]$$

This equation was first derived by Paul Langevin (1872–1946) and carries his name. In order to apply it, important and readily justified approximations are usually made. Since μ is of the order of 10^{-23} A m^2, $s = B\mu/kT \approx 2.5 \times 10^{-3}$ in a magnetic field of one Tesla and the expansion of the exponential functions (see Box 3.1) can be truncated after terms in s^3 giving:

$$\mu_{av} = \mu\left[\frac{\exp\{s\} + \exp\{-s\}}{\exp\{s\} - \exp\{-s\}} - \frac{1}{s}\right] \approx \mu\left[\frac{2 + s^2}{2s + s^3/3} - \frac{1}{s}\right] = \mu\left[\frac{2 + s^2}{2s(1 + s^2/6)} - \frac{1}{s}\right]$$

$$\approx \mu\left[\frac{1}{2s} \cdot (2 + s^2)(1 - s^2/6) - \frac{1}{s}\right] = \mu\left[\frac{s}{3} - \frac{s^3}{12}\right] \approx \frac{B\mu^2}{3kT}$$

In the third step we have used the fact that $(1 + s^2/6)^{-1} \approx (1 - s^2/6)$ if s is small. Thus:

$$\mu_{av} = B\mu^2/3kT$$

and for an assembly of N molecules per unit volume:

$$M = N\mu_{av} = NB\mu^2/3kT$$

Therefore:

$$\chi = M/H = NB\mu^2/3kTH$$

and since $B = \mu_o(H + M)$, when M is small we have:

$$\chi = N\mu_o\mu^2/3kT \equiv \{N\mu_o\mu_B^2/3kT\}\mu_{eff}^2 \qquad (B12.3.1)$$

The identity in Equation (B12.3.1) defines the *Bohr magneton number* (μ_{eff}), which is taken to be positive. It is the effective magnetic moment measured in Bohr magnetons. Comparing the result with Curie's equation, $\chi = C/T$, we obtain an expression for C:

$$C = N\mu_o\mu^2/3k \quad \text{or} \quad \{N\mu_o\mu_B^2/3k\}\mu_{eff}^2 \qquad (B12.3.2)$$

We have derived a classical expression for the magnetic moment (M) induced in a collection of N dipoles by the application of a magnetic field (H). This is the macroscopic, experimental side of the problem, a relationship between the susceptibility (χ), which we can measure, and a property (μ or μ_{eff}) of the individual molecule, which we would like to know. Note that the susceptibility is defined in terms of the applied magnetic field (H), as it must be since the applied magnetic field is something that we can measure. But our dipole (μ) interacts with the resulting flux density (B), which we cannot measure directly. But the approximation $B = \mu_o H$ is a good one when M is small.

We now turn to the microscopic, quantum-mechanical aspect of the problem. How do the individual electronic states of a single atomic/molecular magnet contribute to the total μ for that entity? The important factors are the magnetic moment of each state, which depends upon the orbital and spin angular momentum, and the energy of the state, which determines its population through the Boltzmann distribution. Let these two factors for any state i be μ_i and E_i respectively; the means of calculating them are described in the main body of the text. We set i = 1 for the state which has the lowest energy in the applied field and express the populations (N_i) of all other states in terms of the population N_1. According to Boltzmann we have:

$$N_i = N_1 \exp\{(E_1 - E_i)/kT\} \equiv N_1 e^{(E_1 - E_i)/kT}$$

The total magnetic moment (μ_t) for the molecule or atom is:

$$\mu_t = \sum_i \mu_i N_i$$

To obtain the average moment (μ_{av}) the above expression for the total must be divided by the total population:

$$\mu_{av} = \frac{\sum_i \mu_i N_i}{\sum_i N_i} = \frac{N_1 \sum_i \mu_i e^{(E_1 - E_i)/kT}}{N_1 \sum_i e^{(E_1 - E_i)/kT}} = \frac{N_1 e^{E_1/kT} \sum_i \mu_i e^{-E_i/kT}}{N_1 e^{E_1/kT} \sum_i e^{-E_i/kT}}$$

Cancelling out the factors multiplying both series, the expression for μ simplifies to:

$$\bar{\mu} = \frac{\sum_i \mu_i e^{-E_i/kT}}{\sum_i e^{-E_i/kT}} \tag{B12.3.3}$$

Applications of Equations (B12.3.1), (B12.3.2) and (B12.3.3) can be found in Chapter 12.

PROBLEMS FOR CHAPTER 12

1. *Hückel orbitals for methylenecyclopropene (C_4H_4).* The molecule has the carbon-atom skeleton below:

Following the method described in Section 12.1.4, set up the secular equations for the π-electron system and show that the molecular orbital energies are given by the roots of the equation:

$$x^4 - 4x^2 + 2x + 1 = 0$$

where $x = (\alpha - E)/\beta$.

The evaluation of the Hückel energies and atomic orbital coefficients is rather tedious; they are found to be:

Energy	Coefficients			
$\alpha - 1.481\beta$	0.506	−0.749	0.302	0.302
$\alpha - 1.000\beta$	0.0	0.0	0.707	−0.707
$\alpha + 0.311\beta$	−0.815	−0.254	0.368	0.368
$\alpha + 2.170\beta$	0.282	0.612	0.523	0.523

Draw an energy-level diagram for the molecule and calculate the π-electron charge densities on each atom and the bond orders for all bonds for the neutral molecule.

[Ans: $Q_1 = 1.488$, $Q_2 = 0.877$, $Q_3 = Q_4 = 0.818$, $P_{12} = 0.758$, $P_{34} = 0.818$, $P_{23} = P_{24} = 0.453$.]

What would be the distribution of positive or negative charge in the ions $[C_4H_4]^+$ and $[C_4H_4]^-$?

2. *C–C bond lengths and bond orders*. Complete the following table:

Molecule	π-Bond order	Bond length[a] (pm)
Ethane		153.4
Benzene		139.7
Ethene		133.9
Ethyne		120.4

[a] *Interatomic Distances Supplement*, Special Publication No. 18, The Chemical Society, London, 1965.

[Hints for determining bond order: Ethane has no π bond, ethene has one and ethyne two. The bond order of benzene can be calculated from the molecular orbital coefficients given in Figure 11.5.]

Plot a graph of bond order versus bond length and use it to estimate the lengths of the two C–C bonds in butadiene for which the bond order data can be found in Table 12.4. Compare your results with the experimental values of $r_{1-2} = 133.7$ pm and $r_{2-3} = 148.3$ pm. The prediction is not very good; suggest why.

3. *Hückel orbitals for methanal* $(H_2C{=}O)$.
The hetero-atom, oxygen, can be incorporated into the Hückel scheme using the procedure described at the end of Section 12.1.6. Form the two secular equations for methanal with $h_O = 1.2$ and $k_O = 2.0$ and show that the energy levels are given by the roots of the polynomial $x^2 + 1.2x - 4 = 0$. Calculate the energies and coefficients; you should find:

$$E = \alpha - 1.488\beta \quad C_C = +0.802 \quad C_O = -0.597$$

$$E = \alpha + 2.688\beta \quad C_C = +0.597 \quad C_O = +0.802$$

Calculate the π-electron charge densities and bond orders for the neutral molecule. [$Q_C = 0.713$, $Q_O = 1.286$, $P_{CO} = 0.958$]. Note how the larger value of α assumed for oxygen draws charge to that end of the molecule.

4. Construct a table like Table 12.5 for a single d electron in a strong octahedral crystal field. Assume that spin-orbit coupling is zero and that, because of the large field, only the orbitals d_{xy}, d_{xz} and d_{yz} are occupied. The atomic orbitals having $m_l = 1$ (i.e. d_{xz} and d_{yz}) are not mixed by the field (Chapter 7) and can be used in the form $|m_l, m_s\rangle$ in the calculation. The d_{xy} however must be used in the calculation in the form:

$$|d_{xy}, \pm\tfrac{1}{2}\rangle = \{i/\sqrt{2}\}\{|-2, \pm\tfrac{1}{2}\rangle - |+2, \pm\tfrac{1}{2}\rangle\}$$

and

$$\langle d_{xy}, \pm\tfrac{1}{2}| = -\{i/\sqrt{2}\}\{\langle-2, \pm\tfrac{1}{2}| - \langle+2, \pm\tfrac{1}{2}|\}$$

Show that:

$$\bar{\mu} = \frac{2\exp(2x) - 2\exp(-2x) + \exp(x) - \exp(-x)}{\exp(2x) + \exp(-2x) + \exp(x) + \exp(-x) + 2\exp(0)} = \frac{2\sinh(2x) + \sinh(x)}{\cosh(2x) + \cosh(x) + 1}$$

where $x = B/kT$.

Use a spreadsheet or other plotting program to draw the magnetisation curve like graph 2 in Figure 12.12.

5. Calculate the magnetic energy levels of a system consisting of two interacting metal ions one of which has $S = \frac{1}{2}$, $M_S = \pm\frac{1}{2}$ and the other has $S = 1$, $M_S = 0, \pm1$. Use the coupling Hamiltonian operator ($\hat{\mathcal{H}} = -2K\hat{S}_1 \cdot \hat{S}_2$) and proceed as in Section 12.2.10. You will find that your result has much in common with the energy levels of a single p electron, in which spin and orbital magnetic moments are coupled by the spin-orbit coupling. [The ubiquity of phenomena based on angular momentum is such that the same problems occur again and again.]

6. Show that the wave functions (Ψ_m) defined by Equation (12.3.4) are orthogonal to each other, i.e. show that $\langle\Psi_m|\Psi_n\rangle = 0.0$ if m $\neq$ n.

7. Derive an expression for the general matrix element $\langle\Psi_m|\hat{V}|\Psi_n\rangle$ where $\hat{V}$ is the potential defined by Equation(12.3.3) and Ψ_m are the wave functions defined by Equation (12.3.4).

Use a spreadsheet or other plotting program to draw the magnetization curve like graph 2 in Figure 12.12.

APPENDIX 1

Fundamental Constants and Atomic Units

Fundamental constants[1]

	Constant	Value in SI units	Value in other units
c	vacuum velocity of light	2.99792×10^8 m s^{-1}	2.99792×10^{10} cm s^{-1}
e	elementary charge	1.60218×10^{-19} C	4.80321×10^{-10} esu
k	Boltzmann's constant	1.38066×10^{-23} J K^{-1}	1.38066×10^{-16} erg deg^{-1}
h	Planck's constant	6.62608×10^{-34} J s	6.62608×10^{-27} erg s
$\hbar$	Planck's constant/2π	1.05457×10^{-34} J s	1.05457×10^{-27} erg s
N_A	Avogadro's constant	6.02214×10^{23} mol^{-1}	
ε_0	permittivity of free space	8.85419×10^{-12} J^{-1} C^2 m^{-1} (F m^{-1})	
μ_0	permeability of free space	$4\pi \times 10^{-7}$ J s^2 C^{-2} m^{-1} (N A^{-2})	
m_e	rest mass of electron	9.10939×10^{-31} kg	
m_p	rest mass of proton	1.67262×10^{-27} kg	
μ_B	Bohr magneton	9.27402×10^{-24} A m^2	9.27402×10^{-21} erg gauss^{-1}
μ_N	nuclear magneton	5.05079×10^{-27} A m^2	5.05079×10^{-24} erg gauss^{-1}

Atomic units

The metre, kilogram, etc. are not very convenient units with which to measure atomic and molecular quantities; they are too large. It is easier to work with units which are directly related to the fundamental physical quantities such as electronic charge and mass. Such a system is in widespread use and the units are known as *atomic units*.

Apart from the convenience associated with the appropriate sizes of the atomic units, the system has another, equally important advantage. A result expressed in terms of the fundamental physical constants is always correct algebraically and can be easily brought up to date numerically whenever a more accurate determination of any of the physical

The Quantum in Chemistry   R. Grinter
© 2005 John Wiley & Sons, Ltd

constants is made. The following atomic units occur most widely, a full list is given by Mills *et al.*[1]

The atomic unit of mass is the rest mass of the electron; $m_e = 9.10956 \times 10^{-31}$ kg.

The unit of charge is the elementary charge; $e = 1.60219 \times 10^{-19}$ C.

The atomic unit of length is the Bohr radius or bohr (a_o), the radius of the orbit of lowest energy of the electron of the hydrogen atom in the Bohr theory. The exact value of this quantity depends upon the mass of the nucleus (see Section 5.5). Therefore, the radius is defined for a nucleus of infinite mass; $a_0 = \varepsilon_0 h^2 / m_e e^2 = 5.29177 \times 10^{-11}$ m.

The atomic unit of energy is the Hartree; $E_H = h^2 / 4\pi^2 m_e a_0^2 = 4.35975 \times 10^{-18}$ J. The Hartree is equal to the coulombic repulsion energy of two elementary charges one bohr apart or twice the ionisation energy of the hydrogen atom (assuming an infinite nuclear mass).

The atomic unit of action is Planck's constant divided by 2π. It has no special name and is very frequently denoted by drawing a bar through the h for Planck's constant:

$$h/2\pi \equiv \hbar = 1.05459 \times 10^{-34} \text{ J s.}$$

The atomic unit of time has no special name; it is $\hbar/E_H = 2.41888 \times 10^{-17}$ s.

The atomic unit of velocity has no special name; it may be expressed as $a_0 E_H / \hbar = 2.18769 \times 10^6 \text{ m s}^{-1}$.

In atomic units, the Schrödinger equation for an electron moving in a potential V can be written:

$$\nabla^2 \Psi + 2(E - V)\Psi = 0$$

REFERENCE

1. I. Mills, T. Cvitaš, K. Homann, N. Kallay and K. Kuchitsu, *Quantities, Units and Symbols in Physical Chemistry*, 2nd edn, International Union of Pure and Applied Chemistry, 1993.

APPENDIX 2
The Variation Method and the Secular Equations

The number of problems for which we can solve the Schrödinger equation directly is small. It is therefore very fortunate that the hydrogen atom is one of this type, since the solutions obtained form the starting point of all our quantum-mechanical models of atomic structure and for many descriptions of molecular structure. In the great majority of cases, however, a direct solution of Schrödinger's equation is not possible and we must adopt other methods. This appendix describes one such approach. As a preliminary we need a proof of the variation theorem.

The variation theorem

Given any approximate wavefunction ϕ, which satisfies the boundary conditions of the problem, the expectation value $\overline{P}_\phi$, of the observable P calculated from ϕ with the operator $\hat{\mathcal{P}}$ will always be greater than or equal to the lowest exact eigenvalue of $\hat{\mathcal{P}}$.

Proof

Suppose that we have an operator $\hat{\mathcal{P}}$ the exact eigenvalues and eigenfunctions of which are $p_0, p_1, p_2, \ldots$ and $\psi_0, \psi_1, \psi_2, \ldots$ respectively. Then:

$$\hat{\mathcal{P}}|\psi_i\rangle = p_i|\psi_i\rangle \qquad (A2.1)$$

and the $|\psi_i\rangle$ can always be chosen to be orthogonal and normalised so that:

$$\langle \psi_i|\psi_j\rangle = \delta_{ij} \qquad (A2.2)$$

Suppose now that we have a wave function $|\phi\rangle$, which is not one of the $|\psi_i\rangle$, then the expectation value $\overline{P}_\phi$ of $\hat{\mathcal{P}}$ calculated for the function $|\phi\rangle$ is given (Section 3.9) by:

$$\overline{P}_\phi = \frac{\langle\phi|\hat{\mathcal{P}}|\phi\rangle}{\langle\phi|\phi\rangle} \qquad (A2.3)$$

The Quantum in Chemistry   R. Grinter
© 2005 John Wiley & Sons, Ltd

The functions ψ_i form a complete set and any function ϕ that satisfies the boundary conditions of the problem can be written in terms of them:

$$|\phi\rangle = \sum_i C_i |\psi_i\rangle \tag{A2.4}$$

[Two vectors at right-angles form a complete set and any other vector in the same plane can be written in terms of them].

We can always normalise $|\phi\rangle$ so that:

$$\langle\phi|\phi\rangle = \sum_i C_i^* \sum_j C_j \langle\psi_i|\psi_j\rangle = 1$$

Note that the sums over i and j in the above expression are independent of each other so that if the are n functions ψ_i there are n^2 terms in total.

Now, because of Equation (A2.2):

$$\sum_i C_i^* C_i = 1 \tag{A2.5}$$

Expanding ϕ in Equation (A2.3) using Equation (A2.4) we have:

$$\overline{P}_\phi = \frac{\sum_i C_i^* \sum_j C_j \langle\psi_i|\hat{P}|\psi_j\rangle}{\sum_k C_k^* \sum_l C_l \langle\psi_k|\psi_l\rangle} \tag{A2.6}$$

Using Equation (A2.1) gives:

$$\overline{P}_\phi = \frac{\sum_i C_i^* \sum_j C_j p_j \langle\psi_i|\psi_j\rangle}{\sum_k C_k^* \sum_l C_l \langle\psi_k|\psi_l\rangle}$$

With the help of Equations (A2.2) and (A2.5) we find:

$$\overline{P}_\phi = \sum_i C_i^* C_i p_i \Big/ \sum_k C_k^* C_k = \sum_i C_i^* C_i p_i \tag{A2.7}$$

Subtracting the lowest eigenvalue (p_0) from both sides of Equation (A2.7) we obtain:

$$\overline{P}_\phi - p_0 = \sum_i C_i^* C_i p_i - p_0 = \sum_i C_i^* C_i (p_i - p_0)$$

where we may take p_0 into the bracket because of Equation (A2.5).

But, $p_i \geq p_0$, by definition and $C_i^* C_i$ must always be positive (Appendix 8), therefore:

$$\overline{P}_\phi - p_0 \geq 0 \text{ or } \overline{P}_\phi \geq p_0$$

Q.E.D.

Application of the variation method

Suppose that we have a set of n basis functions $|a\rangle$. They might be atomic electronic states characterised by $|L, M_L, S, M_S\rangle$, spin functions $|\alpha(1)\beta(2)\alpha(3)\rangle$ in an NMR problem or atomic orbitals $|2p_x\rangle$, ... etc. in the molecular orbital theory. We wish to find linear combinations of these basis functions:

$$|\psi\rangle = \sum_{a=1}^{n} C_{\psi,a}|a\rangle$$

which are eigenfunctions (or as near to eigenfunctions as is possible within the limitations of the basis) of some operator, $\hat{\mathcal{P}}$ say. The coefficients $C_{\psi,a}$ are simply numbers which may be real, pure imaginary or complex. In what follows we shall assume that they are real. Ideally, the function $|\psi\rangle$ must be such that:

$$\hat{\mathcal{P}}|\psi\rangle = P_\psi|\psi\rangle$$

The operator may represent any measurable property P and applications in which it is energy are especially common. P_ψ is the eigenvalue of $\hat{\mathcal{P}}$ corresponding to the eigenfunction $|\psi\rangle$. We know from the variation theorem that the expectation value of $\hat{\mathcal{P}}$ with the function $|\psi\rangle$, $\overline{P}_\psi$, will always be greater than the lowest eigenvalue of $\hat{\mathcal{P}}$, or equal to it if we find the exact eigenfunction. We use this as a criterion for determining the values of the coefficients $C_{\psi,a}$ by finding the condition that the expectation value shall be a minimum. We choose to minimise the expectation value rather than the eigenvalue because we cannot be sure, *a priori*, that our set of basis functions will allow us to find an eigenfunction of our problem; they may not form a complete set for example.

The expectation value of $\hat{\mathcal{P}}$ with the function $|\psi\rangle$, $\overline{P}_\psi$, is (Section 3.9):

$$\overline{P}_\psi = \frac{\langle\psi|\hat{\mathcal{P}}|\psi\rangle}{\langle\psi|\psi\rangle} = \frac{\sum_a C_{\psi,a}\sum_b C_{\psi,b}\langle a|\hat{\mathcal{P}}|b\rangle}{\sum_a C_{\psi,a}\sum_b C_{\psi,b}\langle a|b\rangle} \equiv \frac{\sum_a C_{\psi,a}\sum_b C_{\psi,b}P_{ab}}{\sum_a C_{\psi,a}\sum_b C_{\psi,b}S_{ab}}$$

P_{ab} and S_{ab} are simply abbreviations for matrix elements, $\langle a|\hat{\mathcal{P}}|b\rangle$, and overlap integrals, $\langle a|b\rangle$, which we must know or be able to calculate. Therefore:

$$\overline{P}_\psi \sum_a C_{\psi,a} \sum_b C_{\psi,b}S_{ab} = \sum_a C_{\psi,a} \sum_b C_{\psi,b}P_{ab}$$

But, according to the variation theorem, the lowest value of $\overline{P}_\psi$ we can obtain is the best, so we look for the set of coefficients which give lowest point in the surface of the plot of $\overline{P}_\psi$ against the unknown coefficients C. Firstly we differentiate both sides of the above equation with respect to a particular C, $C_{\psi,i}$ say. We must remember that $\overline{P}_\psi$ is itself a function of $C_{\psi,i}$, though P_{ab} and S_{ab} are not. Thus, on the left-hand side we have a product to differentiate:

$$\frac{\partial\overline{P}_\psi}{\partial C_{\psi,i}} \cdot \sum_a C_{\psi,a} \sum_b C_{\psi,b}S_{ab} + \overline{P}_\psi \cdot \frac{\partial\{\sum_a C_{\psi,a}\sum_b C_{\psi,b}S_{ab}\}}{\partial C_{\psi,i}} = \frac{\partial\{\sum_a C_{\psi,a}\sum_b C_{\psi,b}P_{ab}\}}{\partial C_{\psi,i}}$$

For a minimum in P_ψ, $\partial\overline{P}_\psi/\partial C_{\psi,i}$ is zero and we have:

$$\overline{P}_\psi \cdot \frac{\partial\{\sum_a C_{\psi,a}\sum_b C_{\psi,b}S_{ab}\}}{\partial C_{\psi,i}} = \frac{\partial\{\sum_a C_{\psi,a}\sum_b C_{\psi,b}P_{ab}\}}{\partial C_{\psi,i}}$$

To carry out the differentiations we express the double sum as three terms, paying particular attention to terms involving $C_{\psi,i}$:

$$\sum_a C_{\psi,a} \sum_b C_{\psi,b}S_{ab} = \sum_{a\neq i} C_{\psi,a} \sum_{b\neq i} C_{\psi,b}S_{ab} + 2C_{\psi,i}\sum_{b\neq i} C_{\psi,b}S_{ib} + C_{\psi,i}^2 S_{ii}$$

Where we have assumed that $S_{ab} = S_{ba}$ and that all the coefficients C are real. If we wish to recognise the fact that the coefficients C and the overlap integrals S may be complex or pure imaginary we have to proceed slightly differently at this point, but we shall not take that course.

The differentiation is now straightforward and we find:

$$\frac{\partial\{\sum_a C_{\psi,a} \sum_b C_{\psi,b} S_{ab}\}}{\partial C_{\psi,i}} = 0 + 2\sum_{b\neq i} C_{\psi,b} S_{ib} + 2C_{\psi,i} S_{ii} = 2\sum_b C_{\psi,b} S_{ib}$$

The other side of the equation gives a result of exactly the same form and we have:

$$2\overline{P}_\psi \cdot \sum_b C_{\psi,b} S_{ib} = 2\sum_b C_{\psi,b} P_{ib} \Rightarrow \sum_b C_{\psi,b}(P_{ib} - \overline{P}_\psi S_{ib}) = 0$$

We obtain one such equation for every $C_{\psi,i}$, i.e. for every basis function, $|i\rangle$. These are the *secular equations*. Thus, for the expectation value $\overline{P}_\psi$ we have for $i = 1, 2, 3, \ldots$:

$$C_{\psi 1}(P_{11} - \overline{P}_\psi S_{11}) + C_{\psi 2}(P_{12} - \overline{P}_\psi S_{12}) + \cdots + C_{\psi n}(P_{1n} - \overline{P}_\psi S_{1n}) = 0$$

$$C_{\psi 1}(P_{21} - \overline{P}_\psi S_{21}) + C_{\psi 2}(P_{22} - \overline{P}_\psi S_{22}) + \cdots + C_{\psi n}(P_{2n} - \overline{P}_\psi S_{2n}) = 0$$

$$\vdots \qquad\qquad \vdots \qquad\qquad \vdots$$

$$C_{\psi 1}(P_{n1} - \overline{P}_\psi S_{n1}) + C_{\psi 2}(P_{n2} - \overline{P}_\psi S_{n2}) + \cdots + C_{\psi n}(P_{nn} - \overline{P}_\psi S_{nn}) = 0$$

These are *simultaneous, homogeneous* equations in which the unknowns are the coefficients C. They are satisfied by the *trivial* solution $C_{\psi,1} = C_{\psi,2} \ldots = C_{\psi,n} = 0.0$ but, clearly, we are not interested in that. The condition that they have a *non-trivial* solution is that the determinant below is equal to zero:

$$\begin{vmatrix} (P_{11} - \overline{P}_\psi S_{11}) & (P_{12} - \overline{P}_\psi S_{12}) & (P_{13} - \overline{P}_\psi S_{13}) & \cdots & (P_{1n} - \overline{P}_\psi S_{1n}) \\ (P_{21} - \overline{P}_\psi S_{21}) & (P_{22} - \overline{P}_\psi S_{22}) & (P_{23} - \overline{P}_\psi S_{23}) & \cdots & (P_{2n} - \overline{P}_\psi S_{2n}) \\ \vdots & \vdots & \vdots & & \vdots \\ \vdots & \vdots & \vdots & & \vdots \\ (P_{n1} - \overline{P}_\psi S_{n1}) & (P_{n2} - \overline{P}_\psi S_{n2}) & (P_{n3} - \overline{P}_\psi S_{n3}) & \cdots & (P_{nn} - \overline{P}_\psi S_{nn}) \end{vmatrix} = 0$$

Since all the P_{ab} and S_{ab} are known, we can expand the determinant and obtain a polynomial of order n in $\overline{P}_\psi$. We require the lowest root. We also make use of higher roots, as will be described below.

Having found the required value of $\overline{P}_\psi$, we can enter it in the secular equations and solve them for the coefficients C. The fact that the solution gives only the ratios of the coefficients and not their absolute values (because the equations are homogeneous), presents no great problem since we also have the requirement that the eigenfunctions be normalised which gives us one further equation:

$$\sum_a C_{\psi,a} \sum_b C_{\psi,b} S_{ab} = 1.0$$

In the development so far we have made no particular assumptions about the basis functions $|a\rangle$, so if they are not orthogonal to each other terms in S_{ab} appear off the diagonal and this raises extra, but not insurmountable, problems. Fortunately, we can usually find a set of orthogonal and normalised basis functions so that $S_{ab} = \delta_{ab}$ and we then have only diagonal terms in S. The secular equations become:

$$C_{\psi 1}(P_{11} - \overline{P}_\psi) + C_{\psi 2}P_{12} + C_{\psi 3}P_{13} + \cdots + C_{\psi n}P_{1n} = 0$$

$$C_{\psi 1}P_{21} + C_{\psi 2}(P_{22} - \overline{P}_\psi) + C_{\psi 3}P_{23} + \cdots + C_{\psi n}P_{2n} = 0$$

$$\vdots \qquad \vdots \qquad \qquad \vdots \qquad \qquad \vdots$$

$$\vdots \qquad \vdots \qquad \qquad \vdots \qquad \qquad \vdots$$

$$C_{\psi 1}P_{n1} + C_{\psi 2}P_{n2} + C_{\psi 3}P_{n3} + \cdots + C_{\psi n}(P_{nn} - \overline{P}_\psi) = 0$$

The condition for a non-trivial solution of the equations:

$$\begin{vmatrix} (P_{11} - \overline{P}_\psi) & P_{12} & P_{13} & \cdots & P_{1n} \\ P_{21} & (P_{22} - \overline{P}_\psi) & P_{23} & \cdots & P_{2n} \\ \vdots & \vdots & \vdots & & \vdots \\ P_{n1} & P_{n2} & P_{n3} & \cdots & (P_{nn} - \overline{P}_\psi) \end{vmatrix} = 0$$

is of exactly the same form as the procedure for extracting the eigenvalues $\overline{P}_\psi$ of the matrix formed of the elements P_{ab} by evaluating the characteristic equation of the matrix; an nth order polynomial in $\overline{P}_\psi$ (Appendix 3). This illustrates the connection between the two approaches to the problem. Since the basis states are orthonormal the normalisation requirement reduces to:

$$\sum_a C_{\psi,a}^2 = 1.0$$

Our purpose has been to find the lowest value of $\overline{P}_\psi$, but what of the other values which the method inevitably produces? We can distinguish two cases.

Case 1. In some problems, especially in angular momentum, we have a complete set of basis states $|a\rangle$. For example, if we are calculating the spin-orbit coupling of a single p electron then there are only six states characterised by $|l, m_l, s, m_s\rangle$ (three values of m_l and two values of m_s). Under such circumstances the n values of $\overline{P}_\psi$ are the n eigenvalues of our problem and the coefficents C give us the corresponding n eigenfunctions. These types of problem are also easy to handle because the basis states conform rigorously to the condition $S_{ab} = \delta_{ab}$.

Case 2. If the basis set is of necessity incomplete, as in the LCAO-MO theory for example where we can never include all the AOs which contribute to the MOs (Section 6.13), then the higher values of $\overline{P}_\psi$ have no fundamental significance. However, we always regard them as approximations to the higher eigenvalues of the system, though we have no way of knowing how accurate they are except by comparison with experiment. In the Hückel theory (Section 12.1) we also make the drastic approximation $S_{ab} = \delta_{ab}$. Though it is manifestly untrue it simplifies the problem enormously.

Example: The hydrogen atom in a magnetic field

The nucleus and the electron which together form a hydrogen atom each have a magnetic moment which is proportional to their spin angular momentum. These magnetic momenta interact with each other, the Fermi contact interaction, and with an applied magnetic field, the Zeeman effect. The mutual magnetic interaction of nucleus and electron is not the classical dipole-dipole interaction which averages to zero in a free atom. The Fermi contact interaction represents the energy of the nuclear moment in the magnetic field of the electron when the electron is at the nucleus. (For s orbitals the maximum electron density is found at the nucleus, as an examination of atomic orbital wave functions will show.) The Hamiltonian operator for the magnetic interactions of a hydrogen atom in a flux density (B) is:

$$\hat{\mathcal{H}} = g\mu_B B \hat{S}_Z - g_N \mu_N B \hat{I}_Z + a\hat{\mathbf{S}} \cdot \hat{\mathbf{I}}$$

The first two terms are the Zeeman terms and the last the Fermi contact. g and g_N are the electron and nuclear g-factors, μ_B and μ_N are the Bohr and nuclear magnetons, $\hat{\mathbf{I}}$ and $\hat{\mathbf{S}}$ are the operators for the spin angular momenta of the nucleus and electron and $\hat{S}_Z$ and $\hat{I}_Z$ are the operators for the z components of those momenta. $a = (\frac{2}{3})\mu_0 g\mu_B g_N \mu_N |\Psi_{(0)}|^2$ where μ_0 is the vacuum permeability and $|\Psi_{(0)}|^2$ is the electron density at the nucleus. (See Section 12.2 for more on magnetic properties.) The scalar product ($\hat{\mathbf{S}} \cdot \hat{\mathbf{I}}$) can be written in terms of the Cartesian components of the vector operators ($\hat{\mathbf{S}}$ and $\hat{\mathbf{I}}$), or in terms of the corresponding raising and lowering operators (Section 4.7):

$$a\hat{\mathbf{S}} \cdot \hat{\mathbf{I}} = a\{\hat{S}_x \hat{I}_x + \hat{S}_y \hat{I}_y + \hat{S}_z \hat{I}_z\} = a\hat{S}_z \hat{I}_z + (a/2)\{\hat{S}_+ \hat{I}_- + \hat{S}_- \hat{I}_+\}$$

Our basis states are the spin states of the hydrogen atom and they can be defined by writing down all possible combinations of the z-components of the nuclear and electron spin, i.e. $|m_S$ (electron), m_I (nucleus)$\rangle$. Since m_S and m_I can each take values of $+\frac{1}{2}$ and $-\frac{1}{2}$ we have the four basis functions:

$$|a\rangle = |+\tfrac{1}{2}, +\tfrac{1}{2}\rangle, \ |b\rangle = |+\tfrac{1}{2}, -\tfrac{1}{2}\rangle, \ |c\rangle = |-\tfrac{1}{2}, +\tfrac{1}{2}\rangle \text{ and } |d\rangle = |-\tfrac{1}{2}, -\tfrac{1}{2}\rangle$$

The individual spin functions are orthonormal, i.e. $\langle+\frac{1}{2}|+\frac{1}{2}\rangle = \langle-\frac{1}{2}|-\frac{1}{2}\rangle = 1.0$ and $\langle+\frac{1}{2}|-\frac{1}{2}\rangle = \langle-\frac{1}{2}|+\frac{1}{2}\rangle = 0$. Consequently, the functions $|a\rangle - |d\rangle$ are also ortho-normal so that for the overlap of any pair $S_{ij} = \delta_{ij}$.

To set up the secular equations we require all 4×4 terms of the form $\langle i|\hat{\mathcal{H}}|j\rangle$; two examples will suffice to show how they are calculated:

$$\langle b|\hat{\mathcal{H}}|b\rangle = \langle+\tfrac{1}{2}, -\tfrac{1}{2}|\hat{\mathcal{H}}|+\tfrac{1}{2}, -\tfrac{1}{2}\rangle$$

$$= \langle+\tfrac{1}{2}, -\tfrac{1}{2}|g\mu_B B\hat{S}_Z - g_N\mu_N B\hat{I}_Z + a\hat{S}_Z\hat{I}_Z + \tfrac{1}{2}a\{\hat{S}_+\hat{I}_- + \hat{S}_-\hat{I}_+\}|+\tfrac{1}{2}, -\tfrac{1}{2}\rangle$$

Since $\hat{\mathbf{S}}$ and its components operate only on the electron, $\hat{\mathbf{I}}$ and its components operate only on the nucleus and the other symbols are not operators, we can write:

$$\langle b|\hat{\mathcal{H}}|b\rangle = g\mu_B B\langle+\tfrac{1}{2}|\hat{S}_Z|+\tfrac{1}{2}\rangle\langle-\tfrac{1}{2}|-\tfrac{1}{2}\rangle - g_N\mu_N B\langle+\tfrac{1}{2}|+\tfrac{1}{2}\rangle\langle-\tfrac{1}{2}|\hat{I}_Z|-\tfrac{1}{2}\rangle$$

$$+ a\langle+\tfrac{1}{2}|\hat{S}_Z|+\tfrac{1}{2}\rangle\langle-\tfrac{1}{2}|\hat{I}_Z|-\tfrac{1}{2}\rangle$$

$$+ \tfrac{1}{2}a\{\langle+\tfrac{1}{2}|\hat{S}_+|+\tfrac{1}{2}\rangle\langle-\tfrac{1}{2}|\hat{I}_-|-\tfrac{1}{2}\rangle + \langle+\tfrac{1}{2}|\hat{S}_-|+\tfrac{1}{2}\rangle\langle-\tfrac{1}{2}|\hat{I}_+|-\tfrac{1}{2}\rangle\}$$

The results of operating with $\hat{S}_+$, $\hat{S}_-$ and $\hat{S}_Z$ on $|+\frac{1}{2}\rangle$ and $|-\frac{1}{2}\rangle$ are very simple (Box 4.1):

	$|+\frac{1}{2}\rangle$	$|-\frac{1}{2}\rangle$
$\hat{S}_+$	0	$|+\frac{1}{2}\rangle$
$\hat{S}_-$	$|-\frac{1}{2}\rangle$	0
$\hat{S}_Z$	$+\frac{1}{2}|+\frac{1}{2}\rangle$	$-\frac{1}{2}|-\frac{1}{2}\rangle$

and there are corresponding results for $\hat{I}_+$, $\hat{I}_-$ and $\hat{I}_Z$, so that:

$$\langle b|\hat{\mathcal{H}}|b\rangle = +\tfrac{1}{2}g\mu_B B + \tfrac{1}{2}g_N\mu_N B - \tfrac{1}{4}a$$

Similarly:

$$\langle b|\hat{\mathcal{H}}|c\rangle = g\mu_B B\langle +\tfrac{1}{2}|\hat{S}_Z| -\tfrac{1}{2}\rangle\langle -\tfrac{1}{2}| +\tfrac{1}{2}\rangle - g_N\mu_N B\langle -\tfrac{1}{2}| +\tfrac{1}{2}\rangle\langle +\tfrac{1}{2}|\hat{I}_Z| -\tfrac{1}{2}\rangle$$

$$+ a\langle +\tfrac{1}{2}|\hat{S}_Z| -\tfrac{1}{2}\rangle\langle -\tfrac{1}{2}|\hat{I}_Z| +\tfrac{1}{2}\rangle$$

$$+ (a/2)\{\langle +\tfrac{1}{2}|\hat{S}_+| -\tfrac{1}{2}\rangle\langle -\tfrac{1}{2}|\hat{I}_-| +\tfrac{1}{2}\rangle + \langle +\tfrac{1}{2}|\hat{S}_-| -\tfrac{1}{2}\rangle\langle -\tfrac{1}{2}|\hat{I}_+| +\tfrac{1}{2}\rangle\}$$

The first three terms are zero, on account of the orthogonality of the individual spin functions, and so is the fourth because of the properties of the raising and lowering operators. Only the term in $\hat{S}_+\hat{I}_-$ is non-zero so that:

$$\langle b|\hat{\mathcal{H}}|c\rangle = a/2$$

Thus, setting $\Delta_e = g\mu_B B$ and $\Delta_N = g_N\mu_N B$, we find the secular equations:

$$C_1(\tfrac{1}{2}\Delta_e - \tfrac{1}{2}\Delta_N + a/4 - E) + C_2 0 + C_3 0 + C_4 0 = 0$$

$$C_1 0 + C_2(\tfrac{1}{2}\Delta_e + \tfrac{1}{2}\Delta_N - a/4 - E) + C_3(a/2) + C_4 0 = 0$$

$$C_1 0 + C_2(a/2) + C_3(-\tfrac{1}{2}\Delta_e - \tfrac{1}{2}\Delta_N - a/4 - E) + C_4 0 = 0$$

$$C_1 0 + C_2 0 + C_3 0 + C_4(-\tfrac{1}{2}\Delta_e + \tfrac{1}{2}\Delta_N + a/4 - E) = 0$$

The first secular equation contains only C_1 and is therefore independent of the others, so that we have:

$$\tfrac{1}{2}\Delta_e - \tfrac{1}{2}\Delta_N + a/4 - E = 0 \Rightarrow E = \tfrac{1}{2}\Delta_e - \tfrac{1}{2}\Delta_N + a/4 \text{ and } C_1 = 1.0$$

Similarly, the last equation gives $E = -\tfrac{1}{2}\Delta_e + \tfrac{1}{2}\Delta_N + a/4$ and $C_4 = 1.0$.
The secular determinant for equations 2 and 3 is:

$$\begin{vmatrix} \tfrac{1}{2}D - a/4 - E & +a/2 \\ +a/2 & -\tfrac{1}{2}D - a/4 - E \end{vmatrix} = 0$$

Where $D = \Delta_e + \Delta_N$. Multiplying the determinant out, we find that the quadratic equation for E is:

$$E^2 + (a/2)E - \tfrac{1}{4}D^2 - 3(a/4)^2 = 0$$

giving

$$E = -a/4 \pm \tfrac{1}{2}\{a^2 + D^2\}^{\frac{1}{2}}$$

Given numerical values of a and D, it is quite simple to extract the two coefficients, C_2 and C_3, corresponding to these two energies, but it is tedious and not worthwhile to obtain algebraic expressions for them. However, the case of zero magnetic field is of interest and merits further discussion.

In the absence of a magnetic field $D = \Delta_e + \Delta_N = 0$ and we find three degenerate energy levels of $a/4$ and one of $-3a/4$. The separation between these two states of the hydrogen atom, which correspond to the two possible relative orientations of the nuclear and electronic magnetic moment, is $\sim$1,420 MHz. It is the frequency which has been scanned with the intention of detecting intelligent life in the galaxy (Section 4.9). To obtain the coefficients C_2 and C_3 for the zero-field case we substitute the two energies back into the secular equations. When $E = a/4$ we have:

$$C_2(-a/4 - a/4) + C_3(a/2) = 0 \quad \Rightarrow \quad aC_2/2 = aC_3/2$$

and

$$C_2(a/2) + C_3(-a/4 - a/4) = 0 \quad \Rightarrow \quad aC_2/2 = aC_3/2$$

from which it is clear that $C_2 = C_3$ and, since $C_2{}^2 + C_3{}^2 = 1.0$, their absolute values are:

$$C_2 = C_3 = 1/\sqrt{2}$$

When $E = -3a/4$ we have:

$$C_2(-a/4 + 3a/4) + C_3(a/2) = 0 \quad \Rightarrow \quad aC_2/2 = -aC_3/2$$

and

$$C_2(a/2) + C_3(-a/4 + 3a/4) = 0 \quad \Rightarrow \quad aC_2/2 = -aC_3/2$$

giving

$$C_2 = -C_3 = 1/\sqrt{2}$$

Thus, summarising our results for the zero-field case we have:
The singlet

$$E = -3a/4 \qquad \Psi_s = \{1/\sqrt{2}\}\{|+\tfrac{1}{2}, -\tfrac{1}{2}\rangle - |-\tfrac{1}{2}, +\tfrac{1}{2}\rangle\}$$

The triplet

$$E = +a/4 \qquad \Psi_{t+1} = |+\tfrac{1}{2}, +\tfrac{1}{2}\rangle$$

$$E = +a/4 \qquad \Psi_{t0} = \{1/\sqrt{2}\}\{|+\tfrac{1}{2}, -\tfrac{1}{2}\rangle + |-\tfrac{1}{2}, +\tfrac{1}{2}\rangle\}$$

$$E = +a/4 \qquad \Psi_{t-1} = |-\tfrac{1}{2}, -\tfrac{1}{2}\rangle$$

APPENDIX 3

Energies and Wave Functions by Matrix Diagonalisation

To fully understand this subject the reader needs to know how to multiply matrices and to find the inverse of a matrix.

The determination of eigenvalues and eigenfunctions by the solution of secular equations (Appendix 2) plays an important pedagogical role in quantum mechanics but finds very little application today. The advent of the digital computer has made matrix algebra much more potent for the formulation and solution of practical problems in quantum mechanics. We shall use the determination of the Hückel MOs for the propyl radical, C_3H_5, to compare the two methods.

The secular equations for the π-electron system based on three conjugated carbon atoms having AOs ϕ_1, ϕ_2 and ϕ_3 are (Section 12.1):

$$C_1(\alpha - E) + C_2\beta \qquad = 0$$

$$C_1\beta + C_2(\alpha - E) + C_3\beta = 0$$

$$C_2\beta + C_3(\alpha - E) = 0$$

Their solution, by the method described in Section 12.1, leads to the following energies and wave functions; eigenvalues and eigenvectors respectively in matrix terminology:

$$\psi_1 = \tfrac{1}{2}\phi_1 + (1/\sqrt{2})\phi_2 + \tfrac{1}{2}\phi_3 \qquad E_1 = \alpha + \sqrt{2}\beta$$

$$\psi_2 = (1/\sqrt{2})\phi_1 + 0\phi_2 - (1/\sqrt{2})\phi_3 \quad E_2 = \alpha$$

$$\psi_3 = \tfrac{1}{2}\phi_1 - (1/\sqrt{2})\phi_2 + \tfrac{1}{2}\phi_3 \qquad E_3 = \alpha - \sqrt{2}\beta$$

Matrix elements

When using matrix-algebra we first set up a matrix the elements of which are the possible values, nine in this case, of $\langle \phi_i | \hat{\mathcal{H}} | \phi_j \rangle$, where $\hat{\mathcal{H}}$ is the Hamiltonian operator and the ϕ_i, $i = 1 - 3$, are known as the basis functions. In the present case they are the three carbon $2p_z$ atomic orbitals. These are the quantities which we first denoted by H_{ij} in the Hückel

The Quantum in Chemistry   R. Grinter
© 2005 John Wiley & Sons, Ltd

theory and then replaced by α and β in accordance with Hückel's approximations. We obtain the Hamiltonian matrix [H]:

$$
\begin{array}{c|ccc}
\hat{\mathcal{H}} & \phi_1 & \phi_2 & \phi_3 \\
\hline
\phi_1 & \alpha & \beta & 0 \\
\phi_2 & \beta & \alpha & \beta \\
\phi_3 & 0 & \beta & \alpha
\end{array} \equiv [\mathrm{H}]
$$

Note that, following Hückel's method, we set all overlap integrals to zero. If the basis functions are not orthogonal and there are overlap integrals then their inclusion complicates the problem, though it can also be handled perfectly well with matrix methods. However, most chemical applications of quantum mechanics, apart from those arising from MO theory, start from a basis of orthogonal functions and the problems associated with overlap do not therefore arise. The present description of the matrix method of obtaining eigenvalues and eigenfunctions applies only to a set of basis functions which are either truly orthogonal or are assumed to be so.

In a computational solution of our problem we would now insert appropriate numerical values for α and β and use a computer to diagonalise the resulting matrix. However, in order to understand the process of diagonalisation we shall here proceed algebraically. We pose the question: What would the Hamiltonian matrix look like if we used the three functions ψ_i rather than the ϕ_i to construct it? To answer that question we evaluate two matrix elements as examples:

$$
\begin{aligned}
\langle \psi_1 | \hat{\mathcal{H}} | \psi_1 \rangle &= \langle \tfrac{1}{2}\phi_1 + (1/\sqrt{2})\phi_2 + \tfrac{1}{2}\phi_3 | \hat{\mathcal{H}} | \tfrac{1}{2}\phi_1 + (1/\sqrt{2})\phi_2 + \tfrac{1}{2}\phi_3 \rangle \\
&= \tfrac{1}{4}\langle \phi_1 | \hat{\mathcal{H}} | \phi_1 \rangle + \tfrac{1}{2}\langle \phi_2 | \hat{\mathcal{H}} | \phi_2 \rangle + \tfrac{1}{4}\langle \phi_3 | \hat{\mathcal{H}} | \phi_3 \rangle \\
&\quad + 2 \cdot (\tfrac{1}{2}) \cdot (1/\sqrt{2})\langle \phi_1 | \hat{\mathcal{H}} | \phi_2 \rangle + 2 \cdot (\tfrac{1}{2}) \cdot (1/\sqrt{2})\langle \phi_2 | \hat{\mathcal{H}} | \phi_3 \rangle \\
&= \tfrac{1}{4}\alpha + \tfrac{1}{2}\alpha + \tfrac{1}{4}\alpha + (1/\sqrt{2})\beta + (1/\sqrt{2})\beta = \alpha + \sqrt{2}\beta \\
\langle \psi_1 | \hat{\mathcal{H}} | \psi_2 \rangle &= \langle \tfrac{1}{2}\phi_1 + (1/\sqrt{2})\phi_2 + \tfrac{1}{2}\phi_3 | \hat{\mathcal{H}} | (1/\sqrt{2})\phi_1 - (1/\sqrt{2})\phi_3 \rangle \\
&= (1/2\sqrt{2})\langle \phi_1 | \hat{\mathcal{H}} | \phi_1 \rangle - (1/2\sqrt{2})\langle \phi_3 | \hat{\mathcal{H}} | \phi_3 \rangle + \tfrac{1}{2}\langle \phi_2 | \hat{\mathcal{H}} | \phi_1 \rangle - \tfrac{1}{2}\langle \phi_2 | \hat{\mathcal{H}} | \phi_3 \rangle \\
&= (1/2\sqrt{2})\alpha - (1/2\sqrt{2})\alpha + \tfrac{1}{2}\beta - \tfrac{1}{2}\beta = 0
\end{aligned}
$$

When we evaluate the remaining elements we find, as the reader can confirm, that the matrix constructed using the three functions ψ_i is the diagonal matrix [D], in which the diagonal elements are the MO energies, or eigenvalues of our problem:

$$
\begin{array}{c|ccc}
\hat{\mathcal{H}} & \psi_1 & \psi_2 & \psi_3 \\
\hline
\psi_1 & \alpha + \sqrt{2}\beta & 0 & 0 \\
\psi_2 & 0 & \alpha & 0 \\
\psi_3 & 0 & 0 & \alpha - \sqrt{2}\beta
\end{array} \equiv [\mathrm{D}]
$$

The process of diagonalising a matrix is the process of forming combinations of the basis functions such that all the off-diagonal elements of the matrix constructed with the new combinations are zero. The diagonal elements are then the eigenvalues. The formation of the combinations is subject to the requirement that each of the original basis functions is completely 'used up' in the process. From a purely mathematical viewpoint, the procedure involves the construction of a second matrix [V] such that:

$$[V]^{-1} \times [H] \times [V] = [D]$$

where [H] is the Hamiltonian matrix to be diagonalised, $[V]^{-1}$ is the inverse of [V] and [D] is the resulting diagonal matrix. This form of transformation of [H] to [D] is called a *similarity transformation*. The multiplications are, of course, matrix multiplications. The matrix [V] is built up by an iterative process which computers do very well, and many sophisticated subroutines are available for the purpose. When the procedure is complete the required eigenvectors, which are the atomic orbital coefficients, C_i in this particular application, are the elements of the matrix [V]. Thus, diagonalisation is a process whereby those combinations of the basis functions, ϕ_i, which are eigenfunctions of $\hat{\mathcal{H}}$ and their associated eigenvalues, are found simultaneously.

Readers who are familiar with matrix operations may like to confirm that if:

$$[V] = \begin{bmatrix} \frac{1}{2} & \frac{1}{\sqrt{2}} & \frac{1}{2} \\ \frac{1}{\sqrt{2}} & 0 & \frac{-1}{\sqrt{2}} \\ \frac{1}{2} & \frac{-1}{\sqrt{2}} & \frac{1}{2} \end{bmatrix} \text{ then } [V]^{-1} = \begin{bmatrix} \frac{1}{2} & \frac{1}{\sqrt{2}} & \frac{1}{2} \\ \frac{1}{\sqrt{2}} & 0 & \frac{-1}{\sqrt{2}} \\ \frac{1}{2} & \frac{-1}{\sqrt{2}} & \frac{1}{2} \end{bmatrix}$$

(There is no error here. In this particular case, as is often found in applications of quantum mechanics, $[V] = [V]^{-1}$. [V] is called a *unitary* matrix.)

and

$$[V]^{-1}[H][V] = [D] = \begin{bmatrix} \alpha + \sqrt{2}\beta & 0 & 0 \\ 0 & \alpha & 0 \\ 0 & 0 & \alpha - \sqrt{2}\beta \end{bmatrix}$$

The characteristic equation of a matrix

A further example of the underlying similarity between the use of matrix and determinantal methods to find eigenvalues is found in the characteristic equation of a matrix. We return to our Hückel matrix for the propyl radical [H] and subtract E from each diagonal element:

	ϕ_1	ϕ_2	ϕ_3
ϕ_1	$\alpha - E$	β	0
ϕ_2	β	$\alpha - E$	β
ϕ_3	0	β	$\alpha - E$

If we regard the new matrix as a determinant and multiply it out according to the rules of determinant algebra we obtain a cubic equation in E:

$$\{\alpha - E\}\{(\alpha - E)^2 - \beta^2\} - \beta\{\beta(\alpha - E)\} = 0$$

$$\Rightarrow \quad \{\alpha - E\}\{(\alpha - E)^2 - 2\beta^2\} = 0$$

The polynomial we have obtained in E is called the *characteristic equation* of the matrix and its roots are the eigenvalues of the matrix, as we now confirm. The roots are:

$$\alpha - E = 0 \quad \Rightarrow \quad E = \alpha$$

and

$$(\alpha - E)^2 - 2\beta^2 = 0 \Rightarrow \alpha - E = \pm\sqrt{2}\beta \text{ or } E = \alpha \pm \sqrt{2}\beta$$

In principle, the eigenvalues of any matrix can be determined by finding the roots of the characteristic equation. However, apart from in the simplest cases, diagonalisation is by far the more efficient method.

Blocked-out matrices

In quantum-mechanical calculations it is quite common to find matrices where a diagonal element has no off-diagonal element associated with it. One may also find smaller sub-matrices within a larger matrix where there are no off-diagonal elements connecting the sub-matrix with the remainder of the matrix:

$$\begin{bmatrix} a & x & 0 & 0 \\ x & b & 0 & 0 \\ 0 & 0 & c & y \\ 0 & 0 & y & d \end{bmatrix}$$

In the example above, the two 2×2 sub-matrices at top left and bottom right have no connections to each other through off-diagonal elements. Matrices of this form are said to be *blocked out* and each block is independent of the others. Thus the above matrix is actually two independent matrices:

$$\begin{bmatrix} a & x \\ x & b \end{bmatrix} \text{ and } \begin{bmatrix} c & y \\ y & d \end{bmatrix}$$

each of which can be diagonalised separately.

General solution of a 2 × 2 eigenvalue problem

There is a general solution of the 2×2 eigenvalue problem which we use several times in this book. Suppose that we have to determine the eigenvalues and eigenvectors of the following matrix in which the basis functions (ϕ_a and ϕ_b) are normalised and orthogonal:

$$\begin{array}{cc} & \phi_a \quad \phi_b \\ \phi_a & A \quad Z \\ \phi_b & Z \quad B \end{array}$$

We define:

$$\Sigma = \tfrac{1}{2}(A + B), \Delta = \tfrac{1}{2}(A - B) \text{ and } \Omega = \{\Delta^2 + Z^2\}^{\frac{1}{2}}$$

where for Ω we take the root which has the same sign as Z.

We also define an angle (θ) such that:

$$\cos\theta = |\{(\Omega + \Delta)/2\,\Omega\}^{\frac{1}{2}}| \text{ and } \sin\theta = |\{(\Omega - \Delta)/2\,\Omega\}^{\frac{1}{2}}|$$

where we take the positive square root in each case.

The eigenvalues (E) and their normalised, orthogonal eigenvectors (Ψ) are now given by:

$$E_1 = \Sigma + \Omega \quad \Psi_1 = \cos\theta\,\phi_a + \sin\theta\,\phi_b$$

$$E_2 = \Sigma - \Omega \quad \Psi_2 = -\sin\theta\,\phi_a + \cos\theta\,\phi_b$$

If $Z > 0$ then $\Omega > 0$ and the lower eigenvalue has a node while if $Z < 0$ $\Omega < 0$ and the higher eigenvalue has a node. This is the case, for example, when Z represents the bonding interaction between two atomic wave functions (ϕ_a and ϕ_b) and is therefore a negative quantity. Then, Ω is <0, E_1 is the lower energy and Ψ_1 is the bonding molecular orbital. Ψ_2 is the antibonding MO.

APPENDIX 4
Perturbation Theory

Exact solution of Schrödinger's equation is possible for only a very small proportion of the problems of interest in the physical sciences. Great importance therefore attaches to approximate methods of solution and among these methods perturbation theory, which is also extensively used in classical mechanics, occupies a very important place. The technique can be applied where the Hamiltonian can be written as a sum of two parts, a simple part which if present alone would generate a soluble Schrödinger equation, and a second part consisting of one or more relatively small additional terms. The approximate behaviour of the system can then be obtained by considering the soluble part as giving the dominant behaviour and treating the actual behaviour as a relatively minor deviation, or *perturbation*, from this calculable behaviour. The perturbation can be estimated by studying the small, complicating additional terms in the second part of the Hamiltonian.

The analysis of time-independent and time-dependent perturbations is different and we treat only the former type of problem here.

Time-independent perturbation theory

We have a Hamiltonian operator of the form:

$$\hat{\mathcal{H}} = \hat{\mathcal{H}}^{(0)} + \hat{\mathcal{H}}'$$

(A4.1)

where the energy associated with $\hat{\mathcal{H}}^{(0)}$ is large compared with that derived from $\hat{\mathcal{H}}'$. In order to facilitate the algebra we write Equation (A4.1) in the form:

$$\hat{\mathcal{H}} = \hat{\mathcal{H}}^{(0)} + \lambda\hat{\mathcal{H}}'$$

(A4.2)

λ is an arbitrary parameter, which we use to keep track of the *order* of the perturbation, i.e. the degree to which our approximate Hamiltonian $(\hat{\mathcal{H}}^{(0)} + \lambda\hat{\mathcal{H}}')$ approaches the true Hamiltonian $(\hat{\mathcal{H}})$. Once it has performed its labelling duty λ is simply set equal to 1. We seek eigenfunctions $|\psi_k\rangle$ and energies E_k which satisfy the Schrödinger equation:

$$\hat{\mathcal{H}}|\psi_k\rangle = (\hat{\mathcal{H}}^{(0)} + \lambda\hat{\mathcal{H}}')|\psi_k\rangle = E_k|\psi_k\rangle$$

(A4.3)

In addition to the assumption concerning the relative magnitudes of the energies associated with $\hat{\mathcal{H}}^{(0)}$ and $\hat{\mathcal{H}}'$ we also assume the following:

i) $\hat{\mathcal{H}}$ does not depend explicitly on the time.

The Quantum in Chemistry   R. Grinter
© 2005 John Wiley & Sons, Ltd

ii) The complete set of orthonormal eigenfunctions $|\psi_n^{(0)}\rangle$ and energies $E_n^{(0)}$ which satisfy the Schrödinger equation:

$$\hat{\mathcal{H}}^{(0)}|\psi_n^{(0)}\rangle = E_n|\psi_n^{(0)}\rangle \qquad (A4.4)$$

are known.

iii) It is possible to expand the solutions of the perturbed problem as a power series in λ:

$$|\psi_k\rangle = |\psi_k^{(0)}\rangle + \lambda|\psi_k^{(1)}\rangle + \lambda^2|\psi_k^{(2)}\rangle + \cdots$$

$$E_k = E_k^{(0)} + \lambda E_k^{(1)} + \lambda^2 E_k^{(2)} + \cdots \qquad (A4.5)$$

$|\psi_k^{(1)}\rangle$ is the *first order correction* to $|\psi_k\rangle$, $|\psi_k^{(2)}\rangle$ is the *second order correction*, and so on, and it is these functions, together with the corresponding values of $E_k^{(1)}$, $E_k^{(2)}$ etc., which we wish to find. The Expansion (A4.5) requires that the successive absolute values of $E_k^{(n)}$ decrease rapidly and it should be noted that it is difficult to establish this, *a priori*, in any particular case. Therefore, when perturbation calculations are carried out the values of $E_k^{(n)}$ must always be checked to ensure that the required decrease in their absolute values is in fact found.

If we now substitute the Expansions (A4.5) into the Schrödinger Equation (A4.3) we obtain:

$$(\hat{\mathcal{H}}^{(0)} + \lambda\hat{\mathcal{H}}')(|\psi_k^{(0)}\rangle + \lambda|\psi_k^{(1)}\rangle + \lambda^2|\psi_k^{(2)}\rangle + \cdots)$$

$$= (E_k^{(0)} + \lambda E_k^{(1)} + \lambda^2 E_k^{(2)} + \ldots)(|\psi_k^{(0)}\rangle + \lambda|\psi_k^{(1)}\rangle + \lambda^2|\psi_k^{(2)}\rangle + \cdots) \qquad (A4.6)$$

Since λ is an arbitrary parameter, the coefficients of the same power of λ on the two sides of Equation (A4.6) must be equal. Equating these coefficients for increasing powers of λ, i.e. orders of perturbation, we find:

$$\lambda^0 \quad \hat{\mathcal{H}}^{(0)}|\psi_k^{(0)}\rangle = E_k^{(0)}|\psi_k^{(0)}\rangle$$

$$\lambda^1 \quad \hat{\mathcal{H}}^{(0)}|\psi_k^{(1)}\rangle + \hat{\mathcal{H}}'|\psi_k^{(0)}\rangle = E_k^{(0)}|\psi_k^{(1)}\rangle + E_k^{(1)}|\psi_k^{(0)}\rangle$$

$$\lambda^2 \quad \hat{\mathcal{H}}^{(0)}|\psi_k^{(2)}\rangle + \hat{\mathcal{H}}'|\psi_k^{(1)}\rangle = E_k^{(0)}|\psi_k^{(2)}\rangle + E_k^{(1)}|\psi_k^{(1)}\rangle + E_k^{(2)}|\psi_k^{(0)}\rangle$$

etc. $\qquad (A4.7)$

λ has now served its purpose. The first of these equations is simply Equation (A4.4), the Schrödinger equation for the unperturbed system and it tells us what we already knew, namely that the zeroth approximations to $|\psi_k\rangle$ and E_k are $|\psi_k^{(0)}\rangle$ and $E_k^{(0)}$.

The second equation may be written:

$$(\hat{\mathcal{H}}^{(0)} - E_k^{(0)})|\psi_k^{(1)}\rangle + (\hat{\mathcal{H}}' - E_k^{(1)})|\psi_k^{(0)}\rangle = 0 \qquad (A4.8)$$

Two possibilities are open to us here. We may be able to solve this equation directly to obtain the required corrections $|\psi_k^{(1)}\rangle$ and $E_k^{(1)}$, but this is rarely possible. Normally we proceed further by expressing the correction $|\psi_k^{(1)}\rangle$ in terms of the complete set of zeroth order wave functions, $|\psi_n^{(0)}\rangle$, which are assumed to be known (assumption (ii) above):

$$|\psi_k^{(1)}\rangle = \sum_n C_n|\psi_n^{(0)}\rangle \qquad (A4.9)$$

We substitute this expansion into Equation (A4.8), multiply on the left by one of the $\langle\psi_n^{(0)}|$, $\langle\psi_j^{(0)}|$ say, and integrate over all space giving:

$$\langle\psi_j^{(0)}|(\hat{\mathcal{H}}^{(0)} - E_k^{(0)})\sum_n C_n|\psi_n^{(0)}\rangle + \langle\psi_j^{(0)}|(\hat{\mathcal{H}}' - E_k^{(1)})|\psi_k^{(0)}\rangle = 0$$

And using Equation (A4.4):

$$\Rightarrow \sum_n C_n \langle \psi_j^{(0)} | (E_n^{(0)} - E_k^{(0)}) | \psi_n^{(0)} \rangle + \langle \psi_j^{(0)} | (\hat{\mathcal{H}}' - E_k^{(1)}) | \psi_k^{(0)} \rangle = 0$$

$$\Rightarrow \sum_n C_n (E_n^{(0)} - E_k^{(0)}) \langle \psi_j^{(0)} | \psi_n^{(0)} \rangle + \langle \psi_j^{(0)} | (\hat{\mathcal{H}}' - E_k^{(1)}) | \psi_k^{(0)} \rangle = 0. \quad (A4.10)$$

Because the functions $|\psi_n^{(0)}\rangle$ are orthonormal, of the n terms in the sum only that for which $n = j$ is non-zero. Therefore, when $j = k$ the only term in the sum which remains has $n = k$ and is zero because the energy difference is zero. Equation (A4.10) then reduces to:

$$\langle \psi_k^{(0)} | \hat{\mathcal{H}}' - E_k^{(1)} | \psi_k^{(0)} \rangle = 0$$

or

$$\langle \psi_k^{(0)} | \hat{\mathcal{H}}' | \psi_k^{(0)} \rangle = E_k^{(1)} \langle \psi_k^{(0)} | \psi_k^{(0)} \rangle = E_k^{(1)}.$$

This gives us the first correction to the energy ($E_k^{(1)}$), in the form of the matrix element of the unperturbed wave function $|\psi_k^{(0)}\rangle$ with the perturbation $\hat{\mathcal{H}}'$:

$$E_k^{(1)} = \langle \psi_k^{(0)} | \hat{\mathcal{H}}' | \psi_k^{(0)} \rangle \equiv H_{kk} \quad (A4.11)$$

where H_{kk} is simply a convenient symbol for the matrix element.

If $j \neq k$ in Equation (A4.10), the only term in the sum over n which can be non-zero is the one where $n = j$ which gives us an expression for all the coefficients except C_k:

$$C_j (E_j^{(0)} - E_k^{(0)}) \langle \psi_j^{(0)} | \psi_j^{(0)} \rangle + \langle \psi_j^{(0)} | (\hat{\mathcal{H}}' - E_k^{(1)}) | \psi_k^{(0)} \rangle = 0$$

$$\Rightarrow C_j (E_j^{(0)} - E_k^{(0)}) \langle \psi_j^{(0)} | \psi_j^{(0)} \rangle + \langle \psi_j^{(0)} | \hat{\mathcal{H}}' | \psi_k^{(0)} \rangle - E_k^{(1)} \langle \psi_j^{(0)} | \psi_k^{(0)} \rangle = 0$$

$$\Rightarrow C_j E_j^{(0)} - C_j E_k^{(0)} + H_{jk} - 0 = 0$$

Or

$$C_j = H_{jk} / (E_k^{(0)} - E_j^{(0)}) \quad (A4.12)$$

The form of Equation (A4.12) raises two important points:

i) If there is degeneracy such that $E_k^{(0)} = E_j^{(0)}$ the coefficient becomes infinite and the theory breaks down. We return to this point later.

ii) The initial assumption that $\hat{\mathcal{H}}'$, is small compared with $\hat{\mathcal{H}}^{(0)}$ implies that H_{jk} is small compared with $E_k^{(0)}$ and $E_j^{(0)}$. Thus, provided that there is no degeneracy, or near degeneracy, C_j will be small and the perturbed wave function will not differ much from the unperturbed, which is the basic premise of the theory.

By means of Equation (A4.12) we can find all the C_j except C_k which can be found using the requirement that the wave function, up to the order considered (power of $\lambda = 1$ in this case), shall be normalised. According to Equation (A4.5), our wave function correct to first order is:

$$|\psi_k\rangle = |\psi_k^{(0)}\rangle + |\psi_k^{(1)}\rangle$$

so the condition for normalisation is:

$$\langle \psi_k | \psi_k \rangle = \langle \psi_k^{(0)} + \psi_k^{(1)} | \psi_k^{(0)} + \psi_k^{(1)} \rangle = 1.0 \quad (A4.13)$$

From Equation (A4.9) we have:

$$|\psi_k{}^{(1)}\rangle = \sum_{j \neq k} C_j |\psi_j{}^{(0)}\rangle + C_k |\psi_k{}^{(0)}\rangle \tag{A4.14}$$

where C_j can be obtained from Equation (A4.12). Thus the normalisation integral (Equation (A4.13)) becomes:

$$\langle \psi_k{}^{(0)} + \sum_{j \neq k} C_j^* \psi_j{}^{(0)} | \psi_k{}^{(0)} + \sum_{j \neq k} C_j \psi_j{}^{(0)} \rangle \tag{A4.15}$$

where we have, for the moment, omitted the term $C_k |\psi_k{}^{(0)}\rangle$ in $|\psi_k{}^{(1)}\rangle$.

Since the $|\psi_n{}^{(0)}\rangle$ are orthonormal (A4.15) becomes:

$$\langle \psi_k{}^{(0)} | \psi_k{}^{(0)} \rangle + \sum_{j \neq k} C_j^* C_j \langle \psi_j{}^{(0)} | \psi_j{}^{(0)} \rangle = 1.0 + \sum_{j \neq k} C_j^* C_j$$

Thus, the normalisation integral (Equation (A4.15)) is equal to unity plus the term $\sum_{j \neq k} C_j^* C_j$ which is of the order of λ^2 because C_j is of order λ since it is a measure of the small mixing of $|\psi_j{}^{(0)}\rangle$ into $|\psi_k{}^{(0)}\rangle$ induced by the perturbation. Therefore, the wave function $|\psi_k\rangle = |\psi_k{}^{(0)}\rangle + |\psi_k{}^{(1)}\rangle$ is normalised to order λ if we neglect the term $C_k |\psi_k{}^{(0)}\rangle$, i.e. if $C_k = 0$. This is reasonable, $|\psi_k{}^{(0)}\rangle$ cannot be corrected by adding a further small contribution from the same function.

Our wave function, correct to first order and normalised, is therefore:

$$\Psi_k = \Psi_k{}^{(0)} + \sum_{j \neq k} \left\{ \frac{H_{jk}}{E_k{}^{(0)} - E_j{}^{(0)}} \right\} \Psi_j{}^{(0)} \tag{A4.16}$$

To obtain $E_k{}^{(2)}$ we write the third Equation (A4.7) as:

$$(\hat{\mathcal{H}}^{(0)} - E_k{}^{(0)})|\psi_k{}^{(2)}\rangle + (\hat{\mathcal{H}}' - E_k{}^{(1)})|\psi_k{}^{(1)}\rangle - E_k{}^{(2)}|\psi_k{}^{(0)}\rangle = 0 \tag{A4.17}$$

and expand $|\psi_k{}^{(2)}\rangle$ in terms of the $|\psi_m{}^{(0)}\rangle$ as before:

$$|\psi_k{}^{(2)}\rangle = \sum_m B_m |\psi_m{}^{(0)}\rangle \tag{A4.18}$$

Substituting Equations (A4.9) and (A4.18) into Equation (A4.17), multiplying on the left by $\langle \psi_j{}^{(0)} |$ and integrating over all space, we obtain:

$$\langle \psi_j{}^{(0)} | \hat{\mathcal{H}}^{(0)} - E_k{}^{(0)} | \sum_m B_m \psi_m{}^{(0)} \rangle + \langle \psi_j{}^{(0)} | \hat{\mathcal{H}}' - E_k{}^{(1)} | \sum_n C_n \psi_n{}^{(0)} \rangle$$

$$- \langle \psi_j{}^{(0)} | E_k{}^{(2)} | \psi_k{}^{(0)} \rangle = 0$$

$$\Rightarrow \sum_n C_n \langle \psi_j{}^{(0)} | \hat{\mathcal{H}}' | \psi_n{}^{(0)} \rangle - \sum_n C_n E_k{}^{(1)} \langle \psi_j{}^{(0)} | \psi_n{}^{(0)} \rangle - E_k{}^{(2)} \langle \psi_j{}^{(0)} | \psi_k{}^{(0)} \rangle$$

$$+ \sum_m B_m \langle \psi_j{}^{(0)} | \hat{\mathcal{H}}^{(0)} | \psi_m{}^{(0)} \rangle - \sum_m B_m \langle \psi_j{}^{(0)} | E_k{}^{(0)} | \psi_m{}^{(0)} \rangle = 0 \tag{A4.19}$$

If $j = k$ the last two terms of Equation (A4.19) cancel each other and we have:

$$\sum_n C_n \langle \psi_k{}^{(0)} | \hat{\mathcal{H}}' | \psi_n{}^{(0)} \rangle - C_k E_k{}^{(1)} \langle \psi_k{}^{(0)} | \psi_k{}^{(0)} \rangle - E_k{}^{(2)} \langle \psi_k{}^{(0)} | \psi_k{}^{(0)} \rangle = 0$$

or

$$E_k^{(2)} = \sum_n C_n H_{kn} - C_k E_k^{(1)} = \sum_{n \neq k} C_n H_{kn} \tag{A4.20}$$

Using Equations (A4.11) and (A4.12) gives:

$$E_k^{(2)} = \sum_{n \neq k} \left\{ \frac{H_{nk} H_{kn}}{E_k^{(0)} - E_j^{(0)}} \right\} \tag{A4.21}$$

We do not usually pursue perturbation theory to terms of order higher than $E_k^{(1)}$, $E_k^{(2)}$ and $\psi_k^{(1)}$, but such formulae are available. [e.g. J.O. Hirschfelder, W. Byers Brown and S.T. Epstein, *Advances in Quantum Chemistry*, **1**, 256 (1964)]. The results obtained here are summarised in Equations (A4.22) and (A4.23).

Energy correct to second order:

$$E_k = H_{kk} + \sum_{n \neq k} \left\{ \frac{H_{nk} H_{kn}}{E_k^{(0)} - E_n^{(0)}} \right\} \tag{A4.22}$$

Wave function correct to first order:

$$\Psi_k = \Psi_k^{(0)} + \sum_{j \neq k} \left\{ \frac{H_{jk}}{E_k^{(0)} - E_j^{(0)}} \right\} \Psi_j^{(0)} \tag{A4.23}$$

The case of degenerate eigenvalues in the unperturbed solution

Formulae (A4.22) and (A4.23) fail when there are degeneracies. Under such circumstances we adopt a variational approach and proceed as follows, we have:

$$\hat{\mathcal{H}} = \hat{\mathcal{H}}^{(0)} + \hat{\mathcal{H}}' \tag{A4.1}$$

$$\hat{\mathcal{H}} |\psi_k\rangle = E_k |\psi_k\rangle \tag{A4.3}$$

We expand $|\psi_k\rangle$ using Equation (A4.9), multiply on the left by $\langle \psi_j^{(0)}|$ and integrate over all space giving:

$$\langle \psi_j^{(0)} | \hat{\mathcal{H}} | \sum_n C_n \psi_n^{(0)} \rangle - E_k \sum_n C_n \langle \psi_j^{(0)} | \psi_n^{(0)} \rangle$$

$$= \sum_n C_n \{ \langle \psi_j^{(0)} | \hat{\mathcal{H}} | \psi_n^{(0)} \rangle - E_k \langle \psi_j^{(0)} | \psi_n^{(0)} \rangle \}$$

$$\equiv \sum_n C_n \{ \mathcal{H}_{jn} - E_k \delta_{jn} \} \tag{A4.24}$$

Here, $\mathcal{H}_{jn}$ is a matrix element of the full Hamiltonian and δ_{jn} (the Kronecker delta) = 1 if $j = n$ and 0 otherwise. There is an equation of the form of Equation (A4.24) for each degenerate wave function $\langle \psi_j^{(0)}|$, so that we obtain a set of linear, homogeneous equations which are known as the secular equations. Methods of solving secular equations are described in Appendices 2 and 3.

Example

Consider the problem of a particle confined within a one-dimensional box of length L by infinite potential walls at $x = 0$ and $x = L$. The solutions of the Schrödinger equation for the unperturbed problem:

$$\hat{\mathcal{H}}^{(0)}|\psi_n^{(0)}\rangle = E_n^{(0)}|\psi_n^{(0)}\rangle \qquad (A4.4)$$

are (Section 3.6)–eigenvalues: $E_n^{(0)} = n^2h^2/8mL^2$ and eigenfunctions: $\psi_n^{(0)} = \sqrt{(2/L)}$ sin $(n\pi x/L)$, where h is Planck's constant, m the mass of the particle and $n = 1 - \infty$ the quantum number. These results apply where the particle has only kinetic energy. The perturbation which we shall apply is an electrostatic field such that the potential energy of the particle (V) increases linearly from $-v/2$ at $x = 0$ to $+v/2$ at $x = L$. This potential forms our perturbing Hamiltonian, which we can write as:

$$\hat{\mathcal{H}}' = v(x - \tfrac{1}{2}L)/L = vx/L - v/2$$

In order to apply perturbation theory using Equations (A4.22) and (A4.23) we require matrix elements of the form:

$$\langle \psi_j^{(0)}|\hat{\mathcal{H}}'|\psi_k^{(0)}\rangle \equiv H_{jk}$$

and it is convenient to evaluate and tabulate these. In general we have:

$$H_{jk} = \langle \psi_j^{(0)}|vx/L - v/2|\psi_k^{(0)}\rangle = \langle \psi_j^{(0)}|vx/L|\psi_k^{(0)}\rangle - \langle \psi_j^{(0)}|v/2|\psi_k^{(0)}\rangle$$

$$= (v/L)\langle \psi_j^{(0)}|x|\psi_k^{(0)}\rangle - (v/2)\langle \psi_j^{(0)}|\psi_k^{(0)}\rangle$$

Since the eigenfunctions are orthonormal, the last term contributes a constant energy of $-v/2$ to all diagonal elements and zero to all off-diagonal elements. To evaluate the contribution of the first term we write out its full algebraic form and consider first the diagonal elements:

$$\langle \psi_k|V|\psi_k\rangle = \frac{2}{L}\cdot\frac{v}{L}\int_0^L x\cdot\sin^2\left\{\frac{k\pi x}{L}\right\}dx = \frac{v}{L^2}\int_0^L\left[x - x\cos\left\{\frac{2k\pi x}{L}\right\}\right]dx = \frac{v}{2}$$

Thus, the two contributions to the diagonal elements cancel and all $H_{kk} = 0$.
When we consider the off-diagonal elements we find:

$$\langle \psi_j|V|\psi_k\rangle = \frac{2}{L}\cdot\frac{v}{L}\int_0^L \sin\left\{\frac{j\pi x}{L}\right\}\cdot x\cdot\sin\left\{\frac{k\pi x}{L}\right\}dx$$

$$= \frac{v}{L^2}\int_0^L\left\{x\cos\left\{\frac{(j-k)\pi x}{L}\right\} - x\cos\left\{\frac{(j+k)\pi x}{L}\right\}\right\}dx$$

$$= \left[\frac{L^2}{(j-k)^2\pi^2}\cos\left\{\frac{(j-k)\pi x}{L}\right\} + \frac{xL}{(j-k)\pi}\sin\left\{\frac{(j-k)\pi x}{L}\right\}\right]_0^L$$

$$- \left[\frac{L^2}{(j+k)^2\pi^2}\cos\left\{\frac{(j+k)\pi x}{L}\right\} + \frac{xL}{(j+k)\pi}\sin\left\{\frac{(j+k)\pi x}{L}\right\}\right]_0^L$$

It is necessary to consider whether $(j+k)$ and $(j-k)$ are odd or even. When $(j+k)$ and $(j-k)$ are even, inserting the limits of 0 and L shows that in this case the integral is zero and there are therefore no off-diagonal matrix elements when j and k themselves are either both even or both odd. Another way of saying this is to say that there are no off-diagonal elements between states of the same parity and we shall return to this important point below. Where $(j+k)$ and $(j-k)$ are odd, i.e. where j and k themselves are of different parities, we find on inserting limits that:

$$\langle \psi_j | V | \psi_k \rangle = \frac{2v}{\pi^2} \cdot \left\{ \frac{-4jk}{(j^2 - k^2)^2} \right\}$$

Inserting appropriate values for j and k and adding the constant contribution of $-v/2$ to each diagonal element, we can tabulate the matrix elements for the lower quantum numbers:

$\langle \psi_j^{(0)} \| \hat{\mathcal{H}}' \| \psi_k^{(0)} \rangle$	k = 1	2	3	4	5
j = 1	0	$-1.778 \ v/\pi^2$	0	$-0.142 \ v/\pi^2$	0
j = 2	$-1.778 \ v/\pi^2$	0	$-1.920 \ v/\pi^2$	0	$-0.181 \ v/\pi^2$
j = 3	0	$-1.920 \ v/\pi^2$	0	$-1.959 \ v/\pi^2$	0
j = 4	$-0.142 \ v/\pi^2$	0	$-1.959 \ v/\pi^2$	0	$-1.975 \ v/\pi^2$
j = 5	0	$-0.181 \ v/\pi^2$	0	$-1.975 \ v/\pi^2$	0

It is interesting to note how rapidly the magnitude of the off-diagonal matrix elements decreases as the difference between j and k increases; $(\pi^2/v)H_{1k} = -1.778, -0.142, -0.039, -0.016$ for k = 2, 4, 6 and 8 respectively. The matrix elements for $j - k = \pm 1$ decrease slowly to a limiting value of $-2.0v/\pi^2$ for large values of j and k.

The absence of off-diagonal matrix elements between eigenfunctions of the same parity exemplifies a quantum-mechanical principle that is of particular importance. In order that an integral such as $\langle \psi_j^{(0)} | \hat{\mathcal{H}}' | \psi_k^{(0)} \rangle$ be non-zero, the symmetry of each component of the integral $\langle \psi_j^{(0)} |, \hat{\mathcal{H}}'$ and $| \psi_k^{(0)} \rangle$ must be such that their product is of the highest possible symmetry. In the language of group theory we say that the integral must belong to the totally symmetric representation of the symmetry group of the problem. The only symmetry property which we have recognised here is that of parity; the phases of the wave functions can either change sign (odd) or be unchanged (even) on inversion in the centre of the box. Even functions have the highest symmetry, so in order to have a finite value the integral $\langle \psi_j^{(0)} | \hat{\mathcal{H}}' | \psi_k^{(0)} \rangle$ must be even. The phases of the wave functions $| \psi_k^{(0)} \rangle$ (Figure 3.6) determine that the functions are even when k is even and odd when k is odd. And the potential is odd since it changes sign at the centre point of the box. Therefore, when $\langle \psi_j^{(0)} |$ and $| \psi_k^{(0)} \rangle$ are of the same parity the integral as a whole is odd and must therefore be zero as the above numerical results show. But when $\langle \psi_j^{(0)} |$ and $| \psi_k^{(0)} \rangle$ are of opposite parity the integral as a whole is even and a finite result is possible.

We are now in a position to apply perturbation theory to a problem. In Box 3.4 we interpreted the electronic spectra of the cyanine dyes in terms of the energies of an electron in a one-dimensional box. We found there that the experimental data gave a value of 313×10^{-20} J for $h^2/8ml^2$, where l is the length of an individual bond in the cyanine chain. For our example we consider a chain of 16 bonds for which, since L = 16l, $h^2/8mL^2 = (313/16^2) \times 10^{-20}$ J = 1.223×10^{-20} J. Thus, the lower energy values for this molecule are:

$$E_1^{(0)} = h^2/8mL^2 = 1.223 \times 10^{-20} \text{ J}$$

$$E_2^{(0)} = 4h^2/8mL^2 = 4.891 \times 10^{-20} \text{ J}$$

$$E_3^{(0)} = 9h^2/8mL^2 = 11.004 \times 10^{-20} \text{ J}$$

Suppose, for the sake of this example, that $v = 2.0 \times 10^{-20}$ J and we wish to determine the new E_2 and $|\psi_2\rangle$ which result from this perturbation. According to Equation (A4.22):

$$E_2 = H_{22} + \sum_{j \neq 2} \frac{H_{j2}H_{2j}}{E_k^{(0)} - E_k^{(0)}}$$

$H_{22} = 0.0$ and the largest term in the summation is when $j = 1$ and for that term in the summation we find:

$$E_2^{(2)}{}_{(j=1)} = \frac{(1.778 \times 2.0 \times 10^{-20})^2}{\pi^4(3.668 \times 10^{-20})} = \frac{12.65 \times 10^{-20}}{357.3} = 0.035 \times 10^{-20} \text{ J}$$

For $j = 3$:

$$E_2^{(2)}{}_{(j=3)} = \frac{(1.920 \times 2.0 \times 10^{-20})^2}{\pi^4(-6.113 \times 10^{-20})} = \frac{14.75 \times 10^{-20}}{-595.5} = -0.025 \times 10^{-20} \text{ J}$$

Note how interaction with a state of lower energy, $|\psi_1^{(0)}\rangle$ raises the energy of $|\psi_2\rangle$, whereas interaction with a state of higher energy, $|\psi_3^{(0)}\rangle$, depresses it. The contributions to the energy of $|\psi_2\rangle$ from states of energy higher than $E_3^{(0)}$ are negligible and we finally have:

$$E_2 = (4.891 + 0.035 - 0.025) \times 10^{-20} \text{ J} = 4.901 \times 10^{-20} \text{ J}$$

For the corrections to the wave function we use:

$$|\psi_2\rangle = |\psi_2^{(0)}\rangle + \sum_{j \neq 2} \frac{H_{j2}}{E_2^{(0)} - E_j^{(0)}} |\psi_j^{(0)}\rangle$$

$$= |\psi_2^{(0)}\rangle - \frac{1.778 \times 2.0 \times 10^{-20}}{36.202 \times 10^{-20}} |\psi_1^{(0)}\rangle + \frac{1.920 \times 2.0 \times 10^{-20}}{60.333 \times 10^{-20}} |\psi_3^{(0)}\rangle$$

$$= |\psi_2^{(0)}\rangle - 0.098|\psi_1^{(0)}\rangle + 0.064|\psi_3^{(0)}\rangle$$

Finally, it is of interest to consider the first order correction to $|\psi_1^{(0)}\rangle$ which is readily found to be:

$$|\psi_1\rangle = |\psi_1^{(0)}\rangle + 0.098|\psi_2^{(0)}\rangle$$

We note that the small admixture of $|\psi_2^{(0)}\rangle$ has a positive coefficient so that (see Figure 3.6) the amplitude of the wave function $|\psi_1^{(0)}\rangle$ in the region between $x = 0$ and $L/2$ is slightly increased by the perturbation while between $x = L/2$ and L it is reduced. Since the electron density is proportional to the square of the wave function, this means that the perturbation increases the electron density in the lower half of the box and decreases it at the upper end. This is just what we would expect, since the perturbing potential increases linearly with x and the electron will respond to this by moving to lower values of x. In applying perturbation theory it is rather easy to make trivial errors of sign and simple, qualitative considerations such as the above can be a useful aid to the detection of such errors.

APPENDIX 5

The Spherical Harmonics and Hydrogen Atom Wave Functions

The functions of θ and ϕ which form the solutions of the angular part of the hydrogen-atom problem are well known. They are frequently found as the solutions of differential equations arising from the theoretical analysis of problems having spherical symmetry, such as the vibration of a gas inside a sphere or the motion of the water on a flooded planet. They are especially important in atomic physics because, since all atoms are spherically symmetrical, the solutions of Schrödinger's equation for any atom can be formed from the same angular functions as the hydrogen atom; the presence of other electrons and the consequent inter-electronic repulsion changes the radial but not the angular part of the wave function.

Furthermore, since combinations of atomic orbital functions are frequently used to construct orbitals for molecules their use extends throughout theoretical chemistry. We shall not describe the extensive mathematics of these functions, but simply write down the lower members of the group in a form which is useful in chemical applications.

The explicit normalised forms of the function $\Phi_m(\phi)$ for $m = 0, \pm 1, \pm 2$

$$\Phi_0(\phi) = (2\pi)^{-\frac{1}{2}} \qquad\qquad \Phi_0(\phi) = (2\pi)^{-\frac{1}{2}}$$

$$\Phi_1(\phi) = (2\pi)^{-\frac{1}{2}} \exp(i\phi) \qquad \Phi_{1c}(\phi) = (\pi)^{-\frac{1}{2}} \cos(\phi)$$

$$\Phi_{-1}(\phi) = (2\pi)^{-\frac{1}{2}} \exp(-i\phi) \qquad \Phi_{1s}(\phi) = (\pi)^{-\frac{1}{2}} \sin(\phi)$$

$$\Phi_2(\phi) = (2\pi)^{-\frac{1}{2}} \exp(2i\phi) \qquad \Phi_{2c}(\phi) = (\pi)^{-\frac{1}{2}} \cos(2\phi)$$

$$\Phi_{-2}(\phi) = (2\pi)^{-\frac{1}{2}} \exp(-2i\phi) \quad \Phi_{2s}(\phi) = (\pi)^{-\frac{1}{2}} \sin(2\phi)$$

The above functions are normalised in the sense that:

$$\int_0^{2\pi} \Phi_m^*(\phi)\Phi_m(\phi)\,\mathrm{d}\phi = 1.0$$

Note that there is a choice of function. The functions on the left are eigenfunctions of both the energy operator and the angular momentum operator. Those on the right are eigenfunctions of the energy operator only.

The Quantum in Chemistry   R. Grinter
© 2005 John Wiley & Sons, Ltd

The explicit normalised forms of the function $\Theta_{l,m}(\theta)$ for $l = 0, 1$ and 2

$l = 0$ s-functions	$l = 1$ p-functions	$l = 2$ d-functions
$\Theta_{0,0} = +\frac{1}{2}\sqrt{2}$	$\Theta_{1,0} = +\sqrt{(3/2)} \cdot \cos\theta$	$\Theta_{2,0} = +\frac{1}{2}\sqrt{(5/2)} \cdot (3\cos^2\theta - 1)$
	$\Theta_{1,+1} = -\frac{1}{2}\sqrt{3} \cdot \sin\theta$	$\Theta_{2,+1} = -\frac{1}{2}\sqrt{15} \cdot \sin\theta \cdot \cos\theta$
	$\Theta_{1,-1} = +\frac{1}{2}\sqrt{3} \cdot \sin\theta$	$\Theta_{2,-1} = +\frac{1}{2}\sqrt{15} \cdot \sin\theta \cdot \cos\theta$
		$\Theta_{2,+2} = +\frac{1}{4}\sqrt{15} \cdot \sin^2\theta$
		$\Theta_{2,-2} = +\frac{1}{4}\sqrt{15} \cdot \sin^2\theta$

The above functions are normalised in the sense that:

$$\int_0^\pi \Theta_{l,m}^*(\theta)\Theta_{l,m}(\theta) \sin\theta \, d\theta = 1.0$$

Note that the value of m is restricted to integers in the range $+l$ to $-l$. The positive signs have been included to emphasise that there is a rather subtle sign convention in use here. It will not affect our use of the functions in any way, but it is important when they are used in conjunction with advanced vector-coupling methods.

There are two useful recursion formulae for the Θ.

$$\sin\theta \cdot \Theta_{l,|m|-1}(\theta) = [(l + |m| + 1)(l + |m|)/(2l + 3)(2l + 1)]^{\frac{1}{2}}\Theta_{l+1,|m|}(\theta)$$

$$- [(l - |m| + 1)(l - |m|)/(2l + 1)(2l - 1)]^{\frac{1}{2}}\Theta_{l-1,|m|}(\theta)$$

and

$$\cos\theta \cdot \Theta_{l,|m|}(\theta) = [(l + |m| + 1)(l - |m| + 1)/(2l + 2)(2l + 1)]^{\frac{1}{2}}\Theta_{l+1,|m|}(\theta)$$

$$+ [(l + |m|)(l - |m|)/(2l + 1)(2l - 2)]^{\frac{1}{2}}\Theta_{l-1,|m|}(\theta)$$

The spherical harmonics $Y_{l,m}(\theta, \phi)$

Products of the functions $\Theta_{l,m}(\theta)$ and $\Phi_m(\phi)$ are commonly known as the spherical harmonics. Some normalised examples follow:

$l = 0$ s-functions	$Y_{0,0} = +\frac{1}{2}\sqrt{(1/\pi)}$

$l = 1$ p-functions	$l = 2$ d-functions
$Y_{1,0} = +\frac{1}{2}\sqrt{(3/\pi)} \cdot \cos\theta$	$Y_{2,0} = +\frac{1}{4}\sqrt{(5/\pi)} \cdot (3\cos^2\theta - 1)$
$Y_{1,+1} = -\frac{1}{2}\sqrt{(3/2\pi)} \cdot \sin\theta \cdot \exp(i\phi)$	$Y_{2,+1} = -\frac{1}{2}\sqrt{(15/2\pi)} \cdot \sin\theta \cdot \cos\theta \cdot \exp(i\phi)$
$Y_{1,-1} = +\frac{1}{2}\sqrt{(3/2\pi)} \cdot \sin\theta \cdot \exp(-i\phi)$	$Y_{2,-1} = +\frac{1}{2}\sqrt{(15/2\pi)} \cdot \sin\theta \cdot \cos\theta \cdot \exp(-i\phi)$
	$Y_{2,+2} = +\frac{1}{4}\sqrt{(15/2\pi)} \cdot \sin^2\theta \cdot \exp(2i\phi)$
	$Y_{2,-2} = +\frac{1}{4}\sqrt{(15/2\pi)} \cdot \sin^2\theta \cdot \exp(-2i\phi)$

The hydrogen atom wave functions

Although they are not spherical harmonics, the hydrogen atom radial functions ($R_{n,l}$ (r)) and the complete hydrogen wave functions ($\Psi_{n,l,m}$) are included here for the sake of completeness.

The normalised explicit forms of the function $R_{n,l}$ (r) for n = 1, 2 and 3

We define $\rho = \dfrac{2Z}{na_0} \cdot r$

where Z is the nuclear charge in units of $+e$ and a_0 is the Bohr radius.

$$n = 1 \quad l = 0 \quad \mathbf{R}_{1,0}(r) = (Z/a_0)^{\frac{3}{2}} \cdot 2 \exp(-\rho/2)$$

$$n = 2 \quad l = 0 \quad \mathbf{R}_{2,0}(r) = \frac{(Z/a_0)^{\frac{3}{2}}}{2\sqrt{2}}(2 - \rho) \exp(-\rho/2)$$

$$n = 2 \quad l = 1 \quad \mathbf{R}_{2,1}(r) = \frac{(Z/a_0)^{\frac{3}{2}}}{2\sqrt{6}}\rho \exp(-\rho/2)$$

$$n = 3 \quad l = 0 \quad \mathbf{R}_{3,0}(r) = \frac{(Z/a_0)^{\frac{3}{2}}}{9\sqrt{3}}(6 - 6\rho - \rho^2) \exp(-\rho/2)$$

$$n = 3 \quad l = 1 \quad \mathbf{R}_{3,1}(r) = \frac{(Z/a_0)^{\frac{3}{2}}}{9\sqrt{6}}(4 - \rho)\rho \exp(-\rho/2)$$

$$n = 3 \quad l = 2 \quad \mathbf{R}_{3,2}(r) = \frac{(Z/a_0)^{\frac{3}{2}}}{9\sqrt{30}}\rho^2 \exp(-\rho/2)$$

All the above functions of r, θ and ϕ are all individually normalised to unity and mutually orthogonal:

$$\int_0^{2\pi} \Phi_m^*(\phi)\Phi_{m'}(\phi)\, d\phi = \delta_{m,m'};$$

$$\int_0^{\pi} \Theta_{l,m}^*(\theta)\Theta_{l',m'}(\theta) \sin\theta\, d\theta = \delta_{l,l'}\delta_{m,m'}; \quad \int_0^{\infty} \mathbf{R}_{n,l}^*\mathbf{R}_{n',l'}r^2\, dr = \delta_{n,n'}\delta_{l,l'}$$

Therefore, a complete hydrogen atom wave function formed from a product of three of them is also normalised to one. See Appendix 7 concerning the volume element in polar co-ordinates.

Some complete normalised wave functions for the hydrogen atom

$$\Psi_{1s} = \frac{1}{\sqrt{\pi}}\left(\frac{Z}{a_0}\right)^{\frac{3}{2}} \exp(-\rho/2)$$

$$\Psi_{2s} = \frac{1}{4\sqrt{2\pi}} \left(\frac{Z}{a_0}\right)^{\frac{3}{2}} (2 - \rho) \exp(-\rho/2)$$

$$\Psi_{2pz} = \frac{1}{4\sqrt{2\pi}} \left(\frac{Z}{a_0}\right)^{\frac{3}{2}} \cos\theta \cdot \rho \exp(-\rho/2) = \frac{1}{4\sqrt{2\pi}} \left(\frac{Z}{a_0}\right)^{\frac{5}{2}} z \cdot \exp(-\rho/2)$$

Using the functions $\Theta_{1,+1}$, Θ_{1-1}, $\Phi_{1c}(\phi)$ and $\Phi_{1s}(\phi)$, or simply by noting that the three 2p orbitals must be exactly the same apart from their dependence upon x, y or z, we obtain:

$$\Psi_{2px} = \frac{1}{4\sqrt{2\pi}} \left(\frac{Z}{a_0}\right)^{\frac{5}{2}} x \cdot \exp(-\rho/2)$$

$$\Psi_{2py} = \frac{1}{4\sqrt{2\pi}} \left(\frac{Z}{a_0}\right)^{\frac{5}{2}} y \cdot \exp(-\rho/2)$$

APPENDIX 6
Slater Determinants

Suppose that we have a system in which n atomic or molecular orbitals are each doubly occupied by a total of 2n electrons. The wave function for the system might be thought to be:

$$\Omega = \phi_1\alpha(1)\phi_1\beta(2)\phi_2\alpha(3)\phi_2\beta(4)\ldots\phi_{n-1}\beta(2n-2)\phi_n\alpha(2n-1)\phi_n\beta(2n)$$

But it is fundamental to quantum mechanics that electrons are indistinguishable and an equally acceptable wave function could be obtained by interchanging any two electrons, e.g. 1 and 4, giving:

$$\Omega = \phi_1\alpha(4)\phi_1\beta(2)\phi_2\alpha(3)\phi_2\beta(1)\ldots\phi_{n-1}\beta(2n-2)\phi_n\alpha(2n-1)\phi_n\beta(2n)$$

Indeed, any one of the $(2n)!$ permutations[‡] of electrons in orbitals is equally acceptable and all must be included in the complete description of the system. Furthermore, we know that the wave functions describing sets of indistinguishable, fundamental particles must have a clear symmetry with respect to interchange of any two of those particles (see Sections 11.4 and 11.5 for a further discussion of this point). For electrons the wave function must change sign (be antisymmetric) when a pair of electrons are interchanged, which is simply another way of stating the Pauli principle. John Clarke Slater (1900–1976) pointed out that these requirements could be concisely fulfilled by writing the total wave function as a determinant of the form:

$$\Psi = 1/\sqrt{(2n)!}\begin{vmatrix} \phi_1\alpha(1)\phi_1\alpha(2)\ldots\phi_1\alpha(2n-1)\phi_1\alpha(2n) \\ \phi_1\beta(1)\phi_1\beta(2)\ldots\phi_1\beta(2n-1)\phi_1\beta(2n) \\ \phi_2\alpha(1)\phi_2\alpha(2)\ldots\phi_2\alpha(2n-1)\phi_2\alpha(2n) \\ .\qquad .\qquad .\qquad . \\ .\qquad .\qquad .\qquad . \\ .\qquad .\qquad .\qquad . \\ \phi_n\beta(1)\phi_n\beta(2)\ldots\phi_n\beta(2n-1)\phi_n\beta(2n) \end{vmatrix}$$

Determinantal wave functions such as Ψ are frequently represented by the leading diagonal of the complete determinant enclosed between two pairs of vertical bars:

$$\Psi = 1/\sqrt{(2n)!}\;||\phi_1\alpha(1)\phi_1\beta(2)\phi_2\alpha(3)\phi_2\beta(4)\ldots\phi_{n-1}\beta(2n-2)\phi_n\alpha(2n-1)\phi_n\beta(2n)||$$

The determinant includes all possible permutations of electrons in orbitals and when we multiply it out according to the established rules of determinantal algebra each constituent

---

[‡] $(2n)!$ (spoken '2n factorial' or 'factorial 2n') $= 2n \times (2n-1) \times (2n-2) \times \cdots \times 1$.

The Quantum in Chemistry   R. Grinter
© 2005 John Wiley & Sons, Ltd

permutation is generated, with the correct change of sign for interchange of any two electrons; e.g.

$$\Psi = (1/\sqrt{6}) \begin{vmatrix} \phi_1\alpha(1)\phi_1\alpha(2)\phi_1\alpha(3) \\ \phi_1\beta(1)\phi_1\beta(2)\phi_1\beta(3) \\ \phi_2\alpha(1)\phi_2\alpha(2)\phi_2\alpha(3) \end{vmatrix}$$

$$\equiv (1/\sqrt{6})||\phi_1\alpha(1)\phi_1\beta(2)\phi_2\alpha(3)||$$

On multiplying the 3×3 determinant out we obtain the total wave function in the explicit form:

$$(1/\sqrt{6}) \{\phi_1\alpha(1)\phi_1\beta(2)\phi_2\alpha(3) - \phi_1\alpha(1)\phi_1\beta(3)\phi_2\alpha(2) + \phi_1\alpha(2)\phi_1\beta(3)\phi_2\alpha(1)$$

$$- \phi_1\alpha(2)\phi_1\beta(1)\phi_2\alpha(3) + \phi_1\alpha(3)\phi_1\beta(1)\phi_2\alpha(2) - \phi_1\alpha(3)\phi_1\beta(2)\phi_2\alpha(1)\}$$

The example of a three-electron wave function alerts us to a further point about the complete description of many-electron systems. Clearly, the function in which the electron in orbital ϕ_2 has β spin would be equally acceptable and the complete description of the system must therefore be the normalised sum of two Slater determinants:

$$\Psi = (1/\sqrt{12}) \{||\phi_1\alpha(1)\phi_1\beta(2)\phi_2\alpha(3)|| + ||\phi_1\alpha(1)\phi_1\beta(2)\phi_2\beta(3)||\}$$

APPENDIX 7
Spherical Polar Co-ordinates

We require three co-ordinates to specify a position in space. In the Cartesian system we choose three mutually perpendicular axes which we normally designate x, y and z. However, for a spherically symmetrical problem such as the hydrogen atom, the spherical polar co-ordinate system is much more convenient.

The relationships between the Cartesian and polar co-ordinate systems is shown in Figure A7.1. Consider the point P; its projection onto the xy-plane is p. In the Cartesian system the co-ordinates of P are x, y and z; in the polar system they are r, θ and ϕ. r is the distance of P from the origin of the co-ordinate system, O. θ is the angle between OP and the z-axis. ϕ is the angle between Op and the x-axis. Since $Op = r \cdot \sin\theta$, simple trigonometry shows that the relationships between the two co-ordinate systems are:

$$x = r \cdot \sin\theta \cdot \cos\phi \qquad y = r \cdot \sin\theta \cdot \sin\phi \qquad z = r \cdot \cos\theta$$

and

$$r^2 = x^2 + y^2 + z^2$$

The volume element in polar co-ordinates

Many quantum-mechanical calculations involve the evaluation of the integral of a product of two or more wave functions over the whole of co-ordinate space. For example, the

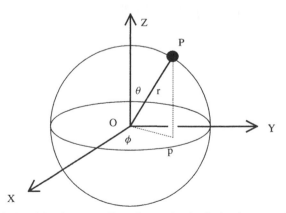

Figure A7.1 The relationships between Cartesian and spherical polar co-ordinates

The Quantum in Chemistry   R. Grinter
© 2005 John Wiley & Sons, Ltd

wave function (ψ) must be normalised, i.e. we must have:

$$\iiint\limits_{\text{all space}} \Psi^*\Psi \, dx \, dy \, dz \equiv \int\limits_{\text{all space}} \Psi^*\Psi \, dv = 1.0$$

In Cartesian co-ordinates the volume element, $dx \, dy \, dz \equiv dv$ is an infinitesimally small cube with sides dx, dy and dz. We obtain it by moving infinitesimally small distances along the x, y and z axes in the vicinity of a point such as P in the figure above. To obtain the volume element in polar co-ordinates we have to do the same thing, but the process and the result is not quite so simple:

1. First we move a distance dr along OP.

2. When we change the angle θ by $d\theta$, P moves along an arc the length of which depends not only upon $d\theta$ but also upon r, the distance of P from O, according to the formula: arc $= r \, d\theta$, where $d\theta$ is measured in radians.

3. A movement along the co-ordinate ϕ brings about a movement of p, the projection of P in the xy-plane. As above, since the length of Op is $r \cdot \sin\theta$, p moves along an arc of length $r \cdot \sin\theta \, d\phi$.

Thus, in polar co-ordinates, the sides of our infinitesimal volume are dr, $r d\theta$ and $r \cdot \sin\theta \, d\phi$, and the new volume element is $r^2 \sin\theta \, dr \, d\theta \, d\phi$. Therefore, the normalisation integral becomes:

$$\iiint\limits_{\text{all space}} \Psi^*\Psi \, dx \, dy \, dz \Rightarrow \int\limits_{\text{all space}} \Psi^*\Psi r^2 \cdot \sin\theta \, dr \, d\theta \, d\phi = 1.0$$

The limits of the three co-ordinates are: $0 \le r \le \infty$, $0 \le \theta \le \pi$ and $0 \le \phi \le 2\pi$. In order to make clear which integration limits belong to each variable, multiple integrals of this sort are often written:

$$\int\limits_0^\infty r^2 \, dr \int\limits_0^\pi \sin\theta \, d\theta \int\limits_0^{2\pi} \Psi^*\Psi \, d\phi = 1.0$$

APPENDIX 8
Numbers: Real, Imaginary and Complex

Any number which can be plotted as a point on a scale extending from $-\infty$ to $+\infty$ is known as a *real* number. Real numbers are divided into two types: *rational* numbers can be expressed in the form p/q, where p and q are integers, and *irrational* numbers, which cannot be expressed in that form; e.g. $1.375 = 11/8$ is a rational number and $\pi = 4(1 - 1/3 + 1/5 - 1/7 + \cdots)$ is an irrational number.

The square root of -1 is denoted by i. This quantity cannot be expressed as a real number and the product of i and any real number is a number which also cannot be placed on the scale of real numbers. It is known as an *imaginary* number. No real number can be equal to an imaginary number; apart from zero which is found on both scales at the point where they cross (Figure A8.1). Note that $i^2 = -1$, $i^3 = -i$, $i^4 = +1$, $i^5 = i$, and that these values recur in cycles for higher powers of i.

A *complex* number is the sum of a real and an imaginary number written in the form $(a + ib)$, where a and b are both real numbers. It can be represented in a diagram (Figure A8.1) where the scale of imaginary numbers is plotted at right angles to the real number scale. This type of representation of complex numbers is known as an *Argand diagram*. When two complex numbers are added (subtracted) the real and imaginary parts are added (subtracted) separately:

$$(a + ib) - (x + iy) = a - x + i(b - y)$$

In an equation involving complex numbers, the real and imaginary parts on the two sides of the equation are separately equal.

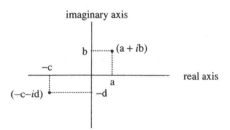

Figure A8.1 The representation of complex numbers on an Argand diagram

The Quantum in Chemistry   R. Grinter
© 2005 John Wiley & Sons, Ltd

The two complex numbers, $(a + ib)$ and $(a - ib)$, have a special relationship and are said to be the *complex conjugates* of each other. When a complex number is multiplied by its complex conjugate a positive real number is always obtained:

$$(a + ib)(a - ib) = a^2 - abi + bai - b^2 i^2 = a^2 + b^2 \text{ which must be positive.}$$

In quantum mechanics the complex conjugate of a complex quantity is usually indicated by an asterisk so that $(a + ib)^* = (a - ib)$ and it is especially important that the product of two such quantities always gives a positive result.

A fraction having a complex number in the denominator can be simplified by multiplying the numerator and the denominator of the fraction by the complex conjugate of the denominator.

The Italian mathematician Girolamo Cardano (1501–1576) appears to have been the first to publish (in 1545) mathematics involving the square roots of negative numbers which he obtained in the process of solving quadratic equations. For many years mathematicians regarded such entities as strange and useless. In 1702, Gottfried Wilhelm Leibnitz (1646–1716), who developed calculus independently of Newton and whose notation we use in modern mathematics, wrote: 'The Divine Spirit found a sublime outlet in that wonder of [mathematical] analysis, that portent of the ideal world, that amphibian between being and non-being, which we call the imaginary root of negative unity'. And indeed, anyone might have been excused for thinking that such an unworldly concept would have no application in science but, on the contrary, it has proved extremely valuable. Complex numbers play a very important role in the theoretical models of many phenomena including alternating currents, diffusion and the refraction of light. The mere existence of this brief account attests to their importance in quantum mechanics.

A particularly interesting and useful result is obtained when the argument of the exponential function contains i. From Box 3.1 we have:

$$\exp(iax) = 1 + iax + (iax)^2/2! + (iax)^3/3! + (iax)^4/4! + \cdots$$

which, using the properties of the powers of i noted above, and the series for sine and cosine given in box 4.1:

$$= 1 + iax - (ax)^2/2! - i(ax)^3/3! + (ax)^4/4! + \cdots = \cos(ax) + i \cdot \sin(ax)$$

Similarly:

$$\exp(-iax) = \cos(ax) - i \cdot \sin(ax)$$

Alternatively, $\cos(ax)$ and $\sin(ax)$ can be expressed in terms of the exponential functions:

$$\cos(ax) = \tfrac{1}{2}\{\exp(+iax) + \exp(-iax)\}$$
$$\sin(ax) = \tfrac{1}{2i}\{\exp(+iax) - \exp(-iax)\}$$

APPENDIX 9
Dipole and Transition Dipole Moments

The dipole moment of a molecule is an important quantity in its own right and the transition dipole moment plays a central role in the description of the transition probability in infrared and electronic spectroscopy. Here we first determine the quantum-mechanical expression for the dipole moment of a molecule. In order to be specific we shall take as our example the lithium hydride molecule and we shall assume that its orientation in space is fixed so that the Li–H bond is aligned along the x-axis. The situation might be crudely represented as in the schematic representation of the lithium hydride molecule shown in Figure A9.1, in which the black dots represent the electrons and the open circles the nuclei.

As an illustration of the principles of the procedure, consider first the determination of the centre of mass of the molecule. If we neglect the mass of the four electrons and consider only the nuclei then the x co-ordinate of the centre of mass (x_0) is given by:

$$x_0 = \frac{m_H \cdot x_H + m_{Li} \cdot x_{Li}}{m_H + m_{Li}} \tag{A9.1}$$

where x_H and x_{Li} are the co-ordinates of the hydrogen and lithium nuclei with respect to any co-ordinate origin which we choose. Rearrangement of Equation (A9.1) gives:

$$m_{Li} \cdot (x_0 - x_{Li}) = m_H \cdot (x_H - x_0)$$

and if we set the origin to be the centre of mass, i.e. $x_0 = 0.0$, then we obtain the more familiar relationship:

$$m_{Li} \cdot x_{Li} + m_H \cdot x_H = 0$$

The centre of charge can be determined in an exactly analogous way, but there are two obvious differences. Firstly, since the charge on the electron is equal in magnitude to that

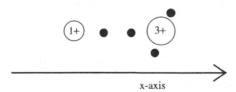

Figure A9.1   A schematic representation of the lithium hydride molecule

The Quantum in Chemistry   R. Grinter
© 2005 John Wiley & Sons, Ltd

of the proton we cannot neglect it and, secondly, we must take account of the different signs of the charges on the nuclei and the electrons. In fact, we can determine a centre of positive and a centre of negative charge. If these two centres are coincident then the molecule has no dipole moment but if they are not coincident then the molecule has a dipole moment (M) and *for a neutral molecule*:

$$M = Q \cdot D \tag{A9.2}$$

where Q is the sum of either the electronic or the nuclear charges and D is the distance between the centres of positive and negative charge.

The centre of positive charge for our lithium hydride molecule might be written:

$$x_0^+ = \frac{3 \cdot x_{Li} + 1 \cdot x_H}{4} = \frac{\sum_N q_N x_N}{4} \tag{A9.3}$$

where q_N is the charge on nucleus (N). But this equation assumes that the nuclei are stationary and we know that they are not. However, by using the vibrational wave function, (ψ_{vib}), which describes the motions of the nuclei in the vibrational ground state, we can replace Equation (A9.3) by the quantum-mechanically exact expression:

$$x_0^+ = \frac{\int \Psi_{vib}^* \hat{M}_{nuc} \Psi_{vib} \partial v}{4} \equiv \frac{\langle \Psi_{vib} | \hat{M}_{nuc} | \Psi_{vib} \rangle}{4} \tag{A9.4}$$

where $\hat{M}_{nuc} = \Sigma_N q_N X_N$ in which the co-ordinate X_N has been replaced by the operator X_N. (We use X for the co-ordinates of the nuclei and x for those of the electrons.)

The remarks above are even more true with regard to the determination of the centre of negative charge because we certainly cannot consider the electrons to be localised point charges. But we do know that the distribution of the electrons is given by the product of a wave function (ψ_{elec}) and its complex conjugate (ψ_{elec}^*) so that:

$$x_0^- = \frac{\int \Psi_{elec}^* \hat{M}_{elec} \Psi_{elec} \partial v}{4} \equiv \frac{\langle \Psi_{elec} | \hat{M}_{elec} | \Psi_{elec} \rangle}{4} \tag{A9.5}$$

where $\hat{M}_{elec} = \Sigma_e q_e x_e$.

But Equations (A9.4) and (A9.5) are also oversimplified in that they assume that we can use wave functions for the nuclei (ψ_{vib}) and for the electrons (ψ_{elec}) independently of each other. But we know that they are inter-related because, for example, the strength of a bond, and hence the force constant and the vibrational frequency, depend upon the electron distribution in the molecule. Therefore, we must write our wave functions as functions of both the electronic and the nuclear co-ordinates and, according to Born and Oppenheimer (Section 6.3), it is a very good approximation to write the total wave function as a simple product:

$$\phi_{total}(x,X) = \psi_{elec}(x,X) \cdot \psi_{vib}(X) \tag{A9.6}$$

In Equation (A9.6) x represents all the co-ordinates (not just the x co-ordinates) of all the electrons described by the electronic wave function and X represents the co-ordinates of the nuclei. It is therefore a general expression that is not limited to diatomic molecules. The presence of both electronic and nuclear co-ordinates following ψ_{elec} reminds us of the fact that the electronic wave function depends upon the positions of the nuclei as well as upon those of the electrons. Thus, the full expression for the molecular dipole moment is:

$$M = 4(X_0^+ - x_0^-) = \langle \phi_{total}(x,X)|\hat{M}_{nuc}|\phi_{total}(x,X)\rangle - \langle \phi_{total}(x,X)|\hat{M}_{elec}|\phi_{total}(x,X)\rangle$$

(A9.7)

This is the quantum-mechanical expression for the dipole moment of a molecule in the state $\phi_{total}(x,X)$. It can be written more concisely in the form:

$$M = \langle \phi_{total}(x,X)|\hat{M}|\phi_{total}(x,X)\rangle$$

(A9.8)

where $\hat{M} = \Sigma_k q_k x_k$ and the sum runs over all particles, electrons and nuclei, with appropriate changes of the sign and magnitude of q_k.

The transition dipole moment

For the determination of the probability that a spectroscopic transition between an initial state (ϕ^i_{total}) and a final state (ϕ^f_{total}) (Section 8.6) will be allowed we require an expression for the *transition* dipole moment. This is a generalisation of Equation (A9.8) in which the wave function to the left of the operator is replaced by one for the excited molecular state (ϕ^f_{total}). The resulting integral represents the transitory dipole moment generated when the molecule changes its state, which is the 'handle' by means of which the oscillating electric field of the radiation is able to couple with the molecule. That is – dropping the (x,X) to simplify the notation – we require:

$$M_{i,f} = \langle \phi^f_{total}|\hat{M}|\phi^i_{total}\rangle = \langle \psi^f_{elec}\psi^f_{vib}|\hat{M}|\psi^i_{elec}\psi^i_{vib}\rangle$$

(A9.9)

Equation (A9.9) is the basic form of the expression for a transition dipole moment. In order to apply it to UV-VIS and IR spectroscopies it is expedient to develop the formula further. This is done in Section 8.7.

Selection rules

In other parts of this book, notably Chapters 5 and 8, selection rules have been discussed, often on the basis of the conservation of energy, angular momentum and parity. This is not the most common approach to the problem. It is more usual to start from Equation (A9.9), which shows that the observation of a spectroscopic transition requires a non-zero value for the transition moment integral ($M_{i,f}$). The fulfilment of that requirement is best investigated by considering the e-m radiation as a wave rather than a particle and to invoke the powerful methods of group theory. Since group theory is not discussed here we shall not use it, but the basic principles of its application can be illustrated by deriving Laporte's atomic spectral selection rule (Section 5.10.2). The integral $\langle \phi^f_{total}|\hat{M}|\phi^i_{total}\rangle$ consists of three parts to each of which a parity can be assigned. The operator ($\hat{M}$) is a property of the e-m radiation or photon and the parity of the dipole photon is u (Section 8.5). The initial and final states can be of either parity and if they are both g the overall parity of the integrand is g × u × g = u. If both states are u the integrand is also u, but if the two states are of different parity then the integrand is g. The crucial argument now goes as follows. The fact that we can classify our operator and states and their product in terms of their parities means that they are composed of pairs of parts which are *exactly equal in magnitude* and of the *same sign* if the function is g but of *opposite sign* if it is u. The process of integration is the summation of these parts, so the integration of a u function is always a sum of equal and opposite parts which has to give zero. In the case of a g function, however, a non-zero result is expected, though a fortuitous cancellation of oppositely signed contributions to the sum can always occur. Thus we deduce that allowed electric dipole transitions in atomic spectroscopy are always u $\leftrightarrow$ g in accord with Laporte's rule.

APPENDIX 10

Wave Functions for the ^{3}F States of d^2 using Shift Operators

The use of the lowering operators ($\hat{L}_-$ and $\hat{l}_-$) to determine the combination of microstates which form the ^{3}F state of the configuration d^2 provides a good example of the application of the raising and lowering or shift operators (Section 4.7).

Notation:

The required ^{3}F states are completely described by the *kets* $|L, M_L, S, M_S\rangle$. However, since we shall concentrate on the component of the triplet which has $S = 1$ and $M_S = +1$, it will be convenient to abbreviate to $|L, M_L\rangle$, i.e. $|3, M_L, 1, +1\rangle$ becomes $|3, M_L\rangle$.

Similarly, to simplify the notation the microstates are written in the form $|m_l^\pm(1), m_l^\pm(2)\rangle$ where the $\pm$ superscripts indicate the m_s value of the electron.

A table of the microstates of the electron configuration d^2 is not an essential prerequisite for this exercise, but it does provide a useful running check. Another useful check is the fact that each combination of microstates, as it is found, is normalised.

The microstates of the configuration d^2

		M_S +1	M_S 0	M_S −1
	+4		$|+2^+(1), +2^-(2)\rangle$	
	+3	$|+2^+(1), +1^+(2)\rangle$	$|+2^+(1), +1^-(2)\rangle|+2^-(1), +1^+(2)\rangle$	$|+2^-(1), +1^-(2)\rangle$
	+2	$|+2^+(1), 0^+(2)\rangle$	$|+2^+(1), 0^-(2)\rangle|+2^-(1), 0^+(2)\rangle$	$|+2^-(1), 0^-(2)\rangle$
M_L			$|+1^+(1), +1^-(2)\rangle$	
	+1	$|+1^+(1), 0^+(2)\rangle$	$|+1^+(1), 0^-(2)\rangle|+1^-(1), 0^+(2)\rangle$	$|+1^-(1), 0^-(2)\rangle$
		$|+2^+(1), -1^+(2)\rangle$	$|+2^+(1), -1^-(2)\rangle|+2^-(1), -1^+(2)\rangle$	$|+2^{-1}(1), -1^-(2)\rangle$
	0	$|+1^+(1), -1^+(2)\rangle$	$|+1^+(1), -1^-(2)\rangle|+1^-(1), -1^+(2)\rangle$	$|+1^-(1), -1^-(2)\rangle$
		$|+2^+(1), -2^+(2)\rangle$	$|+2^+(1), -2^-(2)\rangle|+2^-(1), -2^+(2)\rangle$	$|+2^-(1), -2^-(2)\rangle$
			$|0^+(1), 0^-(2)\rangle$	

The Quantum in Chemistry   R. Grinter
© 2005 John Wiley & Sons, Ltd

plus the 'mirror image' of microstates having M_L values of -1, -2, -3 and -4.

Clearly, the component of the ^{3}F which has the highest z-component of orbital angular momentum may be written in the two equivalent notations:

$$|3, 3\rangle \approx |+2^+(1), +1^+(2)\rangle \tag{A10.1}$$

We now apply the lowering operator:

$$\hat{L}_-|L, M\rangle = \{L(L+1) - M(M-1)\}^{\frac{1}{2}}|L, M-1\rangle \tag{A10.2}$$

to the left-hand side of Equation (10.1) and the operator:

$$\hat{l}_-|l, m\rangle = \{l(l+1) - m(m-1)\}^{\frac{1}{2}}|l, m-1\rangle \tag{A10.3}$$

to each electron on the right-hand side. Since no confusion can arise, subscripts L and l have been dropped to simplify the notation.

We find that:

$$\hat{L}_-|3, 3\rangle = \sqrt{6}|3, 2\rangle \tag{A10.4}$$

while

$$\sum_i \hat{l}_{i-}|+2^+(1), +1^+(2)\rangle = 2|+1^+(1), +1^+(2)\rangle + \sqrt{6}|+2^+(1), 0^+(2)\rangle \tag{A10.5}$$

Note that $\hat{l}_{i-}$ has to be applied to the two electrons, 1 and 2 (indexed with subscript i), in turn and the results added. But the first term on the right is unacceptable, because both electrons have the same quantum numbers, and it must therefore be discarded. The right-hand sides of Equations (A10.4) and (A10.5) must be equal so that:

$$|3, 2\rangle = |+2^+(1), 0^+(2)\rangle$$

We proceed in exactly the same manner to obtain $|3, 1\rangle$ from $|3, 2\rangle$:

$$\hat{L}_-|3, 2\rangle = \sqrt{(10)}|3, 1\rangle$$

and

$$\sum_i \hat{l}_{i-}|+2^+(1), 0^+(2)\rangle = 2|+1^+(1), 0^+(2)\rangle + \sqrt{6}|+2^+(1), -1^+(2)\rangle$$

so that:

$$|3, 1\rangle = \sqrt{(2/5)}|+1^+(1), 0^+(2)\rangle + \sqrt{(3/5)}|+2^+(1), -1^+(2)\rangle$$

Note that the coefficients of the linear combination of microstates are such that the function is normalised. This provides a good check for errors.

Similarly, after applying the lowering operators to both sides of the last equation we find:

$$\sqrt{12}|3, 0\rangle = \sqrt{(12/5)}|0^+(1), 0^+(2)\rangle + \sqrt{(12/5)}|+1^+(1), -1^+(2)\rangle$$
$$+ \sqrt{(12/5)}|+1^+(1), -1^+(2)\rangle + \sqrt{(12/5)}|+2^+(1), -2^+(2)\rangle$$

Rejecting the first term on the right because it contravenes the Pauli principle (both electrons in the same orbital, $m_l = 0$, with the same spin, $m_s = +\frac{1}{2}$) we have:

$$|3, 0\rangle = (2/\sqrt{5})|+1^+(1), -1^+(2)\rangle + (1/\sqrt{5})|+2^+(1), -2^+(2)\rangle$$

The process can be continued until the remaining three components of the ^{3}F state with $M_S = +1$ have been found.

A similar procedure using $\hat{S}_-$ and $\hat{s}_-$ can be used to find the components having $M_S = 0$ and -1. As an illustration we do this for the $|3, 0, 1, +1\rangle$ wave function where we have reintroduced the S and M_S quantum numbers. Since $M_S = S = 1$ we find:

$$\hat{S}_-|3, 0, 1, +1\rangle = \sqrt{2}|3, 0, 1, 0\rangle$$

while

$$\sum_i \hat{s}_{i-}\{(2/\sqrt{5})|+1^+(1), -1^+(2)\rangle + (1/\sqrt{5})|+2^+(1), -2^+(2)\rangle\}$$

$$= (2/\sqrt{5})\{|+1^-(1), -1^+(2)\rangle + |+1^+(1), -1^-(2)\rangle\}$$

$$+ (1/\sqrt{5})\{|+2^-(1), -2^+(2)\rangle + |+2^+(1), -2^-(2)\rangle\}$$

Therefore:

$$|3, 0, 1, 0\rangle = \sqrt{(2/5)}\{|+1^-(1), -1^+(2)\rangle + |+1^+(1), -1^-(2)\rangle\}$$

$$+ (1/\sqrt{10})\{|+2^-(1), -2^+(2)\rangle + |+2^+(1), -2^-(2)\rangle\}$$

Continuing, and noting that an electron spin denoted by a superscript – must be annihilated by $\hat{s}_-$, we find:

$$|3, 0, 1, -1\rangle = (2/\sqrt{5})|+1^-(1), -1^-(2)\rangle + (1/\sqrt{5})|+2^-(1), -2^-(2)\rangle$$

The method described above is ideally suited to problems involving a small number of electrons, but for numbers of electrons larger than three the method is very tedious to apply. For these cases more direct approaches, based on advanced angular momentum theory, are available.

Index

Note: Figures and Tables are indicated by *italic page numbers*, Boxes by **emboldened numbers**, footnotes by suffix 'n'

ab initio calculation of IR spectra 316–17
ab initio MO calculations 167, 317
ab initio VB calculations 167
absolute zero (of temperature) 4–5
absorbance 249, **257**
absorption coefficient **257**
absorption spectroscopy, transitions in 228–9
absorptivity 14
alkali metals, properties explained 107
allowed electronic transition 236–7
alternant hydrocarbons 361
ammonia, bond angle *164*, 165
Amontons, Guillaume 3, 4
angular correlation 341
angular functions (wave functions) of hydrogen
    atom 99–100
    polar diagrams 100–4
angular integral(s)
    calculation of **204**
    values for octahedral complex *186*
angular momentum **31**, 73–93
    addition of 79, 245
    in classical mechanics 73–5
    conservation of 75, 79, 227, 243
    definition 74
    of electron on ring 50–2
    and energy **31**
    examples 84–6
    and magnetic dipole moment 367
    notation 82, 83
    observing transitions 81
    of photon 115, 225
    quantum-mechanical
        addition of 79–80
        conservation of 80
        laws 81
    selection rules 115–16

vector model 77–8
    as vector quantity 75–6
    z components 81, 84
    see also orbital angular momentum;
        spin angular momentum; total
        angular momentum
angular momentum eigenfunctions, Dirac
    notation for 82
angular momentum operator(s)
    commutation rules **87–92**
    eigenfunctions *52*, 54, 82
    for electron on ring 50, 60
angular momentum quantum number 79, 111
anharmonic oscillator 302
anharmonicity 309
anharmonicity constant 303
    carbon dioxide 314–15
annihilation operator **92**
    see also lowering operator
anomalous Zeeman effect 120
antibonding molecular orbital(s), hydrogen
    molecule 145–6
antiferromagnetism 365, 374
Argand diagram *433*
aromatic hydrocarbons
    bonding in 160
    delocalisation energies 359–60
    π-electron systems in 45, 160, 329
    see also benzene
aromaticity 44–6
Arrhenius equation 7
atom
    Bohr's model 24–6
    first (Lucretius) model 2
    Rutherford's model 24
atomic emission spectra 23–6
atomic nuclei, magnetic properties 262–4

atomic orbital overlap 146–9
  δ (delta) overlap 148
  non-bonding overlap 148–9
  π (pi) overlap 147–8
  σ (sigma) overlap 147
atomic orbitals
  and electronic spectroscopy 328
  linear combination of 200, 353
atomic spectroscopy, selection rules in
    115–17, 437, 438
atomic transition probabilities **125–9**
atomic units 401–2
atomic wave functions, classification of 117
aufbau principle 46
  applications 107, 150
Avogadro constant 401

Balmer equation for hydrogen atom emission
    spectrum 25, **33**, **121**
  Schrödinger's interpretation 97
Balmer series (hydrogen atom emission
    spectrum of hydrogen atom) 25, 97, **121**,
    122
band gap(s)
  formation by periodic potential *386*
  insulators 388
  semiconductors 388–9
band theory of solids 378–90
  electron–gas (free-electron) approximation
    381–6
    applications *387*
  tight binding approximation 378–81
    applications *380, 381, 387*
barium, $^1S_0$-to-$^1P_1$ transition 118, *119*
basis functions 404–5, 411
  linear combinations of 405
Becquerel, J. 182
Beer–Lambert–Bouguer law 249–50, **256–8**
Beer law **257**
bent molecules *166*, 167, *167*
benzene
  'breathing' vibration 303
  delocalisation energy, compared with
    enthalpy of combustion 359–60, *359*,
    391
  Kekulé structures, enthalpies/heats of
    combustion 359–60, **391**
  molecular orbitals 341–2
  π-electron energy levels 342–4
  π-electron spectrum 341–4
  protons **281**
beryllium atom, electron repulsion integrals
    175
Bethe, Hans Albrecht 182
black body, meaning of term 14
black-body radiation 13–18

emission spectra *14*
  interpretation by Planck's hypothesis
    16–18
  Planck's hypothesis 15–16
  Rayleigh–Jeans equation *14*, 15
  Stefan's law 13
blocking out of matrices 385, 414
Bohr–Einstein condition (selection rule) 115,
    226–7, 232, 248
  in NMR spectroscopy 264, 266
Bohr–Einstein equation 24, 86, 224, **258**
Bohr magneton 368, 401
Bohr magneton number **396**
Bohr model of hydrogen atom 24–6, **30**,
    31–3, 78, 223, 226, 366
Bohr radius 402
Bohr theory of hydrogen atom structure 24–6
  experimental agreement 26, **33**
  further development of 26
Boltzmann, Ludwig Edward 7, 13
Boltzmann constant 15, 401
Boltzmann law 248
bond angles
  carbon compounds 156
  hydrides *164*, 165
bond order 361
  butadiene *361*
bond pairs (of electrons) 165
  in various molecules *166–167*
bonding models
  hybrid orbital approach 153–62
  molecular orbital theory 139
  valence bond theory 137–9
bonding molecular orbital(s), hydrogen
    molecule 144–5
Born–Haber cycle 132
Born–Oppenheimer approximation 136–7
  applications 234, 297
  basis 136, **253**
  limitations 144
Born's interpretation of eigenfunctions 41–4,
    51
boron isotopes, nuclear spin quantum numbers
    84
Bouguer–Lambert law **257**
  see also Beer–Lambert–Bouguer law
Boyle's law 3
bra–ket notation **205**, 230
  see also kets
Brackett series (emission spectrum of hydrogen
    atom) **121**
'breathing' vibration (of benzene molecule)
    303
bridge analogy (marching soldiers) 238
Brillouin functions 371
bromochloromethane
  IR spectrum 306–7

compared with dibromomethane *307*
4-bromo-2,6-dimethylphenol, NMR spectrum *268*
Bunsen, Robert Wilhelm 23
butadiene
  charge densities *361*
  Hückel orbitals *356*
  π-electron charge densities *361*
  trans isomer *354*
  carbon
  electron configuration 156
  hybridisation of atomic orbitals 156–61
    sp hybrid orbitals *159*, 160
    sp^2 hybrid orbitals *159*, 160
    sp^3 hybrid orbitals 158, *158*

carbon–carbon atomic orbital overlap *161*
carbon compounds, bond angles 156
carbon dioxide molecule
  angle-bending vibration *310*, 311
    effect of Coriolis interaction 315
  asymmetric bond stretching vibration 244, *310*
    effect of Coriolis interaction 312–15
  C=O stretching vibrations 244, 310, *310*, 312
  Coriolis interaction 312–16
  effect of Fermi resonance 310–11, *316*
  geometry 166
  symmetric stretching vibration 244, 310, *310*, 312
  vibrational spectrum 309–11
carbonyl compounds, correlation of C=O bond order with carbonyl stretching frequencies *363*
Cardano, Girolamo 434
Cartesian coordinates, relationship with polar coordinates *431*
centre of charge, determination of 435–6
characteristic equation of matrix 413–14
charge correlation energy 339
  listed for $1s^1 2s^1$ states of helium atom *338*
charge densities 361
  butadiene *361*
charge distribution 153
charge transfer complex, energy curves for *332*
charge transfer (CT) spectra 330–2
charge transfer to solvent (CTTS) transitions 331
charge transfer transitions 196–7, 330–1
charge unit 402
Charles, Jacques 3
chemical bonding theory, questions to be answered 132

chemical shift (in NMR spectroscopy) 267–70
  delta (δ) scale 268–70
    listed for various species *269*
chemically equivalent protons **281**
circle, motion in **30–1**
circularly polarised radiation 221
  left-handed (LCP) 221, 225
  in NMR spectrometer **280**
  propagation of *221*
  right-handed (RCP) 221, 225
Clar, E. 344
classical mechanics
  angular momentum in 73–5
  applicability 12
  development of 9–12
  failure 12–13
  quantisation in 18–20
Clebsch–Gordan coefficients 80
Clebsch–Gordan series 79, **92–3**, 223
closed shells 109
coloured complexes, explanations 188, 196
coloured metals, explanation 390
combination bands (in IR spectroscopy) 303
commutation of operators **34**, 57–8, **68**
  angular momentum operators **87–8**
complete orbital wave functions 104
complex conjugates 51, 230, 434
complex numbers 433–4
Compton, Arthur Holly, on scattering of X-rays by electrons 225, **250–1**
Compton wavelength **251**
computational developments 167–8
conceptual frameworks 2
Condon, Edward Uhler 236, **253**
  see also Franck–Condon principle
conduction band 388
configuration interaction 143
conjugated polyenes, electronic spectra **65–7**, 329
conservation of energy 115, 227, **251**
conservation of mass 224
conservation of momentum 75, 79, 227, 243, **251**
Coolidge, A.S. 143
Coriolis interaction 311–12
  in carbon dioxide molecule 312–16
  and Fermi resonance 315–16
correlation diagram, for d^2 energy levels in octahedral field *191*, 192
cosine function **62**
Coulomb integral **174**, **206**, 337, 375
Coulombic attraction 331
Coulombic repulsion **174**, **206**, 337
Coulson, Charles Alfred 132
coupling, of angular momenta 79–80, *80*
covalency, and ligand field theory 199–203

covalent chemical bond(s)   131–79
  magnetic moments affected by   368
covalent molecules
  spectra   329–30
    n→π^* transitions   330
    π→π^* transitions   329–30
    transition-metal complexes   196–7, 330
creation operator   **92**
  see also raising operator
crystal field (CF) theory   182–7
  compared with molecular orbital theory
    200
  transition-metal complexes   187–96
    inter-electron repulsion in   188, 194
    spin–orbit coupling in   195–6
cubic symmetry, d-electron energy levels   *188*
Curie constant   365, **396**
Curie law   364–5
Curie temperature   365
  for iron   374
Curie–Weiss law   365, 372
cyanine dyes, electronic spectra   **65–7**, 423–4

δ (delta) overlap   148
d^4 complex, in octahedral field   197
d^2 configuration
  energy levels for weak, strong and
    intermediate octahedral fields   191–3
  ^3F states, application of shift operators to
    determine wave functions for   440–1
  microstates   439
  strong-field coupling scheme in octahedral
    field   193–5
  weak-field coupling schemes in octahedral
    field   189–91
d→d transitions   196
d-electron matrix element   184
d-functions   426
d-orbitals   100
  angular part of wave function   103–4
  radial part of wave function   *99*
^1D state (singlet D)   109
  energies for various atoms   *110*
^2D state (doublet D)   108
Davisson, Clinton Joseph   27
de Broglie, Louis   26–8
degenerate eigenvalues, in unperturbed solution
  421
degenerate energy levels   44
degrees of freedom   290
delocalisation energy (DLE)   359
  benzene, compared with enthalpy of
    combustion   *359*, **391**
delocalised electrons, in hydrocarbons   160
density functional theory (DFT)   168, 181

density of states, transition metals   381
descriptive models   2
diagonalisation of matrix   412–13
diamagnetism   364
diamond
  band theory applied   387, *387*
  mechanical properties   390
diatomic molecule
  dumb-bell model   18, 20, 291, *292*
    comparison with real molecule   302–3
    rotation of   **319–20**
    vibration of   291, *292*, **318–19**
  Morse potential energy curve   302, *303*
  rotation of   294–7
dibromomethane, IR spectrum, compared with
  dibromochloromethane   *307*
Dickinson, W.C.   267
dipole–dipole interaction, compared with
  electron–nucleus interaction 200n,   272
dipole moments   435–7
Dirac, Paul   **69**, 78
  notation for angular momentum
    eigenfunctions   82
double-zeta functions   **173**
doublet states   **281–2**
dumb-bell oscillator   18
  as model for diatomic molecule   18, 20,
    291, *292*, **318–20**
    comparison with real molecule   302–3
    rotation of   **319–20**
    vibration of   291, *292*, **318–19**
  potential energy characteristics   19
Dwingeloo 1 galaxy   86

effective nuclear charge   **172**
eigenfunction(s)   29, 39
  Born's interpretation   41–4
  combinations of   58
  determination of
    by matrix algebra   411–15
    by solution of secular equations   406–10
  of different operators   52–3
  and experimental measurements   53–4
  of hydrogen atom   29
  normalisation of   46, **64**
  orthogonality   46, **65**
  physically acceptable   43–4, 47–8
eigenvalue problem, general solution of 2 x 2
  problem   414–15
eigenvalue(s)   29, 39
  angular momentum of electron on ring   *51*
  and experimental measurements   53
  of hydrogen atom   29, 96–7
Einstein, Albert   8, 13
  on light-quanta (photons)   21, 223, 224,
    226

on photoelectric effect 20, 21, 223
   experimental confirmation 22, 223
  relativity theory 13
  see also Bohr–Einstein...
Einstein coefficients 250, **258–9**
electric current, units of measurement 7, 216
electric dipole interaction energy 234
electric dipole radiation 222
  photons
    angular momentum 115, 225
    selection rules 225–6
electric dipole transitions, conservation of
    angular momentum 242
electric quadrupole radiation 222
  photons, selection rules 226
electric quadrupole transitions 116
  conservation of angular momentum 243
electrical anharmonicity 309
electrical conductivity, temperature effects
   388
electromagnetic radiation 216–19
  electric field component 217–19
  forms 222
  magnetic field component 219
  Maxwell's equations 7, 216, 217–19
  see also light
electromagnetic spectrum *217*, 222
electromagnetic units (emu) 7, 216
  ratio to electrostatic units 7, 216
electron
  in Bohr's model of atom 24
  charge on 22
  magnetic properties *262*
  mass at rest 401, 402
  particle–wave duality 27–8
  probability distribution for 42, 51
electron affinity, in charge transfer absorption
   band 331
electron configuration(s) 108
  carbon atom 156
  oxygen molecule 164
electron correlation 106–7
  excited helium atom 341
  hydrogen molecule 143–5
electron g-factor 118, 408
electron in linear box 46–8
  angular momentum eigenvalues *51*
  boundary conditions 48
  eigenfunctions for 47–8
    normalisation of **64**
    orthogonality **65**
  Hamiltonian operator for 46–7
  linear momentum 49–50
electron-mediated coupling 272
electron–neutron interaction 85–7, 200n, 272
electron–nucleus attraction, in helium atom
   332

electron paramagnetic resonance (EPR)
  spectroscopy 199, 261
electron repulsion energy
  singlet state **174**, 337–8
  triplet state **174**, 336–7
electron repulsion integrals **205–6**, 337
  combination for ^{3}P and ^{3}F states of d^2
   190, **207–8**
  listed for d orbitals
   in complex forms **210**
   in real forms **209**
electron on ring 40–4
  acceptable eigenfunctions for 41–4
  angular momentum 50–2
   calculation of expectation value **67–8**
  energy levels for *45*
  Hamiltonian operator for 40–1
  Hückel's rule for aromatic hydrocarbons
   44–6
electron spin 105
  magnetic properties affected by 78, 369
  transition probability affected by 344–6
electron spin correlation 173–5
electron spin quantum number 78
electron spin resonance (ESR) spectroscopy,
  interactions studied 261
electron transition density, $\pi \rightarrow \pi^*$ transition in
  benzene *347*
electron volt 248
electronic energy levels, transition-metal
  complexes 187–96
electronic orbital motion, magnetic effect of
  366–8
electronic spectroscopy 327–48
  applications 328
  and atomic orbitals 328
  classification of spectra 46, 344
  and molecular orbitals 328
  selection rules 344–8
  timescales 246, *246*
  transition-metal complexes 196–7
  see also UV–VIS spectroscopy
electronic states of atom 107–11
electronic transition dipole moment 236
electronic transitions
  allowed 236–7
  compared with IR and NMR transitions
   241
  forbidden 237
  hydrogen atom (1s→2p transition) 238–9
electrostatic field due to ligands 182
  see also crystal field theory
electrostatic units (esu) 7, 216
  ratio to electromagnetic units 7, 216
elementary charge (electron/proton) 401, 402
emission spectra 23–6
emissivity 13

energy
  conservation of   115, 227
  units   248, 402
energy-conversion factors   *249*
energy level diagram(s)
  atomic electronic spectroscopy, sodium D
      lines   242–3
  infrared spectroscopy   *296*
  NMR spectroscopy, first-order spectrum
      275
  π-electron energy levels for butadiene   *357*
  three interacting magnetic centres   *378*
energy operator
  eigenfunctions of   *52*
  for electron in linear box   46–7
  for electron on ring   40–1
    eigenfunctions   *52*
equilibrium internuclear distance   133
  hydrogen chloride   **322**
equivalent nuclei   **280–7**
ethene   *159*
  bonding in   158, 160
  electronic spectroscopy   243–4
  Hückel orbitals   243
ethyne   *160*
  bonding in   160
exchange integral   **175**, **206**, 337, 375
expectation value   54
experiment design   2–3
experimental measurements
  effect on system   55–7
  and eigenfunctions/eigenvalues   53
exponential function   **62**
extinction coefficient   **257**
  see also absorption coefficient

f-orbitals   100
  angular part of wave function   104
^3F state (triplet F), of d^2 in octahedral field
  calculation of matrix elements of
      crystal-field Hamiltonian   **204–5**
  weak-field coupling scheme   189–90
    with ^3P state   190–1
Faraday, Michael   7, 216
Fermi contact interaction 200n,   272, 408
Fermi level   380, *381*
Fermi resonance   309–11
  and Coriolis interaction   315–16
Fermi resonance interaction term   310, 316
ferromagnetism   365, 374
fine structure
  IR spectra   297, 300
  NMR spectra   200
fingerprint (in IR absorption spectrum)   309
Fizeau, Armand   216
Fleming's left-hand rule   **392**

Fock, Vladimir A.   106
forbidden electronic transition   237
force constant   291
  angle-bending, for small molecules   *306*
  bond-stretching, for small molecules   *306*
Foucault, Jean   216
Fourier transform IR spectroscopy   244
Fourier transform NMR spectroscopy   278
Franck–Condon principle   236–7, **253–6**
free induction decay (NMR spectroscopy)
    278
frontier orbitals *see* highest occupied molecular
    orbital; lowest unoccupied molecular
    orbital
fundamental constants (listed)   401

g-factor *see* electron g-factor; orbital g-factor
gas equations
  pressure–volume   3
  pressure–volume–temperature (for 'ideal'
      gases)   5
  temperature–volume   3–4, *4*
Gay-Lussac's law   3–4
German nouns, construction of   29
Germer, Lester Halbert   27
Gillespie, R.J.   164
'good' quantum numbers   80, 81
Goudsmit, Samuel Abraham   78, 105, 118,
    120
group theory   182
gyroscope   75, 265

Haber, Fritz   132
  see also Born–Haber cycle
Hamiltonian formulation   12
Hamiltonian operator   40
  for crystal field   188
    calculation of matrix elements   **204–5**
    matrix   189
  for electron in box   46–7
  for electron on ring   40–1
  for harmonic oscillator   293
  for hydrogen molecule   134–6
  for many-electron atom   105–6
  for nuclear spin system   273, **282**
  for rotation of diatomic molecule   **321**
  for vibration of diatomic molecule   **319**
harmonic oscillator   291
  compared with real diatomic molecule
      02–3
  eigenfunctions   294, *295*, 312, **323–4**
  energy–distance curves   *19*, *292*
  quantisation of energy   293–4
  quantum mechanics first applied   290
Hartree–Fock functions   106

Hartree (unit of energy)   402
Heisenberg, Karl Werner, matrix mechanics   26
Heisenberg uncertainty principle   56–7, 246
  effects   58, 294
Heisenberg uncertainty relationships   56, **68**
Heitler, Walter   132, 137
Heitler and London's calculations for energy of covalent bond   132, 137–8
helium atom
  ground state ($1s^2$) configuration, wave function   332–3, 344–5
  $1s^1 2s^1$ configuration   333–41
    electron repulsion in triplet and singlet states   339–41
    energies of $1s \to 2s$ excited states   335–9
    energies of singlet and triplet states   338–9
    magnetic interactions   375–6
    one-electron energies   335–6
    total wave function   334
    two-electron (repulsion) energy   336–8, 374–5
helium–helium bond   146
Hermite polynomials   **323**
Herschel, (Sir) William   217
Hertz, Heinrich   7, 20, 217
heteronuclear diatomic molecules, molecular orbital theory applied   151–3
high-spin complexes   197
highest occupied molecular orbital (HOMO)   357
  energy difference between HOMO and LUMO, correlation with spectral band energies   329, 358, *358*
Hoffman, Roald, extended Hückel MO theory   203
homonuclear diatomic molecules, molecular orbital theory applied   149–51
Hooke's law   18, 290–1
hot bands   302
Hückel molecular orbital (HMO) theory   352–63
  applications of HMO coefficients   361–2
  applications of HMO energies   357–60
  assumptions   354
  basis   352–3
  delocalisation energies   359
  determination of HMO energies   354–5
  Hoffman's extended version   203
  HOMO and LUMO energies calculated using   329, 357
  LUMO–HOMO energy difference   358
    correlation with energy of *para* band of aromatic hydrocarbons   *359*
  method   353–4

π-electron energy calculations   329, 358–60
π-electron structure of benzene   341
Hückel orbitals
  butadiene   *356*
  ethene   243
Hückel's ($4N + 2$) rule   44–6
Hund–Mulliken bonding model   139
  and electronic spectroscopy   328
Hund–Mulliken function   139
Hund's rules   110–11
  applications   150, 156, 197, 273
  basis   **174**
  data illustrating   *110*
Huygens, Christian   216
hybrid atomic orbitals   154
hybrid orbitals, choice of   161–2
hybrid-orbital bonds, properties   162
hybridisation   153–62
  of carbon AOs   156–60, *161*
  choice of hybrid orbitals   161–2
  Pauling's approach   153–5
  and valence bond theory   156
hydrides, bond angles   *164*, 165
hydrocarbons
  molecular geometry
    type 1 (alkanes)   156, 157–8
    type 2 (alkenes and aromatic hydrocarbons)   156, 158–60
    type 3 (alkynes)   156, 160
hydrogen 1s atomic orbital(s)
  out-of-phase combination of   143, 145–6
  parity operation on   116–17
  polarisation of   140–1
  radial density function   *99*
hydrogen atom
  Bohr's model   24–6, **30**, **31–3**, 78, 223, 226, 366
  characteristic emission (1420 MHz/ 0.2111 m)   86, 410
  eigenfunctions   29
  eigenvalues   29, 96–7
  electronic energy   133
  electronic transition ($1s \to 2p$ transition)   238–9
  interactions in magnetic field   407–10
  mass   20, 38
  parallel-to-antiparallel transition   86
  radial density functions   *99*
  spectral series   **121–2**
  spin properties   85–6
  transition dipole moment   **252–3**
  wave functions   97–100, 427–8
    angular parts   99–104, *104*
    effect of polarisation   140–1
    Heitler–London (valence bond) approach   137–9

hydrogen atom (*continued*)
  Hund–Mulliken (molecular orbital)
      approach 139
  improvements 140–1
  nodes 104, *104*
  nuclear charge and 140
  radial parts 98–9, *104*
hydrogen atom emission spectrum 25, 97, **121–2**
  Lyman series 97, **121**, **122**
hydrogen chloride ($^1H–^{35}Cl$), vibration of **321–3**
hydrogen molecule
  binding energy 133–4
      Heitler and London's calculations 132, 137–8
  bond length 20, 38
  dissociation energy 133
      experimental value 133, *138*
      various calculations *138*, 143–4
  energy terms
      kinetic energy of electrons 134
      kinetic energy of nuclei 134
      potential energy due to interelectronic repulsion 135
      potential energy due to internuclear repulsion 134
      potential energy due to nucleus–electron attraction 134
  energy vs internuclear distance *136*
  equilibrium bond length 133
      experimental value 133, *138*
      various calculations *138*, 143–4
  frequency of oscillation 20, 38
  Hamiltonian operator for 134–6
  molecular orbital energy-level scheme for *146*
  normalisation of wave functions for **169–70**
  quantisation in 20
  spin–spin coupling in 271–2
hydrogen sulfide, bond angle *164*, 165
hydroxyl radical
  charge distribution in 153
  molecular orbital energy-level scheme for *152*
  molecular orbital theory applied 151–3
  O–H bond length 151
hypothesis
  meaning of term 6–7
  testing correctness 3

ideal gas constant 5
ideal gas law 5
imaginary numbers 49, 433

infinitesimal rotations theory **90**
infrared (IR) spectroscopy 289–324
  basis 290
  characteristic group frequencies 308
  compared with NMR spectroscopy 240–1
  electric dipole interaction energy 234
  experimental aspects 240
  selection rules 244, 297–302
  timescales 246, *246*
  vibrations of polyatomic molecules 305–8
infrared transitions 235–6
  compared with UV–VIS/electronic and NMR transitions 238–41
  intensities 301–2
insulators 388
  effect of temperature on electrical conductivity *388*
integration, over 'all space' 54, **63**
inter-electron repulsion
  calculation for some states of d^2 **205–8**
  in crystal field theory 188, 194
  for p^2 configuration **123–4**
  triplet/singlet states **174**
inter-electronic spin–orbit coupling 115
intermediate spin–orbit coupling 113–14
intermolecular charge transfer 330–1
intramolecular charge transfer 330
inverse distance, expressed in terms of spherical harmonics 183
ionic contributions, in covalent bonding models 142
ionic solids
  mechanical properties 390
  tight-binding version of band theory used 387
ionisation of atom 97
ionisation energy, in charge transfer absorption band 331
IR spectra
  ab initio calculations 316–17
  bromochloromethane compared with dibromomethane *307*
  combination bands 303
  fingerprint 309
  overtones 303
IR spectroscopy *see* infrared (IR) spectroscopy
iron
  effective nuclear charge for occupied orbitals **172**
  ferromagnetism 374
irrational numbers 433
isochronous protons **281**

Jahn–Teller effect 199
James, H.M. 143

Jeans, James Hopwood   15
jj coupling   112–13

Kelvin scale (of temperature)   4
Kepler's laws of planetary motion   10–11
keto–enol equilibrium   245, 247
kets   189, 439
    see also bra–ket notation
kinetic energy operator   60
kinetic theory of gases   7
Kirchhoff, Gustav   7, 14, 23, 216

l-type doubling   315
ladder operators   82–3
Lagrangian formulation   12
Landé formula   120
Landé interval rule   195
Langevin relationship   366, **395**
Laporte (selection) rule   117, 196, 437
Larmor frequency   265, 266
Larmor precession   265, *266*
laws
    deduction from experimentation   5–6
    meaning of term   6
lead atom, eigenstates   **128**
Leibnitz, Gottfried Wilhelm   434
Lenard, Philipp Eduard Anton   20
length units   402
Lennard-Jones, J.   **175**
Lewis, Gilbert Newton   223
ligand field (LF) theory   181
    and covalency   199–203
    historical development   182
ligand-to-metal charge transfer (LMCT)   196,
    330
light
    corpuscular model   216
    Einstein's quantum model   22, 223
    velocity of   223, 401
        factors affecting   216
        as limiting velocity   13
        ratio of electrostatic to electromagnetic
            units   7, 216
    wave theory   216
        limitations   21, 223
    see also electromagnetic radiation
linear combination of atomic orbitals (LCAO)
    approach   200, 353
linear molecules   166, *166*, *167*
linear momentum
    conservation of   **251**
    definition   73–4
    of electron in linear box   49–50
    of photon   224–5
linear momentum operator   49, 60

linearly polarised radiation   220
    as sum of left and right circularly polarised
        radiation   **280**
Linnett, J.W.   161
London, Fritz   132, 137
    see also Heitler and London's
        calculations
lone pairs (of electrons)   165
    in coordination complexes   202
    in various molecules   *166*, *167*
lone-pair electrons, n→π transitions   330
low-spin complexes   197
lowering operator   82–3, **89, 90–1, 92**
    applications   **279, 281, 282, 283**
lowest unoccupied molecular orbital (LUMO)
    357
    energy difference between HOMO and
        LUMO, correlation with spectral band
        energies   329, 358, *358*
LS coupling   111–12
Lucretius   2
Lummer, Otto Richard   14
Lyman series (emission spectrum of hydrogen
    atom)   97, **121, 122**

magnetic dipole
    definition   366, **392**
    orientation in magnetic field   **392–3**
magnetic dipole–dipole interaction   374
magnetic dipole interaction energy   234, 374
magnetic dipole radiation   222
    photons, selection rules   225–6
magnetic dipole transitions   116
magnetic field   364, 365–6
    polarised light affected by   7, 118, *119*
    spectral lines affected by   118–20
magnetic interactions
    helium excited state   374–6
    three-centre   377–8
magnetic moment(s)
    for atomic nuclei   262–3
    formulae for   197–8
    listed for various transition-metal ions   *198*
    and magnetic susceptibility   364, **393–7**
magnetic properties
    atomic nuclei   262–4
    electrons   *262*
    protons   *262*
magnetic susceptibility   364–5
    relationship with atomic/molecular magnetic
        moment   **393–7**
    see also volume magnetic susceptibility
magnetically equivalent nuclei   **281**
magnetisation   364
magnetism   363–78
    effect of electron spin   369

magnetism (*continued*)
  origins **392**
  practical applications 370–3
  and spin angular momentum 78, 369
  transition-metal complexes 197–9
  see also nuclear magnetic resonance
    (NMR) spectroscopy
magnetogyric ratio 263
many-electron atom(s) 105–7
  electronic structure 112
  quantum numbers for 108
  selection rules for dipolar radiation
    115–16
many-electron wave functions 332–3
mass
  conservation of 224
  reduced 291–2, **319**
  unit 402
mathematical models 2
matrix diagonalisation 412–13
matrix elements 411–13
matrix mechanics 26
Maxwell's equations 7, 216
mean value
  of energy 137
  of observable 53
measurement, effect on system 55–7, **70**
mechanical anharmonicity 309
mechanical properties, solids 390
metal-surface selection rule 244–5, 306
metal-to-ligand charge transfer (MLCT)
  196–7, 330
metals
  effect of temperature on electrical
    conductivity *388*, 389
  theory of
    free-electron approximation 381–6
    tight-binding approximation 378–81
    Wigner and Seitz's remarks 135, 387
methane
  bonding in 157–8
  magnetically equivalent protons **281**
microstates 109
  for d^2 configuration 439
  for p^2 configuration *109*
  as Slater determinants **122, 207**
Millennium Bridge, London 238
Millikan, Robert Andrews 22
  experiments on Einstein's photon hypothesis
    23, 223
models, categories 2
molecular geometry 163–7
  VSEPR model used 164–7
molecular magnets
  interactions 374–6
  orientation in magnetic field **392–3**
molecular orbital energy-level schemes

hydrogen molecule *146*
hydroxyl radical *152*
  using sp hybrid oxygen orbitals *155*
lines joining AOs and MOs 153
nitrogen molecule *150*
octahedral MH_6 molecule (hypothetical)
  202
oxygen molecule *150*
molecular orbital (MO) theory 139
  compared with crystal field theory 200
  compared with VB theory 156, 163
  computational developments 167
  heteronuclear diatomic molecules 151–3
  homonuclear diatomic molecules 149–51
  unification with valence bond theory
    141–3
molecular orbital (MO) wave functions,
  normalisation of **170**
molecular orbitals
  bonding and antibonding 145–6
  and electronic spectroscopy 328
  imaging of 168
molecular solids
  mechanical properties 390
  tight-binding version of band theory used
    387
moment of inertia 296
momentum *see* angular momentum; linear
  momentum
Morse potential energy curve 302, *303*
Mulliken, Robert Sanderson 139, 331
  see also Hund–Mulliken bonding model
Mulliken population analysis 153
Mulliken wave functions 331
multiple bonds, VSEPR theory 165–7

n-type semiconductor *388*, 389
natural phenomena, observation of 1–2
near-infrared (NIR) spectroscopy 317
near-infrared spectrum *217*
  first discovered 217
nebulae, emission spectra 243
Néel temperature 365
Newton's corpuscular view of light 216
Newton's laws of motion 6, 7, 11–12
nitrogen molecule
  bonding in 150
  molecular orbitals *150*
    imaging of highest occupied MO 168
NMR tomography 267
NMR transitions 264
  compared with UV–VIS and IR transitions
    238–41, 277
  selection rule 266, **278–80**
non-bonding (atomic orbital) overlap 148–9
normal mode of vibration 304

normalisation
  of eigenfunctions   46, **63–64**, 425
  of wave functions
    in Hückel MO calculation   356
    for hydrogen molecule   **169–70**
notation, angular momentum   83
nuclear charge
  effective   **172**
  in hydrogen molecule   140
nuclear g-factor   408
nuclear magnetic resonance (NMR) imaging,
    medical applications   267
nuclear magnetic resonance (NMR)
    spectroscopy   261–87
  A_2 spectrum   275–6
  A_3 spectrum   **285–7**
  ABX spectrum   **285**
  AX spectrum   274–5
  chemical shift   267–70
    delta (δ) scale   268–70
    shielding constant   270
    standard used   269
  early discoveries   261
  equivalent nuclei   **280–7**
  experimental aspects   240, 270
  first order spectra   274–5
  free induction decay   278
  frequency region   264
  intensities of NMR spectral lines   276–7
  interactions studied   261
  magnetic dipole interaction energy   234
  and quantum mechanics   277–8
  relaxation processes   276
  second order spectra   275–6
  signal saturation   276
  timescales   246, *246*
nuclear magneton   401
nuclear spin angular momentum quantum
    number(s)   83
  boron isotopes   84
nuclear spin–spin coupling   270–3
nuclear spin system, energy levels   273–6
Nyholm, R.S.   164

observables   38
observation of natural phenomena   1–2
  prediction based on   2
octahedral complex
  angular integrals listed for   *186*
  Brillouin functions for single d electron
    371, 372
octahedral crystal field
  calculation of matrix elements   182–7
  results   186–7
octahedral MH_6 molecule (hypothetical)   *200*
  atomic orbital overlap in   *201*

molecular orbital energy-level scheme
    202
molecular orbital theory applied   200–2
octahedral species   *166–7*
octahedral symmetry, d-electron energy levels
    188
one-dimensional box
  electron in
    acceptable eigenfunctions   47–8
    boundary conditions   48
    Hamiltonian operator for   46–7
    linear momentum   49–50
  particle in   421–2
  perturbation theory applied   422–4
one-dimensional model
  in band theory   381–6
  see also electron in linear box
one-electron atoms   104–5
  selection rules for dipolar radiation   115
  see also hydrogen atom
one-electron energies, excited states of helium
    atom   335–6
operands, meaning of term   **33**
operators   **33–4**, 39
  angular momentum operator   50, 60
  commutation/non-commutation of   **34**,
    57–8, **68**
  eigenfunctions of different   52–3
  kinetic energy operator   60
  linear momentum operator   49, 60
  position operator   59
  potential energy operator   59–60
Oppenheimer, J. Robert   136
  see also Born–Oppenheimer
    approximation
optical density   **257**
  see also absorbance
optical properties, solids   390
orbital angular momentum   76–8, **87**
  causes   78
  notation   83
orbital angular momentum quantum number
  of rotating diatomic molecule   83
  of single electron   83–4, 108
orbital g-factor   368
orbital hybridisation   153–62
orbital overlap   146–9
Orgel diagram   *191*
  see also Tanabe–Sugano diagram
orthogonality of eigenfunctions   46, **63–5**
oscillating electric field component of light
    217–19
oscillating magnetic field component of light
    219
overlap integrals   **169**
  in Hückel MO theory   356
overtones (in IR absorption spectra)   303

oxygen
  Schumann–Runge bands **256**
  sp hybrid orbital *154*
oxygen molecule
  bonding in 150–1
  electron configuration 164

π bonding, in hydrocarbons 160
π-electron spectra, benzene 341–4
π-electron systems
  in aromatic hydrocarbons 45, 341
  in conjugated polyenes **65**
π (pi) overlap 147–8
$\pi \rightarrow \pi^*$ transitions 329–30
  in benzene, electron transition density *347*
p^2 configuration, terms determined for 109–10
p^2 electron configuration
  energies **122–5**
  spin–orbit coupling in *114*
  terms determined for 109–10
p-orbitals 100
  angular part of wave function 101–3
  radial part of wave function *99*
^3P state (triplet P) 110–111
  energies for various atoms *110*
  weak-field coupling scheme in octahedral field 190–1
^5P state (quintet P) 108
p^2-to-s^1p^1 transition **126–7**
p-functions 426
p-type semiconductor *388*, 389
pairing energies 197
paramagnetism 364
  effect of temperature 364–5
  temperature-independent 365
parity 423
  atomic orbitals 116–17
  in crystal field theory 184
  photons 225
parity selection rule 117, 225
Paschen series (emission spectrum of hydrogen atom) **121**
Pauli, Wolfgang 78, 261
Pauli exclusion principle 45, 78, 105, 138, 333
  applications 107, 113, 150, 440
  effects on valence electron distribution 161–2, **173**
Pauling, Linus Carl 153, 182
2,4-pentane dione, keto–enol equilibrium 245, 247
periodic boundary conditions 48, 382
periodic potential, band gap formed by 382–6
periodic table, quantum-mechanical view 107, 114

perturbation theory 417–24
  examples of 112, 382, 421–4
Pfund series (emission spectrum of hydrogen atom) **121**
phase angle 291
photochemical reactions 146, 338
photoelectric effect 20–3, 223
  Einstein's theory 21
  experimental confirmation 22
photon counting **69**
photon(s) 21, 223
  angular momentum 115, 225
  detection of 242
  energy 224
  linear momentum 224–5
  mass 224
  parity 225
  properties 223–6
  release of electrons caused by 20–3, 55
  term first used 223
  velocity 223–4
photosynthesis 146
physical laws 5–6
Planck, Max 8, 15
  black-body radiator hypothesis 13, 15–16, 24
Planck's constant 15, 24, 401
  experimental determination of 22
plane polarised light 220
  directional interaction 306
  resolution into x and y components *220*
planetary motion 10–11
Platt, J.R. 46, 344
polar coordinates 431
  relationship with Cartesian coordinates *431*
  Schrödinger equation expressed in 97–8, 303–4
  volume element in 431–2
polar diagrams
  d-orbital angular functions 103–4
  p-orbital angular functions 101–3
  s-orbital angular function 101
polarisation
  of atomic orbitals 140–1
  as observable property **70**
polarised light 219–21
  circularly polarised light 221
  effect of magnetic field 7, 118, *119*
  linearly polarised light 220
  and quantum mechanics **69–71**
polyatomic molecules, vibrations 303–9
polynomial, roots of, for Hückel molecular orbital calculation 355
Pople, J.A. **175**
position operator 59
postulate, meaning of term 6–7

potential-energy curves, diatomic molecules **253–255**
potential-energy operator 59–60
Powell, H.M. 164
prediction, approaches to 2
principal quantum number 96
principle, meaning of term 6
Pringsheim, Ernst 14
probability, determination in quantum mechanics 232
Proctor, W.G. 267
progression (of spectral bands) 238
propyl radical, Hückel matrix for 412–13
proteins, NMR spectroscopy 270
proton
  magnetic properties *262*
  mass at rest 401
proton NMR transitions
  compared with electronic and infrared transitions *241*
  see also nuclear magnetic resonance (NMR) spectroscopy

quanta/quantum, term first coined 16
quantisation of energy 16
  in classical mechanics 18–20
quantum electrodynamics (QED) **92**, 247
quantum mechanical harmonic oscillator 293–4
  quantisation of energy 293–4
  vibrational eigenfunctions 294
  zero-point energy 294
quantum mechanics
  addition and conservation of angular momentum 79–80
  applications 37–72
    covalent chemical bond 131–79
    and NMR spectroscopy 277–8
    orbital angular momentum in 76–8
  progression from classical mechanics 13–34
  Schrödinger method 39–40
  of transition probability 227–33
quantum number(s) 16
  for many-electron atom 108
  of single electron 108
quarter-wave plate 221
quartet state **281**
quenching 198, 368

Racah parameters 197, **206**
  A parameter 194, **206**
  B parameter 193, **206**
  C parameter **206**
radial functions 98–9

of hydrogen atom *99*
  Slater's simplification **172**
radiation–matter interaction 216
  time-dependent interaction 230–3, 266–7
  time-independent interaction 230, 233–45, 297
radio astronomy, hydrogen emission used 86
raising operator 82–3, **89–92**
  applications **279–280, 282–3**
Raman spectroscopy 309
  carbon dioxide vibrational spectrum 310, 311
rare earth coupling scheme 188
rational numbers 433
Rayleigh–Jeans equation *14*, 15
real numbers, types 433
reduced mass 291–2, **319**
  hydrogen chloride **321**
Regnault, Charles 4
relativistic mechanics 13
relaxation processes, in NMR spectroscopy 276, 278
resonance energy 359
  see also delocalisation energy
resonance stabilisation, in charge transfer transitions 331
resonance structures 163
  benzene *163*
  carbonate anion *163*
  cyanine dyes **66**
restoring force (in simple harmonic motion) 290
rigid rotator model 297
  eigenfunctions 297
  rotational motions 298–9
  vibrational motions 299–300
Ritter, Johann Wilhelm 217
Rosen, N. 140–1
rotation of diatomic molecules **319–20**
rotational constant 296
  carbon dioxide 314–15
rotational selection rule 298–9
  semi-classical interpretation 301
Russell–Saunders coupling 111–12
Rutherford's model of atom 24
Ryberg, Johannes Robert, on Balmer equation **121**
Ryberg constant **121**
  values listed for various atoms and ions *105*

σ bonding, in hydrocarbons 160
σ (sigma) overlap 147
s-functions 426
s-orbitals 100

s-orbitals (*continued*)
 angular part of wave function 101
 radial part of wave function *99*
^{1}S states (singlet S) 110–11
 energies for various atoms *110*
saturation (magnetism) 371
SCF (self-consistent field) calculations 106
Schrödinger, Erwin 26, 28
Schrödinger equation 28–9
 in atomic units 402
 expression in polar coordinates 97–8, 303–4
  angular functions 99–100, 304
  radial function 98–9, 304
 solution of 41, 53, 96, 100, 135
  for Morse potential energy curve 302–3
  perturbation theory used 417–24
  for polyatomic molecules 304
 for system under perturbation 229
 time-dependent equation 228
 time-independent equation 228
Schrödinger method in quantum mechanics 39–40, 96
'sea' of electrons 381
secular equations 406–7
 for butadiene 355
 for propyl radical 411
 see also variation theorem
Seitz, Frederick 135, 387
selection rules 115–17, 196, 225–7, 437–8
 electronic spectroscopy 344–8
 IR spectroscopy 244, 297–302
 NMR spectroscopy 264–7, **278–80**
 see also Bohr–Einstein condition
self-consistent field 106
 see also SCF calculations
semiconductors 388–9
 effect of temperature on electrical conductivity *388*, 389
SETI (search for extraterrestrial intelligence) 86, 410
shielding constant (σ) in NMR spectroscopy 270
shift operators 82–3
 wave functions for ^{3}F states of d^2 439–41
Sidgwick, N.V. 164
similarity transformation 413
simple harmonic motion 290–3
 see also harmonic oscillator
simultaneous homogeneous equations 406
sine function **62**
singlet states 85, **173**
 electron distributions for excited helium atom 339–41
 electron repulsion energy **174**, 337–8
 lifetimes 338

space wave functions **173–4**, 333, *335*
spin wave functions **173**, 334, *335*
total energies in 1s^{1}2s^1 helium atom 338–9
Slater determinant(s) 189, 348, 429–30
 microstates as **122, 207**
Slater rules for calculation of screening 172
Slater-type orbitals (STOs) **171–3**
 orbital overlap for *161*
sodium D-lines 224, *242*
 Zeeman effect *120, 242*
sodium metal
 band structure 379, *380*
 bonding in 378–9
solids
 band theory 378–89
 electrical conductivity 387–9
 mechanical properties 390
 optical properties 390
Sommerfeld, Arnold 26, 381
sp hybrid orbitals *159*, 160
sp^2 hybrid orbitals *159*, 160
sp^3 hybrid orbitals 158, *158*
 proof of normalisation and orthogonality **171**
space functions **173–4**
space quantisation 77
spectroscopic timescales 245–7
spectroscopic transitions
 comparison of IR, NMR and UV–VIS transitions 238–41
 particle and wave views 242–5
 time-independent and time-dependent parts of wave functions 227–8, 266–7
spectroscopic units and notation 247–9
spectroscopy 215–59
 definition 216
 see also absorption...; atomic...; electronic...; infrared...; nuclear magnetic resonance spectroscopy
spectrum
 energy/frequency/wavelength axis, units used 248–9
 intensity/absorbance axis, units used 249–50
spherical harmonics 426
 applications 100, 183, 297
spherical polar coordinates 431
 see also polar coordinates
spin angular momentum 78, 261
 notation 83
 of nucleus 262
spin-forbidden transitions 197
spin functions 85, **173**
spin–orbit coupling (SOC) 81, 111–15
 in crystal field theory 195–6
 first-order **129**

inter-electronic   115
intermediate coupling   113–14
jj coupling   112–13
    matrix for microstates of p^2 configuration   **124–5**
    Russell–Saunders (*LS*) coupling   111–12
    and state energies   *114*
spin–orbit coupling constant   195
spin–orbit coupling energy   111
spin quantum number
    of nuclei   83, 262
    of single electron   83–4, 108
spin–spin coupling, in NMR spectra   270–3
spin–spin coupling energy   272
spin states
    hydrogen atom   85–6
    single electron   84
    two-electron system   85
spinning bullets/projectiles   75
square planar molecules   *166–7*
Stark effect   299
state energies, effect of spin–orbit coupling   *114*
Stefan–Boltzmann constant   13–4
Stefan's radiation law   13
strong-field coupling scheme   188
    for d^2 in octahedral field   193–5
Strutt, John William   15
sulfur dioxide, molecular geometry   166–7
super-pairs (of electrons)   166
    in various molecules   *167*
superconductors   389
superposition of states   59
surface-adsorbed species, FTIR spectroscopy   244–5
symmetry species   304, 306
synchronised stepping frequency   238

T-shaped molecules   *166*
Tanabe–Sugano diagrams   *191*, 192–3
    limitations   193
temperature-independent paramagnetism (t.i.p.)   365
term energies   110
term symbols   108
    assignment of   108–10
tetragonal antiprismatic crystal field
    states of single d electron
        energy levels   *373*
        magnetic moments   372–3, *373*
tetrahedral complexes, d→d transitions   196
tetrahedral molecules   157–8, *166–7*
tetrahedral symmetry, d-electron energy levels   *188*
tetramethylsilane (TMS), as standard in NMR spectroscopy   269
theory

at end-19th century   7–8
    role in science   1–3
thermal radiation   13
thermodynamics, laws of   6
Thomson, George Paget   27
three-body problem   106
time, atomic unit of   402
time-independent perturbation theory   417–21
tin atom, eigenstates   **128**
total angular momentum   78–81
transition dipole moment   234–8, 437
    in hydrogen atom   **252–3**
transition-metal complexes
    electronic energy levels   187–96
    electronic spectroscopy   196–7, 330
    magnetism   197–9
transition metals
    band structure   *381*
    bonding in   379–81
    density of states   381
    properties explained   107
transition moment (TM) integral, evaluation of   **126–7**
transition moment operator   **125**
transition probability
    effect of electron spin   344–6
    nuclear magnetic resonance   **278–9**
    quantum mechanics   227–33
        confirmation of method   239
    spatial aspects for allowed electronic transition   346–7
    vibrational factor in   347–8
1,1,2-trichloroethene
    ^{1}H NMR spectrum   270–1, *271*
    spin–spin coupling in   272–3
trigonal bipyramidal molecules   *166–167*
trigonal molecules   *166–167*
trigonal pyramidal molecules   *166*
triplet states   85, **173**
    electron distributions for excited helium atom   339–41
    electron repulsion energy   **174**, 336–7
    lifetimes   338, 341
    space wave functions   **173–4**, 333–4, *335*
    spin wave functions   **173**, 334, *335*
    total energies in $1s^12s^1$ helium atom   338–9
two-electron repulsion integrals   **205–6**
    listed for d orbitals
        in complex forms   **210**
        in real forms   **209**
Tycho Brahe   10

Uhlenbeck, George Eugene   78, 105, 118, 120
ultraviolet catastrophe   *14*, 15
ultraviolet spectrum   *217*
    first discovered   217

uncertainty principle   56–7, 246
    effects   58, 77
unification of molecular orbital and valence
        bond models   141–3
unitary matrix   413
units   *248*, 401–2
    conversion factors   *249*
UV–VIS spectroscopy
    electric dipole interaction energy   234
    experimental aspects   240
    terminology   **257**
    timescales   246, *246*
    see also electronic spectroscopy
UV–VIS transitions, compared with IR and
        NMR transitions   238–41

vacuum magnetic susceptibility   364
    relationship with magnetic moment   366,
        395
valence band   387
valence bond (VB) theory   137–9
    compared with MO theory   156, 163
    computational developments   167–8
    and hybridisation   156
    and resonance   163
    unification with molecular orbital theory
        141–3
valence bond (VB) wave functions,
        normalisation of   **169**
valence electron distribution, effects of Pauli
        exclusion principle   161–2
valence electrons/orbitals   107
valence-shell electron-pair repulsion (VSEPR)
        model   164–7
    first developed   164
    and multiple bonds   165–7
    water molecular shape considered   165
Van de Hulst, Hendrick   86
Van der Waals' gas equation   5
Van Vleck, John Hasbrouck   182, 199, 375
vanadium(III) hexacyano complex ion,
        spectrum   194–5
variation theorem   403–10
    application of   404–7
    examples of use   140, 151, 353, 407–10
    proof   403–4
vector quantity   74
    angular momentum as   74–6
velocity, atomic unit of   402
velocity of light   223, 401
    factors affecting   216
    as limiting velocity   13
    ratio of electrostatic to electromagnetic units
        7, 216
vibrational angular momentum, and Coriolis
        interaction   311–16

vibrational circular dichroism (VCD) spectra
        317
vibrational selection rule   299–300
    semi-classical interpretation   301
vibrations of diatomic molecules   291, *292*,
        318–19
vibrations of polyatomic molecules   305–8
    angle bending mode   305–6, *310*
    asymmetric stretching mode   305–6, *307*,
        310
    normal mode   304
    rocking mode   307, *307*, 308
    scissoring mode   307, *307*, 308
    skeletal vibrations   309
    symmetric stretching mode   305–6, *307*,
        310
    wagging mode   307, *307*, 308
vibronic transition   236
virial theorem   97
volume magnetic susceptibility   364
    and Bohr magneton number   **396**
    variation with temperature   364–5
VSEPR model *see* valence-shell electron-pair
        repulsion model

Wang, S.C.   140
water molecule
    bond angle   164
    vibrations   305–6
    VSEPR model   165
wave functions
    of hydrogen atom   97–100
        angular parts   99–104, *104*
        Heitler–London (valence bond) approach
            137–9
        Hund–Mulliken (molecular orbital)
            approach   139
        improvements   140–1
        nodes   104, *104*
        radial part   98–9, *104*
    normalisation of
        in Hückel MO calculation   356
        for hydrogen molecule   **169–70**
    time-dependent part   227
    time-independent part   227–8
wave mechanics   26, 28–9, 50
wave number   248
wave theory of light   216
    limitations   21, 223
weak-field coupling scheme   188
    for d^2 (^{3}F in octahedral field)   189–90
    for d^2 (inclusion of ^{3}P)   190–1
Weinbaum, S.   142
Weiss, Pierre   365
    see also Curie–Weiss law
Wigner, Eugen P.   135, 387

Wilson, Edgar Bright 316
work function (of surface) 21

Young, Thomas 216
Yu, F.C. 267

Zeeman effect 105, 117–20, 408
    anomalous Zeeman effect 120
    normal Zeeman effect 118–20
    practical applications 119–20
zero of energy, definition 96, 133
zero-point energy 16, 47, 58, 134, 294

With thanks to Paul Nash for creation of this index.

Printed and bound by CPI Group (UK) Ltd, Croydon, CR0 4YY

27/10/2024

14580152-0005